Switching, Routing, and Wireless Essentials Companion Guide (CCNAv7)

Cisco Press

Hoboken, New Jersey

Switching, Routing, and Wireless Essentials Companion Guide (CCNAv7)

Published by:
Cisco Press
Hoboken, New Jersey

14 2023

Library of Congress Control Number: 2020936826

ISBN-13: 978-0-13-672935-8
ISBN-10: 0-13-672935-5

Warning and Disclaimer

This book is designed to provide information about the Cisco Networking Academy Switching, Routing, and Wireless Essentials course. Every effort has been made to make this book as complete and as accurate as possible, but no warranty or fitness is implied.

The information is provided on an "as is" basis. The authors, Cisco Press, and Cisco Systems, Inc. shall have neither liability nor responsibility to any person or entity with respect to any loss or damages arising from the information contained in this book or from the use of the discs or programs that may accompany it.

The opinions expressed in this book belong to the author and are not necessarily those of Cisco Systems, Inc.

This book is part of the Cisco Networking Academy® series from Cisco Press. The products in this series support and complement the Cisco Networking Academy curriculum. If you are using this book outside the Networking Academy, then you are not preparing with a Cisco trained and authorized Networking Academy provider.

CISCO.

For more information on the Cisco Networking Academy or to locate a Networking Academy, Please visit www.netacad.com

Editor-in-Chief
Mark Taub

Alliances Manager, Cisco Press
Arezou Gol

Director, ITP Product Management
Brett Bartow

Senior Editor
James Manly

Managing Editor
Sandra Schroeder

Development Editor
Marianne Bartow

Senior Project Editor
Tonya Simpson

Copy Editor
Barbara Hacha

Technical Editor
Rick Graziani

Editorial Assistant
Cindy Teeters

Cover Designer
Chuti Prasertsith

Composition
codeMantra

Indexer
Cheryl Ann Lenser

Proofreader
Abigail Manheim

Trademark Acknowledgments

All terms mentioned in this book that are known to be trademarks or service marks have been appropriately capitalized. Cisco Press or Cisco Systems, Inc., cannot attest to the accuracy of this information. Use of a term in this book should not be regarded as affecting the validity of any trademark or service mark.

Special Sales

For information about buying this title in bulk quantities, or for special sales opportunities (which may include electronic versions; custom cover designs; and content particular to your business, training goals, marketing focus, or branding interests), please contact our corporate sales department at corpsales@pearsoned.com or (800) 382-3419.

For government sales inquiries, please contact governmentsales@pearsoned.com.

For questions about sales outside the U.S., please contact intlcs@pearson.com.

Feedback Information

At Cisco Press, our goal is to create in-depth technical books of the highest quality and value. Each book is crafted with care and precision, undergoing rigorous development that involves the unique expertise of members from the professional technical community.

Readers' feedback is a natural continuation of this process. If you have any comments regarding how we could improve the quality of this book, or otherwise alter it to better suit your needs, you can contact us through email at feedback@ciscopress.com. Please make sure to include the book title and ISBN in your message.

We greatly appreciate your assistance.

Americas Headquarters
Cisco Systems, Inc.
San Jose, CA

Asia Pacific Headquarters
Cisco Systems (USA) Pte. Ltd.
Singapore

Europe Headquarters
Cisco Systems International BV Amsterdam,
The Netherlands

Cisco has more than 200 offices worldwide. Addresses, phone numbers, and fax numbers are listed on the Cisco Website at **www.cisco.com/go/offices.**

About the Contributing Authors

Bob Vachon is a professor at Cambrian College (Sudbury, Ontario, Canada) and Algonquin College (Ottawa, Ontario, Canada). He has more than 30 years of teaching experience in computer networking and information technology. He has also collaborated on many Cisco Networking Academy courses, including CCNA, CCNA Security, CCNP, and Cybersecurity as team lead, lead author, and subject matter expert. Bob enjoys family, friends, and being outdoors playing guitar by a campfire.

Allan Johnson entered the academic world in 1999 after 10 years as a business owner/operator to dedicate his efforts to his passion for teaching. He holds both an MBA and an M.Ed. in training and development. He taught CCNA courses at the high school level for seven years and has taught both CCNA and CCNP courses at Del Mar College in Corpus Christi, Texas. In 2003, Allan began to commit much of his time and energy to the CCNA Instructional Support Team providing services to Networking Academy instructors worldwide and creating training materials. He now works full time for Cisco Networking Academy as Curriculum Lead.

Contents at a Glance

Reader Services

Register your copy at www.ciscopress.com/title/9780136729358 for convenient access to downloads, updates, and corrections as they become available. To start the registration process, go to www.ciscopress.com/register and log in or create an account*. Enter the product ISBN 9780136729358 and click Submit. When the process is complete, you will find any available bonus content under Registered Products.

*Be sure to check the box that you would like to hear from us to receive exclusive discounts on future editions of this product.

Contents

Command Syntax Conventions

The conventions used to present command syntax in this book are the same conventions used in the IOS Command Reference. The Command Reference describes these conventions as follows:

- **Boldface** indicates commands and keywords that are entered literally as shown. In actual configuration examples and output (not general command syntax), boldface indicates commands that are manually input by the user (such as a **show** command).
- *Italic* indicates arguments for which you supply actual values.
- Vertical bars (|) separate alternative, mutually exclusive elements.
- Square brackets ([ ]) indicate an optional element.
- Braces ({ }) indicate a required choice.
- Braces within brackets ([{ }]) indicate a required choice within an optional element.

Introduction

Switching, Routing, and Wireless Essentials Companion Guide (CCNAv7) is the official supplemental textbook for the Cisco Network Academy CCNA Switching, Routing, and Wireless Essentials version 7 course. Cisco Networking Academy is a comprehensive program that delivers information technology skills to students around the world. The curriculum emphasizes real-world practical application while providing opportunities for you to gain the skills and hands-on experience needed to design, install, operate, and maintain networks in small- to medium-sized businesses as well as enterprise and service provider environments.

As a textbook, this book provides a ready reference to explain the same networking concepts, technologies, protocols, and devices as the online curriculum. This book emphasizes key topics, terms, and activities and provides some alternate explanations and examples compared with the course. You can use the online curriculum as directed by your instructor and then use this Companion Guide's study tools to help solidify your understanding of all the topics.

Who Should Read This Book

The book, as well as the course, is designed as an introduction to data network technology for those pursuing careers as network professionals as well as those who need only an introduction to network technology for professional growth. Topics are presented concisely, starting with the most fundamental concepts and progressing to a comprehensive understanding of network communication. The content of this text provides the foundation for additional Cisco Networking Academy courses and preparation for the CCNA certification.

Book Features

The educational features of this book focus on supporting topic coverage, readability, and practice of the course material to facilitate your full understanding of the course material.

Topic Coverage

The following features give you a thorough overview of the topics covered in each chapter so that you can make constructive use of your study time:

- **Objectives:** Listed at the beginning of each chapter, the objectives reference the core concepts covered in the chapter. The objectives match the objectives stated in the corresponding chapters of the online curriculum; however, the question format in the Companion Guide encourages you to think about finding the answers as you read the chapter.

- **Notes:** These are short sidebars that point out interesting facts, timesaving methods, and important safety issues.
- **Chapter summaries:** At the end of each chapter is a summary of the chapter's key concepts. It provides a synopsis of the chapter and serves as a study aid.
- **Practice:** At the end of each chapter is a full list of all the labs, class activities, and Packet Tracer activities to refer back to for study time.

Readability

The following features assist your understanding of the networking vocabulary:

- **Key terms:** Each chapter begins with a list of key terms, along with a page-number reference from inside the chapter. The terms are listed in the order in which they are explained in the chapter. This handy reference allows you to find a term, flip to the page where the term appears, and see the term used in context. The Glossary defines all the key terms.
- **Glossary:** This book contains an all-new Glossary with more than 300 terms.

Practice

Practice makes perfect. This Companion Guide offers you ample opportunities to put what you learn into practice. You will find the following features valuable and effective in reinforcing the instruction that you receive:

- **Check Your Understanding questions and answer key:** Review questions are presented at the end of each chapter as a self-assessment tool. These questions match the style of questions that you see in the online course. Appendix A, "Answers to the 'Check Your Understanding' Questions," provides an answer key to all the questions and includes an explanation of each answer.

Interactive Graphic

Video

- **Labs and activities:** Throughout each chapter, you will be directed back to the online course to take advantage of the activities created to reinforce concepts. In addition, at the end of each chapter is a practice section that collects a list of all the labs and activities to provide practice with the topics introduced in the chapter.
- **Page references to online course:** After headings, you will see, for example, (1.1.2). This number refers to the page number in the online course so that you can easily jump to that spot online to view a video, practice an activity, perform a lab, or review a topic.

About Packet Tracer Software and Activities

Packet Tracer
☐ Activity

Interspersed throughout the chapters you'll find a few Cisco Packet Tracer activities. Packet Tracer allows you to create networks, visualize how packets flow in the network, and use basic testing tools to determine whether the network would work. When you see this icon, you can use Packet Tracer with the listed file to perform a task suggested in this book. The activity files are available in the course. Packet Tracer software is available only through the Cisco Networking Academy website. Ask your instructor for access to Packet Tracer.

How This Book Is Organized

This book corresponds closely to the Cisco Networking Academy Switching, Routing, and Wireless Essentials course and is divided into 16 chapters, one appendix, and a glossary of key terms:

- **Chapter 1, "Basic Device Configuration":** This chapter explains how to configure devices using security best practices. Included are initial switch and router configuration, switch port configuration, remote access configuration, and how to verify connectivity between two networks.
- **Chapter 2, "Switching Concepts":** This chapter explains how switches forward data. Included are frame forwarding methods and collision and broadcast domain comparison.
- **Chapter 3, "VLANs":** This chapter explains how to implement VLANs and trunking in a switched network. Included are explanations of the purpose of VLANs, how VLANs forward frames in a multiswitched environment, VLAN port assignments, trunk configuration, and DTP configuration.
- **Chapter 4, "Inter-VLAN Routing":** This chapter explains how to implement inter-VLAN routing. Included are descriptions of inter-VLAN routing options, router-on-a-stick configuration, Layer 3 switch inter-VLAN routing, and troubleshooting common inter-VLAN routing configuration issues.
- **Chapter 5, "STP Concepts":** This chapter explains how STP enables redundancy in a Layer 3 network. Included are explanations of common problems in redundant Layer 2 networks, STP operation, and Rapid PVST+ operation.
- **Chapter 6, "EtherChannel":** This chapter explains how to implement EtherChannel on switched links. Included are descriptions of EtherChannel technology, EtherChannel configuration, and troubleshooting EtherChannel.
- **Chapter 7, "DHCPv4":** This chapter explains how to implement DHCPv4 for multiple LANs. Included is an explanation of DHCPv4 operation, as well as configuring a router as a DHCPv4 server or DHCPv4 client.

- **Chapter 8, "SLAAC and DHCPv6":** This chapter explains how to implement dynamic address allocation in an IPv6 network. Included are explanations of how an IPv6 host acquires its addressing, SLAAC operation, DHCPv6 operation, and configuring a router as a stateful or stateless DHCPv6 server.
- **Chapter 9, "FHRP Concepts":** This chapter explains how FHRPs provide default gateway services in a redundant network. Included are explanations of the purpose of FHRPs and HSRP operation.
- **Chapter 10, "LAN Security Concepts":** This chapter explains how vulnerabilities compromise LAN security. Included are explanations of how to use endpoint security, how AAA and 802.1X are used to authenticate, Layer 2 vulnerabilities, MAC address table attacks, and LAN attacks.
- **Chapter 11, "Switch Security Configuration":** This chapter explains how to configure switch security to mitigate LAN attacks. Included is port security implementation as well as mitigating VLAN, DHCP, ARP, and STP attacks.
- **Chapter 12, "WLAN Concepts":** This chapter explains how WLANs enable network connectivity for wireless devices. Included are explanations of WLAN technology, WLAN components, and WLAN operation. In addition, the chapter discusses how CAPWAP is used to manage multiple APs for a WLC. WLAN channel management is discussed. The chapter concludes with a discussion of threats to WLANs and how to secure WLANs.
- **Chapter 13, "WLAN Configuration":** This chapter explains how to implement a WLAN using a wireless router and a WLC. Included are explanations of wireless router configuration and WLC WLAN configuration for both WPA2 PSK and WPA2 Enterprise authentication. The chapter concludes with a discussion of how to troubleshoot common wireless configuration issues.
- **Chapter 14, "Routing Concepts":** This chapter explains how routers use information in packets to make forwarding decisions. Included are explanations of path determination, packet forwarding, basic router configuration, routing table structure, and static and dynamic routing concepts.
- **Chapter 15, "IP Static Routing":** This chapter explains how to implement IPv4 and IPv6 static routes. Included are static route syntax, static and default routing configuration, floating static routing configuration, and static host route configuration.
- **Chapter 16, "Troubleshoot Static and Default Routes":** This chapter explains how to troubleshoot static and default route implementations. Included are explanations of how a router processes packets when a static route is configured, and how to troubleshoot command static and default route configuration issues.

- **Appendix A, "Answers to the 'Check Your Understanding' Questions":** This appendix lists the answers to the "Check Your Understanding" review questions that are included at the end of each chapter.
- **Glossary:** The Glossary provides you with definitions for all the key terms identified in each chapter.

Figure Credits

Figure 1-6, screenshot of Telnet Session Capture © Wireshark

Figure 1-7, screenshot of SSH Session Capture © Wireshark

Figure 7-8, screenshot of Configuring a Home Router as a DHCPv4 Client © 2020 Belkin International, Inc.

Figure 8-1, screenshot of Manual Configuration of an IPv6 Windows Host © Microsoft 2020

Figure 8-2, screenshot of Automatic Configuration of an IPv6 Windows Host © Microsoft 2020

Figure 10-1, screenshot of WannaCry Ransomware © Lazarus Group

Figure 10-29, screenshot of Wireshark Capture of a CDP Frame © Wireshark

Figure 12-38, screenshot of Disabling SSID Broadcast (SSID Cloaking) on a Wireless Router © 2020 Belkin International, Inc.

Figure 12-39, screenshot of Configuring MAC Address Filtering on a Wireless Router © 2020 Belkin International, Inc.

Figure 12-41, screenshot of Selecting the Authentication Method on a Wireless Router © 2020 Belkin International, Inc.

Figure 12-42, screenshot of Setting the Encryption Method on a Wireless Router © 2020 Belkin International, Inc.

Figure 12-43, screenshot of Configuring WPA2 Enterprise Authentication on a Wireless Router © 2020 Belkin International, Inc.

Figure 13-3, screenshot of Connecting to a Wireless Router Using a Browser © 2020 Belkin International, Inc.

Figure 13-4, screenshot of Basic Network Setup - Step 1 © 2020 Belkin International, Inc.

Figure 13-5, screenshot of Basic Network Setup - Step 2 © 2020 Belkin International, Inc.

CHAPTER 1

Basic Device Configuration

Objectives

Upon completion of this chapter, you will be able to answer the following questions:

- How do you configure initial settings on a Cisco switch?
- How do you configure switch ports to meet network requirements?
- How do you configure secure management access on a switch?
- How do you configure basic settings on a router to route between two directly connected networks, using CLI?
- How do you verify connectivity between two networks that are directly connected to a router?

Key Terms

This chapter uses the following key terms. You can find the definitions in the Glossary.

power-on self-test (POST) Page 2

CPU subsystem Page 2

boot loader software Page 2

BOOT environment variable Page 3

Mode button Page 3

System LED Page 4

Redundant Power System (RPS) LED Page 4

full-duplex Page 11

half-duplex Page 11

autonegotiate Page 12

automatic medium-dependent interface crossover (auto-MDIX) Page 13

input errors Page 17

runts Page 17

giants Page 17

CRC Page 17

output errors Page 17

late collisions Page 17

duplex mismatch Page 20

dual-stack topology Page 27

High-Speed WAN Interface Card (HWIC) Page 27

loopback interface Page 28

Introduction (1.0)

Welcome to the first module in CCNA Switching, Routing, and Wireless Essentials! You know that switches and routers come with some built-in configuration, so why would you need to learn to further configure switches and routers?

Imagine that you purchased a model train set. After you had set it up, you realized that the track was just a simple oval shape and that the train cars ran only clockwise. You might want the track to be a figure-eight shape with an overpass. You might want to have two trains that operate independently of each other and are able to move in different directions. How could you make that happen? You would need to reconfigure the track and the controls. It is the same with network devices. As a network administrator you need detailed control of the devices in your network. This means precisely configuring switches and routers so that your network does what you want it to do. This module has many Syntax Checker and Packet Tracer activities to help you develop these skills. Let's get started!

Configure a Switch with Initial Settings (1.1)

Switches interconnect devices. Unlike a router, which must be initially configured to be operational in a network, switches can be deployed out of the box without initially being configured. However, for management and security reasons, switches should always be manually configured to better meet the needs of the network.

In this section, you learn how to configure initial settings on a Cisco switch.

Switch Boot Sequence (1.1.1)

Before you can configure a switch, you need to turn it on and allow it to go through the five-step boot sequence. This topic covers the basics of configuring a switch and includes a lab at the end.

After a Cisco switch is powered on, it goes through the following five-step boot sequence:

Step 1. The switch loads a *power-on self-test (POST)* program stored in read-only memory (ROM). POST checks the central processing unit *(CPU) subsystem*. It tests the CPU, dynamic random-access memory (DRAM), and the portion of the flash device that makes up the flash file system.

Step 2. The switch loads the *boot loader software*. The boot loader is a small program stored in ROM that is run immediately after POST successfully completes.

Step 3. The boot loader performs low-level CPU initialization. It initializes the CPU registers, which control where physical memory is mapped, the quantity of memory, and its speed.

Step 4. The boot loader initializes the flash file system on the system board.

Step 5. Finally, the boot loader locates and loads a default IOS operating system software image into memory and gives control of the switch over to the IOS.

The boot system Command (1.1.2)

The switch attempts to automatically boot by using information in the *BOOT environment variable*. If this variable is not set, the switch attempts to load and execute the first executable file it can find. On Catalyst 2960 Series switches, the image file is normally contained in a directory that has the same name as the image file (excluding the .bin file extension).

The IOS operating system then initializes the interfaces using the Cisco IOS commands found in the startup-config file. The startup-config file is called config.text and is located in flash.

In the following snippet, the BOOT environment variable is set using the **boot system** global configuration mode command. Notice that the IOS is located in a distinct folder and the folder path is specified. Use the command **show boot** to see what the current IOS boot file is set to.

```
S1(config)# boot system flash./c2960-lanbasek9-mz.150-2.SE/c2960-lanbasek9-mz.150-2.
   SE.bin
```

Table 1-1 defines each part of the **boot system** command.

Table 1-1 The **boot system** Command Syntax

Command	Definition
boot system	The main command
flash.	The storage device
c2960-lanbasek9-mz.150-2.SE/	The path to the file system
c2960-lanbasek9-mz.150-2.SE.bin	The IOS file name

Switch LED Indicators (1.1.3)

Cisco Catalyst switches have several status LED indicator lights. You can use the switch LEDs to quickly monitor switch activity and performance. Switches of different models and feature sets will have different LEDs and their placement on the front panel of the switch may also vary.

Figure 1-1 shows the switch LEDs and the *Mode button* for a Cisco Catalyst 2960 switch.

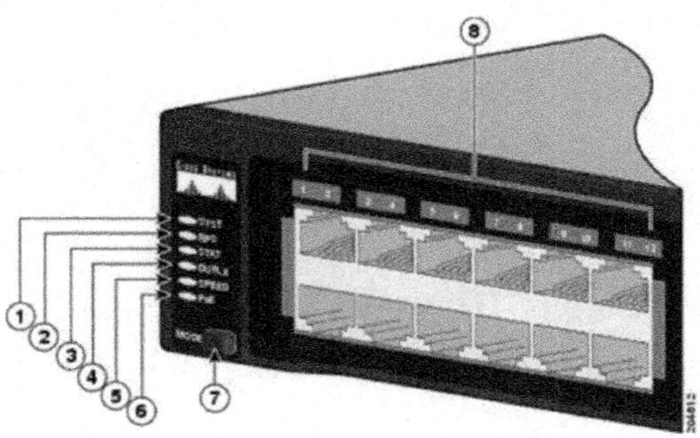

Figure 1-1 Cisco Catalyst 2960 LEDs and Mode Button

The Mode button (7 in Figure 1-1) is used to toggle through port status, port duplex, port speed, and if supported, the Power over Ethernet (PoE) status of the port LEDs (8 in Figure 1-1).

Table 1-2 describes the purpose of the LED indicators (1 through 6 in Figure 1-1), and the meaning of their colors.

Table 1-2 LED Indicators

LED Label	Name	Description
1 SYST	*System LED*	■ Shows whether the system is receiving power and is functioning properly. ■ If the LED is off, it means the system is not powered on. ■ If the LED is green, the system is operating normally. ■ If the LED is amber, the system is receiving power but is not functioning properly.
2 RPS	*Redundant Power System (RPS) LED*	■ Shows the RPS status. ■ If the LED is off, the RPS is off, or it is not properly connected. ■ If the LED is green, the RPS is connected and ready to provide backup power. ■ If the LED is blinking green, the RPS is connected but is unavailable because it is providing power to another device. ■ If the LED is amber, the RPS is in standby mode, or in a fault condition. ■ If the LED is blinking amber, the internal power supply in the switch has failed, and the RPS is providing power.

LED Label	Name	Description
3 STAT	Port Status LED	■ Indicates that the port status mode is selected when the LED is green. (This is the default mode.) ■ When selected, the port LEDs will display colors with different meanings. ■ If the LED is off, there is no link, or the port was administratively shut down. ■ If the LED is green, a link is present. ■ If the LED is blinking green, there is activity and the port is sending or receiving data. ■ If the LED is alternating green-amber, there is a link fault. ■ If the LED is amber, the port is blocked to ensure that a loop does not exist in the forwarding domain and is not forwarding data (typically, ports will remain in this state for the first 30 seconds after being activated). ■ If the LED is blinking amber, the port is blocked to prevent a possible loop in the forwarding domain.
4 DUPLX	Port Duplex LED	■ Indicates that the port duplex mode is selected when the LED is green. ■ When selected, port LEDs that are off are in half-duplex mode. ■ If the port LED is green, the port is in full-duplex mode.
5 SPEED	Port Speed LED	■ Indicates that the port speed mode is selected. ■ When selected, the port LEDs will display colors with different meanings. ■ If the LED is off, the port is operating at 10 Mbps. ■ If the LED is green, the port is operating at 100 Mbps. ■ If the LED is blinking green, the port is operating at 1000 Mbps.
6 PoE	Power over Ethernet (PoE) Mode LED	■ If PoE is supported, a PoE mode LED will be present. ■ If the LED is off, it indicates the PoE mode is not selected and that none of the ports have been denied power or placed in a fault condition. ■ If the LED is blinking amber, the PoE mode is not selected but at least one of the ports has been denied power or has a PoE fault.

LED Label	Name	Description
		■ If the LED is green, it indicates the PoE mode is selected and the port LEDs will display colors with different meanings. ■ If the port LED is off, the PoE is off. ■ If the port LED is green, the PoE is on. ■ If the port LED is alternating green-amber, PoE is denied because providing power to the powered device will exceed the switch power capacity. ■ If the LED is blinking amber, PoE is off because of a fault. ■ If the LED is amber, PoE for the port has been disabled.

Recovering from a System Crash (1.1.4)

The boot loader provides access into the switch if the operating system cannot be used because of missing or damaged system files. The boot loader has a command-line that provides access to the files stored in flash memory.

The boot loader can be accessed through a console connection following these steps:

Step 1. Connect a PC by console cable to the switch console port. Configure terminal emulation software to connect to the switch.

Step 2. Unplug the switch power cord.

Step 3. Reconnect the power cord to the switch and, within 15 seconds, press and hold down the Mode button while the System LED is still flashing green.

Step 4. Continue pressing the Mode button until the System LED turns briefly amber and then solid green; then release the Mode button.

Step 5. The boot loader switch: prompt appears in the terminal emulation software on the PC.

Type **help** or **?** at the boot loader prompt to view a list of available commands.

By default, the switch attempts to automatically boot by using information in the BOOT environment variable. To view the path of the switch BOOT environment variable type the **set** command. Then, initialize the flash file system using the **flash_init** command to view the current files in flash, as shown in Example 1-1.

Example 1-1 Initialize Flash File System

```
switch: set
BOOT=flash:/c2960-lanbasek9-mz.122-55.SE7/c2960-lanbasek9-mz.122-55.SE7.bin
(output omitted)
switch: flash_init
Initializing Flash...
flashfs[0]: 2 files, 1 directories
flashfs[0]: 0 orphaned files, 0 orphaned directories
flashfs[0]: Total bytes: 32514048
flashfs[0]: Bytes used: 11838464
flashfs[0]: Bytes available: 20675584
flashfs[0]: flashfs fsck took 10 seconds.
...done Initializing Flash.
```

After flash has finished initializing you can enter the **dir flash:** command to view the directories and files in flash, as shown in Example 1-2.

Example 1-2 Display Flash Directory

```
switch: dir flash:
Directory of flash:/
    2  -rwx  11834846                    c2960-lanbasek9-mz.150-2.SE8.bin
    3  -rwx  2072                        multiple-fs
```

Enter the **BOOT=flash** command to change the BOOT environment variable path the switch uses to load the new IOS in flash. To verify the new BOOT environment variable path, issue the **set** command again. Finally, to load the new IOS type the **boot** command without any arguments, as shown in Example 1-3.

Example 1-3 Configure the Boot Image

```
switch: BOOT=flash:c2960-lanbasek9-mz.150-2.SE8.bin
switch: set
BOOT=flash:c2960-lanbasek9-mz.150-2.SE8.bin
(output omitted)
switch: boot
```

The boot loader commands support initializing flash, formatting flash, installing a new IOS, changing the BOOT environment variable, and recovery of lost or forgotten passwords.

Switch Management Access (1.1.5)

To prepare a switch for remote management access, the switch must have a switch virtual interface (SVI) configured with an IPv4 address and subnet mask or an IPv6 address and a prefix length for IPv6. The SVI is a virtual interface, not a physical port on the switch. Keep in mind that to manage the switch from a remote network, the switch must be configured with a default gateway. This is very similar to configuring the IP address information on host devices.

In Figure 1-2, the switch virtual interface (SVI) on S1 should be assigned an IP address. A console cable is used to connect to a PC so that the switch can be initially configured. The indicated IP addresses display the default gateway of PC1 and S1.

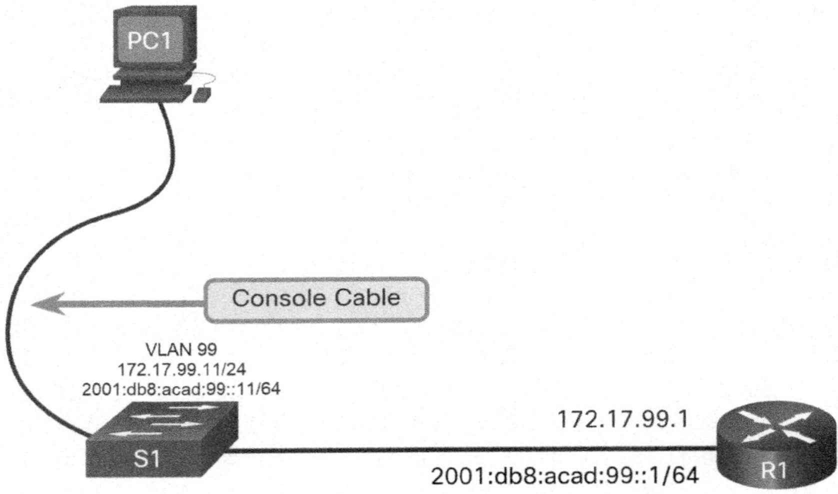

Figure 1-2 Console Connection to a Switch

Switch SVI Configuration Example (1.1.6)

By default, the switch is configured to have its management controlled through VLAN 1. All ports are assigned to VLAN 1 by default. For security purposes, it is considered a best practice to use a VLAN other than VLAN 1 for the management VLAN, such as VLAN 99.

Step 1. **Configure the Management Interface on S1.** From VLAN interface configuration mode, an IPv4 address and subnet mask is applied to the management SVI of the switch. Specifically, SVI VLAN 99 will be assigned the 172.17.99.11/24 IPv4 address and the 2001:db8:acad:99::11/64 IPv6 address as shown in Table 1-3.

Note

The SVI for VLAN 99 will not appear as "up/up" until VLAN 99 is created and there is a device connected to a switch port associated with VLAN 99.

Note

The switch may need to be configured for IPv6. For example, before you can configure IPv6 addressing on a Cisco Catalyst 2960 running IOS version 15.0, you will need to enter the global configuration command **sdm prefer dual-ipv4-and-ipv6 default** and then **reload** the switch.

Table 1-3 Configure IPv4 and IPv6 Addressing for the Management Interface

Task	IOS Commands
Enter global configuration mode.	`S1#` **`configure terminal`**
Enter interface configuration mode for the SVI.	`S1(config)#` **`interface vlan 99`**
Configure the management IPv4 address.	`S1(config-if)#` **`ip address 172.17.99.11 255.255.255.0`**
Configure the management IPv6 address.	`S1(config-if)#` **`ipv6 address 2001:db8:acad:99::11/64`**
Enable the management interface.	`S1(config-if)#` **`no shutdown`**
Return to the privileged EXEC mode.	`S1(config-if)#` **`end`**
Save the running config to the startup config.	`S1#` **`copy running-config startup-config`**

Step 2. **Configure the Default Gateway.** The switch should be configured with a default gateway if it will be managed remotely from networks that are not directly connected, as shown in Table 1-4.

Note

Because it will receive its default gateway information from a router advertisement (RA) message, the switch does not require an IPv6 default gateway.

Table 1-4 Configure the Default Gateway

Task	IOS Commands
Enter global configuration mode.	`S1# configure terminal`
Configure the default gateway for the switch.	`S1(config)# ip default-gateway 172.17.99.1`
Return to the privileged EXEC mode.	`S1(config)# end`
Save the running config to the startup config.	`S1# copy running-config startup-config`

Step 3. **Verify Configuration.** The **show ip interface brief** and **show ipv6 interface brief** commands are useful for determining the status of both physical and virtual interfaces. The output shown in Example 1-4 confirms that interface VLAN 99 has been configured with an IPv4 and IPv6 address.

Note

An IP address applied to the SVI is only for remote management access to the switch; this does not allow the switch to route Layer 3 packets.

Example 1-4 Verify IP Configuration

```
S1# show ip interface brief
Interface        IP-Address       OK? Method    Status      Protocol
Vlan99           172.17.99.11     YES manual    down        down
(output omitted)
S1#
S1# show ipv6 interface brief
Vlan99                        [down/down]
     FE80::C27B:BCFF:FEC4:A9C1
     2001:DB8:ACAD:99::11
(output omitted)
```

Lab - Basic Switch Configuration (1.1.7)

In this lab, you will complete the following objectives:

Part 1: Cable the Network and Verify the Default Switch Configuration

Part 2: Configure Basic Network Device Settings

Part 3: Verify and Test Network Connectivity

Part 4: Manage the MAC Address Table

Configure Switch Ports (1.2)

In this section, you learn how to configure switch ports to meet network requirements.

Duplex Communication (1.2.1)

The ports of a switch can be configured independently for different needs. This section covers how to configure switch ports, how to verify your configurations, common errors, and how to troubleshoot switch configuration issues.

Full-duplex communication increases bandwidth efficiency by allowing both ends of a connection to transmit and receive data simultaneously. This is also known as bidirectional communication and it requires microsegmentation. A microsegmented LAN is created when a switch port has only one device connected and is operating in full-duplex mode. There is no collision domain associated with a switch port operating in full-duplex mode.

Unlike full-duplex communication, *half-duplex* communication is unidirectional. Half-duplex communication creates performance issues because data can flow in only one direction at a time, often resulting in collisions. Half-duplex connections are typically seen in older hardware, such as hubs. Half-duplex hubs have been replaced by switches that use full-duplex communication by default.

Figure 1-3 illustrates full-duplex and half-duplex communication.

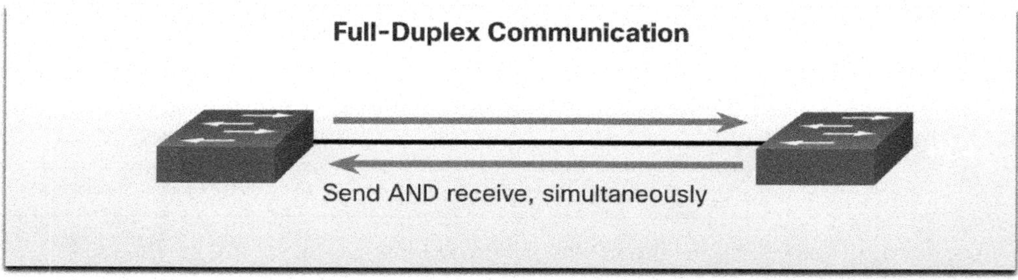

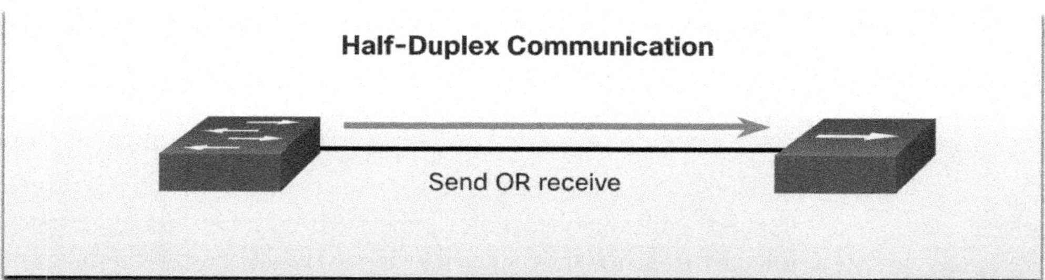

Figure 1-3 Duplex Communications Between Two Switches

Gigabit Ethernet and 10 Gb/s NICs require full-duplex connections to operate. In full-duplex mode, the collision detection circuit on the network interface card (NIC) is disabled. Full-duplex offers 100 percent efficiency in both directions (transmitting and receiving). This results in a doubling of the potential use of the stated bandwidth.

Configure Switch Ports at the Physical Layer (1.2.2)

Switch ports can be manually configured with specific duplex and speed settings. Use the **duplex** interface configuration mode command to manually specify the duplex mode for a switch port. Use the **speed** interface configuration mode command to manually specify the speed. For example, both switches in Figure 1-4 should always operate in full-duplex at 100 Mbps.

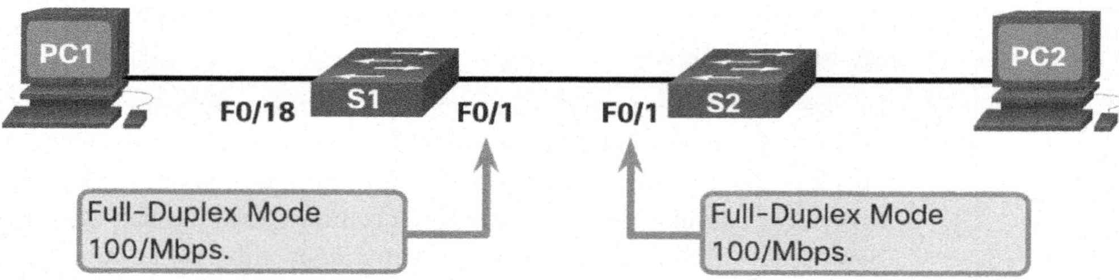

Figure 1-4 Connected Switches Operating at Full-Duplex and 100 Mbps

Table 1-5 shows the commands for S1. The same commands can be applied to S2.

Table 1-5 Duplex and Speed Commands

Task	IOS Commands
Enter global configuration mode.	`S1# configure terminal`
Enter interface configuration mode.	`S1(config)# interface FastEthernet 0/1`
Configure the interface duplex.	`S1(config-if)# duplex full`
Configure the interface speed.	`S1(config-if)# speed 100`
Return to the privileged EXEC mode.	`S1(config-if)# end`
Save the running config to the startup config.	`S1# copy running-config startup-config`

The default setting for both duplex and speed for switch ports on Cisco Catalyst 2960 and 3560 switches is auto. The 10/100/1000 ports operate in either half- or full-duplex mode when they are set to 10 or 100 Mbps and operate only in full-duplex mode when it is set to 1000 Mbps (1 Gbps). *Autonegotiation* is useful when the speed and duplex settings of the device connecting to the port are unknown or may change. When connecting to known devices such as servers, dedicated workstations, or network devices, a best practice is to manually set the speed and duplex settings.

When troubleshooting switch port issues, it is important that the duplex and speed settings should be checked.

Note

Mismatched settings for the duplex mode and speed of switch ports can cause connectivity issues. Autonegotiation failure creates mismatched settings.

All fiber-optic ports, such as 1000BASE-SX ports, operate only at one preset speed and are always full-duplex.

Auto-MDIX (1.2.3)

Until recently, certain cable types (straight-through or crossover) were required when connecting devices. Switch-to-switch or switch-to-router connections required using different Ethernet cables. Using the *automatic medium-dependent interface crossover (auto-MDIX)* feature on an interface eliminates this problem. When auto-MDIX is enabled, the interface automatically detects the required cable connection type (straight-through or crossover) and configures the connection appropriately. When connecting to switches without the auto-MDIX feature, straight-through cables must be used to connect to devices such as servers, workstations, or routers. Crossover cables must be used to connect to other switches or repeaters.

With auto-MDIX enabled, either type of cable can be used to connect to other devices, and the interface automatically adjusts to communicate successfully. On newer Cisco switches, the **mdix auto** interface configuration mode command enables the feature. When using auto-MDIX on an interface, the interface speed and duplex must be set to **auto** so that the feature operates correctly.

The command to enable auto-MDIX is issued in interface configuration mode on the switch as shown:

```
S1(config-if)# mdix auto
```

Note

The auto-MDIX feature is enabled by default on Catalyst 2960 and Catalyst 3560 switches but is not available on the older Catalyst 2950 and Catalyst 3550 switches.

To examine the auto-MDIX setting for a specific interface, use the **show controllers ethernet-controller** command with the **phy** keyword. To limit the output to lines referencing auto-MDIX, use the **include Auto-MDIX** filter. In Example 1-5, Auto-MDIX is on.

Example 1-5 Verifying Auto-MDIX Setting

```
S1# show controllers ethernet-controller fa0/1 phy | include MDIX
 Auto-MDIX            :  On    [AdminState=1    Flags=0x00052248]
```

Switch Verification Commands (1.2.4)

Table 1-6 summarizes some of the more useful switch verification commands.

Table 1-6 Switch Verification Commands

Task	IOS Commands
Display interface status and configuration.	S1# **show interfaces** [*interface-id*]
Display current startup configuration.	S1# **show startup-config**
Display current running configuration.	S1# **show running-config**
Display information about flash file system.	S1# **show flash**
Display system hardware and software status.	S1# **show version**
Display history of command entered.	S1# **show history**
Display IP information about an interface.	S1# **show ip interface** [*interface-id*] OR S1# **show ipv6 interface** [*interface-id*]
Display the MAC address table.	S1# **show mac-address-table** OR S1# **show mac address-table**

Verify Switch Port Configuration (1.2.5)

The **show running-config** command can be used to verify that the switch has been correctly configured. Some important information is shown in Example 1-6:

- Fast Ethernet 0/18 interface is configured with the management VLAN 99.
- VLAN 99 is configured with an IPv4 address of 172.17.99.11 255.255.255.0.
- The default gateway is set to 172.17.99.1.

Example 1-6 Switch Port Configuration Verification

```
S1# show running-config
Building configuration...
Current configuration : 1466 bytes
!
interface FastEthernet0/18
 switchport access vlan 99
 switchport mode access
!
(output omitted)
!
interface Vlan99
 ip address 172.17.99.11 255.255.255.0
 ipv6 address 2001:DB8:ACAD:99::1/64
!
ip default-gateway 172.17.99.1
```

The **show interfaces** command is another commonly used command, which displays status and statistics information on the network interfaces of the switch. The **show interfaces** command is frequently used when configuring and monitoring network devices.

The first line of the output for the **show interfaces fastEthernet 0/18** command in Example 1-7 indicates that the FastEthernet 0/18 interface is up/up, meaning that it is operational. Further down, the output shows that the duplex is full and the speed is 100 Mbps.

Example 1-7 Interface Verification

```
S1# show interfaces fastEthernet 0/18
FastEthernet0/18 is up, line protocol is up (connected)
  Hardware is Fast Ethernet, address is 0025.83e6.9092 (bia 0025.83e6.9092)
  MTU 1500 bytes, BW 100000 Kbit/sec, DLY 100 usec,
     reliability 255/255, txload 1/255, rxload 1/255
  Encapsulation ARPA, loopback not set
  Keepalive set (10 sec)
  Full-duplex, 100Mb/s, media type is 10/100BaseTX
```

Network Access Layer Issues (1.2.6)

The output from the **show interfaces** command is useful for detecting common media issues. One of the most important parts of this output is the display of the line and data link protocol status, as shown Example 1-8.

Example 1-8 Checking Protocol Status

```
S1# show interfaces fastEthernet 0/18
FastEthernet0/18 is up, line protocol is up (connected)
Hardware is Fast Ethernet, address is 0025.83e6.9092 (bia 0025.83e6.9092)MTU 1500
  bytes, BW 100000 Kbit/sec, DLY 100 usec,
```

The first parameter (FastEthernet0/18 is up) refers to the hardware layer and indicates whether the interface is receiving a carrier detect signal. The second parameter (line protocol is up) refers to the data link layer and indicates whether the data link layer protocol keepalives are being received.

Based on the output of the **show interfaces** command, possible problems can be fixed as follows:

- *Interface* is up and line protocol is down: This indicates a problem exists. There could be an encapsulation type mismatch, the interface on the other end could be error-disabled, or there could be a hardware problem.
- *Interface* is down and line protocol is down: This is an indication that a cable is not attached, or some other problem exists. For example, in a back-to-back connection, the other end of the connection may be administratively down.
- *Interface* is administratively down: This indicates that the interface is manually disabled (the **shutdown** command has been issued) in the active configuration.

The **show interfaces** command output displays counters and statistics for the FastEthernet0/18 interface, as highlighted in Example 1-9.

Example 1-9 Interface Counters and Statistics

```
S1# show interfaces fastEthernet 0/18
FastEthernet0/18 is up, line protocol is up (connected)
  Hardware is Fast Ethernet, address is 0025.83e6.9092 (bia 0025.83e6.9092)
  MTU 1500 bytes, BW 100000 Kbit/sec, DLY 100 usec,
     reliability 255/255, txload 1/255, rxload 1/255
  Encapsulation ARPA, loopback not set
  Keepalive set (10 sec)
 Full-duplex, 100Mb/s, media type is 10/100BaseTX
  input flow-control is off, output flow-control is unsupported
  ARP type: ARPA, ARP Timeout 04:00:00
  Last input never, output 00:00:01, output hang never
  Last clearing of "show interface" counters never
  Input queue: 0/75/0/0 (size/max/drops/flushes); Total output drops: 0
  Queueing strategy: fifo
  Output queue: 0/40 (size/max)
  5 minute input rate 0 bits/sec, 0 packets/sec
  5 minute output rate 0 bits/sec, 0 packets/sec
```

```
2295197 packets input, 305539992 bytes, 0 no buffer
Received 1925500 broadcasts (74 multicasts)
0 runts, 0 giants, 0 throttles
3 input errors, 3 CRC, 0 frame, 0 overrun, 0 ignored
0 watchdog, 74 multicast, 0 pause input
0 input packets with dribble condition detected
3594664 packets output, 436549843 bytes, 0 underruns
8 output errors, 1790 collisions, 10 interface resets
0 unknown protocol drops
0 babbles, 235 late collision, 0 deferred
```

Some media errors are not severe enough to cause the circuit to fail but do cause network performance issues. Table 1-7 explains some of these common errors, which can be detected using the **show interfaces** command.

Table 1-7 Common Errors Detected by the **show interface** Command

Error Type	Description
Input Errors	Total number of errors and includes runts, giants, no buffer, CRC, frame, overrun, and ignored counts.
Runts	Frames that are discarded because they are smaller than the minimum packet size for the medium. For instance, any Ethernet frame that is less than 64 bytes is considered a runt.
Giants	May be forwarded or discarded depending on the model of the switch. A giant frame is an Ethernet frame greater than 1,518 bytes.
CRC	CRC errors are generated when the calculated checksum is not the same as the checksum received.
Output Errors	Sum of all errors that prevented the final transmission of datagrams out of the interface that is being examined.
Collisions	Number of messages retransmitted because of an Ethernet collision.
Late Collisions	A collision that occurs after 512 bits of the frame have been transmitted.

Interface Input and Output Errors (1.2.7)

"Input errors" is the sum of all errors in datagrams that were received on the interface being examined. This includes runts, giants, cyclic redundancy check (CRC), no buffer, frame, overrun, and ignored counts. The reported input errors from the **show interfaces** command include the following:

- **Runt Frames:** Ethernet frames that are shorter than the 64-byte minimum allowed length are called runts. Malfunctioning NICs are the usual cause of excessive runt frames, but they can also be caused by collisions.
- **Giants:** Ethernet frames that are larger than the maximum allowed size are called giants. Giants may be forwarded or discarded depending on the model of the switch.
- **CRC errors:** On Ethernet and serial interfaces, CRC errors usually indicate a media or cable error. Common causes include electrical interference, loose or damaged connections, or incorrect cabling. If you see many CRC errors, there is too much noise on the link and you should inspect the cable. You should also search for and eliminate noise sources.

"Output errors" is the sum of all errors that prevented the final transmission of datagrams out the interface that is being examined. The reported output errors from the **show interfaces** command include the following:

- **Collisions:** Collisions in half-duplex operations are normal. However, you should never see collisions on an interface configured for full-duplex communication.
- **Late collisions:** A late collision refers to a collision that occurs after 512 bits of the frame have been transmitted. Excessive cable lengths are the most common cause of late collisions. Another common cause is duplex misconfiguration. For example, you could have one end of a connection configured for full-duplex and the other for half-duplex. You would see late collisions on the interface that is configured for half-duplex. In that case, you must configure the same duplex setting on both ends. A properly designed and configured network should never have late collisions.

Troubleshooting Network Access Layer Issues (1.2.8)

Most issues that affect a switched network are encountered during the original implementation. Theoretically, after it is installed, a network continues to operate without problems. However, cabling gets damaged, configurations change, and new devices are connected to the switch that require switch configuration changes. Ongoing maintenance and troubleshooting of the network infrastructure is required.

To troubleshoot scenarios involving no connection, or a bad connection, between a switch and another device, follow the general process shown in Figure 1-5.

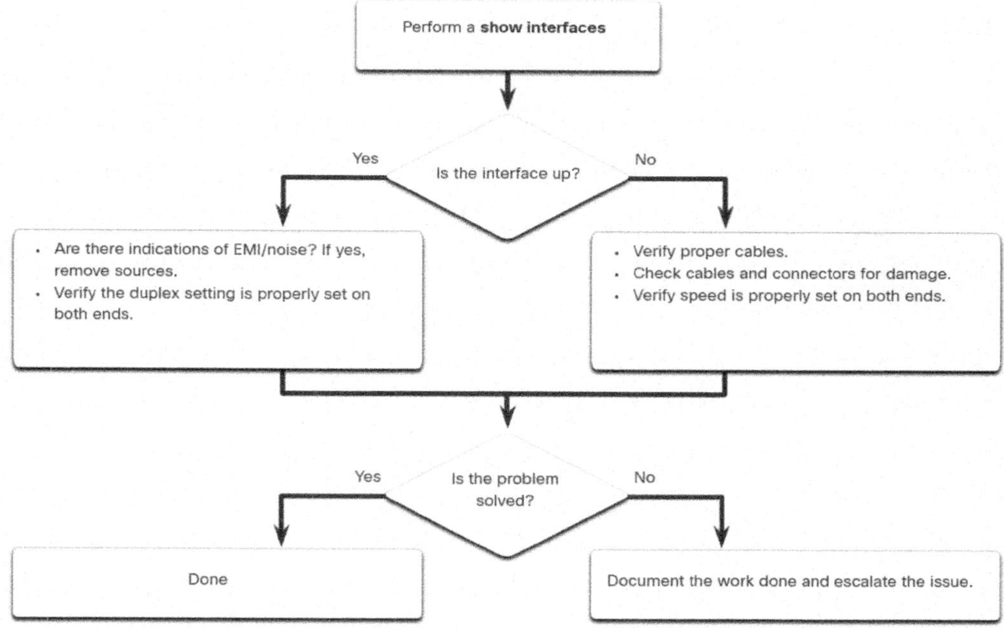

Figure 1-5 Troubleshooting Process for Network Layer Issues

Use the **show interfaces** command to check the interface status.

If the interface is down:

- Check to make sure that the proper cables are being used. Additionally, check the cable and connectors for damage. If a bad or incorrect cable is suspected, replace the cable.
- If the interface is still down, the problem may be due to a mismatch in speed setting. The speed of an interface is typically autonegotiated; therefore, even if it is manually applied to one interface, the connecting interface should autonegotiate accordingly. If a speed mismatch does occur through misconfiguration, or a hardware or software issue, then that may result in the interface going down. Manually set the same speed on both connection ends if a problem is suspected.

If the interface is up but issues with connectivity are still present:

- Using the **show interfaces** command, check for indications of excessive noise. Indications may include an increase in the counters for runts, giants, and CRC errors. If there is excessive noise, first find and remove the source of the noise, if possible. Also, verify that the cable does not exceed the maximum cable length and check the type of cable that is used.

- If noise is not an issue, check for excessive collisions. If there are collisions or late collisions, verify the duplex settings on both ends of the connection. Much like the speed setting, the duplex setting is usually autonegotiated. If there does appear to be a *duplex mismatch*, manually set the duplex to full on both ends of the connection.

Interactive Graphic

Syntax Checker—Configure Switch Ports (1.2.9)

Configure a switch interface based on the specified requirements.

Refer to the online course to complete this activity.

Secure Remote Access (1.3)

In this section, you learn how to configure the management virtual interface on a switch.

Telnet Operation (1.3.1)

You might not always have direct access to your switch when you need to configure it. You need to be able to access it remotely, and it is imperative that your access is secure. This section discusses how to configure Secure Shell (SSH) for remote access. A Packet Tracer activity gives you the opportunity to try this yourself.

Telnet uses Transmission Control Protocol (TCP) port 23. It is an older protocol that uses unsecure plain text transmission of both the login authentication (username and password) and the data transmitted between the communicating devices. A threat actor can monitor packets using Wireshark. For example, in Figure 1-6, the threat actor captured the username **admin** and password **ccna** from a Telnet session.

SSH Operation (1.3.2)

Secure Shell (SSH) is a secure protocol that uses TCP port 22. It provides a secure (encrypted) management connection to a remote device. SSH should replace Telnet for management connections. SSH provides security for remote connections by providing strong encryption when a device is authenticated (username and password) and also for the transmitted data between the communicating devices.

For example, Figure 1-7 shows a Wireshark capture of an SSH session. The threat actor can track the session using the IP address of the administrator device. However, unlike Telnet, with SSH the username and password are encrypted.

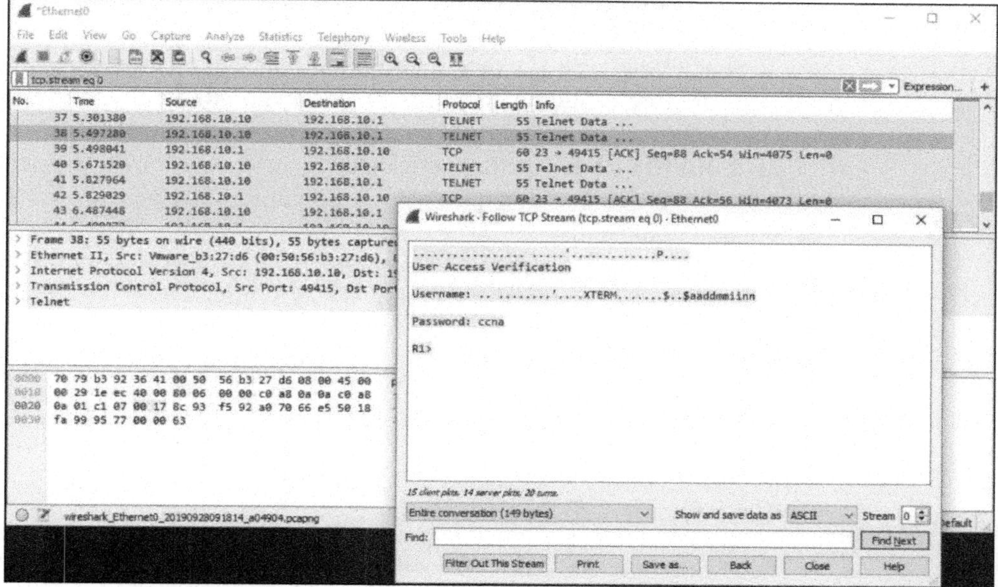

Figure 1-6 Telnet Session Capture

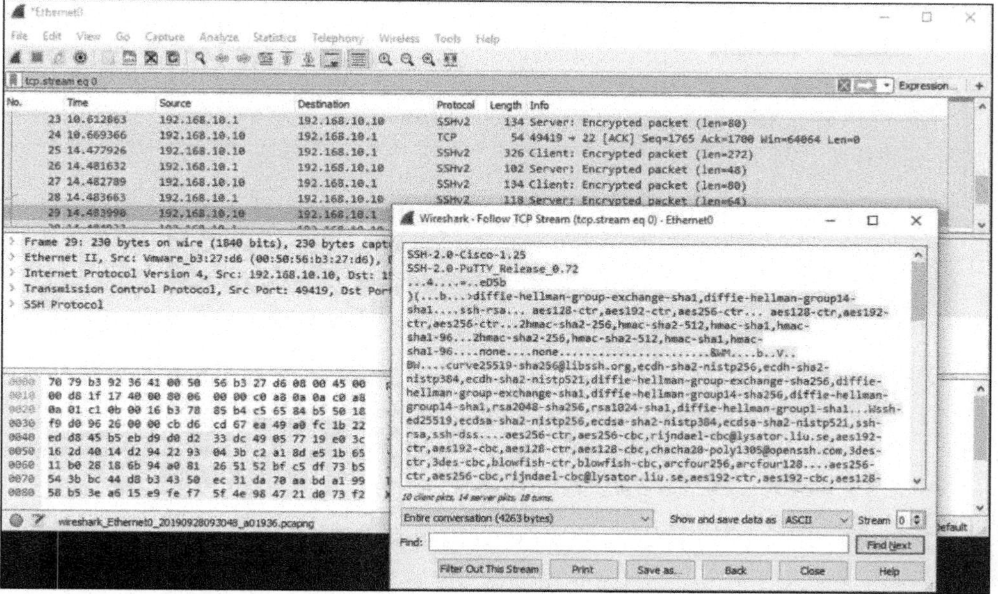

Figure 1-7 SSH Session Capture

Verify the Switch Supports SSH (1.3.3)

To enable SSH on a Catalyst 2960 switch, the switch must be using a version of the IOS software including cryptographic (encrypted) features and capabilities. Use the **show version** command on the switch to see which IOS the switch is currently running. An IOS filename that includes the combination "k9" supports cryptographic (encrypted) features and capabilities. Example 1-10 shows the output of the **show version** command.

Example 1-10 Checking Whether the IOS Supports SSH

```
S1# show version
Cisco IOS Software, C2960 Software (C2960-LANBASEK9-M), Version 15.0(2)SE7,
  RELEASE SOFTWARE (fc1)
```

Configure SSH (1.3.4)

Before configuring SSH, the switch must be minimally configured with a unique hostname and the correct network connectivity settings.

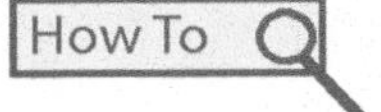

Step 1. **Verify SSH support.** Use the **show ip ssh** command, as shown in Example 1-11, to verify that the switch supports SSH. If the switch is not running an IOS that supports cryptographic features, this command is unrecognized.

Example 1-11 Verify SSH Support

```
S1# show ip ssh
```

Step 2. **Configure the IP domain.** Configure the IP domain name of the network using the **ip domain-name** *domain-name* global configuration mode command. In Example 1-12, the *domain-name* value is cisco.com.

Example 1-12 Configure the IP Domain

```
S1(config)# ip domain-name cisco.com
```

Step 3. **Generate RSA key pairs.** Not all versions of the IOS default to SSH version 2, and SSH version 1 has known security flaws. To configure SSH version 2, issue the **ip ssh version 2** global configuration mode command. Generating a Rivest, Shamir, Adleman (RSA) key pair automatically enables SSH. Use the **crypto key generate rsa** global configuration mode command to enable the SSH server on the switch and generate an RSA key pair. When generating RSA keys, the administrator is prompted to enter a modulus length. The sample configuration in Example 1-13 uses a modulus size of 1,024 bits. A longer modulus length is more secure, but it takes longer to generate and to use.

Note

To delete the RSA key pair, use the **crypto key zeroize rsa** global configuration mode command. After the RSA key pair is deleted, the SSH server is automatically disabled.

Example 1-13 Generate RSA Key Pairs

```
S1(config)# crypto key generate rsa
How many bits in the modulus [512]: 1024
```

Step 4. **Configure user authentication.** The SSH server can authenticate users locally or using an authentication server. To use the local authentication method, create a username and password pair using the **username** *username* **secret** *password* global configuration mode command. In Example 1-14, the user **admin** is assigned the password **ccna.**

Example 1-14 Configure User Authentication

```
S1(config)# username admin secret ccna
```

Step 5. **Configure the vty lines.** Enable the SSH protocol on the vty lines by using the **transport input ssh** line configuration mode command, as shown in Example 1-15. The Catalyst 2960 has vty lines ranging from 0 to 15. This configuration prevents non-SSH (such as Telnet) connections and limits the switch to accept only SSH connections. Use the **line vty** global configuration mode command and then the **login local** line configuration mode command to require local authentication for SSH connections from the local username database.

Example 1-15 Configure the VTY Lines

```
S1(config)# line vty 0 15
S1(config-line)# transport input ssh
S1(config-line)# login local
S1(config-line)# exit
```

Step 6. **Enable SSH version 2.** By default, SSH supports both versions 1 and 2. When supporting both versions, this is shown in the **show ip ssh** output as supporting version 2. Enable SSH version using the **ip ssh version 2** global configuration command, as shown in Example 1-16.

Example 1-16 Enable SSH Version 2

```
S1(config)# ip ssh version 2
```

Verify SSH Is Operational (1.3.5)

On a PC, an SSH client such as PuTTY is used to connect to an SSH server. For example, assume the following is configured:

- SSH is enabled on switch S1.
- Interface VLAN 99 (SVI) with IPv4 address 172.17.99.11 on switch S1.
- PC1 with IPv4 address 172.17.99.21.

Figure 1-8 shows the PuTTY settings for PC1 to initiate an SSH connection to the SVI VLAN IPv4 address of S1.

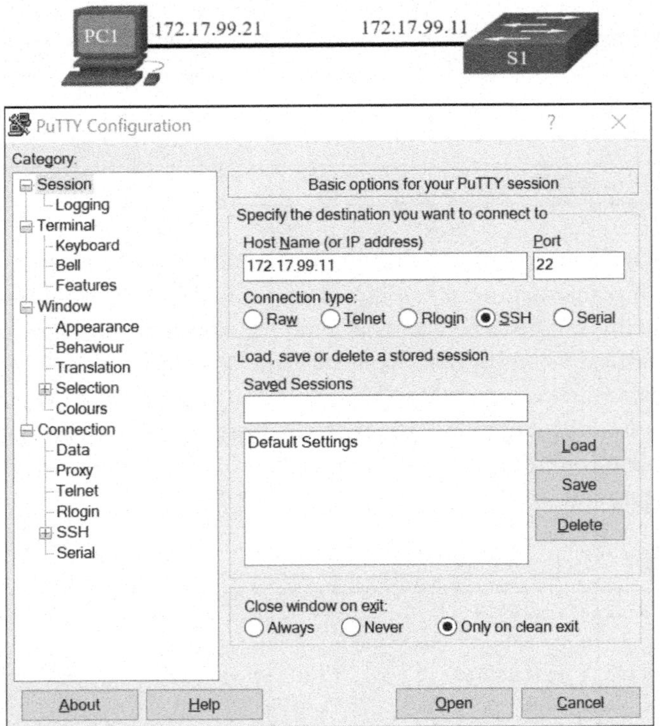

Figure 1-8 PuTTY Settings

When connected, the user is prompted for a username and password, as shown in Example 1-17. Using the configuration from Example 1-14, the username **admin** and password **ccna** are entered. After entering the correct combination, the user is connected via SSH to the command line interface (CLI) on the Catalyst 2960 switch.

Example 1-17 Login Via SSH

```
Login as: admin
Using keyboard-interactive
Authentication.
Password:
S1> enable
Password:
S1#
```

To display the version and configuration data for SSH on the device that you configured as an SSH server, use the **show ip ssh** command. In Example 1-18, SSH version 2 is enabled.

Example 1-18 Verify SSH

```
S1# show ip ssh
SSH Enabled - version 2.0
Authentication timeout: 120 secs; Authentication retries: 3
To check the SSH connections to the device, use the show ssh command as shown.
S1# show ssh
%No SSHv1 server connections running.
Connection Version Mode Encryption  Hmac                  State           Username
0            2.0     IN   aes256-cbc  hmac-sha1   Session started         admin
0            2.0     OUT  aes256-cbc  hmac-sha1   Session started         admin
S1#
```

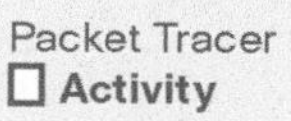

Packet Tracer—Configure SSH (1.3.6)

SSH should replace Telnet for management connections. Telnet uses insecure plain text communications. SSH provides security for remote connections by providing strong encryption of all transmitted data between devices. In this activity, you secure a remote switch with password encryption and SSH.

Basic Router Configuration (1.4)

Every network has unique settings that must be configured on a router. This section introduces basic IOS commands that are required to configure a router.

Configure Basic Router Settings (1.4.1)

Up to now, this module has covered only switches. If you want devices to be able to send and receive data outside of your network, you have to configure routers. This section teaches you basic router configuration and provides two Syntax Checkers and a Packet Tracer activity so you can practice these skills.

Cisco routers and Cisco switches have many similarities. They support a similar modal operating system, similar command structures, and many of the same commands. In addition, both devices have similar initial configuration steps. For example, the following configuration tasks should always be performed. Name the device to distinguish it from other routers and configure passwords, as shown in Example 1-19.

Example 1-19 Basic Router Configuration

```
Router# configure terminal
Enter configuration commands, one per line.  End with CNTL/Z.
Router(config)# hostname R1
R1(config)# enable secret class
R1(config)# line console 0
R1(config-line)# password cisco
R1(config-line)# login
R1(config-line)# exit
R1(config)# line vty 0 4
R1(config-line)# password cisco
R1(config-line)# login
R1(config-line)# exit
R1(config)# service password-encryption
R1(config)#
```

Configure a banner to provide legal notification of unauthorized access, as shown Example 1-20.

Example 1-20 Configure a Banner

```
R1(config)# banner motd $ Authorized Access Only! $
R1(config)#
```

Save the changes on a router, as shown in Example 1-21.

Example 1-21 Save the Configuration

```
R1# copy running-config startup-config
Destination filename [startup-config]?
Building configuration...
[OK]
```

Interactive Graphic

Syntax Checker—Configure Basic Router Settings (1.4.2)

In this Syntax Checker activity, you configure basic settings for R2.

Refer to the online course to complete this activity.

Dual Stack Topology (1.4.3)

One distinguishing feature between switches and routers is the type of interfaces supported by each. For example, Layer 2 switches support LANs; therefore, they have multiple FastEthernet or Gigabit Ethernet ports. The *dual stack topology* in Figure 1-9 is used to demonstrate the configuration of router IPv4 and IPv6 interfaces.

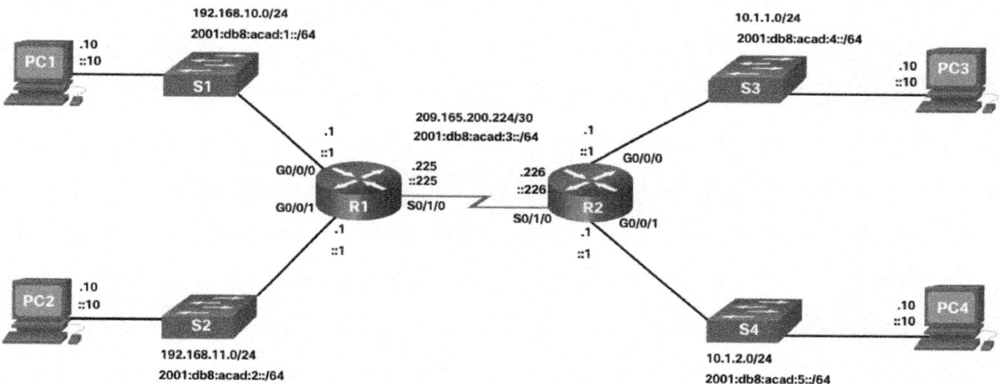

Figure 1-9 Dual Stack Topology

Configure Router Interfaces (1.4.4)

Routers support LANs and WANs and can interconnect different types of networks; therefore, they support many types of interfaces. For example, G2 ISRs have one or two integrated Gigabit Ethernet interfaces and *High-Speed WAN Interface Card (HWIC)* slots to accommodate other types of network interfaces, including serial, DSL, and cable interfaces.

To be available, an interface must be

- **Configured with at least one IP address:** Use the **ip address** *ip-address subnet-mask* and the **ipv6 address** *ipv6-address/prefix* interface configuration commands.
- **Activated:** By default, LAN and WAN interfaces are not activated (**shutdown**). To enable an interface, it must be activated using the **no shutdown** command.

(This is similar to powering on the interface.) The interface must also be connected to another device (a hub, a switch, or another router) for the physical layer to be active.

- **Description:** Optionally, the interface could also be configured with a short description of up to 240 characters. It is good practice to configure a description on each interface. On production networks, the benefits of interface descriptions are quickly realized as they are helpful in troubleshooting and in identifying a third-party connection and contact information.

Example 1-22 shows the configuration for the interfaces on R1.

Example 1-22 R1 Dual Stack Configuration

```
R1(config)# interface gigabitethernet 0/0/0
R1(config-if)# ip address 192.168.10.1 255.255.255.0
R1(config-if)# ipv6 address 2001:db8:acad:1::1/64
R1(config-if)# description Link to LAN 1
R1(config-if)# no shutdown
R1(config-if)# exit
R1(config)# interface gigabitethernet 0/0/1
R1(config-if)# ip address 192.168.11.1 255.255.255.0
R1(config-if)# ipv6 address 2001:db8:acad:2::1/64
R1(config-if)# description Link to LAN 2
R1(config-if)# no shutdown
R1(config-if)# exit
R1(config)# interface serial 0/0/0
R1(config-if)# ip address 209.165.200.225 255.255.255.252
R1(config-if)# ipv6 address 2001:db8:acad:3::225/64
R1(config-if)# description Link to R2
R1(config-if)# no shutdown
R1(config-if)# exit
R1(config)#
```

Interactive Graphic

Syntax Checker—Configure Router Interfaces (1.4.5)

In this Syntax Checker activity, you will configure R2 with its IPv4 and IPv6 interfaces.

Refer to the online course to complete this activity.

IPv4 Loopback Interfaces (1.4.6)

Another common configuration of Cisco IOS routers is enabling a *loopback interface*.

The loopback interface is a logical interface that is internal to the router. It is not assigned to a physical port and can never be connected to any other device. It is considered a software interface that is automatically placed in an "up" state, as long as the router is functioning.

The loopback interface is useful in testing and managing a Cisco IOS device because it ensures that at least one interface will always be available. For example, it can be used for testing purposes, such as testing internal routing processes, by emulating networks behind the router.

Loopback interfaces are also commonly used in lab environments to create additional interfaces. For example, you can create multiple loopback interfaces on a router to simulate more networks for configuration practice and testing purposes. In this curriculum, we often use a loopback interface to simulate a link to the Internet.

Enabling and assigning a loopback address is simple:

```
Router(config)# interface loopback number
Router(config-if)# ip address ip-address subnet-mask
```

Multiple loopback interfaces can be enabled on a router. The IPv4 address for each loopback interface must be unique and unused by any other interface, as shown in Example 1-23, configuration of loopback interface 0 on R1.

Example 1-23 Loopback Interface Configuration

```
R1(config)# interface loopback 0
R1(config-if)# ip address 10.0.0.1 255.255.255.0
R1(config-if)# exit
R1(config)#
%LINEPROTO-5-UPDOWN: Line protocol on Interface Loopback0, changed state to up
```

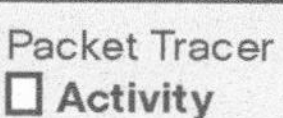

Packet Tracer—Configure Router Interfaces (1.4.7)

In this Packet Tracer activity, you configure routers with IPv4 and IPv6 addressing.

Configure Router Interfaces.

Verify Directly Connected Networks (1.5)

It is always important to know how to troubleshoot and verify whether a device is configured correctly. The focus of this section is on how to verify connectivity between two networks that are directly connected to a router.

Interface Verification Commands (1.5.1)

There is no point in configuring your router unless you verify the configuration and connectivity. This section covers the commands to use to verify directly connected networks. It includes two Syntax Checkers and a Packet Tracer.

There are several **show** commands that can be used to verify the operation and configuration of an interface. The topology in the previous Figure 1-9 is used to demonstrate the verification of router interface settings.

The following commands are especially useful to quickly identify the status of an interface:

- **show ip interface brief** and **show ipv6 interface brief**: These display a summary for all interfaces including the IPv4 or IPv6 address of the interface and current operational status.
- **show running-config interface** *interface-id*: This displays the commands applied to the specified interface.
- **show ip route** and **show ipv6 route**: These display the contents of the IPv4 or IPv6 routing table stored in RAM. In Cisco IOS 15, active interfaces should appear in the routing table with two related entries identified by the code 'C' (Connected) or 'L' (Local). In previous IOS versions, only a single entry with the code 'C' will appear.

Verify Interface Status (1.5.2)

The output of the **show ip interface brief** and **show ipv6 interface brief** commands can be used to quickly reveal the status of all interfaces on the router. You can verify that the interfaces are active and operational as indicated by the Status of "up" and Protocol of "up", as shown in Example 1-24. A different output would indicate a problem with either the configuration or the cabling.

Example 1-24 Verify Interface Status

```
R1# show ip interface brief
Interface              IP-Address      OK? Method Status                Protocol
GigabitEthernet0/0/0   192.168.10.1    YES manual up                    up
GigabitEthernet0/0/1   192.168.11.1    YES manual up                    up
Serial0/1/0            209.165.200.225 YES manual up                    up
Serial0/1/1            unassigned      YES unset  administratively down down
R1# show ipv6 interface brief
GigabitEthernet0/0/0   [up/up]
    FE80::7279:B3FF:FE92:3130
    2001:DB8:ACAD:1::1
```

```
GigabitEthernet0/0/1     [up/up]
    FE80::7279:B3FF:FE92:3131
    2001:DB8:ACAD:2::1
Serial0/1/0              [up/up]
    FE80::7279:B3FF:FE92:3130
    2001:DB8:ACAD:3::1
Serial0/1/1              [down/down]     Unassigned
```

Verify IPv6 Link Local and Multicast Addresses (1.5.3)

The output of the **show ipv6 interface brief** command displays two configured IPv6 addresses per interface. One address is the IPv6 global unicast address that was manually entered. The other address, which begins with FE80, is the link-local unicast address for the interface. A link-local address is automatically added to an interface whenever a global unicast address is assigned. An IPv6 network interface is required to have a link-local address, but not necessarily a global unicast address.

The **show ipv6 interface gigabitethernet 0/0/0** command displays the interface status and all the IPv6 addresses belonging to the interface. Along with the link local address and global unicast address, the output includes the multicast addresses assigned to the interface, beginning with prefix FF02, as shown in Example 1-25.

Example 1-25 Verify IPv6 Link Local and Multicast Addresses

```
R1# show ipv6 interface gigabitethernet 0/0/0
GigabitEthernet0/0/0 is up, line protocol is up
  IPv6 is enabled, link-local address is FE80::7279:B3FF:FE92:3130
  No Virtual link-local address(es):
  Global unicast address(es):
    2001:DB8:ACAD:1::1, subnet is 2001:DB8:ACAD:1::/64
  Joined group address(es):
    FF02::1
    FF02::1:FF00:1
    FF02::1:FF92:3130
  MTU is 1500 bytes
  ICMP error messages limited to one every 100 milliseconds
  ICMP redirects are enabled
  ICMP unreachables are sent
  ND DAD is enabled, number of DAD attempts: 1
  ND reachable time is 30000 milliseconds (using 30000)
  ND advertised reachable time is 0 (unspecified)
  ND advertised retransmit interval is 0 (unspecified)
  ND router advertisements are sent every 200 seconds
  ND router advertisements live for 1800 seconds
  ND advertised default router preference is Medium
```

Verify Interface Configuration (1.5.4)

The output of the **show running-config interface** command displays the current commands applied to the specified interface, as shown in Example 1-26.

Example 1-26 Verify Interface Configuration

```
R1# show running-config interface gigabitethernet 0/0/0
Building configuration...
Current configuration : 158 bytes
!
interface GigabitEthernet0/0/0
 description Link to LAN 1
 ip address 192.168.10.1 255.255.255.0
 negotiation auto
 ipv6 address 2001:DB8:ACAD:1::1/64
end
R1#
```

The following two commands are used to gather more detailed interface information:

- **show interfaces:** Displays interface information and packet flow count for all interfaces on the device.
- **show ip interface** and **show ipv6 interface:** Displays the IPv4 and IPv6 related information for all interfaces on a router.

Verify Routes (1.5.5)

The output of the **show ip route** and **show ipv6 route** commands reveal the three directly connected network entries and the three local host route interface entries, as shown in Example 1-27.

Example 1-27 Verify Routes

```
R1# show ip route
Codes: L - local, C - connected, S - static, R - RIP, M - mobile, B - BGP

Gateway of last resort is not set
      192.168.10.0/24 is variably subnetted, 2 subnets, 2 masks
C        192.168.10.0/24 is directly connected, GigabitEthernet0/0/0
L        192.168.10.1/32 is directly connected, GigabitEthernet0/0/0
      192.168.11.0/24 is variably subnetted, 2 subnets, 2 masks
C        192.168.11.0/24 is directly connected, GigabitEthernet0/0/1
L        192.168.11.1/32 is directly connected, GigabitEthernet0/0/1
      209.165.200.0/24 is variably subnetted, 2 subnets, 2 masks
C        209.165.200.224/30 is directly connected, Serial0/1/0
L        209.165.200.225/32 is directly connected, Serial0/1/0
```

```
R1#
R1# show ipv6 route
IPv6 Routing Table - default - 7 entries
Codes: C - Connected, L - Local, S - Static, U - Per-user Static route

C    2001:DB8:ACAD:1::/64 [0/0]
      via GigabitEthernet0/0/0, directly connected
L    2001:DB8:ACAD:1::1/128 [0/0]
      via GigabitEthernet0/0/0, receive
C    2001:DB8:ACAD:2::/64 [0/0]
      via GigabitEthernet0/0/1, directly connected
L    2001:DB8:ACAD:2::1/128 [0/0]
      via GigabitEthernet0/0/1, receive
C    2001:DB8:ACAD:3::/64 [0/0]
      via Serial0/1/0, directly connected
L    2001:DB8:ACAD:3::1/128 [0/0]
      via Serial0/1/0, receive
L    FF00::/8 [0/0]
      via Null0, receive
R1#
```

The local host route has an administrative distance of 0. It also has a /32 mask for IPv4, and a /128 mask for IPv6. The local host route is for routes on the router that owns the IP address. It is used to allow the router to process packets destined to that IP.

The IPv6 global unicast address applied to the interface is also installed in the routing table as a local route. The local route has a /128 prefix. Local routes are used by the routing table to efficiently process packets with the interface address of the router as the destination.

A 'C' next to a route within the routing table indicates that this is a directly connected network. When the router interface is configured with a global unicast address and is in the "up/up" state, the IPv6 prefix and prefix length are added to the IPv6 routing table as a connected route. The administrative distance of a directly connected network is 0.

The **ping** command for IPv6 is identical to the command used with IPv4 except that an IPv6 address is used. As shown in Example 1-28, the **ping** command is used to verify Layer 3 connectivity between R1 and PC1.

Example 1-28 Ping an IPv6 Address

```
R1# ping 2001:db8:acad:1::10
Type escape sequence to abort.
Sending 5, 100-byte ICMP Echos to 2001:DB8:ACAD:1::10, timeout is 2 seconds:
!!!!!
Success rate is 100 percent (5/5), round-trip min/avg/max = 1/1/1 ms
```

Filter Show Command Output (1.5.6)

Commands that generate multiple screens of output are, by default, paused after 24 lines. At the end of the paused output, the --More-- text displays. Pressing Enter displays the next line and pressing the spacebar displays the next set of lines. Use the **terminal length** command to specify the number of lines to be displayed. A value of 0 (zero) prevents the router from pausing between screens of output.

Another very useful feature that improves the user experience in the CLI is the filtering of **show** output. Filtering commands can be used to display specific sections of output. To enable the filtering command, enter a pipe (|) character after the **show** command and then enter a filtering parameter and a filtering expression.

There are four filtering parameters that can be configured after the pipe:

The section Filter

The **section** filter shows the entire section that starts with the filtering expression, as shown in Example 1-29.

Example 1-29 The **section** Filter

```
R1# show running-config | section line vty
line vty 0 4
 password 7 110A1016141D
 login
 transport input all
```

Note

Output filters can be used in combination with any **show** command.

The include Filter

The **include** filter includes all output lines that match the filtering expression, as shown in Example 1-30.

Example 1-30 The **include** Filter

```
R1# show ip interface brief
Interface              IP-Address      OK? Method Status                Protocol
GigabitEthernet0/0/0   192.168.10.1    YES manual up                    up
GigabitEthernet0/0/1   192.168.11.1    YES manual up                    up
Serial0/1/0            209.165.200.225 YES manual up                    up
Serial0/1/1            unassigned      NO  unset  down                  down
```

```
R1#
R1# show ip interface brief | include up
GigabitEthernet0/0/0    192.168.10.1     YES manual up                    up
GigabitEthernet0/0/1    192.168.11.1     YES manual up                    up
Serial0/1/0             209.165.200.225  YES manual up                    up
```

The exclude Filter

The **exclude** filter excludes all output lines that match the filtering expression, as shown in Example 1-31.

Example 1-31 The **exclude** Filter

```
R1# show ip interface brief
Interface               IP-Address       OK? Method Status                Protocol
GigabitEthernet0/0/0    192.168.10.1     YES manual up                    up
GigabitEthernet0/0/1    192.168.11.1     YES manual up                    up
Serial0/1/0             209.165.200.225  YES manual up                    up
Serial0/1/1             unassigned       NO  unset  down                  down
R1#
R1# show ip interface brief | exclude unassigned
Interface               IP-Address       OK? Method Status                Protocol
GigabitEthernet0/0/0    192.168.10.1     YES manual up                    up
GigabitEthernet0/0/1    192.168.11.1     YES manual up                    up
Serial0/1/0             209.165.200.225  YES manual up                    up
```

The begin Filter

The **begin** filter shows all the output lines from a certain point, starting with the line that matches the filtering expression, as shown in Example 1-32.

Example 1-32 The **begin** Filter

```
R1# show ip route | begin Gateway
Gateway of last resort is not set
      192.168.10.0/24 is variably subnetted, 2 subnets, 2 masks
C        192.168.10.0/24 is directly connected, GigabitEthernet0/0/0
L        192.168.10.1/32 is directly connected, GigabitEthernet0/0/0
      192.168.11.0/24 is variably subnetted, 2 subnets, 2 masks
C        192.168.11.0/24 is directly connected, GigabitEthernet0/0/1
L        192.168.11.1/32 is directly connected, GigabitEthernet0/0/1
      209.165.200.0/24 is variably subnetted, 2 subnets, 2 masks
C        209.165.200.224/30 is directly connected, Serial0/1/0
L        209.165.200.225/32 is directly connected, Serial0/1/0
```

Interactive Graphic

Syntax Checker—Filter Show Command Output (1.5.7)

In this Syntax Checker activity, you filter output for **show** commands.

Refer to the online course to complete this activity.

Command History Feature (1.5.8)

The command history feature is useful because it temporarily stores the list of executed commands to be recalled.

To recall commands in the history buffer, press CTRL+P or the Up Arrow key. The command output begins with the most recent command. Repeat the key sequence to recall successively older commands. To return to more recent commands in the history buffer, press CTRL+N or the Down Arrow key. Repeat the key sequence to recall successively more recent commands.

By default, command history is enabled and the system captures the last 10 command lines in its history buffer. Use the **show history** privileged EXEC command to display the contents of the buffer.

It is also practical to increase the number of command lines that the history buffer records during the current terminal session only. Use the **terminal history size** user EXEC command to increase or decrease the size of the buffer.

An example of the **terminal history size** and **show history** commands is shown in Example 1-33.

Example 1-33 Setting and Displaying Command History

```
R1# terminal history size 200
R1# show history
   show ip int brief
   show interface g0/0/0
   show ip route
   show running-config
   show history
   terminal history size 200
```

Interactive Graphic

Syntax Checker—Command History Features (1.5.9)

In this Syntax Checker activity, you use the command history feature.

Refer to the online course to complete this activity.

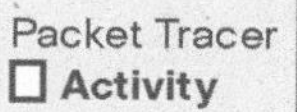

Packet Tracer—Verify Directly Connected Networks (1.5.10)

In this Packet Tracer activity, routers R1 and R2 each have two LANs. Your task is to verify the addressing on each device and verify connectivity between the LANs.

Interactive Graphic

Check Your Understanding—Verify Directly Connected Networks (1.5.11)

Refer to the online course to complete this activity.

Summary (1.6)

The following is a summary of each section in the chapter:

Configure a Switch with Initial Settings

After a Cisco switch is powered on, it goes through a five-step boot sequence. The BOOT environment variable is set using the **boot system** global configuration mode command. The IOS is located in a distinct folder, and the folder path is specified. Use the switch LEDs to monitor switch activity and performance: SYST, RPS, STAT, DUPLX, SPEED, and PoE. The boot loader provides access into the switch if the operating system cannot be used because of missing or damaged system files. The boot loader has a command line that provides access to the files stored in flash memory. To prepare a switch for remote management access, the switch must be configured with an IP address and a subnet mask. To manage the switch from a remote network, the switch must be configured with a default gateway. To configure the switch SVI, you must first configure the management interface, then configure the default gateway, and finally, verify your configuration.

Configure Switch Ports

Full-duplex communication increases effective bandwidth by allowing both ends of a connection to transmit and receive data simultaneously. Half-duplex communication is unidirectional. Switch ports can be manually configured with specific duplex and speed settings. Use autonegotiation when the speed and duplex settings of the device connecting to the port are unknown or may change. When auto-MDIX is enabled, the interface automatically detects the required cable connection type (straight-through or crossover) and configures the connection appropriately. There are several **show** commands to use when verifying switch configurations. Use the **show running-config** command and the **show interfaces** command to verify a switch port configuration. The output from the **show interfaces** command is also useful for detecting common network access layer issues because it displays the line and data link protocol status. The reported input errors from the **show interfaces** command include runt frames, giants, CRC errors, along with collisions and late collisions. Use **show interfaces** to determine whether your network has no connection or a bad connection between a switch and another device.

Secure Remote Access

Telnet (using TCP port 23) is an older protocol that uses unsecure plain text transmission of both the login authentication (username and password) and the data transmitted between the communicating devices. SSH (using TCP port 22) is a secure protocol that provides an encrypted management connection to a remote device. SSH provides security for remote connections by providing strong encryption when a device is authenticated (username and password) and also for the transmitted data between the communicating devices. Use the **show version** command on the switch to see which IOS the switch is currently running. An IOS filename that includes the combination "k9" supports cryptographic features and capabilities. To configure SSH you must verify that the switch supports it, configure the IP domain, generate RSA key pairs, configure use authentication, configure the VTY lines, and enable SSH version 2. To verify that SSH is operational, use the **show ip ssh** command to display the version and configuration data for SSH on the device.

Basic Router Configuration

The following initial configuration tasks should always be performed. Name the device to distinguish it from other routers and configure passwords, configure a banner to provide legal notification of unauthorized access, and save the changes on a router. One distinguishing feature between switches and routers is the type of interfaces supported by each. For example, Layer 2 switches support LANs and, therefore, have multiple FastEthernet or GigabitEthernet ports. The dual stack topology is used to demonstrate the configuration of router IPv4 and IPv6 interfaces. Routers support LANs and WANs and can interconnect different types of networks; therefore, they support many types of interfaces. For example, G2 ISRs have one or two integrated Gigabit Ethernet interfaces and High-Speed WAN Interface Card (HWIC) slots to accommodate other types of network interfaces, including serial, DSL, and cable interfaces. The IPv4 loopback interface is a logical interface that is internal to the router. It is not assigned to a physical port and can never be connected to any other device.

Verify Directly Connected Networks

Use the following commands to quickly identify the status of an interface. **show ip interface brief** and **show ipv6 interface brief** to see a summary of all interfaces (IPv4 and IPv6 addresses and operational status), **show running-config interface** *interface-id* to see the commands applied to a specified interface, and **show ip route** and **show ipv6 route** to see the contents of the IPv4 or IPv6 routing table stored in RAM. The output of the **show ip interface brief** and **show ipv6 interface brief** commands can be used to quickly reveal the status of all interfaces on the router. The **show ipv6 interface** *interface-id* command displays the interface status and all the IPv6 addresses belonging to the interface. Along with the link local address and global unicast address, the output includes the multicast addresses assigned to the interface. The output of the **show running-config interface** command displays the current commands applied to a specified interface. The **show interfaces** command displays interface information and packet flow count for all interfaces on the device. Verify interface configuration using the **show ip interface** and **show ipv6 interface** commands, which display the IPv4 and IPv6 related information for all interfaces on a router. Verify routes using the **show ip route** and **show ipv6 route** commands. Filter **show** command output using the pipe (|) character. Use filter expressions: section, include, exclude, and begin. By default, command history is enabled, and the system captures the last 10 command lines in its history buffer. Use the **show history** privileged EXEC command to display the contents of the buffer.

Packet Tracer Activity

Packet Tracer—Implement a Small Network (1.6.1)

In this Packet Tracer activity, routers R1 and R2 each have two LANs. Your task is to verify the addressing on each device and verify connectivity between the LANs.

Lab—Configure Basic Router Settings (1.6.2)

This is a comprehensive lab to review previously covered IOS router commands. You will cable the equipment and complete basic configurations and IPv4 interface settings on the router. You will then use SSH to connect to the router remotely and use IOS commands to retrieve information from the device to answer questions about the router.

Practice

The following activities provide practice with the topics introduced in this chapter. The Labs are available in the companion *Switching, Routing, and Wireless Essentials Labs and Study Guide (CCNAv7)* (ISBN 9780136634386). The Packet Tracer Activity instructions are also in the Labs & Study Guide. The PKA files are found in the online course.

Labs

Lab 1.1.7: Basic Switch Configuration

Lab 1.6.2: Configure Basic Router Settings

Packet Tracer Activities

Packet Tracer 1.3.6: Configure SSH

Packet Tracer 1.4.7: Configure Router Interfaces

Packet Tracer 1.5.10: Verify Directly Connected Networks

Check Your Understanding Questions

Complete all the review questions listed here to test your understanding of the topics and concepts in this chapter. The appendix "Answers to the 'Check Your Understanding' Questions" lists the answers.

1. Which interface is used by default to manage a Cisco Catalyst 2960 switch?

    A. The FastEthernet 0/1 interface

    B. The GigabitEthernet 0/1 interface

    C. The VLAN 1 interface

    D. The VLAN 99 interface

2. A production switch is reloaded and finishes with a Switch> prompt. What two facts can be determined? (Choose two.)

    A. A full version of the Cisco IOS was located and loaded.

    B. POST occurred normally.

    C. The boot process was interrupted.

    D. There is not enough RAM or flash on this router.

    E. The switch did not locate the Cisco IOS in flash, so it defaulted to ROM.

3. Which two statements are true about using full-duplex Fast Ethernet? (Choose two.)

    A. Full-duplex Fast Ethernet offers 100 percent efficiency in both directions.

    B. Latency is reduced because the NIC processes frames faster.

    C. Nodes operate in full-duplex with unidirectional data flow.

    D. Performance is improved because the NIC is able to detect collisions.

    E. Performance is improved with bidirectional data flow.

4. Which statement describes the port speed LED on the Cisco Catalyst 2960 switch?

    A. If the LED is amber, the port is operating at 1000 Mbps.

    B. If the LED is blinking green, the port is operating at 10 Mbps.

    C. If the LED is green, the port is operating at 100 Mbps.

    D. If the LED is off, the port is not operating.

5. What is a function of the switch boot loader?

   A. To control how much RAM is available to the switch during the boot process
   B. To provide an environment to operate in when the switch operating system cannot be found
   C. To provide security for the vulnerable state when the switch is booting
   D. To speed up the boot process

6. In which situation would a technician use the **show interfaces** command?

   A. To determine whether remote access is enabled
   B. To determine the MAC address of a directly attached network device on a particular interface
   C. When packets are being dropped from a particular directly attached host
   D. When an end device can reach local devices, but not remote devices

7. What is one difference between using Telnet or SSH to connect to a network device for management purposes?

   A. Telnet does not provide authentication, whereas SSH provides authentication.
   B. Telnet sends a username and password in plain text, whereas SSH encrypts the username and password.
   C. Telnet supports a host GUI, whereas SSH supports only a host CLI.
   D. Telnet uses UDP as the transport protocol, whereas SSH uses TCP.

8. What is a characteristic of an IPv4 loopback interface on a Cisco IOS router?

   A. It is a logical interface internal to the router.
   B. It is assigned to a physical port and can be connected to other devices.
   C. Only one loopback interface can be enabled on a router.
   D. The **no shutdown** command is required to place this interface in an up state.

9. What two pieces of information are displayed in the output of the **show ip interface brief** command? (Choose two.)

   A. Interface descriptions
   B. IPv4 addresses
   C. Layer 1 statuses
   D. MAC addresses
   E. Next-hop addresses
   F. Speed and duplex settings

10. What type of cable would be used to connect a router to a switch when neither supports the auto-MDIX feature?

    A. Coaxial
    B. Crossover
    C. Rollover
    D. Straight-through

11. Which statement regarding a loopback interface is true?

    A. It is an internal virtual interface used for testing purposes.
    B. It is used to loop back traffic to an interface.
    C. It must be enabled using the **no shutdown** command.
    D. Only one loopback interface can be created on a device.

12. You are implementing remote access to the VTY lines of a switch using SSH and the **login local** line vty command. Which other command must be entered to avoid being locked out of the switch?

    A. **enable secret** *password*
    B. **password** *password*
    C. **service-password encryption**
    D. **username** *username* **secret** *password*

CHAPTER 2

Switching Concepts

Objectives

Upon completion of this chapter, you will be able to answer the following questions:

- How are frames forwarded in a switched network?
- What are collision domains and broadcast domains?

Key Terms

This chapter uses the following key terms. You can find the definitions in the Glossary.

public switched telephone network (PSTN) Page 46

ingress Page 46

egress Page 46

egress port Page 46

ingress port Page 46

MAC address table Page 47

content addressable memory (CAM) Page 47

application-specific integrated circuits (ASICs) Page 48

store-and-forward switching Page 49

cut-through switching Page 49

frame check sequence (FCS) Page 49

buffer Page 49

rapid frame switching Page 49

fragment free switching Page 50

high-performance computing (HPC) applications Page 50

collision domains Page 51

broadcast domains Page 51

high port density Page 54

Introduction (2.0)

You can connect and configure switches, that's great! But even a network with the newest technology develops its own problems eventually. If you have to troubleshoot your network, you need to know how switches work. This module gives you the fundamentals of switches and switch operation. Luckily, switch operation is easy to understand!

Frame Forwarding (2.1)

In this section, you learn how frames are forwarded in a switched network.

Switching in Networking (2.1.1)

The concept of switching and forwarding frames is universal in networking and telecommunications. Various types of switches are used in LANs, WANs, and in the *public switched telephone network (PSTN)*.

The decision on how a switch forwards traffic is made based on the flow of that traffic. Two terms are associated with frames entering and leaving an interface:

- *Ingress*: This is used to describe the port where a frame enters the device.
- *Egress*: This is used to describe the port that frames will use when leaving the device.

A LAN switch maintains a table that is referenced when forwarding traffic through the switch. The only intelligence of a LAN switch is its ability to use its table to forward traffic. A LAN switch forwards traffic based on the ingress port and the destination MAC address of an Ethernet frame, as shown in Figure 2-1.

In Figure 2-1, the switch receives a frame on port 1 with EA as the destination address. The switch looks up the port for EA and forwards the frame out port 4.

With a LAN switch, there is only one master switching table that describes a strict association between MAC addresses and ports; therefore, an Ethernet frame with a given destination address always exits the same *egress port*, regardless of the *ingress port* it enters.

Note

An Ethernet frame will never be forwarded out the same port it was on which it was received.

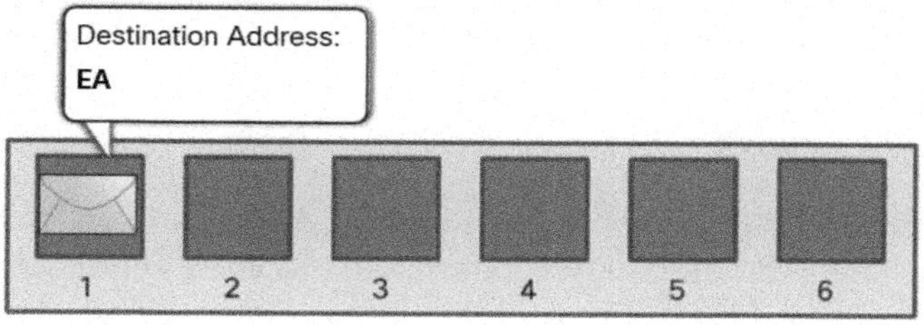

Port Table

Destination Addresses	Port
EE	1
AA	2
BA	3
EA	4
AC	5
AB	6

Figure 2-1 Switch Receives an Ingress Frame

The Switch MAC Address Table (2.1.2)

A switch is made up of integrated circuits and the accompanying software that controls the data paths through the switch. Switches use destination MAC addresses to direct network communications through the switch, out the appropriate port, toward the destination.

For a switch to know which port to use to transmit a frame, it must first learn which devices exist on each port. As the switch learns the relationship of ports to devices, it builds a table called a *MAC address table*. This table is stored in *content addressable memory (CAM)*, which is a special type of memory used in high-speed searching applications. For this reason, the MAC address table is sometimes also called the CAM table.

LAN switches determine how to handle incoming data frames by maintaining the MAC address table. A switch populates its MAC address table by recording the source MAC address of each device connected to each of its ports. The switch references the information in the MAC address table to send frames destined for a specific device out of the port that has been assigned to that device.

The Switch Learn and Forward Method (2.1.3)

The following two-step process is performed on every Ethernet frame that enters a switch.

Step 1. **Learn—Examining the Source MAC Address.** Every frame that enters a switch is checked for new information to learn. It does this by examining the source MAC address of the frame and port number where the frame entered the switch:

- If the source MAC address does not exist in the MAC address table, the MAC address and incoming port number are added to the table.
- If the source MAC address does exist, the switch updates the refresh timer for that entry. By default, most Ethernet switches keep an entry in the table for five minutes. If the source MAC address does exist in the table but on a different port, the switch treats this as a new entry. The entry is replaced using the same MAC address, but with the more current port number.

Step 2. **Forward—Examining the Destination MAC Address.** If the destination MAC address is a unicast address, the switch looks for a match between the destination MAC address of the frame and an entry in its MAC address table:

- If the destination MAC address is in the table, it forwards the frame out of the specified port.
- If the destination MAC address is not in the table, the switch forwards the frame out all ports except the incoming port. This is called an unknown unicast.
- If the destination MAC address is a broadcast or a multicast, the frame is also flooded out all ports except the incoming port.

Video

Video—MAC Address Tables on Connected Switches (2.1.4)

Refer to the online course to view this video.

Switching Forwarding Methods (2.1.5)

Switches make Layer 2 forwarding decisions very quickly. This is because of software on *application-specific-integrated circuits (ASICs)*. ASICs reduce the frame-handling time within the device and allow the device to manage an increased number of frames without degrading performance.

Layer 2 switches use one of two methods to switch frames:

- *Store-and-forward switching*: This method makes a forwarding decision on a frame after it has received the entire frame and checked the frame for errors using a mathematical error-checking mechanism known as a cyclic redundancy check (CRC). Store-and-forward switching is Cisco's primary LAN switching method.
- *Cut-through switching*: This method begins the forwarding process after the destination MAC address of an incoming frame and the egress port have been determined.

Store-and-Forward Switching (2.1.6)

Store-and-forward switching, as distinguished from cut-through switching, has the following two primary characteristics:

- **Error checking:** After receiving the entire frame on the ingress port, the switch compares the *frame check sequence (FCS)* value in the last field of the datagram against its own FCS calculations. The FCS is an error-checking process that helps to ensure that the frame is free of physical and data-link errors. If the frame is error-free, the switch forwards the frame. Otherwise, the frame is dropped.
- **Automatic buffering:** The ingress port buffering process used by store-and-forward switches provides the flexibility to support any mix of Ethernet speeds. For example, handling an incoming frame traveling into a 100 Mbps Ethernet port that must be sent out a 1 Gbps interface would require using the store-and-forward method. With any mismatch in speeds between the ingress and egress ports, the switch stores the entire frame in a *buffer*, computes the FCS check, forwards it to the egress port buffer, and then sends it.

Figure 2-2 illustrates how store-and-forward makes a decision based on the Ethernet frame.

Cut-Through Switching (2.1.7)

The store-and-forward switching method drops frames that do not pass the FCS check. Therefore, it does not forward invalid frames.

By contrast, the cut-through switching method may forward invalid frames because no FCS check is performed. However, cut-through switching has the ability to perform *rapid frame switching*. This means the switch can make a forwarding decision as soon as it has looked up the destination MAC address of the frame in its MAC address table, as shown in Figure 2-3.

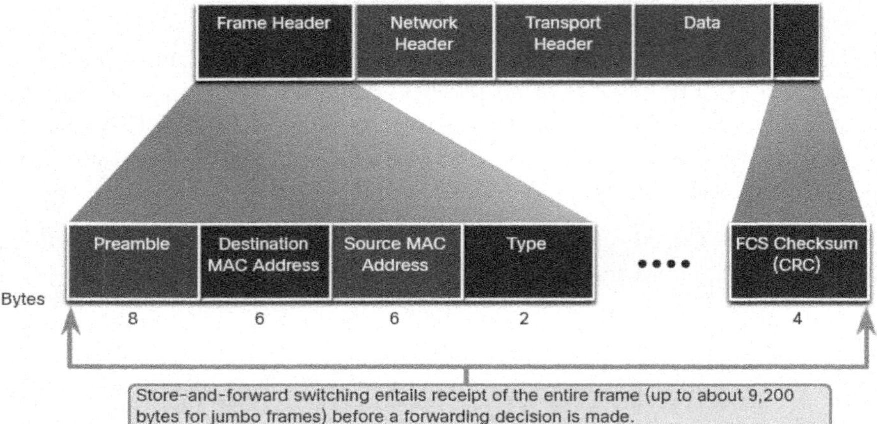

Figure 2-2 Store-and-Forward Method

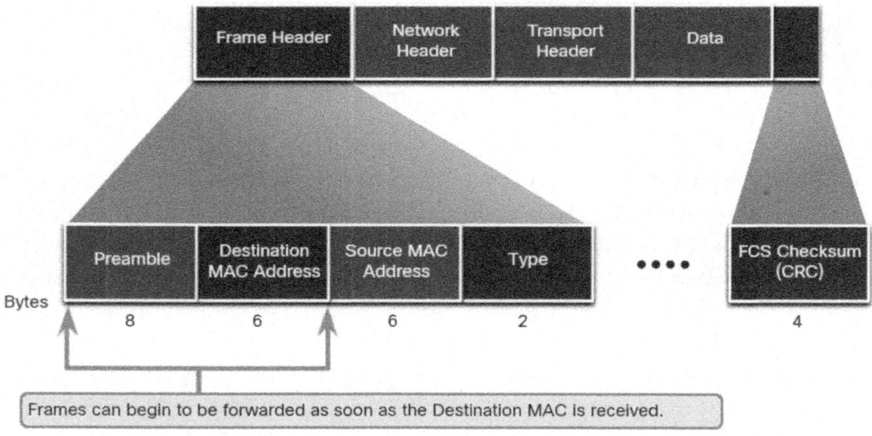

Figure 2-3 Cut-Through Method

The switch does not have to wait for the rest of the frame to enter the ingress port before making its forwarding decision.

Fragment free switching is a modified form of cut-through switching in which the switch only starts forwarding the frame after it has read the Type field. Fragment free switching provides better error checking than cut-through, with practically no increase in latency.

The lower latency speed of cut-through switching makes it more appropriate for extremely demanding, *high-performance computing (HPC) applications* that require process-to-process latencies of 10 microseconds or less.

The cut-through switching method can forward frames with errors. If there is a high error rate (invalid frames) in the network, cut-through switching can have a negative impact on bandwidth, thereby clogging up bandwidth with damaged and invalid frames.

Interactive Graphic

Activity—Switch It! (2.1.8)

Use this activity to check your understanding of how a switch learns and forwards frames.

Refer to the online course to complete this activity.

Collision and Broadcast Domains (2.2)

In this section, you learn about *collision domains* and *broadcast domains*.

Collision Domains (2.2.1)

In the previous section, you gained a better understanding of what a switch is and how it operates. This section discusses how switches work with each other and with other devices to eliminate collisions and reduce network congestion. The terms *collisions* and *congestion* are used here in the same way that you use them in street traffic.

In legacy hub-based Ethernet segments, network devices competed for the shared medium. The network segments that share the same bandwidth between devices are known as collision domains. When two or more devices within the same collision domain try to communicate at the same time, a collision will occur.

If an Ethernet switch port is operating in half-duplex, each segment is in its own collision domain. There are no collision domains when switch ports are operating in full-duplex. However, there could be a collision domain if a switch port is operating in half-duplex.

By default, Ethernet switch ports will autonegotiate full-duplex when the adjacent device can also operate in full-duplex. If the switch port is connected to a device operating in half-duplex, such as a legacy hub, the switch port will operate in half-duplex. In the case of half-duplex, the switch port will be part of a collision domain.

As shown in Figure 2-4, full-duplex is chosen if both devices have the capability along with their highest common bandwidth.

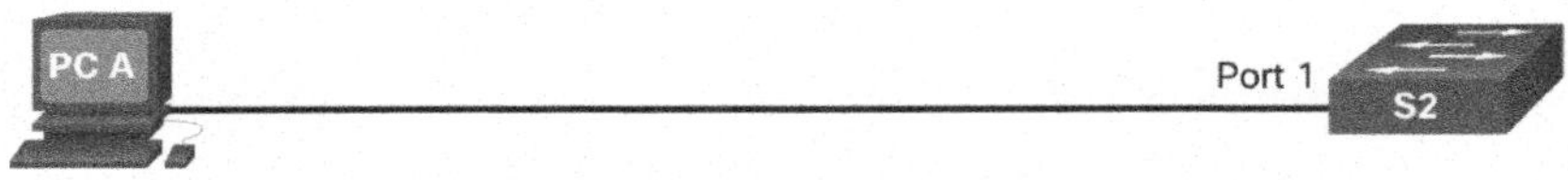

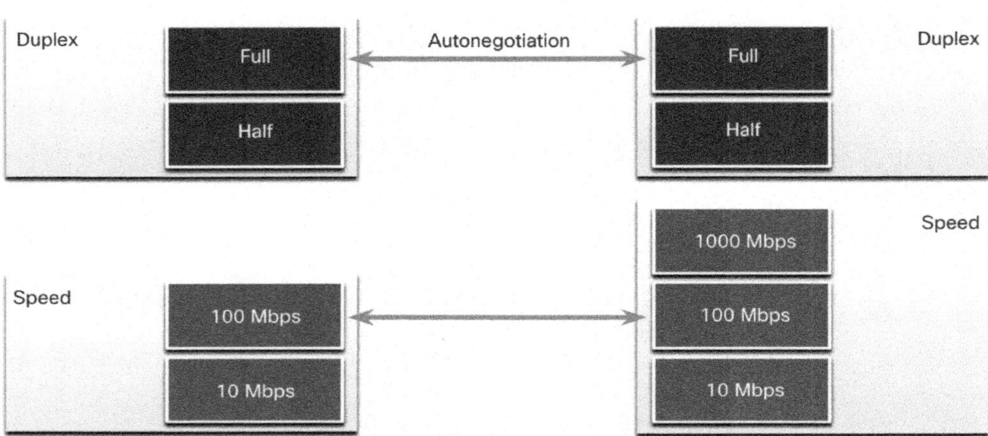

Figure 2-4 Switch Port Auto Negotiating Duplex and Speed

Broadcast Domains (2.2.2)

A collection of interconnected switches forms a single broadcast domain. Only a network layer device, such as a router, can divide a Layer 2 broadcast domain. Routers are used to segment broadcast domains but will also segment a collision domain.

When a device sends a Layer 2 broadcast, the destination MAC address in the frame is set to all binary ones.

The Layer 2 broadcast domain is referred to as the MAC broadcast domain. The MAC broadcast domain consists of all devices on the LAN that receive broadcast frames from a host.

When a switch receives a broadcast frame, it forwards the frame out each of its ports, except the ingress port where the broadcast frame was received. Each device connected to the switch receives a copy of the broadcast frame and processes it. For example, in Figure 2-5, the server sends a broadcast frame to S1. S1 then forwards the frame out all other ports.

Broadcasts are sometimes necessary for initially locating other devices and network services, but they also reduce network efficiency. Network bandwidth is used to propagate the broadcast traffic. Too many broadcasts and a heavy traffic load on a network can result in congestion, which slows down network performance.

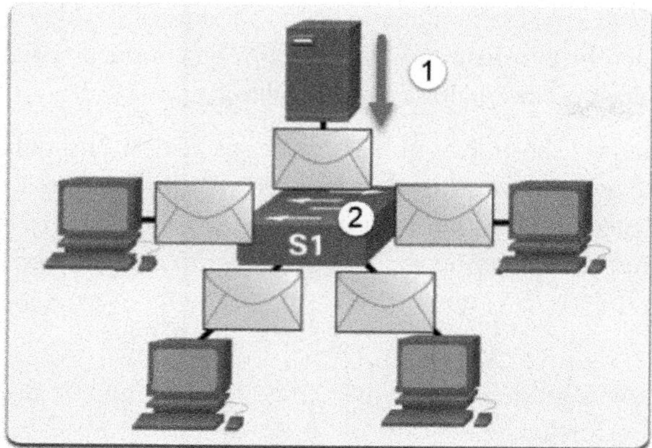

Figure 2-5 Broadcast Domain with One Switch

When two switches are connected together, as shown in Figure 2-6, the broadcast domain is increased—in this case, a broadcast frame from the server forwarded to all connected ports on switch S1. Switch S1 sends the broadcast from to switch S2. The frame is then also propagated to all devices connected to switch S2.

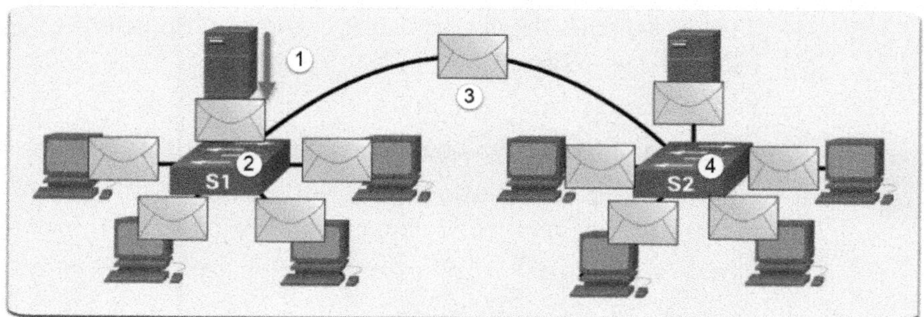

Figure 2-6 Broadcast Domain with Two Switches

Alleviate Network Congestion (2.2.3)

LAN switches have special characteristics that help them alleviate network congestion. By default, interconnected switch ports attempt to establish a link in full-duplex, therefore eliminating collision domains. Each full-duplex port of the switch provides the full bandwidth to the device or devices that are connected to that port. Full-duplex connections have dramatically increased LAN network performance and are required for 1 Gbps Ethernet speeds and higher.

Switches interconnect LAN segments, use a MAC address table to determine egress ports, and can lessen or eliminate collisions entirely. Characteristics of switches that alleviate network congestion include the following:

- **Fast port speeds:** Ethernet switch port speeds vary by model and purpose. For instance, most access layer switches support 100 Mbps and 1 Gbps port speeds. Distribution layer switches support 100 Mbps, 1 Gbps, and 10 Gbps port speeds and core layer and data center switches may support 100 Gbps, 40 Gbps, and 10 Gbps port speeds. Switches with faster port speeds cost more but can reduce congestion.
- **Fast internal switching:** Switches use a fast internal bus or shared memory to provide high performance.
- **Large frame buffers:** Switches use large memory buffers to temporarily store more received frames before having to start dropping them. This enables ingress traffic from a faster port (for example, 1 Gbps) to be forwarded to a slower (for example, 100 Mbps) egress port without losing frames.
- *High port density*: A high port density switch lowers overall costs because it reduces the number of switches required. For instance, if 96 access ports were required, it would be less expensive to buy two 48-port switches instead of four 24-port switches. High port density switches also help keep traffic local, which helps alleviate congestion.

Interactive Graphic

Check Your Understanding—Switching Domains (2.2.4)

Refer to the online course to complete this activity.

Summary (2.3)

The following is a summary of each section in the chapter:

Frame Forwarding

The decision on how a switch forwards traffic is based on the flow of that traffic. The term *ingress* describes the port where a frame enters a device. The term *egress* describes the port that frames use when leaving the device. An Ethernet frame will never be forwarded out the port where it entered. For a switch to know which port to use to transmit a frame, it must first learn which devices exist on each port. As the switch learns the relationship of ports to devices, it builds a table called a MAC address table. Every frame that enters a switch is checked for new information to learn by examining the source MAC address of the frame and port number where the frame entered the switch. If the destination MAC address is a unicast address, the switch will look for a match between the destination MAC address of the frame and an entry in its MAC address table. Switch forwarding methods include store-and-forward and cut-through. Store-and-forward uses error-checking and automatic buffering. Cut-through does not error check. Instead it performs rapid frame switching. This means the switch can make a forwarding decision as soon as it has looked up the destination MAC address of the frame in its MAC address table.

Switching Domains

If an Ethernet switch port is operating in half-duplex, each segment is in its own collision domain. There are no collision domains when switch ports are operating in full-duplex. By default, Ethernet switch ports will autonegotiate full-duplex when the adjacent device can also operate in full-duplex. A collection of interconnected switches forms a single broadcast domain. Only a network layer device, such as a router, can divide a Layer 2 broadcast domain. The Layer 2 broadcast domain is referred to as the MAC broadcast domain. The MAC broadcast domain consists of all devices on the LAN that receive broadcast frames from a host. When a switch receives a broadcast frame, it forwards the frame out each of its ports, except the ingress port where the broadcast frame was received. Each device connected to the switch receives a copy of the broadcast frame and processes it. Switches can do the following: interconnect LAN segments, use a MAC address table to determine egress ports, and lessen or eliminate collisions entirely. Characteristics of switches that alleviate network congestion are fast port speeds, fast internal switching, large frame buffers, and high port density.

Check Your Understanding Questions

Complete all the review questions listed here to test your understanding of the sections and concepts in this chapter. The appendix "Answers to 'Check Your Understanding' Questions" lists the answers.

1. What is one function of a Layer 2 switch?

   A. Determines which interface is used to forward a frame based on the destination MAC address

   B. Duplicates the electrical signal of each frame to every port

   C. Forwards data based on logical addressing

   D. Learns the port assigned to a host by examining the destination MAC address

2. What criteria is used by a Cisco LAN switch to decide how to forward Ethernet frames?

   A. Destination IP address

   B. Destination MAC address

   C. Egress port

   D. Path cost

3. Which type of address does a switch use to build the MAC address table?

   A. Destination IP address

   B. Destination MAC address

   C. Source IP address

   D. Source MAC address

4. What are two reasons a network administrator would segment a network with a Layer 2 switch? (Choose two.)

   A. To create fewer collision domains

   B. To create more broadcast domains

   C. To eliminate virtual circuits

   D. To enhance user bandwidth

   E. To isolate ARP request messages from the rest of the network

   F. To isolate traffic between segments

5. A switch has received a frame on an ingress port. What will the switch do if the unicast destination MAC address is in the MAC address table?

   A. It will drop the frame.

   B. It will forward the frame out all ports.

   C. It will forward the frame out all ports except the incoming port.

   D. It will forward the frame out of the specified port in the MAC address table.

6. A switch has received a frame on an ingress port. What will the switch do if the unicast destination MAC address is not in the MAC address table?

   A. It will drop the frame.

   B. It will forward the frame out all ports.

   C. It will forward the frame out all ports except the incoming port.

   D. It will forward the frame out of the specified port in the MAC address table.

7. A switch has received a frame on an ingress port. What will the switch do if the destination MAC address is a broadcast address?

   A. It will drop the frame.

   B. It will forward the frame out all ports.

   C. It will forward the frame out all ports except the incoming port.

   D. It will forward the frame out of the specified port in the MAC address table.

8. Which switching method makes use of the FCS value?

   A. Broadcast

   B. Cut-though

   C. Large frame buffer

   D. Store-and-forward

9. Which switching method forwards the frame immediately after examining the destination MAC address?

   A. Broadcast

   B. Cut-though

   C. Large frame buffer

   D. Store-and-forward

10. Which statement about half-duplex and full-duplex communication is true?

A. Gigabit Ethernet and 10 Gb/s NICs can operate in full-duplex or half-duplex.

B. Full-duplex communication is bidirectional.

C. Half-duplex communication allows both ends to transmit and receive simultaneously.

D. Half-duplex communication is unidirectional, or one direction at a time.

CHAPTER 3

VLANs

Objectives

Upon completion of this chapter, you will be able to answer the following questions:

- What is the purpose of VLANs in a switched network?
- How does a switch forward frames based on VLAN configuration in a multiswitch environment?
- How do you configure a switch port to be assigned to a VLAN based on requirements?
- How do you configure a trunk port on a LAN switch?
- How do you configure Dynamic Trunking Protocol (DTP)?

Key Terms

This chapter uses the following key terms. You can find the definitions in the Glossary.

Virtual LAN (VLAN) Page 60

hierarchical network addressing Page 61

default VLAN Page 63

native VLAN Page 63

data VLANs Page 64

tagged Page 64

tagged traffic Page 64

802.1Q Page 64

management VLAN Page 64

voice VLAN Page 65

VLAN trunks Page 66

IEEE 802.1Q header Page 69

VLAN tag field Page 69

Canonical Format Identifier Page 69

tag protocol ID (TPID) Page 69

user priority Page 69

untagged frame Page 70

class of service (CoS) Page 72

CoS priority value Page 72

normal range VLANs Page 73

extended range VLANs Page 74

vlan.dat Page 74

VLAN trunking protocol (VTP) Page 74

dynamic trunking protocol (DTP) Page 87

Introduction (3.0)

Imagine that you are in charge of a very large conference. There are people from all over who share a common interest and some who also have special expertise. Imagine if each expert who wanted to present information to a smaller audience had to do that in the same large room with all the other experts and their smaller audiences. Nobody would be able to hear anything. You would have to find separate rooms for all the experts and their smaller audiences. The *Virtual LAN (VLAN)* does something similar in a network. VLANs are created at Layer 2 to reduce or eliminate broadcast traffic. VLANs are how you break up your network into smaller networks so that the devices and people within a single VLAN are communicating with each other and not having to manage traffic from other networks. The network administrator can organize VLANs by location, who is using them, the type of device, or whatever category is needed. You know you want to learn how to do this, so don't wait!

Overview of VLANs (3.1)

In this section, you learn about the purpose of VLANs in a switched network.

VLAN Definitions (3.1.1)

Organizing your network into smaller networks is not as simple as separating screws and putting them into jars. But it will make your network easier to manage. Virtual LANs (VLANs) provide segmentation and organizational flexibility in a switched network. A group of devices within a VLAN communicates as if each device was attached to the same cable. VLANs are based on logical connections instead of physical connections.

As shown in Figure 3-1, VLANs in a switched network enable users in various departments (such as IT, HR, and Sales) to connect to the same network regardless of the physical switch being used or location in a campus LAN.

VLANs allow an administrator to segment networks based on factors such as function, team, or application, without regard for the physical location of the users or devices. Each VLAN is considered a separate logical network. Devices within a VLAN act as if they are in their own independent network, even if they share a common infrastructure with other VLANs. Any switch port can belong to a VLAN.

Unicast, broadcast, and multicast packets are forwarded and flooded only to end devices within the VLAN where the packets are sourced. Packets destined for devices that do not belong to the VLAN must be forwarded through a device that supports routing.

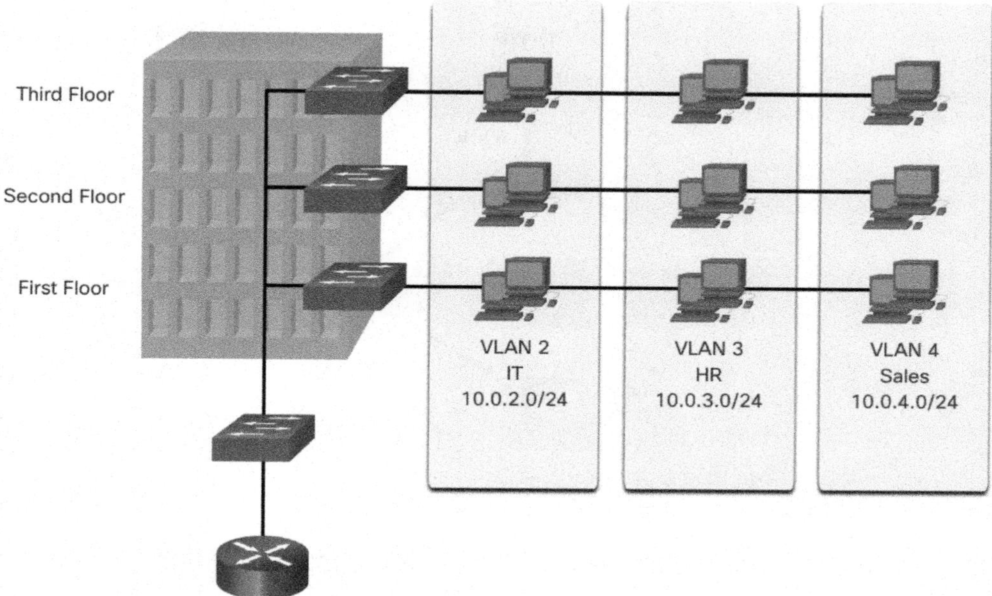

Figure 3-1 Defining VLAN Groups

Multiple IP subnets can exist on a switched network, without the use of multiple VLANs. However, the devices will be in the same Layer 2 broadcast domain. This means that any Layer 2 broadcasts, such as an ARP request, will be received by all devices on the switched network, even by those not intended to receive the broadcast.

A VLAN creates a logical broadcast domain that can span multiple physical LAN segments. VLANs improve network performance by separating large broadcast domains into smaller ones. If a device in one VLAN sends a broadcast Ethernet frame, all devices in the VLAN receive the frame, but devices in other VLANs do not.

Using VLANs, network administrators can implement access and security policies according to specific groupings of users. Each switch port can be assigned to only one VLAN (except for a port connected to an IP phone or to another switch).

Benefits of a VLAN Design (3.1.2)

Each VLAN in a switched network corresponds to an IP network. Therefore, VLAN design must take into consideration the implementation of a hierarchical network-addressing scheme. *Hierarchical network addressing* means that IP network numbers are applied to network segments or VLANs in a way that takes the network as a whole into consideration. Blocks of contiguous network addresses are reserved for and configured on devices in a specific area of the network, as shown in Figure 3-2.

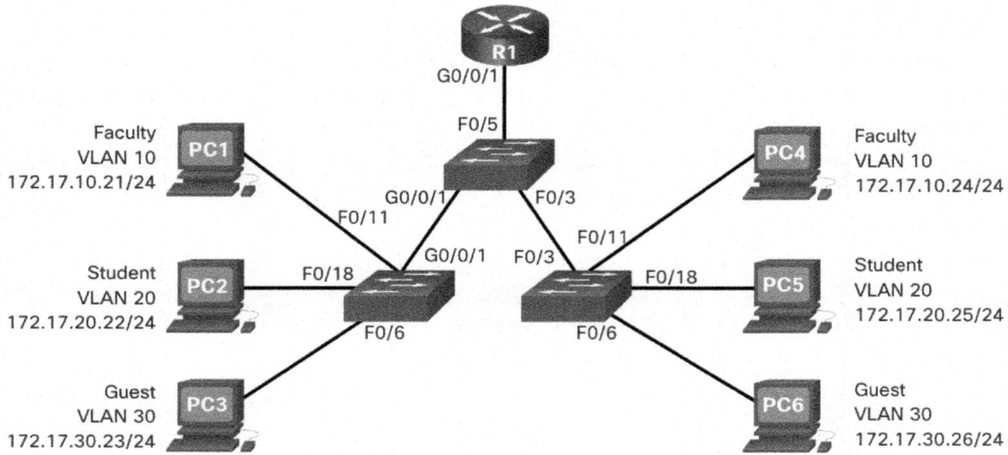

Figure 3-2 VLAN Topology

Table 3-1 lists the benefits of designing a network with VLANs.

Table 3-1 VLAN Benefits

Benefit	Description
Smaller broadcast domains	■ Dividing a network into VLANs reduces the number of devices in the broadcast domain. ■ In Figure 3-1, there are six computers in the network but only three broadcast domains (that is, Faculty, Student, and Guest).
Improved security	■ Only users in the same VLAN can communicate together. ■ For inter-VLAN communication, a router or Layer 3 switch is required. ■ Additional security can be provided using access control lists to restrict traffic between VLANs. ■ In Figure 3-2, faculty network traffic on VLAN 10 is completely separated and secured from users on other VLANs.
Improved IT efficiency	■ VLANs simplify network management because users with similar network requirements can be configured on the same VLAN. ■ VLANs can be named to make them easier to identify. ■ In Figure 3-2, VLAN 10 was named "Faculty," VLAN 20 "Student," and VLAN 30 "Guest."

Benefit	Description
Reduced cost	■ VLANs reduce the need for expensive network upgrades and use the existing bandwidth and uplinks more efficiently, resulting in cost savings.
Better performance	■ Smaller broadcast domains reduce unnecessary traffic on the network and improve performance.
Simpler project and application management	■ VLANs aggregate users and network devices to support business or geographic requirements. ■ Having separate functions makes managing a project or working with a specialized application easier; an example of such an application is an e-learning development platform for faculty.

Types of VLANs (3.1.3)

VLANs are used for different reasons in modern networks. Some VLAN types are defined by traffic classes. Other types of VLANs are defined by the specific function that they serve.

Default VLAN

The *default VLAN* on a Cisco switch is VLAN 1. Therefore, all switch ports are on VLAN 1 unless it is explicitly configured to be on another VLAN. By default, all Layer 2 control traffic is associated with VLAN 1.

Important facts to remember about VLAN 1 include the following:

- All ports are assigned to VLAN 1 by default.
- The *native VLAN* is VLAN 1 by default.
- The management VLAN is VLAN 1 by default.
- VLAN 1 cannot be renamed or deleted.

For instance, in the **show vlan brief** output in Example 3-1, all ports are currently assigned to the default VLAN 1. No native VLAN is explicitly assigned, and no other VLANs are active; therefore, the network is designed with the native VLAN the same as the management VLAN. This is considered a security risk.

Example 3-1 VLAN 1 Default Port Assignments

```
Switch# show vlan brief
VLAN Name                 Status    Ports
---- ---------------- -------   --------------------
1    default              active    Fa0/1, Fa0/2, Fa0/3, Fa0/4
                                    Fa0/5, Fa0/6, Fa0/7, Fa0/8
                                    Fa0/9, Fa0/10, Fa0/11, Fa0/12
                                    Fa0/13, Fa0/14, Fa0/15, Fa0/16
                                    Fa0/17, Fa0/18, Fa0/19, Fa0/20
                                    Fa0/21, Fa0/22, Fa0/23, Fa0/24
                                    Gi0/1, Gi0/2
1002 fddi-default                         act/unsup
1003 token-ring-default                   act/unsup
1004 fddinet-default                      act/unsup
1005 trnet-default                        act/unsup
```

Data VLAN

Data VLANs are VLANs configured to separate user-generated traffic. They are referred to as user VLANs because they separate the network into groups of users or devices. A modern network would have many data VLANs depending on organizational requirements. Note that voice and network management traffic should not be permitted on data VLANs.

Native VLAN

User traffic from a VLAN must be *tagged* with its VLAN ID when it is sent to another switch. Trunk ports are used between switches to support the transmission of *tagged traffic*. Specifically, an *802.1Q* trunk port inserts a 4-byte tag in the Ethernet frame header to identify the VLAN to which the frame belongs.

A switch may also have to send untagged traffic across a trunk link. Untagged traffic is generated by a switch and may also come from legacy devices. The 802.1Q trunk port places untagged traffic on the native VLAN. The native VLAN on a Cisco switch is VLAN 1 (i.e., the default VLAN).

It is a best practice to configure the native VLAN as an unused VLAN, distinct from VLAN 1 and other VLANs. In fact, it is not unusual to dedicate a fixed VLAN to serve the role of the native VLAN for all trunk ports in the switched domain.

Management VLAN

A *management VLAN* is a data VLAN configured specifically for network management traffic, including Secure Shell (SSH), Telnet, Hypertext Transfer Protocol

Secure (HTTPS), Hypertext Transfer Protocol (HTTP), and Simple Network Management Protocol (SNMP). By default, VLAN 1 is configured as the management VLAN on a Layer 2 switch.

Voice VLAN

A separate VLAN called a *voice VLAN* is needed to support Voice over IP (VoIP). VoIP traffic requires the following:

- Assured bandwidth to ensure voice quality
- Transmission priority over other types of network traffic
- Ability to be routed around congested areas on the network
- Delay of less than 150 ms across the network

To meet these requirements, the entire network has to be designed to support VoIP.

In Figure 3-3, VLAN 150 is designed to carry voice traffic. The student computer PC5 is attached to the Cisco IP phone, and the phone is attached to switch S3. PC5 is in VLAN 20, which is used for student data.

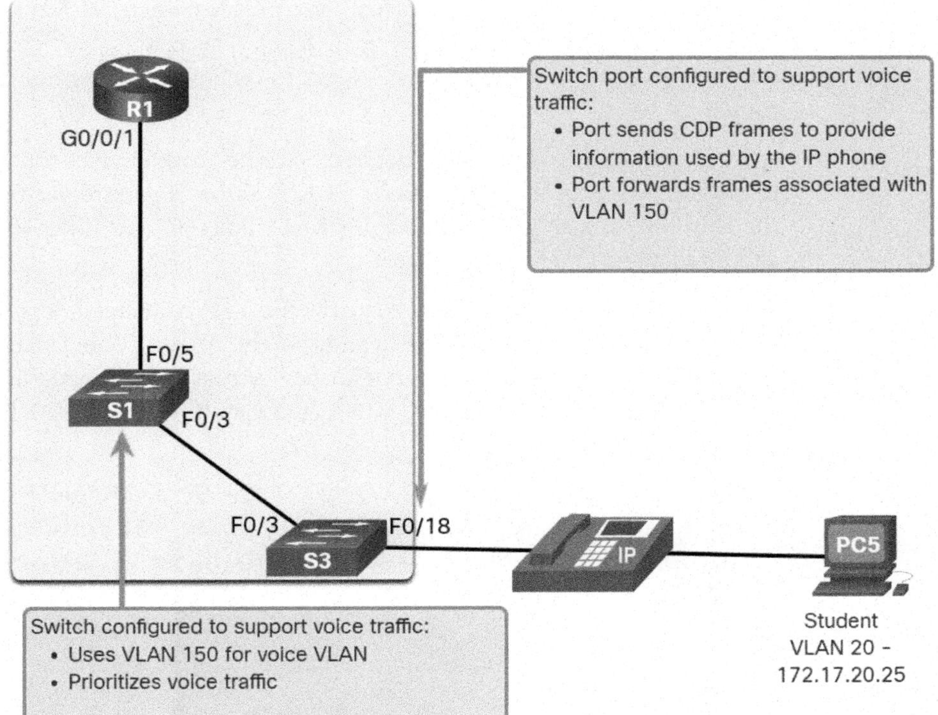

Figure 3-3 Voice VLAN

Packet Tracer—Who Hears the Broadcast? (3.1.4)

In this Packet Tracer activity, you will complete the following objectives:

- Part 1: Observe Broadcast Traffic in a VLAN Implementation
- Part 2: Complete Review Questions

Interactive Graphic

Check Your Understanding—Overview of VLANs (3.1.5)

Refer to the online course to complete this activity.

VLANs in a Multi-Switched Environment (3.2)

In this section, you learn how a switch forwards frames based on VLAN configuration in a multiswitch environment.

Defining VLAN Trunks (3.2.1)

VLANs would not be very useful without *VLAN trunks*. VLAN trunks allow all VLAN traffic to propagate between switches. This enables devices connected to different switches but in the same VLAN to communicate without going through a router.

A trunk is a point-to-point link between two network devices that carries more than one VLAN. A VLAN trunk extends VLANs across an entire network. Cisco supports the IEEE 802.1Q standard for coordinating trunks on Fast Ethernet, Gigabit Ethernet, and 10-Gigabit Ethernet interfaces.

A VLAN trunk does not belong to a specific VLAN. Instead, it is a conduit for multiple VLANs between switches and routers. A trunk could also be used between a network device and server or another device that is equipped with an appropriate 802.1Q-capable NIC. By default, on a Cisco Catalyst switch, all VLANs are supported on a trunk port.

In Figure 3-4, the links between switches S1 and S2, and S1 and S3, are configured to transmit traffic coming from VLANs 10, 20, 30, and 99 (that is, native VLAN) across the network. This network could not function without VLAN trunks.

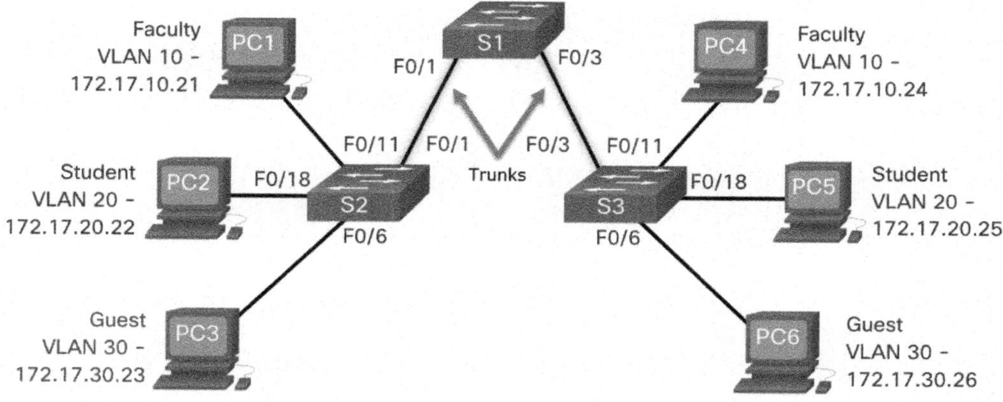

Figure 3-4 VLAN Trunks

Network Without VLANs (3.2.2)

When a switch receives a broadcast frame on one of its ports, it forwards the frame out all other ports except the port where the broadcast was received. As illustrated in Figure 3-5, the entire network is configured in the same subnet (172.17.40.0/24) and no VLANs are configured. As a result, when the faculty computer (PC1) sends out a broadcast frame, switch S2 sends that broadcast frame out all of its ports. Eventually the entire network receives the broadcast because the network is one broadcast domain.

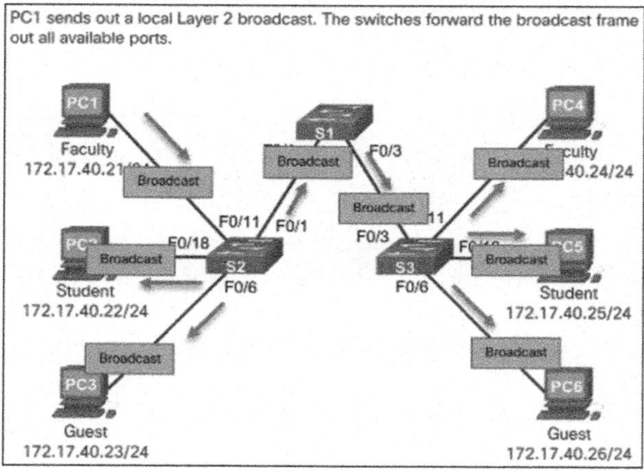

Figure 3-5 Broadcast Domain with No VLANs

Network with VLANs (3.2.3)

VLANs are associated with and configured on individual switch ports. Devices attached to those ports have no concept of VLANs. However, these devices are configured with IP addressing and are members of a specific IP network. This is where the connection between VLAN and IP network is apparent. A VLAN is the equivalent to an IP network (or subnet). VLANs are configured on the switch, whereas IP addressing is configured on the device.

In Figure 3-6, the network has been segmented using two VLANs. Faculty devices are assigned to VLAN 10 and student devices are assigned to VLAN 20. When a broadcast frame is sent from the faculty computer, PC1, to switch S2, Because arrows highlight the trunk lines in this figure, it makes it appear that PC1 is forwarding only to VLAN 20 members not VLAN 10.

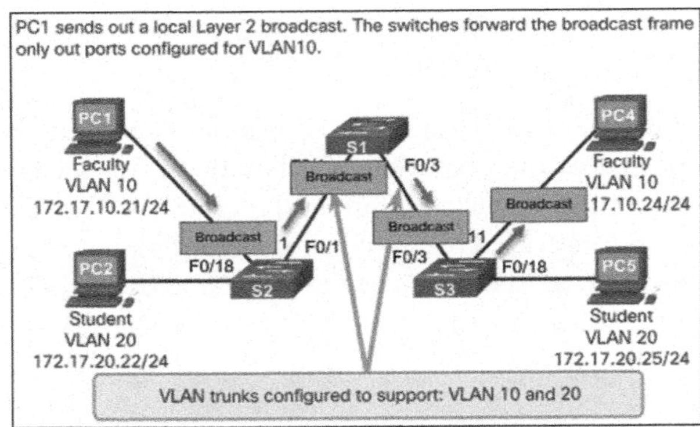

Figure 3-6 VLANs Segment Broadcast Domains

The ports that compose the connection between switches S2 and S1 (ports F0/1), and between S1 and S3 (ports F0/3), are trunks and have been configured to support all the VLANs in the network.

When S1 receives the broadcast frame on port F0/1, S1 forwards that broadcast frame out of the only other port configured to support VLAN 10, which is port F0/3. When S3 receives the broadcast frame on port F0/3, it forwards that broadcast frame out the only other port configured to support VLAN 10, which is port F0/11. The broadcast frame arrives at the only other computer in the network configured in VLAN 10, which is faculty computer PC4.

When VLANs are implemented on a switch, the transmission of unicast, multicast, and broadcast traffic from a host in a particular VLAN are restricted to the devices that are in that VLAN.

VLAN Identification with a Tag (3.2.4)

The standard Ethernet frame header does not contain information about the VLAN to which the frame belongs. Therefore, when Ethernet frames are placed on a trunk, information about the VLANs to which they belong must be added. This process, called *tagging*, is accomplished by using the *IEEE 802.1Q header*, specified in the IEEE 802.1Q standard. The 802.1Q header includes a 4-byte tag inserted within the original Ethernet frame header, specifying the VLAN to which the frame belongs.

When the switch receives a frame on a port configured in access mode and assigned a VLAN, the switch inserts a VLAN tag in the frame header, recalculates the Frame Check Sequence (FCS), and sends the tagged frame out of a trunk port.

VLAN Tag Field Details

The *VLAN tag field* is shown in Figure 3-7.

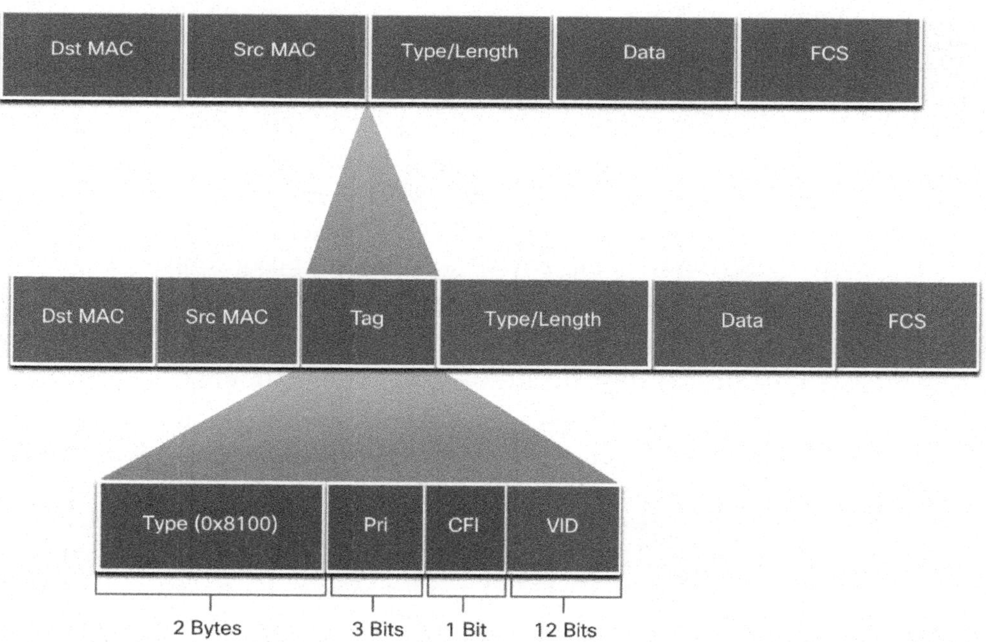

Figure 3-7 Fields in an Ethernet 802.1Q Frame

The VLAN tag control information field consists of a Type field, a Priority field, a *Canonical Format Identifier* field, and VLAN ID field:

- **Type:** A 2-byte value called the *tag protocol ID (TPID)* value. For Ethernet, it is set to hexadecimal 0x8100.
- *User priority*: A 3-bit value that supports level or service implementation.

- **Canonical Format Identifier (CFI):** A 1-bit identifier that enables Token Ring frames to be carried across Ethernet links.
- **VLAN ID (VID):** A 12-bit VLAN identification number that supports up to 4,096 VLAN IDs.

After the switch inserts the tag control information fields, it recalculates the FCS values and inserts the new FCS into the frame.

Native VLANs and 802.1Q Tagging (3.2.5)

The IEEE 802.1Q standard specifies a native VLAN for trunk links, which defaults to VLAN 1. When an *untagged frame* arrives on a trunk port, it is assigned to the native VLAN. Management frames that are sent between switches are an example of traffic that is typically untagged. If the link between two switches is a trunk, the switch sends the untagged traffic on the native VLAN.

Tagged Frames on the Native VLAN

Some devices that support trunking add a VLAN tag to native VLAN traffic. Control traffic sent on the native VLAN should not be tagged. If an 802.1Q trunk port receives a tagged frame with the VLAN ID that is the same as the native VLAN, it drops the frame. Consequently, when configuring a switch port on a Cisco switch, configure devices so that they do not send tagged frames on the native VLAN. Devices from other vendors that support tagged frames on the native VLAN include IP phones, servers, routers, and non-Cisco switches.

Untagged Frames on the Native VLAN

When a Cisco switch trunk port receives untagged frames (which are unusual in a well-designed network), it forwards those frames to the native VLAN. If there are no devices associated with the native VLAN (which is not unusual) and there are no other trunk ports (which is not unusual), then the frame is dropped. The default native VLAN is VLAN 1. When configuring an 802.1Q trunk port, a default Port VLAN ID (PVID) is assigned the value of the native VLAN ID. All untagged traffic coming in or out of the 802.1Q port is forwarded based on the PVID value. For example, if VLAN 99 is configured as the native VLAN, the PVID is 99, and all untagged traffic is forwarded to VLAN 99. If the native VLAN has not been reconfigured, the PVID value is set to VLAN 1.

In Figure 3-8, PC1 is connected by a hub to an 802.1Q trunk link.

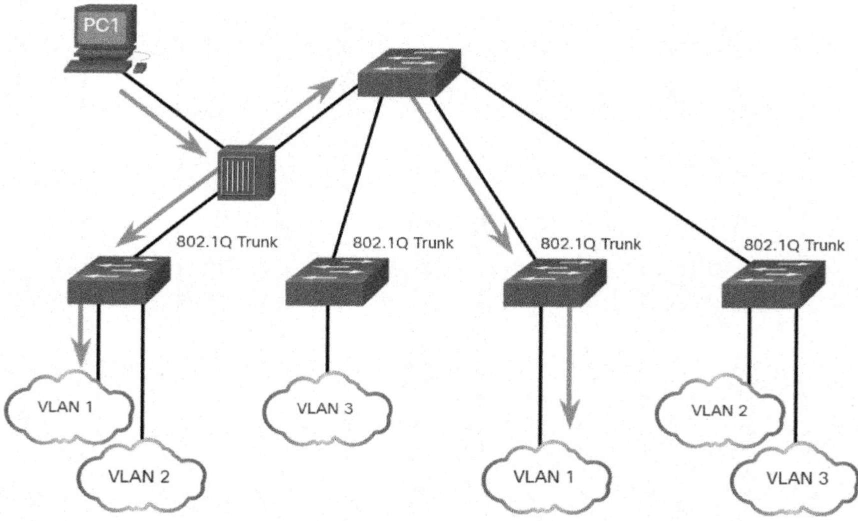

Figure 3-8 Untagged Frames on the Native VLAN

PC1 sends untagged traffic, which the switches associate with the native VLAN configured on the trunk ports, and forward accordingly. Tagged traffic on the trunk received by PC1 is dropped. This scenario reflects poor network design for several reasons: it uses a hub, it has a host connected to a trunk link, and it implies that the switches have access ports assigned to the native VLAN. It also illustrates the motivation for the IEEE 802.1Q specification for native VLANs as a means of handling legacy scenarios.

Voice VLAN Tagging (3.2.6)

A separate voice VLAN is required to support VoIP. This enables quality of service (QoS) and security policies to be applied to voice traffic.

A Cisco IP phone connects directly to a switch port. An IP host can connect to the IP phone to gain network connectivity as well. The access port connected to the Cisco IP phone can be configured to use two separate VLANs. One VLAN is for voice traffic, and the other is a data VLAN to support the host traffic. The link between the switch and the IP phone simulates a trunk link to carry both voice VLAN traffic and data VLAN traffic.

Specifically, the Cisco IP Phone contains an integrated three-port 10/100 switch. The ports provide dedicated connections to the following devices:

- Port 1 connects to the switch or other VoIP device.
- Port 2 is an internal 10/100 interface that carries the IP phone traffic.
- Port 3 (access port) connects to a PC or other device.

The switch access port sends CDP packets instructing the attached IP phone to send voice traffic in one of three ways. The method used varies based on the type of traffic:

- Voice VLAN traffic must be tagged with an appropriate Layer 2 *class of service (CoS)* priority value.
- Access VLAN traffic can also be tagged with a Layer 2 *CoS priority value*.
- Access VLAN is not tagged (no Layer 2 CoS priority value).

In Figure 3-9, the student computer PC5 is attached to a Cisco IP phone, and the phone is attached to switch S3. VLAN 150 is designed to carry voice traffic, while PC5 is in VLAN 20, which is used for student data.

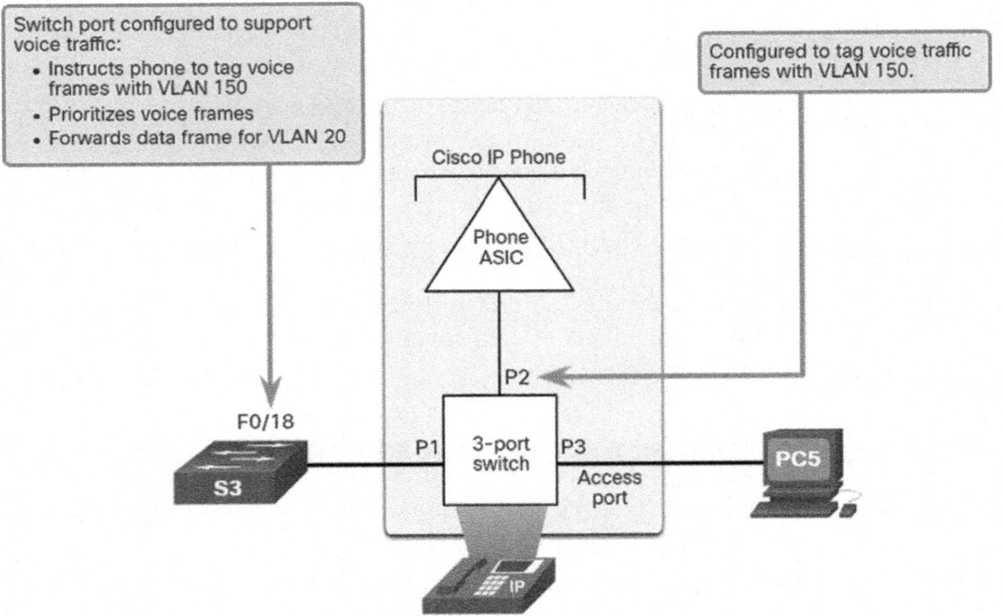

Figure 3-9 Voice VLAN Tagging

Voice VLAN Verification Example (3.2.7)

Example 3-2 shows output for the **show interface fa0/18 switchport** command. The highlighted areas in the sample output show the F0/18 interface configured with a VLAN that is configured for data (VLAN 20) and a VLAN configured for voice (VLAN 150).

Example 3-2 Verifying a Voice VLAN Configuration

```
S1# show interfaces fa0/18 switchport
Name: Fa0/18
Switchport: Enabled
Administrative Mode: static access
Operational Mode: static access
Administrative Trunking Encapsulation: negotiate
Operational Trunking Encapsulation: native
Negotiation of Trunking: Off
Access Mode VLAN: 20 (student)
Trunking Native Mode VLAN: 1 (default)
Administrative Native VLAN tagging: enabled
Voice VLAN: 150 (voice)
```

Packet Tracer
☐ Activity

Packet Tracer—Investigate a VLAN Implementation (3.2.8)

In this activity, you will observe how broadcast traffic is forwarded by the switches when VLANs are configured and when VLANs are not configured.

Interactive Graphic

Check Your Understanding—VLANs in a Multiswitch Environment (3.2.9)

Refer to the online course to complete this activity.

VLAN Configuration (3.3)

In this section, you will configure a switch port to be assigned to a VLAN based on requirements.

VLAN Ranges on Catalyst Switches (3.3.1)

Creating VLANs, like most other aspects of networking, is a matter of entering the appropriate commands. This section details how to configure and verify different types of VLANs.

Different Cisco Catalyst switches support various numbers of VLANs. The number of supported VLANs is large enough to accommodate the needs of most organizations. For example, the Catalyst 2960 and 3650 Series switches support more than 4,000 VLANs. *Normal range VLANs* on these switches are numbered 1 to 1,005, and *extended range VLANs* are numbered 1,006 to 4,094. Example 3-3 illustrates the default VLANs on a Catalyst 2960 switch running Cisco IOS Release 15.x.

Example 3-3 Normal Range VLANs

```
Switch# show vlan brief

VLAN Name                 Status    Ports
---- -------------------- -------   --------------------
1    default              active    Fa0/1, Fa0/2, Fa0/3, Fa0/4
                                    Fa0/5, Fa0/6, Fa0/7, Fa0/8
                                    Fa0/9, Fa0/10, Fa0/11, Fa0/12
                                    Fa0/13, Fa0/14, Fa0/15, Fa0/16
                                    Fa0/17, Fa0/18, Fa0/19, Fa0/20
                                    Fa0/21, Fa0/22, Fa0/23, Fa0/24
                                    Gi0/1, Gi0/2
1002 fddi-default                           act/unsup
1003 token-ring-default                     act/unsup
1004 fddinet-default                        act/unsup
1005 trnet-default                          act/unsup
```

Normal Range VLANs

The following are characteristics of normal range VLANs:

- They are used in all small- and medium-sized business and enterprise networks.
- They are identified by a VLAN ID between 1 and 1005.
- IDs 1002 through 1005 are reserved for legacy network technologies (that is, Token Ring and Fiber Distributed Data Interface).
- IDs 1 and 1002 to 1005 are automatically created and cannot be removed.
- Configurations are stored in the switch flash memory in a VLAN database file called *vlan.dat*.
- When configured, *VLAN trunking protocol (VTP)* helps synchronize the VLAN database between switches.

Extended Range VLANs

The following are characteristics of extended range VLANs:

- They are used by service providers to service multiple customers and by global enterprises large enough to need extended range VLAN IDs.

- They are identified by a VLAN ID between 1006 and 4094.
- Configurations are saved, by default, in the running configuration.
- They support fewer VLAN features than normal range VLANs.
- Requires VTP transparent mode configuration to support extended range VLANs.

Note

4096 is the upper boundary for the number of VLANs available on Catalyst switches because there are 12 bits in the VLAN ID field of the IEEE 802.1Q header.

VLAN Creation Commands (3.3.2)

When configuring normal range VLANs, the configuration details are stored in flash memory on the switch in a file called vlan.dat. Flash memory is persistent and does not require the **copy running-config startup-config** command. However, because other details are often configured on a Cisco switch at the same time that VLANs are created, it is good practice to save running configuration changes to the startup configuration.

Table 3-2 displays the Cisco IOS command syntax used to add a VLAN to a switch and give it a name. Naming each VLAN is considered a best practice in switch configuration.

Table 3-2 Command Syntax for VLAN Creation

Task	IOS Command
Enter global configuration mode.	`Switch#` **`configure terminal`**
Create a VLAN with a valid ID number.	`Switch(config)#` **`vlan`** *`vlan-id`*
Specify a unique name to identify the VLAN.	`Switch(config-vlan)#` **`name`** *`vlan-name`*
Return to the privileged EXEC mode.	`Switch(config-vlan)#` **`end`**

VLAN Creation Example (3.3.3)

In the topology in Figure 3-10, the student computer (PC2) has not been associated with a VLAN yet, but it does have an IP address of 172.17.20.22, which belongs to VLAN 20.

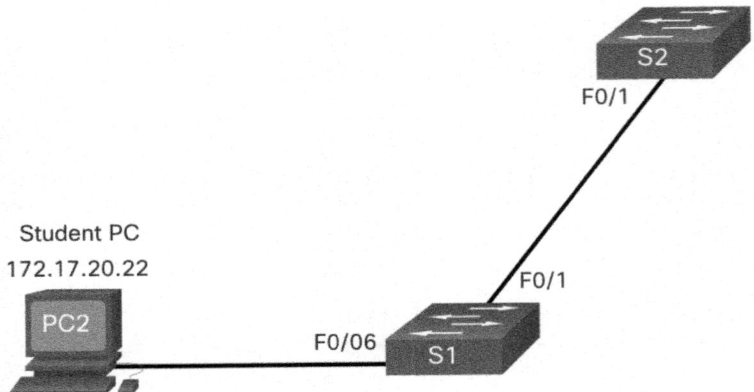

Figure 3-10 VLAN Configuration Example

Example 3-4 shows how the student VLAN (VLAN 20) is configured on switch S1.

Example 3-4 VLAN 20 Configuration

```
S1# configure terminal
S1(config)# vlan 20
S1(config-vlan)# name student
S1(config-vlan)# end
```

Note

In addition to entering a single VLAN ID, a series of VLAN IDs can be entered separated by commas, or a range of VLAN IDs separated by hyphens using the **vlan** *vlan-id* command. For example, entering the **vlan 100,102,105-107** global configuration command would create VLANs 100, 102, 105, 106, and 107.

VLAN Port Assignment Commands (3.3.4)

After creating a VLAN, the next step is to assign ports to the VLAN.

Table 3-3 displays the syntax for defining a port to be an access port and assigning it to a VLAN. The **switchport mode access** command is optional, but strongly recommended as a security best practice. With this command, the interface changes to strictly access mode. Access mode indicates that the port belongs to a single VLAN and will not negotiate to become a trunk link.

Table 3-3 Command Syntax for VLAN Port Assignment

Task	IOS Command
Enter global configuration mode.	`Switch#` **`configure terminal`**
Enter interface configuration mode.	`Switch(config)#` **`interface`** *`interface-id`*
Set the port to access mode.	`Switch(config-if)#` **`switchport mode access`**
Assign the port to a VLAN.	`Switch(config-if)#` **`switchport access vlan`** *`vlan-id`*
Return to the privileged EXEC mode.	`Switch(config-if)#` **`end`**

Note

Use the **interface range** command to simultaneously configure multiple interfaces.

VLAN Port Assignment Example (3.3.5)

In Figure 3-11, port F0/6 on switch S1 is configured as an access port and assigned to VLAN 20. Any device connected to that port will be associated with VLAN 20. Therefore, in our example, PC2 is in VLAN 20.

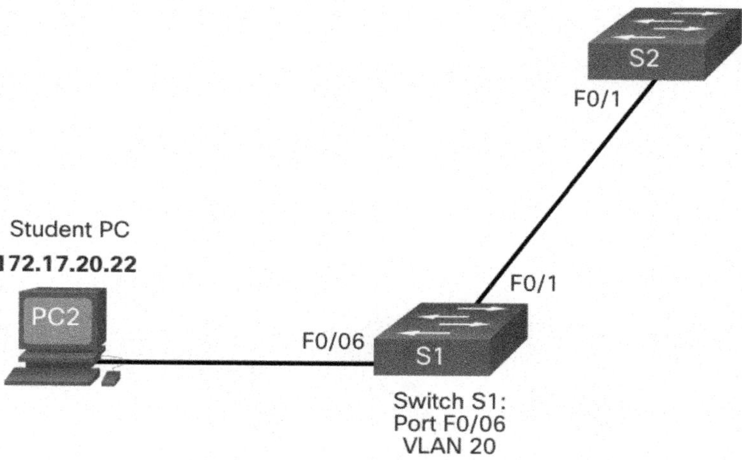

Figure 3-11 Assigning Ports Configuration Example

Example 3-5 shows the configuration for S1 to assign F0/6 to VLAN 20.

Example 3-5 Assigning a Port to a VLAN

```
S1# configure terminal
S1(config)# interface fa0/6
S1(config-if)# switchport mode access
S1(config-if)# switchport access vlan 20
S1(config-if)# end
```

VLANs are configured on the switch port and not on the end device. PC2 is configured with an IPv4 address and subnet mask that is associated with the VLAN, which is configured on the switch port. In this example, it is VLAN 20. When VLAN 20 is configured on other switches, the network administrator must configure the other student computers to be in the same subnet as PC2 (172.17.20.0/24).

Data and Voice VLANs (3.3.6)

An access port can belong to only one data VLAN at a time. However, a port can also be associated to a voice VLAN. For example, a port connected to an IP phone and an end device would be associated with two VLANs: one for voice and one for data.

Consider the topology in Figure 3-12. PC5 is connected to the Cisco IP phone, which in turn is connected to the FastEthernet 0/18 interface on S3. To implement this configuration, a data VLAN and a voice VLAN are created.

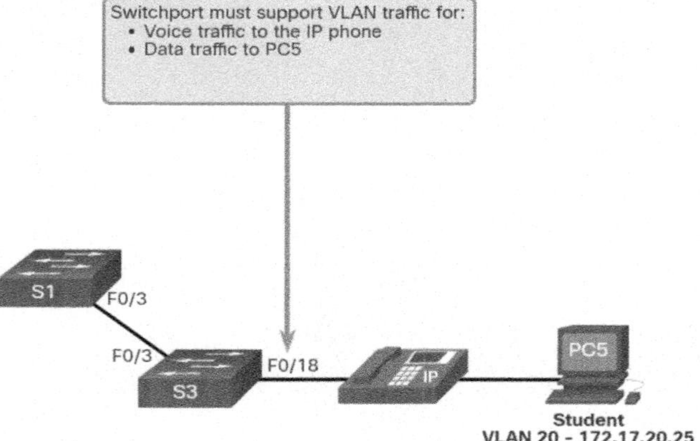

Figure 3-12 Data and Voice VLAN Topology

Data and Voice VLAN Example (3.3.7)

Use the **switchport voice vlan** *vlan-id* interface configuration command to assign a voice VLAN to a port.

LANs supporting voice traffic typically also have quality of service (QoS) enabled. Voice traffic must be labeled as trusted as soon as it enters the network. Use the **mls qos trust [cos | device cisco-phone | dscp | ip-precedence]** interface configuration command to set the trusted state of an interface and to indicate which fields of the packet are used to classify traffic.

The configuration in Example 3-6 creates the two VLANs (that is, VLAN 20 and VLAN 150) and then assigns the F0/18 interface of S3 as a switchport in VLAN 20. It also assigns voice traffic to VLAN 150 and enables Quality of Service (QoS) classification based on the class of service (CoS) assigned by the IP phone.

Example 3-6 Configuring Data and Voice VLANs

```
S3(config)# vlan 20
S3(config-vlan)# name student
S3(config-vlan)# vlan 150
S3(config-vlan)# name VOICE
S3(config-vlan)# exit
S3(config)#
S3(config)# interface fa0/18
S3(config-if)# switchport mode access
S3(config-if)# switchport access vlan 20
S3(config-if)# mls qos trust cos
S3(config-if)# switchport voice vlan 150
S3(config-if)# end
S3#
```

Note

The implementation of QoS is beyond the scope of this course.

The **switchport access vlan** command forces the creation of a VLAN if it does not already exist on the switch. For example, VLAN 30 is not present in the **show vlan brief** output of the switch. If the **switchport access vlan 30** command is entered on any interface with no previous configuration, the switch displays the following:

```
% Access VLAN does not exist. Creating vlan 30
```

Verify VLAN Information (3.3.8)

After a VLAN is configured, VLAN configurations can be validated using Cisco IOS **show** commands.

The **show vlan** command displays a list of all configured VLANs. The **show vlan** command can also be used with options. The complete syntax is **show vlan [brief | id** *vlan-id* | **name** *vlan-name* | **summary]**.

Table 3-4 describes the **show vlan** command options.

Table 3-4 Options for the **show vlan** Command

Task	Command Option
Display VLAN name, status, and its ports one VLAN per line.	`brief`
Display information about the identified VLAN ID number. For *vlan-id*, the range is 1 to 4094.	`id` *`vlan-id`*
Display information about the identified VLAN name. The *vlan-name* is an ASCII string from 1 to 32 characters.	`name` *`vlan-name`*
Display VLAN summary information.	`summary`

The **show vlan summary** command displays the count of all configured VLANs, as shown in Example 3-7.

Example 3-7 The **show vlan summary** Command

```
S1# show vlan summary
Number of existing VLANs                 : 7
Number of existing VTP VLANs             : 7
Number of existing extended VLANS        : 0
```

Other useful commands are the **show interfaces** *interface-id* **switchport** and the **show interfaces vlan** *vlan-id* command. For example, the **show interfaces fa0/18 switchport** command can be used to confirm that the FastEthernet 0/18 port has been correctly assigned to data and voice VLANs, as shown in Example 3-8.

Example 3-8 Verifying an Interface Has the Correct VLAN Assignments

```
S1# show interfaces fa0/18 switchport
Name: Fa0/18
Switchport: Enabled
Administrative Mode: static access
Operational Mode: static access
Administrative Trunking Encapsulation: dot1q
Operational Trunking Encapsulation: native
Negotiation of Trunking: Off
Access Mode VLAN: 20 (student)
Trunking Native Mode VLAN: 1 (default)
Voice VLAN: 150
Administrative private-vlan host-association: none
(Output omitted)
```

Change VLAN Port Membership (3.3.9)

There are a number of ways to change VLAN port membership.

If the switch access port has been incorrectly assigned to a VLAN, simply reenter the **switchport access vlan** *vlan-id* interface configuration command with the correct VLAN ID. For instance, assume Fa0/18 was incorrectly configured to be on the default VLAN 1 instead of VLAN 20. To change the port to VLAN 20, enter **switchport access vlan 20**.

To change the membership of a port back to the default VLAN 1, use the **no switchport access vlan** interface configuration mode command as shown.

In Example 3-9, Fa0/18 is configured to be on the default VLAN 1, as confirmed by the **show vlan brief** command.

Example 3-9 Remove VLAN Assignment Configuration

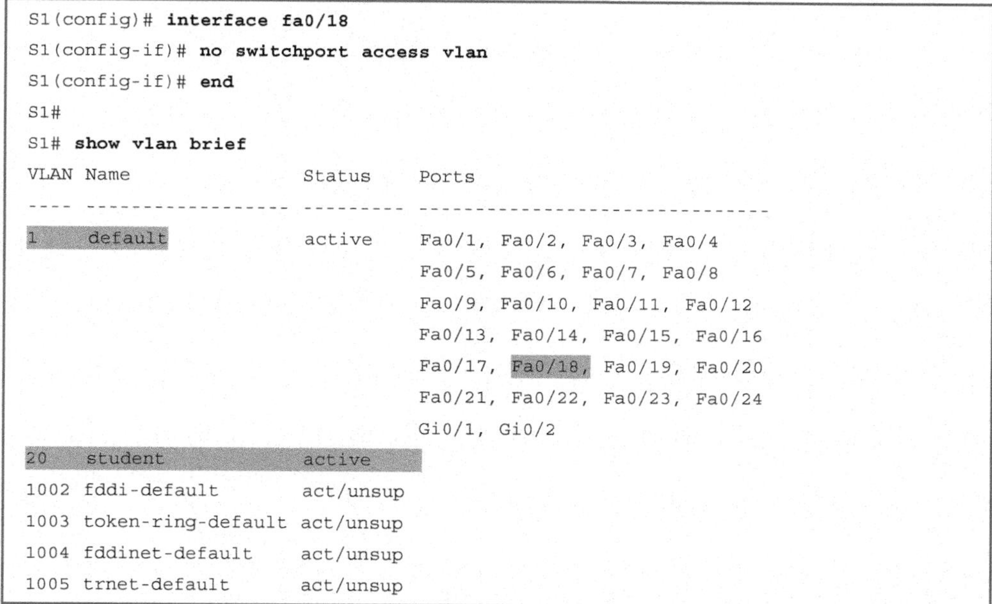

```
S1(config)# interface fa0/18
S1(config-if)# no switchport access vlan
S1(config-if)# end
S1#
S1# show vlan brief
VLAN Name                 Status    Ports
---- ------------------ --------- -------------------------------
1    default              active    Fa0/1, Fa0/2, Fa0/3, Fa0/4
                                    Fa0/5, Fa0/6, Fa0/7, Fa0/8
                                    Fa0/9, Fa0/10, Fa0/11, Fa0/12
                                    Fa0/13, Fa0/14, Fa0/15, Fa0/16
                                    Fa0/17, Fa0/18, Fa0/19, Fa0/20
                                    Fa0/21, Fa0/22, Fa0/23, Fa0/24
                                    Gi0/1, Gi0/2
20   student              active
1002 fddi-default         act/unsup
1003 token-ring-default act/unsup
1004 fddinet-default      act/unsup
1005 trnet-default        act/unsup
```

Notice that VLAN 20 is still active, even though no ports are assigned to it.

The **show interfaces f0/18 switchport** output can also be used to verify that the access VLAN for interface F0/18 has been reset to VLAN 1, as shown in Example 3-10.

Example 3-10 Verify VLAN Is Removed

```
S1# show interfaces fa0/18 switchport
Name: Fa0/18
Switchport: Enabled
Administrative Mode: static access
Operational Mode: static access
Administrative Trunking Encapsulation: negotiate
Operational Trunking Encapsulation: native
Negotiation of Trunking: Off
Access Mode VLAN: 1 (default)
Trunking Native Mode VLAN: 1 (default)
```

Delete VLANs (3.3.10)

The **no vlan** *vlan-id* global configuration mode command is used to remove a VLAN from the switch vlan.dat file.

Caution

Before deleting a VLAN, reassign all member ports to a different VLAN first. Any ports that are not moved to an active VLAN are unable to communicate with other hosts after the VLAN is deleted and until they are assigned to an active VLAN.

The entire vlan.dat file can be deleted using the **delete flash:vlan.dat** privileged EXEC mode command. The abbreviated command version (**delete vlan.dat**) can be used if the vlan.dat file has not been moved from its default location. After issuing this command and reloading the switch, any previously configured VLANs are no longer present. This effectively places the switch into its factory default condition with regard to VLAN configurations.

Note

To restore a Catalyst switch to its factory default condition, unplug all cables except the console and power cable from the switch. Then enter the **erase startup-config** privileged EXEC mode command followed by the **delete vlan.dat** command.

Interactive Graphic

Syntax Checker—VLAN Configuration (3.3.11)

In this Syntax Checker activity, you implement and verify a VLAN configuration for switch interfaces based on the specified requirements.

Refer to the online course to complete this activity.

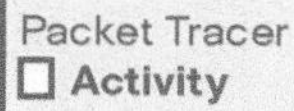

Packet Tracer—VLAN Configuration (3.3.12)

In this Packet Tracer activity, you perform the following:

- Verify the Default VLAN Configuration
- Configure VLANs
- Assign VLANs to Ports

VLAN Trunks (3.4)

In this section, you learn how to configure a trunk port on a LAN switch.

Trunk Configuration Commands (3.4.1)

Now that you have configured and verified VLANs, it is time to configure and verify VLAN trunks. A VLAN trunk is a Layer 2 link between two switches that carries traffic for all VLANs (unless the allowed VLAN list is restricted manually or dynamically).

To enable trunk links, configure the interconnecting ports with the set of interface configuration commands shown in Table 3-5.

Table 3-5 Command Syntax for Trunk Configuration Commands

Task	IOS Command
Enter global configuration mode.	`Switch# configure terminal`
Enter interface configuration mode.	`Switch(config)# interface interface-id`
Set the port to permanent trunking mode.	`Switch(config-if)# switchport mode trunk`
Sets the native VLAN to something other than VLAN 1.	`Switch(config-if)# switchport trunk native vlan vlan-id`
Specify the list of VLANs to be allowed on the trunk link.	`Switch(config-if)# switchport trunk allowed vlan vlan-list`
Return to the privileged EXEC mode.	`Switch(config-if)# end`

Trunk Configuration Example (3.4.2)

In Figure 3-13, VLANs 10, 20, and 30 support the Faculty, Student, and Guest computers (PC1, PC2, and PC3). The F0/1 port on switch S1 is configured as a trunk port and forwards traffic for VLANs 10, 20, and 30. VLAN 99 is configured as the native VLAN.

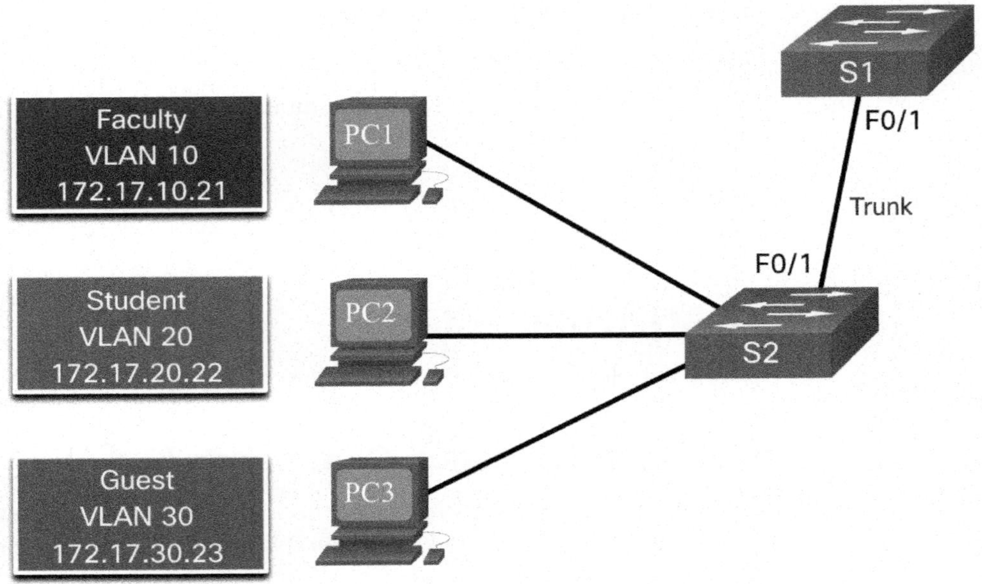

The subnets associated with each VLAN are:

- VLAN 10 - Faculty/Staff - 172.17.10.0/24
- VLAN 20 - Students - 172.17.20.0/24
- VLAN 30 - Guests - 172.17.30.0/24
- VLAN 99 - Native - 172.17.99.0/24

Figure 3-13 Trunk Configuration Topology

Example 3-11 shows the configuration of port F0/1 on switch S1 as a trunk port. The native VLAN is changed to VLAN 99, and the allowed VLAN list is restricted to 10, 20, 30, and 99.

Example 3-11 Trunk Configuration Example

```
S1(config)# interface fastEthernet 0/1
S1(config-if)# switchport mode trunk
S1(config-if)# switchport trunk native vlan 99
S1(config-if)# switchport trunk allowed vlan 10,20,30,99
S1(config-if)# end
```

Note

This configuration assumes the use of Cisco Catalyst 2960 switches, which automatically use 802.1Q encapsulation on trunk links. Other switches may require manual configuration of the encapsulation. Always configure both ends of a trunk link with the same native VLAN. If 802.1Q trunk configuration is not the same on both ends, Cisco IOS Software reports errors.

Verify Trunk Configuration (3.4.3)

The switch output displays the configuration of switch port F0/1 on switch S1. The configuration is verified with the **show interfaces** *interface-ID* **switchport** command, as shown in Example 3-12.

Example 3-12 Verifying Trunk Configuration on S1 Fa0/1

```
S1# show interfaces fa0/1 switchport
Name: Fa0/1
Switchport: Enabled
Administrative Mode: trunk
Operational Mode: trunk
Administrative Trunking Encapsulation: dot1q
Operational Trunking Encapsulation: dot1q
Negotiation of Trunking: On
Access Mode VLAN: 1 (default)
Trunking Native Mode VLAN: 99 (VLAN0099)
Administrative Native VLAN tagging: enabled
Voice VLAN: none
Administrative private-vlan host-association: none
Administrative private-vlan mapping: none
Administrative private-vlan trunk native VLAN: none
Administrative private-vlan trunk Native VLAN tagging: enabled
Administrative private-vlan trunk encapsulation: dot1q
Administrative private-vlan trunk normal VLANs: none
Administrative private-vlan trunk associations: none
Administrative private-vlan trunk mappings: none
Operational private-vlan: none
Trunking VLANs Enabled: ALL
Pruning VLANs Enabled: 2-1001
(output omitted)
```

The top highlighted area shows that port F0/1 has its administrative mode set to **trunk**. The port is in trunking mode. The next highlighted area verifies that the native VLAN is VLAN 99. Further down in the output, the bottom highlighted area shows that VLANs 10, 20, 30, and 99 are enabled on the trunk.

Note

Another useful command for verifying trunk interfaces is the **show interface trunk** command.

Reset the Trunk to the Default State (3.4.4)

Use the **no switchport trunk allowed vlan** and the **no switchport trunk native vlan** commands to remove the allowed VLANs and reset the native VLAN of the trunk. When it is reset to the default state, the trunk allows all VLANs and uses VLAN 1 as the native VLAN. Example 3-13 shows the commands used to reset all trunking characteristics of a trunking interface to the default settings.

Example 3-13 Remove Allowed VLANs and Reset the Native VLAN

```
S1(config)# interface fa0/1
S1(config-if)# no switchport trunk allowed vlan
S1(config-if)# no switchport trunk native vlan
S1(config-if)# end
```

The **show interfaces f0/1 switchport** command in Example 3-14 reveals that the trunk has been reconfigured to a default state.

Example 3-14 Verify Trunk Is Now in Its Default State

```
S1# show interfaces fa0/1 switchport
Name: Fa0/1
Switchport: Enabled
Administrative Mode: trunk
Operational Mode: trunk
Administrative Trunking Encapsulation: dot1q
Operational Trunking Encapsulation: dot1q
Negotiation of Trunking: On
Access Mode VLAN: 1 (default)
Trunking Native Mode VLAN: 1 (default)
Administrative Native VLAN tagging: enabled
Voice VLAN: none
Administrative private-vlan host-association: none
Administrative private-vlan mapping: none
Administrative private-vlan trunk native VLAN: none
Administrative private-vlan trunk Native VLAN tagging: enabled
Administrative private-vlan trunk encapsulation: dot1q
Administrative private-vlan trunk normal VLANs: none
Administrative private-vlan trunk associations: none
Administrative private-vlan trunk mappings: none
Operational private-vlan: none
Trunking VLANs Enabled: ALL
Pruning VLANs Enabled: 2-1001
(output omitted)
```

Example 3-15 shows the commands used to remove the trunk feature from the F0/1 switch port on switch S1. The **show interfaces f0/1 switchport** command reveals that the F0/1 interface is now in static access mode.

Example 3-15 Reset the Port to Access Mode

```
S1(config)# interface fa0/1
S1(config-if)# switchport mode access
S1(config-if)# end
S1# show interfaces fa0/1 switchport
Name: Fa0/1
Switchport: Enabled
Administrative Mode: static access
Operational Mode: static access
Administrative Trunking Encapsulation: dot1q
Operational Trunking Encapsulation: native
Negotiation of Trunking: Off
Access Mode VLAN: 1 (default)
Trunking Native Mode VLAN: 1 (default)
Administrative Native VLAN tagging: enabled
(output omitted)
```

Packet Tracer ☐ Activity

Packet Tracer - Configure Trunks (3.4.5)

In this Packet Tracer activity, you will perform the following:

- Verify VLANs
- Configure Trunks

Lab—Configure VLANs and Trunking (3.4.6)

In this lab, you will perform the following:

- Build the Network and Configure Basic Device Settings
- Create VLANs and Assign Switch Ports
- Maintain VLAN Port Assignments and the VLAN Database
- Configure an 802.1Q Trunk Between the Switches
- Delete the VLAN Database

Dynamic Trunking Protocol (3.5)

In this section, you learn how to configure *dynamic trunking protocol (DTP)* on a LAN switch.

Introduction to DTP (3.5.1)

Some Cisco switches have a proprietary protocol that lets them automatically negotiate trunking with a neighboring device. This protocol is called Dynamic Trunking Protocol (DTP). DTP can speed up the configuration process for a network administrator. Ethernet trunk interfaces support different trunking modes. An interface can be set to trunking or nontrunking, or to negotiate trunking with the neighbor interface. Trunk negotiation is managed by DTP, which operates on a point-to-point basis only, between network devices.

DTP is a Cisco proprietary protocol that is automatically enabled on Catalyst 2960 and Catalyst 3650 Series switches. DTP manages trunk negotiation only if the port on the neighbor switch is configured in a trunk mode that supports DTP. Switches from other vendors do not support DTP.

Note

Some internetworking devices might forward DTP frames improperly, which can cause misconfigurations. To avoid this, turn off DTP on Cisco switch interfaces that are connected to devices that do not support DTP.

The default DTP configuration for Cisco Catalyst 2960 and 3650 switches is dynamic auto.

To enable trunking from a Cisco switch to a device that does not support DTP, use the **switchport mode trunk** and **switchport nonegotiate** interface configuration mode commands. This causes the interface to become a trunk, but it will not generate DTP frames.

```
S1(config-if)# switchport mode trunk
S1(config-if)# switchport nonegotiate
```

To reenable dynamic trunking protocol, use the **switchport mode dynamic auto** command.

```
S1(config-if)# switchport mode dynamic auto
```

If the ports connecting two switches are configured to ignore all DTP advertisements with the **switchport mode trunk** and the **switchport nonegotiate** commands, the ports will stay in trunk port mode. If the connecting ports are set to dynamic auto, they will not negotiate a trunk and will stay in the access mode state, creating an inactive trunk link.

When configuring a port to be in trunk mode, use the **switchport mode trunk** command. Then there is no ambiguity about which state the trunk is in; it is always on.

Negotiated Interface Modes (3.5.2)

The **switchport mode** command has additional options for negotiating the interface mode. The full command syntax is the following:

```
Switch(config-if)# switchport mode { access | dynamic { auto | desirable } | trunk }
```

The options are described in Table 3-6.

Table 3-6 Options for the **switchport mode** Command

Option	Description
`access`	■ Puts the interface (access port) into permanent nontrunking mode and negotiates to convert the link into a nontrunk link. ■ The interface becomes a nontrunk interface, regardless of whether the neighboring interface is a trunk interface.
`dynamic auto`	■ Makes the interface able to convert the link to a trunk link. ■ The interface becomes a trunk interface if the neighboring interface is set to trunk or desirable mode. ■ The default switchport mode for all Ethernet interfaces is **dynamic auto.**
`dynamic desirable`	■ Makes the interface actively attempt to convert the link to a trunk link. ■ The interface becomes a trunk interface if the neighboring interface is set to trunk, desirable, or dynamic auto mode.
`Trunk`	■ Puts the interface into permanent trunking mode and negotiates to convert the neighboring link into a trunk link. ■ The interface becomes a trunk interface even if the neighboring interface is not a trunk interface.

Use the **switchport nonegotiate** interface configuration command to stop DTP negotiation. The switch does not engage in DTP negotiation on this interface. You can use this command only when the interface switchport mode is **access** or **trunk**. You must manually configure the neighboring interface as a trunk interface to establish a trunk link.

Results of a DTP Configuration (3.5.3)

Table 3-7 illustrates the results of the DTP configuration options on opposite ends of a trunk link connected to Catalyst 2960 switch ports. Best practice is to configure trunk links statically whenever possible.

Table 3-7 DTP Configuration Results

	Dynamic Auto	Dynamic Desirable	Trunk	Access
Dynamic Auto	Access	Trunk	Trunk	Access
Dynamic Desirable	Trunk	Trunk	Trunk	Access
Trunk	Trunk	Trunk	Trunk	Limited connectivity
Access	Access	Access	Limited connectivity	Access

Verify DTP Mode (3.5.4)

The default DTP mode is dependent on the Cisco IOS Software version and on the platform. To determine the current DTP mode, issue the **show dtp interface** command, as shown in Example 3-16.

Example 3-16 Verify the DTP Mode

```
S1# show dtp interface fa0/1
DTP information for FastEthernet0/1:
TOS/TAS/TNS: ACCESS/AUTO/ACCESS
TOT/TAT/TNT: NATIVE/NEGOTIATE/NATIVE
Neighbor address 1: C80084AEF101
Neighbor address 2: 000000000000
Hello timer expiration (sec/state): 11/RUNNING
Access timer expiration (sec/state): never/STOPPED
Negotiation timer expiration (sec/state): never/STOPPED
Multidrop timer expiration (sec/state): never/STOPPED
FSM state: S2:ACCESS
# times multi & trunk 0
Enabled: yes
In STP: no
```

Note

A general best practice is to set the interface to **trunk** and **nonegotiate** when a trunk link is required. On links where trunking is not intended, DTP should be turned off.

Packet Tracer—Configure DTP (3.5.5)

In this Packet Tracer activity, you configure and verify DTP.

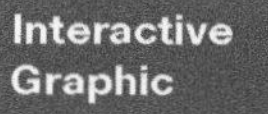

Check Your Understanding—Dynamic Trunking Protocol (3.5.6)

Refer to the online course to complete this activity.

Summary (3.6)

The following is a summary of each section in the module:

Overview of VLANs

Virtual LANs (VLANs) are a group of devices that can communicate as if each device was attached to the same cable. VLANs are based on logical instead of physical connections. Administrators use VLANs to segment networks based on factors such as function, team, or application. Each VLAN is considered a separate logical network. Any switch port can belong to a VLAN. A VLAN creates a logical broadcast domain that can span multiple physical LAN segments. VLANs improve network performance by separating large broadcast domains into smaller ones. Each VLAN in a switched network corresponds to an IP network; therefore, VLAN design must use a hierarchical network-addressing scheme. Types of VLANs include the default VLAN, data VLANs, the native VLAN, management VLANs, and voice VLANs.

VLANs in a Multi-Switched Environment

A VLAN trunk does not belong to a specific VLAN. It is a conduit for multiple VLANs between switches and routers. A VLAN trunk is a point-to-point link between two network devices that carries more than one VLAN. A VLAN trunk extends VLANs across an entire network. When VLANs are implemented on a switch, the transmission of unicast, multicast, and broadcast traffic from a host in a particular VLAN are restricted to the devices that are in that VLAN. VLAN tag fields include the type, user priority, CFI, and VID. Some devices add a VLAN tag to native VLAN traffic. If an 802.1Q trunk port receives a tagged frame with the VID that is the same as the native VLAN, it drops the frame. A separate voice VLAN is required to support VoIP. QoS and security policies can be applied to voice traffic. Voice VLAN traffic must be tagged with an appropriate Layer 2 CoS priority value.

VLAN Configuration

Different Cisco Catalyst switches support various numbers of VLANs, including normal range VLANs and extended range VLANs. When configuring normal range VLANs, the configuration details are stored in flash memory on the switch in a file called vlan.dat. Although it is not required, it is good practice to save running configuration changes to the startup configuration. After creating a VLAN, the next step is to assign ports to the VLAN. There are several commands for defining a port to be an access port and assigning it to a VLAN. VLANs are configured on the switch port and not on the end device. An access port can belong to only one data VLAN at a time. However, a port can also be associated to a voice VLAN. For example, a port connected to an IP phone and an end device would be associated with two VLANs: one for voice and one for data. After a VLAN is configured, VLAN configurations can be validated using Cisco IOS **show** commands. If the switch access port has been incorrectly assigned to a VLAN, reenter the **switchport access vlan** *vlan-id* interface configuration command with the correct VLAN ID. The **no vlan** *vlan-id* global configuration mode command is used to remove a VLAN from the switch vlan.dat file.

VLAN Trunks

A VLAN trunk is an OSI Layer 2 link between two switches that carries traffic for all VLANs. There are several commands to configure the interconnecting ports. To verify VLAN trunk configuration, use the **show interfaces** *interface-ID* **switchport** command. Use the **no switchport trunk allowed vlan** and the **no switchport trunk native vlan** commands to remove the allowed VLANs and reset the native VLAN of the trunk.

Dynamic Trunking Protocol

An interface can be set to trunking or nontrunking, or to negotiate trunking with the neighbor interface. Trunk negotiation is managed by the Dynamic Trunking Protocol (DTP), which operates on a point-to-point basis only, between network devices. DTP is a Cisco proprietary protocol that manages trunk negotiation only if the port on the neighbor switch is configured in a trunk mode that supports DTP. To enable trunking from a Cisco switch to a device that does not support DTP, use the **switchport mode trunk** and **switchport nonegotiate** interface configuration mode commands. The **switchport mode** command has additional options for negotiating the interface mode, including access, dynamic auto, dynamic desirable, and trunk. To verify the current DTP mode, issue the **show dtp interface** command.

Practice

The following activities provide practice with the topics introduced in this chapter. The Labs are available in the companion *Switching, Routing, and Wireless Essentials Labs and Study Guide (CCNAv7)* (ISBN 9780136634386). The Packet Tracer Activity instructions are also in the Labs & Study Guide. The PKA files are found in the online course.

Labs

Lab 3.4.6: Configure VLANs and Trunking

Lab 3.6.2: Implement VLANs and Trunking

Packet Tracer Activity

Packet Tracer Activities

Packet Tracer 3.1.4: Who Hears the Broadcast

Packet Tracer 3.2.8: Investigate a VLAN Implementation

Packet Tracer 3.3.12: VLAN Configuration

Packet Tracer 3.4.5: Configure Trunks

Packet Tracer 3.5.5: Configure DTP

Packet Tracer 3.6.1: Implement VLANs and Trunking

Check Your Understanding Questions

Complete all the review questions listed here to test your understanding of the sections and concepts in this chapter. The appendix "Answers to the 'Check Your Understanding' Questions" lists the answers.

1. Which three statements accurately describe VLAN types? (Choose three).
   A. A data VLAN is used to carry VLAN management data and user-generated traffic.
   B. A management VLAN is any VLAN that is configured to access management features of the switch.
   C. After the initial boot of an unconfigured switch, all ports are members of the default VLAN.
   D. An 802.1Q trunk port, with a native VLAN assigned, supports both tagged and untagged traffic.
   E. Voice VLANs are used to support user phone and email traffic on a network.
   F. VLAN 1 is always used as the management VLAN.

2. Which type of VLAN is used to designate which traffic is untagged when crossing a trunk port?
   A. Data
   B. Default
   C. Native
   D. Management
   E. VLAN 1

3. What are two primary benefits of using VLANs? (Choose two.)

    A. A reduction in the number of trunk links
    B. Cost reduction
    C. Improved IT staff efficiency
    D. No required configuration
    E. Reduced security

4. Which command displays the encapsulation type, the voice VLAN ID, and the access mode VLAN for the Fa0/1 interface?

    A. **show interfaces Fa0/1 switchport**
    B. **show interfaces trunk**
    C. **show mac address-table interface Fa0/1**
    D. **show vlan brief**

5. What must the network administrator do to remove FastEthernet 0/1 from VLAN 2 and assign it to VLAN 3?

    A. Enter the **no shutdown** interface config command on Fa0/1.
    B. Enter the **no vlan 2** and the **vlan 3** global config commands.
    C. Enter the **switchport access vlan 3** interface config command on Fa0/1.
    D. Enter the **switchport trunk native vlan 3** interface config command on Fa0/1.

6. A Cisco Catalyst switch has been added to support the use of multiple VLANs as part of an enterprise network. The network technician finds it necessary to clear all VLAN information from the switch in order to incorporate a new network design. What should the technician do to accomplish this task?

    A. Delete the IP address that is assigned to the management VLAN and reboot the switch.
    B. Delete the startup configuration and the vlan.dat file in the flash memory of the switch and reboot the switch.
    C. Erase the running configuration and reboot the switch.
    D. Erase the startup configuration and reboot the switch.

7. Which two characteristics match extended range VLANs? (Choose two.)

    A. CDP can be used to learn and store these VLANs.
    B. They are commonly used in small networks.
    C. They are saved in the running-config file by default.
    D. VLAN IDs exist between 1006 to 4094.
    E. VLANs are initialized from flash memory.

8. What happens to switch ports after the VLAN to which they are assigned is deleted?

   A. The ports are assigned to VLAN 1, the default VLAN.

   B. The ports are disabled and must be re-enabled using the **no shutdown** command.

   C. The ports are placed in trunk mode.

   D. The ports stop communicating with the attached devices.

9. You must configure a trunk link between a Cisco Catalyst 2960 switch to another vendor Layer 2 switch. Which two commands should be configured to enable the trunk link? (Choose two.)

   A. **switchport mode access**

   B. **switchport mode dynamic auto**

   C. **switchport mode dynamic desirable**

   D. **switchport mode trunk**

   E. **switchport nonegotiate**

CHAPTER 4

Inter-VLAN Routing

Objectives

Upon completion of this chapter, you will be able to answer the following questions:

- What are the options for configuring inter-VLAN routing?
- How do you configure router-on-a-stick inter-VLAN routing?
- How do you configure inter-VLAN routing using Layer 3 switching?
- How do you troubleshoot common inter-VLAN configuration issues?

Key Terms

This chapter uses the following key terms. You can find the definitions in the Glossary.

inter-VLAN routing *Page 98*

legacy inter-VLAN routing *Page 98*

router-on-a-stick *Page 98*

subinterfaces *Page 100*

switched virtual interface (SVI) *Page 112*

routed port *Page 112*

Introduction (4.0)

Now you know how to segment and organize your network into VLANs. Hosts can communicate with other hosts in the same VLAN, and you no longer have hosts sending out broadcast messages to every other device in your network, eating up needed bandwidth. But what if a host in one VLAN needs to communicate with a host in a different VLAN? If you are a network administrator, you know that people will want to communicate with other people outside of your network. This is where inter-VLAN routing can help you. Inter-VLAN routing uses a Layer 3 device, such as a router or a Layer 3 switch. Let's take your VLAN expertise and combine it with your network layer skills and put them to the test!

Inter-VLAN Routing Operation (4.1)

In this section, you learn about two options for configuring for inter-VLAN routing.

What Is Inter-VLAN Routing? (4.1.1)

VLANs are used to segment switched Layer 2 networks for a variety of reasons. Regardless of the reason, hosts in one VLAN cannot communicate with hosts in another VLAN unless there is a router or a Layer 3 switch to provide routing services.

Inter-VLAN routing is the process of forwarding network traffic from one VLAN to another VLAN.

There are three inter-VLAN routing options:

- *Legacy Inter-VLAN routing*: This is a legacy solution. It does not scale well.
- *Router-on-a-Stick*: This is an acceptable solution for a small- to medium-sized network.
- **Layer 3 switch using switched virtual interfaces (SVIs):** This is the most scalable solution for medium to large organizations.

Legacy Inter-VLAN Routing (4.1.2)

The first inter-VLAN routing solution relied on using a router with multiple Ethernet interfaces. Each router interface was connected to a switch port in different VLANs. The router interfaces served as the default gateways to the local hosts on the VLAN subnet.

For example, refer to the topology in Figure 4-1 where R1 has two interfaces connected to switch S1.

Note

The IPv4 addresses of PC1, PC2, and R1 all have a /24 subnet mask.

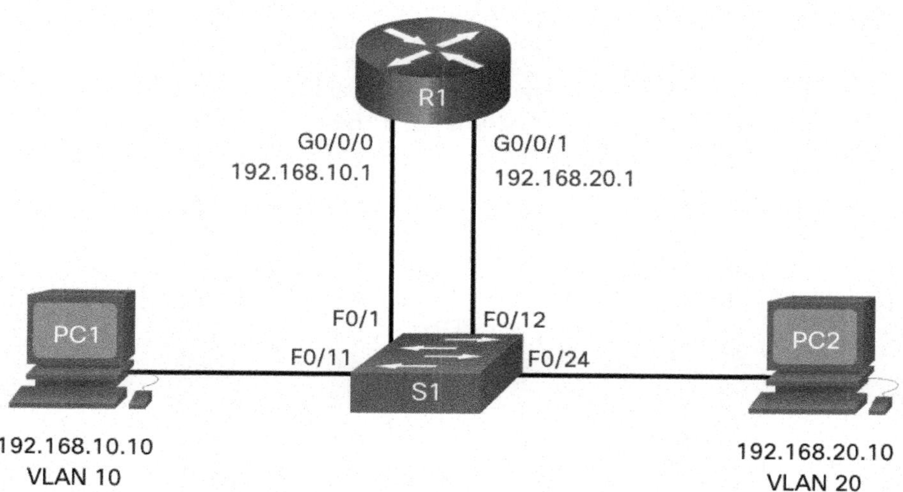

Figure 4-1 Legacy Inter-VLAN Routing Example

As shown in Table 4-1, the example MAC address table of S1 is populated as follows:

- Fa0/1 port is assigned to VLAN 10 and is connected to the R1 G0/0/0 interface.
- Fa0/11 port is assigned to VLAN 10 and is connected to PC1.
- Fa0/12 port is assigned to VLAN 20 and is connected to the R1 G0/0/1 interface.
- Fa0/11 port is assigned to VLAN 20 and is connected to PC2.

Table 4-1 MAC Address Table for S1

Port	MAC Address	VLAN
F0/1	R1 G0/0/0 MAC	10
F0/11	PC1 MAC	10
F0/12	R1 G0/0/1 MAC	20
F0/24	PC2 MAC	20

When PC1 sends a packet to PC2 on another network, it forwards it to its default gateway 192.168.10.1. R1 receives the packet on its G0/0/0 interface and examines the destination address of the packet. R1 then routes the packet out its G0/0/1 interface to the F0/12 port in VLAN 20 on S1. Finally, S1 forwards the frame to PC2.

Legacy inter-VLAN routing using physical interfaces works, but it has a significant limitation. It is not reasonably scalable because routers have a limited number of physical interfaces. Requiring one physical router interface per VLAN quickly exhausts the physical interface capacity of a router.

In our example, R1 required two separate Ethernet interfaces to route between VLAN 10 and VLAN 20. What if there were six (or more) VLANs to interconnect? A separate interface would be required for each VLAN. Obviously, this solution is not scalable.

Note

This method of inter-VLAN routing is no longer implemented in switched networks and is included for explanation purposes only.

Router-on-a-Stick Inter-VLAN Routing (4.1.3)

The "router-on-a-stick" inter-VLAN routing method overcomes the limitation of the legacy inter-VLAN routing method. It requires only one physical Ethernet interface to route traffic between multiple VLANs on a network.

A Cisco IOS router Ethernet interface is configured as an 802.1Q trunk and connected to a trunk port on a Layer 2 switch. Specifically, the router interface is configured using *subinterfaces* to identify routable VLANs.

The configured subinterfaces are software-based virtual interfaces. Each is associated with a single physical Ethernet interface. Subinterfaces are configured in software on a router. Each subinterface is independently configured with an IP address and VLAN assignment. Subinterfaces are configured for different subnets that correspond to their VLAN assignment. This facilitates logical routing.

When VLAN-tagged traffic enters the router interface, it is forwarded to the VLAN subinterface. After a routing decision is made based on the destination IP network address, the router determines the exit interface for the traffic. If the exit interface is configured as an 802.1Q subinterface, the data frames are VLAN-tagged with the new VLAN and sent back out the physical interface.

Figure 4-2 shows an example of router-on-a-stick inter-VLAN routing. PC1 on VLAN 10 is communicating with PC3 on VLAN 30 through router R1 using a single, physical router interface.

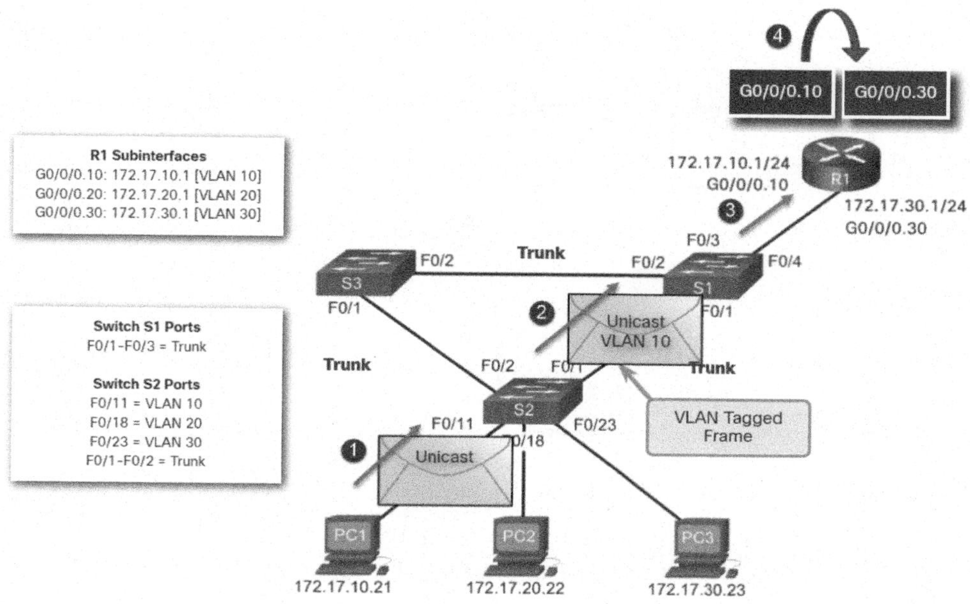

Figure 4-2 Unicast from VLAN 10 Is Route to VLAN 30

Figure 4-2 illustrates the following steps:

Step 1. PC1 sends its unicast traffic to switch S2.

Step 2. Switch S2 tags the unicast traffic as originating on VLAN 10 and forwards the unicast traffic out its trunk link to switch S1.

Step 3. Switch S1 forwards the tagged traffic out the other trunk interface on port F0/3 to the interface on router R1.

Step 4. Router R1 accepts the tagged unicast traffic on VLAN 10 and routes it to VLAN 30 using its configured subinterfaces.

In Figure 4-3, R1 routes the traffic to the correct VLAN.

Figure 4-3 illustrates the following steps:

Step 5. The unicast traffic is tagged with VLAN 30 as it is sent out the router interface to switch S1.

Step 6. Switch S1 forwards the tagged unicast traffic out the other trunk link to switch S2.

Step 7. Switch S2 removes the VLAN tag of the unicast frame and forwards the frame out to PC3 on port F0/23.

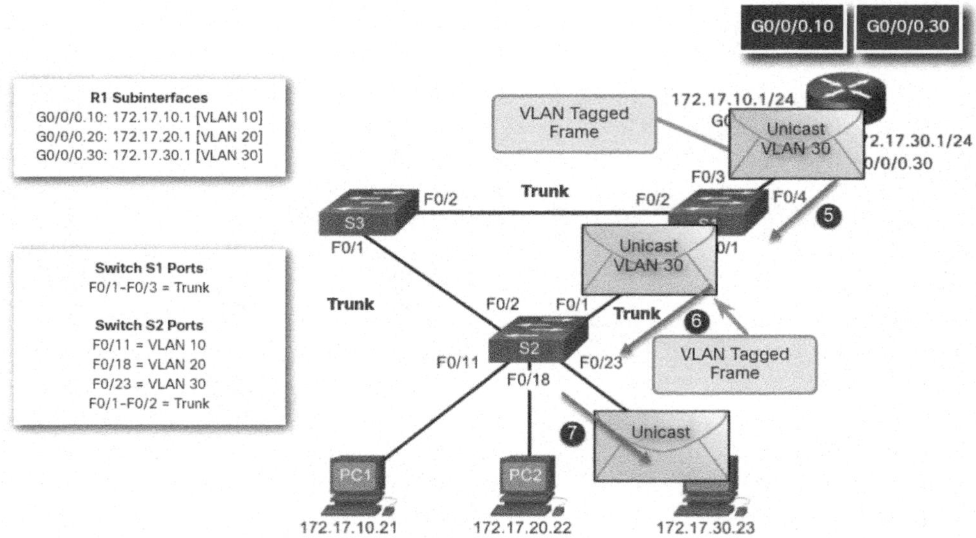

Figure 4-3 Router Tags Unicast Frame with VLAN 30

Note

The router-on-a-stick method of inter-VLAN routing does not scale beyond 50 VLANs.

Inter-VLAN Routing on a Layer 3 Switch (4.1.4)

The modern method of performing inter-VLAN routing is to use Layer 3 switches and switched virtual interfaces (SVI). An SVI is a virtual interface that is configured on a Layer 3 switch, as shown in Figure 4-4.

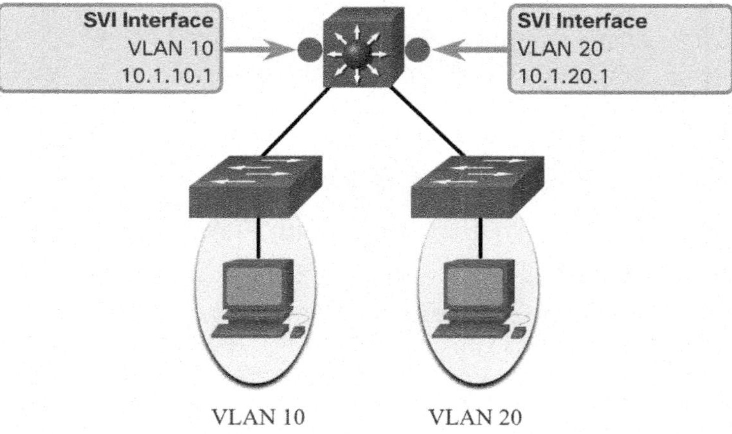

Figure 4-4 Layer 3 Switch Inter-VLAN Routing Example

Note

A Layer 3 switch is also called a multilayer switch because it operates at Layer 2 and Layer 3. However, in this course we use the term Layer 3 switch.

Inter-VLAN SVIs are created the same way that the management VLAN interface is configured. The SVI is created for a VLAN that exists on the switch. Although virtual, the SVI performs the same functions for the VLAN as a router interface would. Specifically, it provides Layer 3 processing for packets that are sent to or from all switch ports associated with that VLAN.

The following are advantages of using Layer 3 switches for inter-VLAN routing:

- They are much faster than router-on-a-stick because everything is hardware switched and routed.
- There is no need for external links from the switch to the router for routing.
- They are not limited to one link because Layer 2 EtherChannels can be used as trunk links between the switches to increase bandwidth.
- Latency is much lower because data does not need to leave the switch to be routed to a different network.
- They are more commonly deployed in a campus LAN than routers.

The only disadvantage is that Layer 3 switches are more expensive than Layer 2 switches, but they can be less expensive than a separate Layer 2 switch and router.

Interactive Graphic

Check Your Understanding—Inter-VLAN Routing Operation (4.1.5)

Refer to the online course to complete this activity.

Router-on-a-Stick Inter-VLAN Routing (4.2)

In this section, you configure router-on-a-stick inter-VLAN routing.

Router-on-a-Stick Scenario (4.2.1)

In the previous section, three ways to create inter-VLAN routing were listed, and legacy inter-VLAN routing was detailed. This section details how to configure router-on-a-stick inter-VLAN routing. You can see in the figure that the router is not in the center of the topology but instead appears to be on a stick near the border, hence the name.

In Figure 4-5, the R1 GigabitEthernet 0/0/1 interface is connected to the S1 FastEthernet 0/5 port. The S1 FastEthernet 0/1 port is connected to the S2 FastEthernet 0/1 port. These are trunk links that are required to forward traffic within and between VLANs.

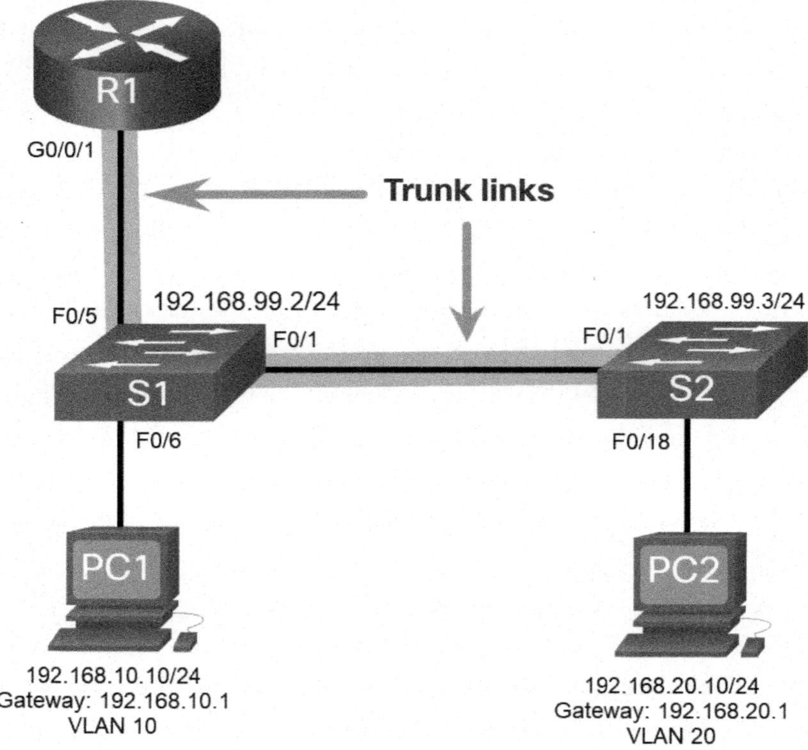

Figure 4-5 Router-on-a-Stick Topology

To route between VLANs, the R1 GigabitEthernet 0/0/1 interface is logically divided into three subinterfaces, as shown in Table 4-2. The table also shows the three VLANs that will be configured on the switches.

Table 4-2 Router R1 Subinterfaces

Subinterface	VLAN	IP Address
G0/0/1.10	10	192.168.10.1/24
G0/0/1.20	20	192.168.20.1/24
G0/0/1.30	99	192.168.99.1/24

Assume that R1, S1, and S2 have initial basic configurations. Currently, PC1 and PC2 cannot **ping** each other because they are on separate networks. Only S1 and S2 can **ping** each other, but they but are unreachable by PC1 or PC2 because they are also on different networks.

To enable devices to ping each other, the switches must be configured with VLANs and trunking, and the router must be configured for inter-VLAN routing.

S1 VLAN and Trunking Configuration (4.2.2)

Complete the following steps to configure S1 with VLANs and trunking:

Step 1. **Create and name the VLANs.** First, the VLANs are created and named, as shown in Example 4-1. VLANs are created only after you exit out of VLAN subconfiguration mode.

Example 4-1 Create and Name VLANs

```
S1(config)# vlan 10
S1(config-vlan)# name LAN10
S1(config-vlan)# exit
S1(config)# vlan 20
S1(config-vlan)# name LAN20
S1(config-vlan)# exit
S1(config)# vlan 99
S1(config-vlan)# name Management
S1(config-vlan)# exit
S1(config)#
```

Step 2. **Create the management interface.** Next, the management interface is created on VLAN 99 along with the default gateway of R1, as shown in Example 4-2.

Example 4-2 Create the Management Interface

```
S1(config)# interface vlan 99
S1(config-if)# ip add 192.168.99.2 255.255.255.0
S1(config-if)# no shut
S1(config-if)# exit
S1(config)# ip default-gateway 192.168.99.1
S1(config)#
```

Step 3. **Configure access ports.** Next, port Fa0/6 connecting to PC1 is configured as an access port in VLAN 10, as shown in Example 4-3. Assume PC1 has been configured with the correct IP address and default gateway.

Example 4-3 Configure Access Ports

```
S1(config)# interface fa0/6
S1(config-if)# switchport mode access
S1(config-if)# switchport access vlan 10
S1(config-if)# no shut
S1(config-if)# exit
S1(config)#
```

Step 4. Configure trunking ports. Finally, ports Fa0/1 connecting to S2 and Fa05 connecting to R1 are configured as trunk ports, as shown in Example 4-4.

Example 4-4 Configure Trunking Ports

```
S1(config)# interface fa0/1
S1(config-if)# switchport mode trunk
S1(config-if)# no shut
S1(config-if)# exit
S1(config)# interface fa0/5
S1(config-if)# switchport mode trunk
S1(config-if)# no shut
S1(config-if)# end
*Mar  1 00:23:43.093: %LINEPROTO-5-UPDOWN: Line protocol on Interface
  FastEthernet0/1, changed state to up
*Mar  1 00:23:44.511: %LINEPROTO-5-UPDOWN: Line protocol on Interface
  FastEthernet0/5, changed state to up
```

S2 VLAN and Trunking Configuration (4.2.3)

The configuration for S2 is similar to S1, as shown in Example 4-5.

Example 4-5 S2 Configuration

```
S2(config)# vlan 10
S2(config-vlan)# name LAN10
S2(config-vlan)# exit
S2(config)# vlan 20
S2(config-vlan)# name LAN20
S2(config-vlan)# exit
S2(config)# vlan 99
S2(config-vlan)# name Management
S2(config-vlan)# exit
S2(config)#
S2(config)# interface vlan 99
S2(config-if)# ip add 192.168.99.3 255.255.255.0
```

```
S2(config-if)# no shut
S2(config-if)# exit
S2(config)# ip default-gateway 192.168.99.1
S2(config)# interface fa0/18
S2(config-if)# switchport mode access
S2(config-if)# switchport access vlan 20
S2(config-if)# no shut
S2(config-if)# exit
S2(config)# interface fa0/1
S2(config-if)# switchport mode trunk
S2(config-if)# no shut
S2(config-if)# exit
S2(config-if)# end
*Mar  1 00:23:52.137: %LINEPROTO-5-UPDOWN: Line protocol on Interface
  FastEthernet0/1, changed state to up
```

R1 Subinterface Configuration (4.2.4)

The router-on-a-stick method requires you to create a subinterface for each VLAN to be routed.

A subinterface is created using the **interface** *interface_id.subinterface_id* global configuration mode command. The subinterface syntax is the physical interface followed by a period and a subinterface number. Although not required, it is customary to match the subinterface number with the VLAN number.

Each subinterface is then configured with the following two commands:

- **encapsulation dot1q** *vlan_id* [**native**]: This command configures the subinterface to respond to 802.1Q encapsulated traffic from the specified *vlan-id*. The **native** keyword option is only appended to set the native VLAN to something other than VLAN 1.
- **ip address** *ip-address subnet-mask*: This command configures the IPv4 address of the subinterface. This address typically serves as the default gateway for the identified VLAN.

Repeat the process for each VLAN to be routed. Each router subinterface must be assigned an IP address on a unique subnet for routing to occur.

When all subinterfaces have been created, enable the physical interface using the **no shutdown** interface configuration command. If the physical interface is disabled, all subinterfaces are disabled.

In the configuration in Example 4-6, the R1 G0/0/1 subinterfaces are configured for VLANs 10, 20, and 99.

Example 4-6 R1 Subinterface Configuration

```
R1(config)# interface G0/0/1.10
R1(config-subif)# description Default Gateway for VLAN 10
R1(config-subif)# encapsulation dot1Q 10
R1(config-subif)# ip add 192.168.10.1 255.255.255.0
R1(config-subif)# exit
R1(config)#
R1(config)# interface G0/0/1.20
R1(config-subif)# description Default Gateway for VLAN 20
R1(config-subif)# encapsulation dot1Q 20
R1(config-subif)# ip add 192.168.20.1 255.255.255.0
R1(config-subif)# exit
R1(config)#
R1(config)# interface G0/0/1.99
R1(config-subif)# description Default Gateway for VLAN 99
R1(config-subif)# encapsulation dot1Q 99
R1(config-subif)# ip add 192.168.99.1 255.255.255.0
R1(config-subif)# exit
R1(config)#
R1(config)# interface G0/0/1
R1(config-if)# description Trunk link to S1
R1(config-if)# no shut
R1(config-if)# end
R1#
*Sep 15 19:08:47.015: %LINK-3-UPDOWN: Interface GigabitEthernet0/0/1, changed
  state to down
*Sep 15 19:08:50.071: %LINK-3-UPDOWN: Interface GigabitEthernet0/0/1, changed
  state to up
*Sep 15 19:08:51.071: %LINEPROTO-5-UPDOWN: Line protocol on Interface
  GigabitEthernet0/0/1, changed state to up
R1#
```

Verify Connectivity Between PC1 and PC2 (4.2.5)

The router-on-a-stick configuration is complete after the switch trunk and the router subinterfaces have been configured. The configuration can be verified from the hosts, router, and switch.

From a host, verify connectivity to a host in another VLAN using the **ping** command. It is a good idea to first verify the current host IP configuration using the **ipconfig** Windows host command, as shown in Example 4-7.

Example 4-7 Verify Windows Host Configuration

```
C:\Users\PC1> ipconfig
Windows IP Configuration
Ethernet adapter Ethernet0:
   Connection-specific DNS Suffix  . :
   Link-local IPv6 Address           : fe80::5c43:ee7c:2959:da68%6
   IPv4 Address                      : 192.168.10.10
   Subnet Mask                       : 255.255.255.0
   Default Gateway                   : 192.168.10.1
C:\Users\PC1>
```

The output confirms the IPv4 address and default gateway of PC1. Next, use **ping** to verify connectivity with PC2 and S1, as shown in Figure 4-5. The **ping** output successfully confirms that inter-VLAN routing is operating, as shown in Example 4-8.

Example 4-8 Verify Inter-VLAN Routing by Pinging from PC1

```
C:\Users\PC1> ping 192.168.20.10
Pinging 192.168.20.10 with 32 bytes of data:
Reply from 192.168.20.10: bytes=32 time<1ms TTL=127
Reply from 192.168.20.10: bytes=32 time<1ms TTL=127
Reply from 192.168.20.10: bytes=32 time<1ms TTL=127
Reply from 192.168.20.10: bytes=32 time<1ms TTL=127
Ping statistics for 192.168.20.10:
    Packets: Sent = 4, Received = 4, Lost = 0 (0% loss).
Approximate round trip times in milli-seconds:
    Minimum = 0ms, Maximum = 0ms, Average = 0ms
C:\Users\PC1>
C:\Users\PC1> ping 192.168.99.2
Pinging 192.168.99.2 with 32 bytes of data:
Request timed out.
Request timed out.
Reply from 192.168.99.2: bytes=32 time=2ms TTL=254
Reply from 192.168.99.2: bytes=32 time=1ms TTL=254
Ping statistics for 192.168.99.2:
    Packets: Sent = 4, Received = 2, Lost = 2 (50% loss).
Approximate round trip times in milli-seconds:
    Minimum = 1ms, Maximum = 2ms, Average = 1ms
C:\Users\PC1>
```

Router-on-a-Stick Inter-VLAN Routing Verification (4.2.6)

In addition to using **ping** between devices, the following **show** commands can be used to verify and troubleshoot the router-on-a-stick configuration.

- **show ip route**
- **show ip interface brief**
- **show interfaces**
- **show interfaces trunk**

As shown in Example 4-9, verify that the subinterfaces are appearing in the routing table of R1 by using the **show ip route** command. Notice that there are three connected routes (C) and their respective exit interfaces for each routable VLAN. The output confirms that the correct subnets, VLANs, and subinterfaces are active.

Example 4-9 Verify Subinterfaces Are in Routing Table

```
R1# show ip route | begin Gateway
Gateway of last resort is not set
      192.168.10.0/24 is variably subnetted, 2 subnets, 2 masks
C        192.168.10.0/24 is directly connected, GigabitEthernet0/0/1.10
L        192.168.10.1/32 is directly connected, GigabitEthernet0/0/1.10
      192.168.20.0/24 is variably subnetted, 2 subnets, 2 masks
C        192.168.20.0/24 is directly connected, GigabitEthernet0/0/1.20
L        192.168.20.1/32 is directly connected, GigabitEthernet0/0/1.20
      192.168.99.0/24 is variably subnetted, 2 subnets, 2 masks
C        192.168.99.0/24 is directly connected, GigabitEthernet0/0/1.99
L        192.168.99.1/32 is directly connected, GigabitEthernet0/0/1.99
R1#
```

Another useful router command is **show ip interface brief**, as shown in Example 4-10. The output confirms that the subinterfaces have the correct IPv4 address configured, and that they are operational.

Example 4-10 Verify Subinterface IP Addresses and Status

```
R1# show ip interface brief | include up
GigabitEthernet0/0/1   unassigned      YES unset  up                    up
Gi0/0/1.10             192.168.10.1    YES manual up                    up
Gi0/0/1.20             192.168.20.1    YES manual up                    up
Gi0/0/1.99             192.168.99.1    YES manual up                    up
R1#
```

Subinterfaces can be verified using the **show interfaces** *subinterface-id* command, as shown in Example 4-11.

Example 4-11 Verify Details of the Subinterface

```
R1# show interfaces g0/0/1.10
GigabitEthernet0/0/1.10 is up, line protocol is up
  Hardware is ISR4221-2x1GE, address is 10b3.d605.0301 (bia 10b3.d605.0301)
  Description: Default Gateway for VLAN 10
  Internet address is 192.168.10.1/24
  MTU 1500 bytes, BW 100000 Kbit/sec, DLY 100 usec,
     reliability 255/255, txload 1/255, rxload 1/255
  Encapsulation 802.1Q Virtual LAN, Vlan ID  10.
  ARP type: ARPA, ARP Timeout 04:00:00
  Keepalive not supported
  Last clearing of "show interface" counters never
R1#
```

The misconfiguration could also be on the trunking port of the switch. Therefore, it is also useful to verify the active trunk links on a Layer 2 switch by using the **show interfaces trunk** command, as shown in Example 4-12. The output confirms that the link to R1 is trunking for the required VLANs.

Note

Although VLAN 1 was not explicitly configured, it was automatically included because control traffic on trunk links will always be forwarded on VLAN 1.

Example 4-12 Verify Trunk Link Status

```
S1# show interfaces trunk
Port        Mode             Encapsulation  Status        Native vlan
Fa0/1       on               802.1q         trunking      1
Fa0/5       on               802.1q         trunking      1
Port        Vlans allowed on trunk
Fa0/1       1-4094
Fa0/5       1-4094
Port        Vlans allowed and active in management domain
Fa0/1       1,10,20,99
Fa0/5       1,10,20,99
Port        Vlans in spanning tree forwarding state and not pruned
Fa0/1       1,10,20,99
Fa0/5       1,10,20,99
S1#
```

Packet Tracer
Activity

Packet Tracer—Configure Router-on-a-Stick Inter-VLAN Routing (4.2.7)

In this Packet Tracer activity, you check for connectivity prior to implementing inter-VLAN routing. Then you configure VLANs and inter-VLAN routing. Finally, you enable trunking and verify connectivity between VLANs.

Lab—Configure Router-on-a-Stick Inter-VLAN Routing (4.2.8)

In this lab, you complete the following objectives:

- Part 1: Build the Network and Configure Basic Device Settings
- Part 2: Configure Switches with VLANs and Trunking
- Part 3: Configure Trunk-Based Inter-VLAN Routing

Inter-VLAN Routing using Layer 3 Switches (4.3)

In this section, you configure inter-VLAN routing using Layer 3 switches.

Layer 3 Switch Inter-VLAN Routing (4.3.1)

Modern enterprise networks rarely use router-on-a-stick because it does not scale easily to meet requirements. In these very large networks, network administrators use Layer 3 switches to configure inter-VLAN routing.

Inter-VLAN routing using the router-on-a-stick method is simple to implement for a small- to medium-sized organization. However, a large enterprise requires a faster, much more scalable method to provide inter-VLAN routing.

Enterprise campus LANs use Layer 3 switches to provide inter-VLAN routing. Layer 3 switches use hardware-based switching to achieve higher-packet processing rates than routers. Layer 3 switches are also commonly implemented in enterprise distribution layer wiring closets.

Capabilities of a Layer 3 switch include the ability to do the following:

- Route from one VLAN to another using multiple *switched virtual interfaces (SVIs)*.
- Convert a Layer 2 switchport to a Layer 3 interface (that is, a *routed port*). A routed port is similar to a physical interface on a Cisco IOS router.

To provide inter-VLAN routing, Layer 3 switches use SVIs. SVIs are configured using the same **interface vlan** *vlan-id* command used to create the management SVI on a Layer 2 switch. A Layer 3 SVI must be created for each of the routable VLANs.

Layer 3 Switch Scenario (4.3.2)

In Figure 4-6, the Layer 3 switch, D1, is connected to two hosts on different VLANs. PC1 is in VLAN 10, and PC2 is in VLAN 20, as shown. The Layer 3 switch will provide inter-VLAN routing services to the two hosts.

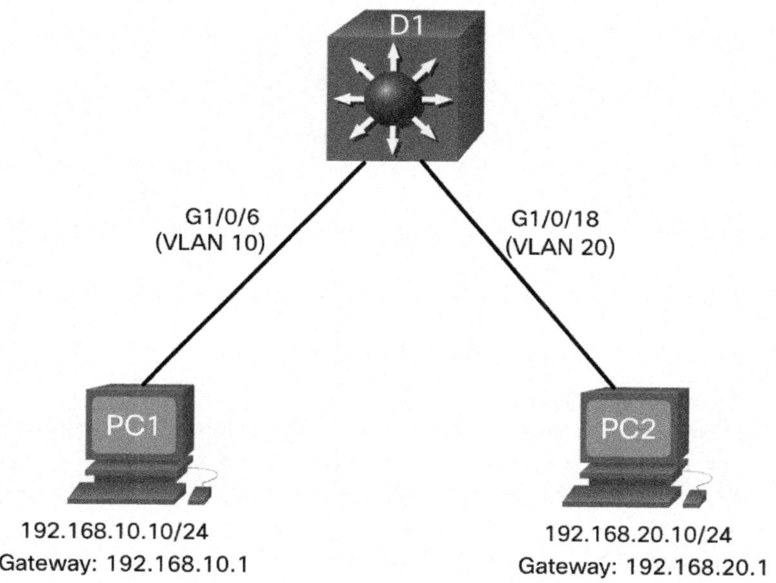

Figure 4-6 Layer 3 Switch Inter-VLAN Routing Topology

Table 4-3 shows the IP addresses for each VLAN.

Table 4-3 D1 VLAN IP Addresses

VLAN Interface	IP Address
10	192.168.10.1/24
20	192.168.20.1/24

Layer 3 Switch Configuration (4.3.3)

Complete the following steps to configure S1 with VLANs and trunking:

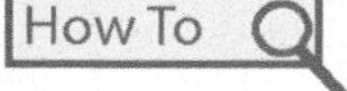

Step 1. **Create the VLANs.** First, create the two VLANs as shown in Example 4-13.

Example 4-13 Create the VLANs

```
D1(config)# vlan 10
D1(config-vlan)# name LAN10
D1(config-vlan)# vlan 20
D1(config-vlan)# name LAN20
D1(config-vlan)# exit
D1(config)#
```

Step 2. **Create the SVI VLAN interfaces.** Configure the SVI for VLANs 10 and 20, as shown in Example 4-14. The IP addresses that are configured will serve as the default gateways to the hosts in the respective VLANs. Notice the informational messages showing the line protocol on both SVIs changed to up.

Example 4-14 Create the SVI VLAN Interfaces

```
D1(config)# interface vlan 10
D1(config-if)# description Default Gateway SVI for 192.168.10.0/24
D1(config-if)# ip add 192.168.10.1 255.255.255.0
D1(config-if)# no shut
D1(config-if)# exit
D1(config)#
D1(config)# int vlan 20
D1(config-if)# description Default Gateway SVI for 192.168.20.0/24
D1(config-if)# ip add 192.168.20.1 255.255.255.0
D1(config-if)# no shut
D1(config-if)# exit
D1(config)#
*Sep 17 13:52:16.053: %LINEPROTO-5-UPDOWN: Line protocol on Interface Vlan10,
  changed state to up
*Sep 17 13:52:16.160: %LINEPROTO-5-UPDOWN: Line protocol on Interface Vlan20,
  changed state to up
```

Step 3. **Configure access ports.** Next, configure the access ports connecting to the hosts and assign them to their respective VLANs, as shown in Example 4-15.

Example 4-15 Configure Access Ports

```
D1(config)# interface GigabitEthernet1/0/6
D1(config-if)# description Access port to PC1
D1(config-if)# switchport mode access
D1(config-if)# switchport access vlan 10
D1(config-if)# exit
D1(config)#
D1(config)# interface GigabitEthernet1/0/18
D1(config-if)# description Access port to PC2
D1(config-if)# switchport mode access
D1(config-if)# switchport access vlan 20
D1(config-if)# exit
```

Step 4. **Enable IP routing.** Finally, enable IPv4 routing with the **ip routing** global configuration command to allow traffic to be exchanged between VLANs 10 and 20, as shown in Example 4-16. This command must be configured to enable inter-VAN routing on a Layer 3 switch for IPv4.

Example 4-16 Enable IP Routing

```
D1(config)# ip routing
D1(config)#
```

Layer 3 Switch Inter-VLAN Routing Verification (4.3.4)

Inter-VLAN routing using a Layer 3 switch is simpler to configure than the router-on-a-stick method. After the configuration is complete, the configuration can be verified by testing connectivity between the hosts.

From a host, verify connectivity to a host in another VLAN using the **ping** command. It is a good idea to first verify the current host IP configuration using the **ipconfig** Windows host command. The output in Example 4-17 confirms the IPv4 address and default gateway of PC1.

Example 4-17 Verify Windows Host Configuration

```
C:\Users\PC1> ipconfig
Windows IP Configuration
Ethernet adapter Ethernet0:
   Connection-specific DNS Suffix  .  :
   Link-local IPv6 Address            : fe80::5c43:ee7c:2959:da68%6
   IPv4 Address                       : 192.168.10.10
   Subnet Mask                        : 255.255.255.0
   Default Gateway                    : 192.168.10.1
C:\Users\PC1>
```

Next, verify connectivity with PC2 using the **ping** Windows host command, as shown in Example 4-18. The **ping** output successfully confirms that inter-VLAN routing is operating.

Example 4-18 Verify Inter-VLAN Routing by Pinging from PC1

```
C:\Users\PC1> ping 192.168.20.10
Pinging 192.168.20.10 with 32 bytes of data:
Reply from 192.168.20.10: bytes=32 time<1ms TTL=127
Reply from 192.168.20.10: bytes=32 time<1ms TTL=127
Reply from 192.168.20.10: bytes=32 time<1ms TTL=127
Reply from 192.168.20.10: bytes=32 time<1ms TTL=127
Ping statistics for 192.168.20.10:
    Packets: Sent = 4, Received = 4, Lost = 0 (0% loss),
Approximate round trip times in milli-seconds:
    Minimum = 0ms, Maximum = 0ms, Average = 0ms
C:\Users\PC1>
```

Routing on a Layer 3 Switch (4.3.5)

If VLANs are to be reachable by other Layer 3 devices, they must be advertised using static or dynamic routing. To enable routing on a Layer 3 switch, a routed port must be configured.

A routed port is created on a Layer 3 switch by disabling the switchport feature on a Layer 2 port that is connected to another Layer 3 device. Specifically, configuring the **no switchport** interface configuration command on a Layer 2 port converts it into a Layer 3 interface. Then the interface can be configured with an IPv4 configuration to connect to a router or another Layer 3 switch.

Routing Scenario on a Layer 3 Switch (4.3.6)

In Figure 4-7, the previously configured D1 Layer 3 switch is now connected to R1. R1 and D1 are both in an Open Shortest Path First (OSPF) routing protocol domain. Assume inter-VLAN has been successfully implemented on D1. The G0/0/1 interface of R1 has also been configured and enabled. Additionally, R1 is using OSPF to advertise its two networks, 10.10.10.0/24 and 10.20.20.0/24.

Note

OSPF routing configuration is covered in another course. In this module, OSPF configuration commands will be given to you in all activities and assessments. It is not required that you understand the configuration in order to enable OSPF routing on the Layer 3 switch.

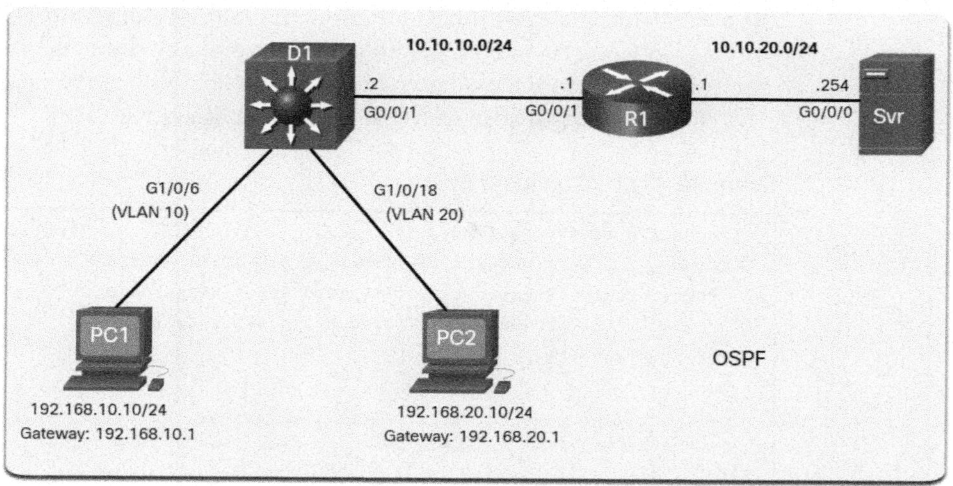

Figure 4-7 Routing Scenario on a Layer 3 Switch Topology

Routing Configuration on a Layer 3 Switch (4.3.7)

Complete the following steps to configure D1 to route with R1:

Step 1. Configure the routed port. Configure G1/0/1 to be a routed port, assign it an IPv4 address, and enable it, as shown in Example 4-19.

Example 4-19 Configure the Routed Port

```
D1(config)# interface GigabitEthernet1/0/1
D1(config-if)# description routed Port Link to R1
D1(config-if)# no switchport
D1(config-if)# ip address 10.10.10.2 255.255.255.0
D1(config-if)# no shut
D1(config-if)# exit
D1(config)#
```

Step 2. Enable routing, as shown in Example 4-20. Ensure IPv4 routing is enabled with the **ip routing** global configuration command.

Example 4-20 Enable Routing

```
D1(config)# ip routing
D1(config)#
```

Step 3. Configure routing. Configure the OSPF routing protocol to advertise the VLAN 10 and VLAN 20 networks, along with the network that is connected to R1, as shown in Example 4-21. Notice the message informing you that an adjacency has been established with R1.

Example 4-21 Configure Routing

```
D1(config)# router ospf 10
D1(config-router)# network 192.168.10.0 0.0.0.255 area 0
D1(config-router)# network 192.168.20.0 0.0.0.255 area 0
D1(config-router)# network 10.10.10.0 0.0.0.3 area 0
D1(config-router)# ^Z
D1#
*Sep 17 13:52:51.163: %OSPF-5-ADJCHG: Process 10, Nbr 10.20.20.1 on
  GigabitEthernet1/0/1 from LOADING to FULL, Loading Done
D1#
```

Step 4. Verify routing. Verify the routing table on D1, as shown in Example 4-22. Notice that D1 now has a route to the 10.20.20.0/24 network.

Example 4-22 Verify Routing

```
D1# show ip route | begin Gateway
Gateway of last resort is not set
      10.0.0.0/8 is variably subnetted, 3 subnets, 3 masks
C        10.10.10.0/30 is directly connected, GigabitEthernet1/0/1
L        10.10.10.2/32 is directly connected, GigabitEthernet1/0/1
O        10.20.20.0/24 [110/2] via 10.10.10.1, 00:00:06, GigabitEthernet1/0/1
      192.168.10.0/24 is variably subnetted, 2 subnets, 2 masks
C        192.168.10.0/24 is directly connected, Vlan10
L        192.168.10.1/32 is directly connected, Vlan10
      192.168.20.0/24 is variably subnetted, 2 subnets, 2 masks
C        192.168.20.0/24 is directly connected, Vlan20
L        192.168.20.1/32 is directly connected, Vlan20
D1#
```

Step 5. Verify connectivity. At this time, PC1 and PC2 are able to ping the server connected to R1, as shown in Example 4-23.

Example 4-23 Verify Connectivity

```
C:\Users\PC1> ping 10.20.20.254
Pinging 10.20.20.254 with 32 bytes of data:
Request timed out.
Reply from 10.20.20.254: bytes=32 time<1ms TTL=127
Reply from 10.20.20.254: bytes=32 time<1ms TTL=127
```

```
Reply from 10.20.20.254: bytes=32 time<1ms TTL=127
Ping statistics for 10.20.20.254:
    Packets: Sent = 4, Received = 3, Lost = 1 (25% loss).
Approximate round trip times in milli-seconds:
    Minimum = 1ms, Maximum = 2ms, Average = 1ms
C:\Users\PC1>
!==================================================
C:\Users\PC2> ping 10.20.20.254
Pinging 10.20.20.254 with 32 bytes of data:
Reply from 10.20.20.254: bytes=32 time<1ms TTL=127
Reply from 10.20.20.254: bytes=32 time<1ms TTL=127
Reply from 10.20.20.254: bytes=32 time<1ms TTL=127
Reply from 10.20.20.254: bytes=32 time<1ms TTL=127
Ping statistics for 10.20.20.254:
    Packets: Sent = 4, Received = 4, Lost = 0 (0% loss).
Approximate round trip times in milli-seconds:
    Minimum = 1ms, Maximum = 2ms, Average = 1ms
C:\Users\PC2>
```

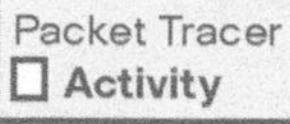

Packet Tracer—Configure Layer 3 Switching and Inter-VLAN Routing (4.3.8)

In this Packet Tracer activity, you configure Layer 3 switching and Inter-VLAN routing on a Cisco 3560 switch.

Troubleshoot Inter-VLAN Routing (4.4)

In this section, you learn how to troubleshoot issues in an inter-VLAN routing environment.

Common Inter-VLAN Issues (4.4.1)

By now, you know that when you configure and verify, you must also be able to troubleshoot. This section discusses some common network problems associated with inter-VLAN routing.

There are a number of reasons why an inter-VAN configuration may not work. All are related to connectivity issues. First, check the physical layer to resolve any issues where a cable might be connected to the wrong port. If the connections are correct, use the list in Table 4-4 for other common reasons why inter-VLAN connectivity may fail.

Table 4-4 Common Inter-VLAN Issues

Issue Type	How to Fix	How to Verify
Missing VLANs	■ Create (or re-create) the VLAN if it does not exist. ■ Ensure host port is assigned to the correct VLAN.	**show vlan** [brief] **show interfaces switchport** **ping**
Switch Trunk Port Issues	■ Ensure trunks are configured correctly. ■ Ensure port is a trunk port and enabled.	**show interfaces trunk** **show running-config**
Switch Access Port Issues	■ Assign correct VLAN to access port. ■ Ensure port is an access port and enabled. ■ Host is incorrectly configured in the wrong subnet.	**show interfaces switchport** **show running-config interface** **ipconfig**
Router Configuration Issues	■ Router subinterface IPv4 address is incorrectly configured. ■ Router subinterface is assigned to the VLAN ID.	**show ip interface brief** **show interfaces**

Troubleshoot Inter-VLAN Routing Scenario (4.4.2)

Next, examples of some of these inter-VLAN routing problems are covered in more detail.

The topology in Figure 4-8 will be used for all of these issues.

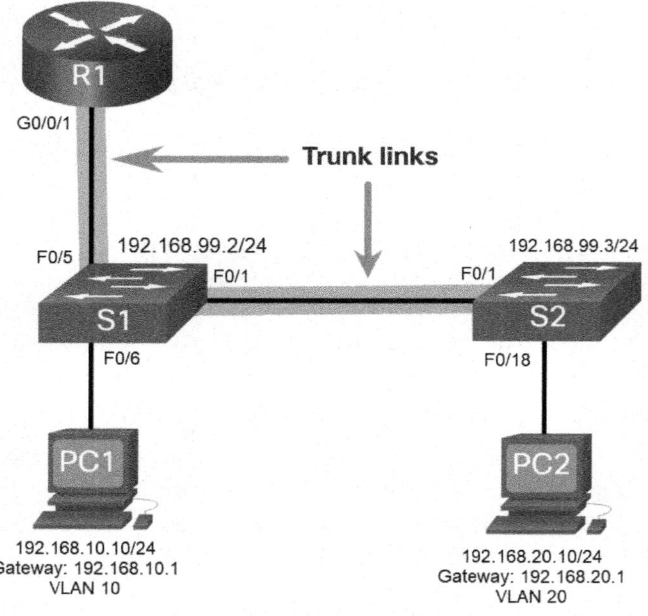

Figure 4-8 Inter-VLAN Routing Troubleshooting Topology

The VLAN and IPv4 addressing information for R1 is shown in Table 4-5.

Table 4-5 Router R1 Subinterfaces

Subinterface	VLAN	IP Address
G0/0/0.10	10	192.168.10.1/24
G0/0/0.20	20	192.168.20.1/24
G0/0/0.30	99	192.168.99.1/24

Missing VLANs (4.4.3)

An inter-VLAN connectivity issue could be caused by a missing VLAN. The VLAN could be missing if it was not created, it was accidently deleted, or it is not allowed on the trunk link.

For example, PC1 is currently connected to VLAN 10, as shown in the **show vlan brief** command output in Example 4-24.

Example 4-24 Verify VLAN for PC1

```
S1# show vlan brief
VLAN Name                             Status    Ports
---- -------------------------------- --------- -------------------------------
1    default                          active    Fa0/2, Fa0/3, Fa0/4, Fa0/7
                                                Fa0/8, Fa0/9, Fa0/10, Fa0/11
                                                Fa0/12, Fa0/13, Fa0/14, Fa0/15
                                                Fa0/16, Fa0/17, Fa0/18, Fa0/19
                                                Fa0/20, Fa0/21, Fa0/22, Fa0/23
                                                Fa0/24, Gi0/1, Gi0/2
10   LAN10                            active    Fa0/6
20   LAN20                            active
99   Management                       active
1002 fddi-default                     act/unsup
1003 token-ring-default               act/unsup
1004 fddinet-default                  act/unsup
1005 trnet-default                    act/unsup
S1#
```

Now assume that VLAN 10 is accidently deleted, as shown in Example 4-25.

Example 4-25 VLAN 10 Is Deleted

```
S1(config)# no vlan 10
S1(config)# do show vlan brief
VLAN Name                             Status    Ports
---- -------------------------------- --------- -------------------------------
1    default                          active    Fa0/2, Fa0/3, Fa0/4, Fa0/7
                                                Fa0/8, Fa0/9, Fa0/10, Fa0/11
                                                Fa0/12, Fa0/13, Fa0/14, Fa0/15
                                                Fa0/16, Fa0/17, Fa0/18, Fa0/19
                                                Fa0/20, Fa0/21, Fa0/22, Fa0/23
                                                Fa0/24, Gi0/1, Gi0/2
20   LAN20                            active
99   Management                       active
1002 fddi-default                     act/unsup
1003 token-ring-default               act/unsup
1004 fddinet-default                  act/unsup
1005 trnet-default                    act/unsup
S1(config)#
```

Notice that VLAN 10 is now missing from the output in Example 4-25. Also notice that port Fa0/6 has not been reassigned to the default VLAN. The reason is because when you delete a VLAN, any ports assigned to that VLAN become inactive. They remain associated with the VLAN (and thus inactive) until you assign them to a new VLAN or re-create the missing VLAN.

Use the **show interface** *interface-id* **switchport** command to verify the VLAN membership, as shown in Example 4-26.

Example 4-26 Verify an Interface's VLAN Membership

```
S1(config)# do show interface fa0/6 switchport
Name: Fa0/6
Switchport: Enabled
Administrative Mode: static access
Operational Mode: static access
Administrative Trunking Encapsulation: dot1q
Operational Trunking Encapsulation: native
Negotiation of Trunking: Off
Access Mode VLAN: 10 (Inactive)
Trunking Native Mode VLAN: 1 (default)
Administrative Native VLAN tagging: enabled
Voice VLAN: none
(Output omitted)
```

Re-creating the missing VLAN would automatically reassign the hosts to it, as shown in Example 4-27.

Example 4-27 Attempt to Re-create and Verify VLAN 10

```
S1(config)# vlan 10
S1(config-vlan)# do show vlan brief
VLAN Name                             Status    Ports
---- -------------------------------- --------- -------------------------------
1    default                          active    Fa0/2, Fa0/3, Fa0/4, Fa0/7
                                                Fa0/8, Fa0/9, Fa0/10, Fa0/11
                                                Fa0/12, Fa0/13, Fa0/14, Fa0/15
                                                Fa0/16, Fa0/17, Fa0/18, Fa0/19
                                                Fa0/20, Fa0/21, Fa0/22, Fa0/23
                                                Fa0/24, Gi0/1, Gi0/2
20   LAN20                            active
99   Management                       active
1002 fddi-default                     act/unsup
1003 token-ring-default               act/unsup
1004 fddinet-default                  act/unsup
1005 trnet-default                    act/unsup
S1(config-vlan)#
```

Notice that the VLAN has not been created as expected. The reason is because you must exit from VLAN sub-configuration mode to create the VLAN, as shown in Example 4-28.

Example 4-28 Exit VLAN Configuration Mode and Then Re-create and Verify VLAN

```
S1(config-vlan)# exit
S1(config)# vlan 10
S1(config)# do show vlan brief
VLAN Name                             Status    Ports
---- -------------------------------- --------- -------------------------------
1    default                          active    Fa0/2, Fa0/3, Fa0/4, Fa0/7
                                                Fa0/8, Fa0/9, Fa0/10, Fa0/11
                                                Fa0/12, Fa0/13, Fa0/14, Fa0/15
                                                Fa0/16, Fa0/17, Fa0/18, Fa0/19
                                                Fa0/20, Fa0/21, Fa0/22, Fa0/23
                                                Fa0/24, Gi0/1, Gi0/2
10   VLAN0010                         active    Fa0/6
20   LAN20                            active
99   Management                       active
1002 fddi-default                     act/unsup
1003 token-ring-default               act/unsup
1004 fddinet-default                  act/unsup
1005 trnet-default                    act/unsup
S1(config)#
```

Now notice that the VLAN is included in the list and that the host connected to Fa0/6 is on VLAN 10.

Switch Trunk Port Issues (4.4.4)

Another issue for inter-VLAN routing includes misconfigured switch ports. In a legacy inter-VLAN solution, this could be caused when the connecting router port is not assigned to the correct VLAN.

However, with a router-on-a-stick solution, the most common cause is a misconfigured trunk port.

For example, assume PC1 was able to connect to hosts in other VLANs until recently. A quick look at maintenance logs revealed that the S1 Layer 2 switch was recently accessed for routine maintenance. Therefore, you suspect the problem may be related to that switch.

On S1, verify that the port connecting to R1 (i.e., F0/5) is correctly configured as a trunk link using the **show interfaces trunk** command, as shown in Example 4-29.

Example 4-29 Verify Trunking

```
S1# show interfaces trunk
Port        Mode              Encapsulation  Status        Native vlan
Fa0/1       on                802.1q         trunking      1
Port        Vlans allowed on trunk
Fa0/1       1-4094
Port        Vlans allowed and active in management domain
Fa0/1       1,10,20,99
Port        Vlans in spanning tree forwarding state and not pruned
Fa0/1       1,10,20,99
S1#
```

The Fa0/5 port connecting to R1 is mysteriously missing from the output. Verify the interface configuration using the **show running-config interface fa0/5** command, as shown in Example 4-30.

Example 4-30 Verify Interface Configuration

```
S1# show running-config interface fa0/5
Building configuration...
Current configuration : 96 bytes
!
interface FastEthernet0/5
 description Trunk link to R1
 switchport mode trunk
 shutdown
end
S1#
```

As you can see, the port was accidently shut down. To correct the problem, reenable the port and verify the trunking status, as shown in Example 4-31.

Example 4-31 Reenable and Verify the Port

```
S1(config)# interface fa0/5
S1(config-if)# no shut
S1(config-if)#
*Mar  1 04:46:44.153: %LINK-3-UPDOWN: Interface FastEthernet0/5, changed state to
  up
S1(config-if)#
*Mar  1 04:46:47.962: %LINEPROTO-5-UPDOWN: Line protocol on Interface
  FastEthernet0/5, changed state to up
S1(config-if)# do show interface trunk
Port        Mode                 Encapsulation  Status         Native vlan
Fa0/1       on                   802.1q         trunking       1
Fa0/5       on                   802.1q         trunking       1
Port        Vlans allowed on trunk
Fa0/1       1-4094
Fa0/5       1-4094
Port        Vlans allowed and active in management domain
Fa0/1       1,10,20,99
Fa0/5       1,10,20,99
Port        Vlans in spanning tree forwarding state and not pruned
Fa0/1       1,10,20,99
Fa0/1       1,10,20,99
S1(config-if)#
```

To reduce the risk of a failed inter-switch link disrupting inter-VLAN routing, redundant links and alternate paths should be part of the network design.

Switch Access Port Issues (4.4.5)

When a problem is suspected with a switch access port configuration, use verification commands to examine the configuration and identify the problem.

Assume PC1 has the correct IPv4 address and default gateway but is not able to **ping** its own default gateway. PC1 is supposed to be connected to a VLAN 10 port.

Verify the port configuration on S1 using the **show interfaces** *interface-id* **switchport** command, as shown in Example 4-32.

Example 4-32 Verify the Port Configuration

```
S1# show interface fa0/6 switchport
Name: Fa0/6
Switchport: Enabled
Administrative Mode: static access
Operational Mode: static access
Administrative Trunking Encapsulation: dot1q
Operational Trunking Encapsulation: native
Negotiation of Trunking: Off
Access Mode VLAN: 1 (default)
Trunking Native Mode VLAN: 1 (default)
Administrative Native VLAN tagging: enabled
Voice VLAN: none
```

The Fa0/6 port has been configured as an access port, as indicated by "static access". However, it appears that it has not been configured to be in VLAN 10. Verify the configuration of the interface, as shown in Example 4-33.

Example 4-33 Verify the Port Configuration in the Running-Config

```
S1# show running-config interface fa0/6
Building configuration...
Current configuration : 87 bytes
!
interface FastEthernet0/6
 description PC-A access port
 switchport mode access
end
S1#
```

Assign port Fa0/6 to VLAN 10 and verify the port assignment, as shown in Example 4-34.

Example 4-34 Assign the VLAN to the Port and Verify the Configuration

```
S1# configure terminal
S1(config)# interface fa0/6
S1(config-if)# switchport access vlan 10
S1(config-if)#
S1(config-if)# do show interface fa0/6 switchport
Name: Fa0/6
Switchport: Enabled
Administrative Mode: static access
Operational Mode: static access
```

```
Administrative Trunking Encapsulation: dot1q
Operational Trunking Encapsulation: native
Negotiation of Trunking: Off
Access Mode VLAN: 10 (VLAN0010)
Trunking Native Mode VLAN: 1 (default)
Administrative Native VLAN tagging: enabled
Voice VLAN: none
(Output omitted)
```

PC1 is now able to communicate with hosts on other VLANs.

Router Configuration Issues (4.4.6)

Router-on-a-stick configuration problems are usually related to subinterface misconfigurations. For instance, an incorrect IP address was configured or the wrong VLAN ID was assigned to the subinterface.

For example, R1 should be providing inter-VLAN routing for users in VLANs 10, 20, and 99. However, users in VLAN 10 cannot reach any other VLAN.

You verified the switch trunk link and all appears to be in order. Verify the subinterface status using the **show ip interface brief** command, as shown in Example 4-35.

Example 4-35 Verify the Status of the Subinterfaces

```
R1# show ip interface brief
Interface              IP-Address      OK? Method Status                Protocol
GigabitEthernet0/0/0   unassigned      YES unset  administratively down down
GigabitEthernet0/0/1   unassigned      YES unset  up                    up
Gi0/0/1.10             192.168.10.1    YES manual up                    up
Gi0/0/1.20             192.168.20.1    YES manual up                    up
Gi0/0/1.99             192.168.99.1    YES manual up                    up
Serial0/1/0            unassigned      YES unset  administratively down down
Serial0/1/1            unassigned      YES unset  administratively down down
R1#
```

The subinterfaces have been assigned the correct IPv4 addresses, and they are operational.

Verify which VLANs each of the subinterfaces is on. To do so, the **show interfaces** command is useful, but it generates a great deal of additional unrequired output.

The command output can be reduced using IOS command filters as shown in Example 4-36.

Example 4-36 Verify the VLANs Configured on Each Subinterface

```
R1# show interfaces | include Gig|802.1Q
GigabitEthernet0/0/0 is administratively down, line protocol is down
GigabitEthernet0/0/1 is up, line protocol is up
  Encapsulation 802.1Q Virtual LAN, Vlan ID  1., loopback not set
GigabitEthernet0/0/1.10 is up, line protocol is up
  Encapsulation 802.1Q Virtual LAN, Vlan ID  100.
GigabitEthernet0/0/1.20 is up, line protocol is up
  Encapsulation 802.1Q Virtual LAN, Vlan ID  20.
GigabitEthernet0/0/1.99 is up, line protocol is up
  Encapsulation 802.1Q Virtual LAN, Vlan ID  99.
R1#
```

The pipe symbol (|) along with some select keywords is a useful method to help filter command output. In this example, the keyword **include** was used to identify that only lines containing the letters "Gig" or "802.1Q" will be displayed. Because of the way the **show interface** output is naturally listed, using these filters produces a condensed list of interfaces and their assigned VLANs.

Notice that the G0/0/1.10 interface has been incorrectly assigned to VLAN 100 instead of VLAN 10. This is confirmed by looking at the configuration of the R1 GigabitEthernet 0/0/1.10 subinterface, as shown in Example 4-37.

Example 4-37 Verify the Configuration of the Subinterface in the Running-Config

```
R1# show running-config interface g0/0/1.10
Building configuration...
Current configuration : 146 bytes
!
interface GigabitEthernet0/0/1.10
 description Default Gateway for VLAN 10
 encapsulation dot1Q 100
 ip address 192.168.10.1 255.255.255.0
end
R1#
```

To correct this problem, configure subinterface G0/0/1.10 to be on the correct VLAN using the **encapsulation dot1q 10** subinterface configuration mode command, as shown in Example 4-38.

Example 4-38 Correct and Verify the Subinterface Configuration

```
R1# conf t
Enter configuration commands, one per line.  End with CNTL/Z.
R1(config)# interface gigabitEthernet 0/0/1.10
R1(config-subif)# encapsulation dot1Q 10
R1(config-subif)# end
R1#
R1# show interfaces | include Gig|802.1Q
GigabitEthernet0/0/0 is administratively down, line protocol is down
GigabitEthernet0/0/1 is up, line protocol is up
  Encapsulation 802.1Q Virtual LAN, Vlan ID  1., loopback not set
GigabitEthernet0/0/1.10 is up, line protocol is up
  Encapsulation 802.1Q Virtual LAN, Vlan ID  10.
GigabitEthernet0/0/1.20 is up, line protocol is up
  Encapsulation 802.1Q Virtual LAN, Vlan ID  20.
GigabitEthernet0/0/1.99 is up, line protocol is up
R1#
```

When the subinterface has been assigned to the correct VLAN, it is accessible by devices on that VLAN, and the router can perform inter-VLAN routing.

With verification, router configuration problems are quickly addressed, allowing inter-VLAN routing to function properly.

Interactive Graphic

Check Your Understanding—Troubleshoot Inter-VLAN Routing (4.4.7)

Refer to the online course to complete this activity.

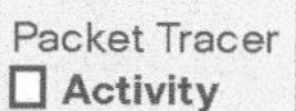

Packet Tracer—Troubleshoot Inter-VLAN Routing (4.4.8)

In this Packet Tracer activity, you complete the following objectives:

- Part 1: Locate Network Problems
- Part 2: Implement the Solution
- Part 3: Verify Network Connectivity

Lab—Troubleshoot Inter-VLAN Routing (4.4.9)

In this lab, you complete the following objectives:

- Part 1: Build the Network and Load Device Configurations
- Part 2: Troubleshoot the Inter-VLAN Routing Configuration
- Part 3: Verify VLAN Configuration, Port Assignment, and Trunking
- Part 4: Test Layer 3 Connectivity

Summary (4.5)

The following is a summary of each section in the chapter:

Inter-VLAN Routing Operation

Hosts in one VLAN cannot communicate with hosts in another VLAN unless there is a router or a Layer 3 switch to provide routing services. Inter-VLAN routing is the process of forwarding network traffic from one VLAN to another VLAN. Three options include legacy, router-on-a-stick, and a Layer 3 switch using SVIs. Legacy used a router with multiple Ethernet interfaces. Each router interface was connected to a switch port in different VLANs. Requiring one physical router interface per VLAN quickly exhausts the physical interface capacity of a router. The router-on-a-stick inter-VLAN routing method requires only one physical Ethernet interface to route traffic between multiple VLANs on a network. A Cisco IOS router Ethernet interface is configured as an 802.1Q trunk and connected to a trunk port on a Layer 2 switch. The router interface is configured using subinterfaces to identify routable VLANs. The configured subinterfaces are software-based virtual interfaces associated with a single physical Ethernet interface. The modern method is Inter-VLAN routing on a Layer 3 switch using SVIs. The SVI is created for a VLAN that exists on the switch. The SVI performs the same functions for the VLAN as a router interface. It provides Layer 3 processing for packets being sent to or from all switch ports associated with that VLAN.

Router-on-a-Stick Inter-VLAN Routing

To configure a switch with VLANs and trunking, complete the following steps: create and name the VLANs, create the management interface, configure access ports, and configure trunking ports. The router-on-a-stick method requires a subinterface to be created for each VLAN to be routed. A subinterface is created using the **interface** *interface_id.subinterface_id* global configuration mode command. Each router subinterface must be assigned an IP address on a unique subnet for routing to occur. When all subinterfaces have been created, the physical interface must be enabled using the **no shutdown** interface configuration command. From a host, verify connectivity to a host in another VLAN using the **ping** command. Use **ping** to verify connectivity with the host and the switch. To verify and troubleshoot, use the **show ip route**, **show ip interface brief**, **show interfaces**, and **show interfaces trunk** commands.

Inter-VLAN Routing Using Layer 3 Switches

Enterprise campus LANs use Layer 3 switches to provide inter-VLAN routing. Layer 3 switches use hardware-based switching to achieve higher-packet processing rates than routers. Capabilities of a Layer 3 switch include routing from one VLAN to another using multiple switched virtual interfaces (SVIs) and converting a Layer 2 switch port to a Layer 3 interface (that is, a routed port). To provide inter-VLAN routing, Layer 3 switches use SVIs. SVIs are configured using the same **interface vlan** *vlan-id* command used to create the management SVI on a Layer 2 switch. A Layer 3 SVI must be created for each of the routable VLANs. To configure a switch with VLANS and trunking, complete the following steps: create the VLANS, create the SVI VLAN interfaces, configure access ports, and enable IP routing. From a host, verify connectivity to a host in another VLAN using the **ping** command. Next, verify connectivity with the host using the **ping** Windows host command. VLANs must be advertised using static or dynamic routing. To enable routing on a Layer 3 switch, a routed port must be configured. A routed port is created on a Layer 3 switch by disabling the switch port feature on a Layer 2 port that is connected to another Layer 3 device. The interface can be configured with an IPv4 configuration to connect to a router or another Layer 3 switch. To configure a Layer 3 switch to route with a router, follow these steps: configure the routed port, enable routing, configure routing, verify routing, and verify connectivity.

Troubleshoot Inter-VLAN Routing

There are a number of reasons why an inter-VAN configuration may not work. All are related to connectivity issues such as missing VLANs, switch trunk port issues, switch access port issues, and router configuration issues. A VLAN could be missing if it was not created, it was accidently deleted, or it is not allowed on the trunk link. Another issue for inter-VLAN routing includes misconfigured switch ports. In a legacy inter-VLAN solution, a misconfigured switch port could be caused when the connecting router port is not assigned to the correct VLAN. With a router-on-a-stick solution, the most common cause is a misconfigured trunk port. When a problem is suspected with a switch access port configuration, use **ping** and **show interfaces** *interface-id* **switchport** commands to identify the problem. Router configuration problems with router-on-a-stick configurations are usually related to subinterface misconfigurations. Verify the subinterface status using the **show ip interface brief** command.

Packet Tracer—Inter-VLAN Routing Challenge (4.5.1)

In this activity, you demonstrate and reinforce your ability to implement inter-VLAN routing, including configuring IP addresses, VLANs, trunking, and subinterfaces.

Lab—Implement Inter-VLAN Routing (4.5.2)

In this lab, you complete the following objectives:

- Part 1: Build the Network and Configure Basic Device Settings
- Part 2: Create VLANs and Assign Switch Ports

- Part 3: Configure an 802.1Q Trunk between the Switches
- Part 4: Configure Inter-VLAN Routing on the S1 Switch
- Part 5: Verify Inter-VLAN Routing is Working

Practice

The following activities provide practice with the topics introduced in this chapter. The Labs are available in the companion *Switching, Routing, and Wireless Essentials Labs and Study Guide (CCNAv7)* (ISBN 9780136634386). The Packet Tracer Activity instructions are also in the Labs & Study Guide. The PKA files are found in the online course.

Labs

Lab 4.2.8: Configure Router-on-a-Stick Inter-VLAN Routing

Lab 4.4.9: Troubleshoot Inter-VLAN Routing

Lab 4.5.2: Implement Inter-VLAN Routing

Packet Tracer Activity

Packet Tracer Activities

Packet Tracer 4.2.7: Configure Router-on-a-Stick Inter-VLAN Routing

Packet Tracer 4.3.8: Configure Layer 3 Switching and Inter-VLAN Routing

Packet Tracer 4.4.8: Troubleshoot Inter-VLAN Routing

Packet Tracer 4.5.1: Inter-VLAN Routing Challenge

Check Your Understanding Questions

Complete all the review questions listed here to test your understanding of the sections and concepts in this chapter. The appendix "Answers to the 'Check Your Understanding' Questions" lists the answers.

1. A router has two FastEthernet interfaces and needs to connect to four VLANs in the local network. How can this be accomplished using the fewest number of physical interfaces without unnecessarily decreasing network performance?

   A. Add a second router to handle the inter-VLAN traffic.

   B. Implement a router-on-a-stick configuration.

   C. Interconnect the VLANs via the two additional FastEthernet interfaces.

   D. Use a hub to connect the four VLANS with a FastEthernet interface on the router.

2. What distinguishes traditional legacy inter-VLAN routing from router-on-a-stick?

   A. Traditional routing is able to use only a single switch interface, whereas a router-on-a-stick can use multiple switch interfaces.

   B. Traditional routing requires a routing protocol, whereas a router-on-a-stick only needs to route directly connected networks.

   C. Traditional routing uses one port per logical network, whereas a router-on-a-stick uses subinterfaces to connect multiple logical networks to a single router port.

   D. Traditional routing uses multiple paths to the router and therefore requires STP, whereas router-on-a-stick does not provide multiple connections and therefore eliminates the need for STP.

3. Subinterface G0/1.10 on R1 must be configured as the default gateway for the VLAN 10 192.168.10.0/24 network. Which command should be configured on the subinterface to enable inter-VLAN routing for VLAN 10?

   A. **encapsulation dot1q 10**

   B. **encapsulation vlan 10**

   C. **switchport mode access**

   D. **switchport mode trunk**

4. What is important to consider while configuring the subinterfaces of a router when implementing inter-VLAN routing?

   A. The IP address of each subinterface must be the default gateway address for each VLAN subnet.

   B. The **no shutdown** command must be given on each subinterface.

   C. The physical interface must have an IP address configured.

   D. The subinterface numbers must match the VLAN ID number.

5. What are the steps that must be completed in order to enable inter-VLAN routing using router-on-a-stick?

   A. Configure the physical interfaces on the router and enable a routing protocol.

   B. Create the VLANs on the router and define the port membership assignments on the switch.

   C. Create the VLANs on the switch to include port membership assignment and enable a routing protocol on the router.

   D. Create the VLANs on the switch to include port membership assignment and configure subinterfaces on the router matching the VLANs.

6. What two statements are true regarding the use of subinterfaces for inter-VLAN routing? (Choose two.)

   A. Fewer router Ethernet ports required than in traditional inter-VLAN routing

   B. Less complex physical connection than in traditional inter-VLAN routing

   C. More switch ports required than in traditional inter-VLAN routing

   D. Simpler Layer 3 troubleshooting than with traditional inter-VLAN routing

   E. Subinterfaces have no contention for bandwidth

7. Which router-on-a-stick command and prompt on R1 correctly encapsulates 802.1Q traffic for VLAN 20?

   A. R1(config-if)# **encapsulation 802.1q 20**

   B. R1(config-if)# **encapsulation dot1q 20**

   C. R1(config-subif)# **encapsulation 802.1q 20**

   D. R1(config-subif)# **encapsulation dot1q 20**

8. What are two disadvantages of using the router-on-a-stick inter-VLAN routing method in a large network? (Choose two.)

   A. A dedicated router is required.

   B. It does not scale well.

   C. It requires multiple physical interfaces on a router.

   D. It requires subinterfaces to be configured on the same subnets.

   E. Multiple SVIs are needed.

9. What is a characteristic of a routed port on a Layer 3 switch? (Choose two.)

   A. It requires the **switchport mode access interface** config command.

   B. It requires the **no switchport interface** config command.

   C. It requires the **switchport access vlan** *vlan-id* interface config command.

   D. It supports trunking.

10. What are two advantages of using a Layer 3 switch with SVIs for inter-VLAN routing? (Choose two.)

    A. A router is not required.

    B. It switches packets faster than using the router-on-a-stick method.

    C. SVIs can be bundled into EtherChannels.

    D. SVIs can be divided using subinterfaces.

    E. SVIs eliminate the need for a default gateway in the hosts.

CHAPTER 5

STP Concepts

Objectives

Upon completion of this chapter, you will be able to answer the following questions:

- What are common problems in a redundant, L2 switched network?
- How does STP operate in a simple switched network?
- How does Rapid PVST+ operate?

Key Terms

This chapter uses the following key terms. You can find the definitions in the Glossary.

Layer 2 loops Page 139

802.1D Page 140

blocking state Page 141

Time to Live (TTL) Page 141

hop limit Page 141

broadcast storm Page 143

Spanning Tree Algorithm (STA) Page 145

root bridge Page 145

Bridge Protocol Data Unit (BPDU) Page 149

root ports Page 149

designated ports Page 149

alternate ports Page 149

bridge ID (BID) Page 149

extended system ID Page 149

bridge priority Page 149

Rapid Spanning Tree Protocol (RSTP) Page 149

spanning tree instance Page 151

root path cost Page 152

default port costs Page 152

802.1w Page 152

listening state Page 158

learning state Page 158

STP diameter Page 159

forwarding state Page 159

disabled state Page 159

Multiple Spanning Tree Protocol (MSTP) Page 161

STP Page 161

Common Spanning Tree (CST) Page 161

PVST+ Page 162

Per-VLAN Spanning Tree (PVST+) Page 162

PortFast Page 162

BPDU guard Page 162

BPDU filter Page 162

root guard Page 162

loop guard Page 162

RSTP Page 162

Rapid PVST+ Page 162

Multiple Spanning Tree (MST) Page 162

discarding state Page 163

Introduction (5.0)

A well-designed Layer 2 network will have redundant switches and paths to ensure that if one switch goes down, another path to a different switch is available to forward data. Users of the network would not experience any disruption of service. Redundancy in a hierarchical network design fixes the problem of a single point of failure, yet it can create a different kind of problem called *Layer 2 loops*.

What is a loop? Imagine that you are at a concert. The singer's microphone and the amplified loudspeaker can, for a variety of reasons, create a feedback loop. What you hear is an amplified signal from the microphone that comes out of the loudspeaker which is then picked up again by the microphone, amplified further, and passed again through the loudspeaker. The sound quickly becomes very loud, unpleasant, and makes it impossible to hear any actual music. This continues until the connection between the microphone and the loudspeaker is cut.

A Layer 2 loop creates similar chaos in a network. It can happen very quickly and make it impossible to use the network. There are a few common ways that a Layer 2 loop can be created and propagated. Spanning Tree Protocol (STP) is designed specifically to eliminate Layer 2 loops in your network. This module discusses causes of loops and the various types of spanning tree protocols. It includes a video and a Packet Tracer activity to help you understand STP concepts.

Purpose of STP (5.1)

In this section, you learn how to build a simple switched network with redundant links.

Redundancy in Layer 2 Switched Networks (5.1.1)

This section covers the causes of loops in a Layer 2 network and briefly explains how Spanning Tree Protocol works. Redundancy is an important part of the hierarchical design for eliminating single points of failure and preventing disruption of network services to users. Redundant networks require the addition of physical paths, but logical redundancy must also be part of the design. Having alternate physical paths for data to traverse the network makes it possible for users to access network resources, despite path disruption. However, redundant paths in a switched Ethernet network may cause both physical and logical Layer 2 loops.

Ethernet LANs require a loop-free topology with a single path between any two devices. A loop in an Ethernet LAN can cause continued propagation of Ethernet frames until a link is disrupted and breaks the loop.

Spanning Tree Protocol (5.1.2)

Spanning Tree Protocol (STP) is a loop-prevention network protocol that allows for redundancy while creating a loop-free Layer 2 topology. *IEEE 802.1D* (that is, Spanning Tree Protocol) is the original IEEE MAC Bridging standard for STP.

STP ensures that there is only one logical path between all destinations on the network by intentionally blocking redundant paths that could cause a loop. A port is considered blocked when user data is prevented from entering or leaving that port. The physical paths still exist to provide redundancy, but these paths are disabled to prevent the loops from occurring.

The following steps illustrate how a converged STP-enabled LAN operates.

Step 1. Figure 5-1 illustrates normal STP operation when all switches have STP enabled.

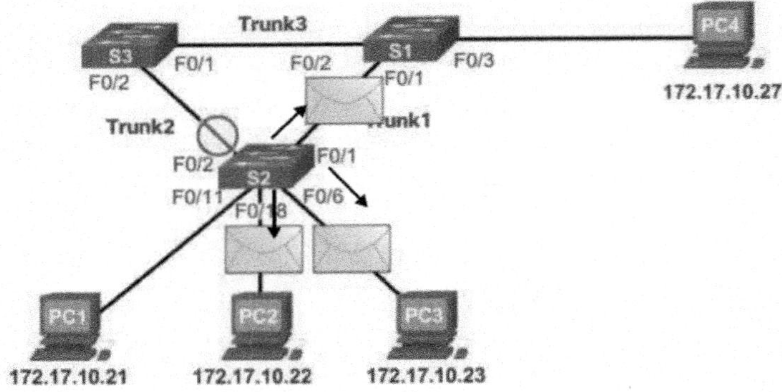

Figure 5-1 Normal STP Operation

Figure 5-2 shows how STP recalculates the path when a failure occurs.

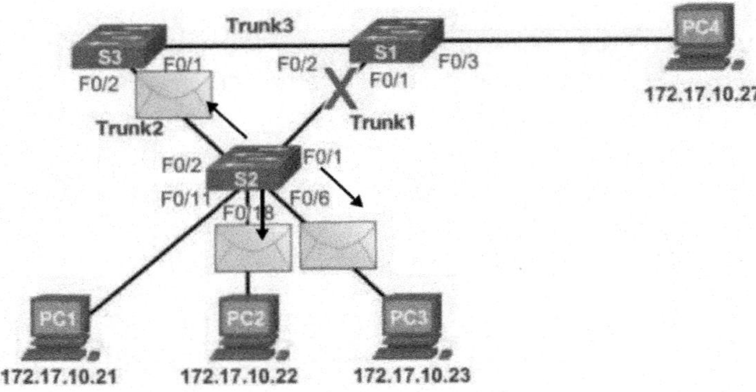

Figure 5-2 STP Compensates for Network Failure

Step 2. S2 is configured with STP and has set the port for Trunk2 to a *blocking state*. The blocking state prevents ports from being used to forward user data, which prevents a loop from occurring. S2 forwards a broadcast frame out all of the switch ports, except the originating port from PC1 and the F0/2 port for Trunk2.

Step 3. S1 receives the broadcast frame from S2 and forwards it out all of its switch ports, except the receiving port, where it reaches PC4 and S3. S3 forwards the frame out the port for Trunk2 and S2 drops the frame. The Layer 2 loop is prevented.

STP Recalculation (5.1.3)

What happens when a link fails? If the path is ever needed to compensate for a network cable or switch failure, STP unblocks the necessary ports to allow the redundant path to become active.

The following steps illustrate how an STP-enabled LAN can provide redundancy and keep traffic flowing.

Step 1. PC1 sends a broadcast out onto the network.

Step 2. The broadcast is then forwarded around the network.

Step 3. As shown in Figure 5-2, the trunk link between S2 and S1 fails, resulting in the previous path being disrupted.

Step 4. S2 unblocks the previously blocked port for Trunk2 and allows the broadcast traffic to traverse the alternate path around the network, permitting communication to continue. If the link comes back up, STP reconverges, and the port on S2 is again blocked.

Issues with Redundant Switch Links (5.1.4)

Path redundancy provides multiple network services by eliminating the possibility of a single point of failure. When multiple paths exist between two devices on an Ethernet network, and there is no spanning tree implementation on the switches, a Layer 2 loop occurs. A Layer 2 loop can result in MAC address table instability, link saturation, and high CPU utilization on switches and end devices, resulting in the network becoming unusable.

Unlike the Layer 3 protocols, IPv4 and IPv6, Layer 2 Ethernet does not include a mechanism to recognize and eliminate endlessly looping frames. Both IPv4 and IPv6 include a mechanism that limits the number of times a Layer 3 networking device can retransmit a packet. A router will decrement the *Time to Live (TTL)* in every IPv4 packet and the *Hop Limit* field in every IPv6 packet. When these fields are decremented to 0, a router drops the packet. Ethernet and Ethernet switches have no comparable mechanism for limiting the number of times a switch retransmits a Layer 2 frame. STP was developed specifically as a loop prevention mechanism for Layer 2 Ethernet.

Layer 2 Loops (5.1.5)

Without STP enabled, Layer 2 loops can form, causing broadcast, multicast, and unknown unicast frames to loop endlessly. This can bring down a network within a very short amount of time, sometimes in just a few seconds. For example, broadcast frames, such as an Address Resolution Protocol (ARP) Request are forwarded out all the switch ports, except the original ingress port. This ensures that all devices in a broadcast domain are able to receive the frame. If there is more than one path for the frame to be forwarded out of, an endless loop can result. When a loop occurs, the MAC address table on a switch will constantly change with the updates from the broadcast frames, which results in MAC database instability. This can cause high CPU utilization, which makes the switch unable to forward frames.

Broadcast frames are not the only type of frames that are affected by loops. Unknown unicast frames sent onto a looped network can result in duplicate frames arriving at the destination device. An unknown unicast frame is when the switch does not have the destination MAC address in its MAC address table and must forward the frame out all ports, except the ingress port.

The following sequence of events demonstrates the MAC database instability issue:

1. PC1 sends a broadcast frame to S2. S2 receives the broadcast frame on F0/11. When S2 receives the broadcast frame, it updates its MAC address table to record that PC1 is available on port F0/11.
2. Because it is a broadcast frame, S2 forwards the frame out all ports, including Trunk1 and Trunk2. When the broadcast frame arrives at S3 and S1, the switches update their MAC address tables to indicate that PC1 is available out port F0/1 on S1 and out port F0/2 on S3.
3. Because it is a broadcast frame, S3 and S1 forward the frame out all ports, except the ingress port. S3 sends the broadcast frame from PC1 to S1. S1 sends the broadcast frame from PC1 to S3. Each switch updates its MAC address table with the incorrect port for PC1.
4. Each switch forwards the broadcast frame out all of its ports, except the ingress port, which results in both switches forwarding the frame to S2.
5. When S2 receives the broadcast frames from S3 and S1, the MAC address table is updated with the last entry received from the other two switches.
6. S2 forwards the broadcast frame out all ports except the last received port. The cycle starts again.

Note

To view an animation of these sequences, refer to the online course.

Figure 5-3 shows a snapshot during sequence 6. Notice that S2 now thinks PC1 is reachable out the F0/1 interface.

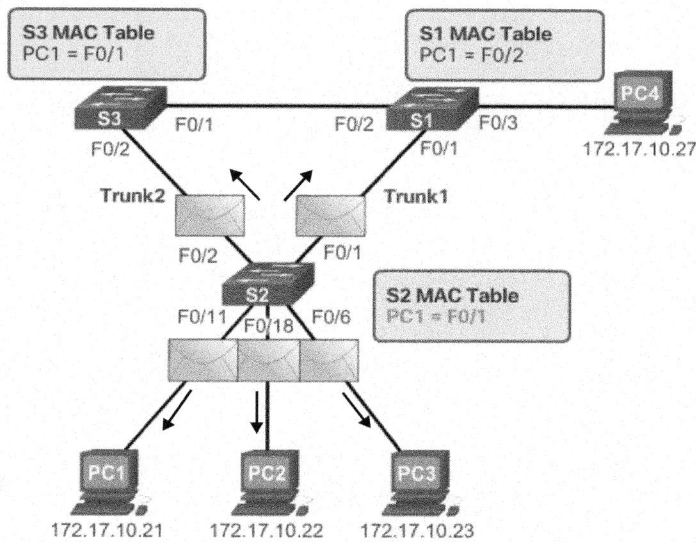

Figure 5-3 MAC Database Instability Example

Broadcast Storm (5.1.6)

A *broadcast storm* is an abnormally high number of broadcasts overwhelming the network during a specific amount of time. Broadcast storms can disable a network within seconds by overwhelming switches and end devices. Broadcast storms can be caused by a hardware problem, such as a faulty network interface card (NIC), or from a Layer 2 loop in the network.

Layer 2 broadcasts in a network, such as ARP Requests, are very common. A Layer 2 loop is likely to have immediate and disabling consequences on the network. Layer 2 multicasts are typically forwarded the same way as a broadcast by the switch. So, although IPv6 packets are never forwarded as a Layer 2 broadcast, Internet Control Message Protocol version 6 (ICMPv6) Neighbor Discovery uses Layer 2 multicasts.

The following sequence of events demonstrates the broadcast storm issue:

1. PC1 sends a broadcast frame out onto the looped network.
2. The broadcast frame loops between all the interconnected switches on the network.
3. PC4 also sends a broadcast frame out onto the looped network.
4. The PC4 broadcast frame gets caught in the loop between all the interconnected switches, just like the PC1 broadcast frame.
5. As more devices send broadcasts over the network, more traffic is caught in the loop and consumes resources. This eventually creates a broadcast storm that causes the network to fail.
6. When the network is fully saturated with broadcast traffic that is looping between the switches, new traffic is discarded by the switch because it is unable to process it. Figure 5-4 displays the resulting broadcast storm.

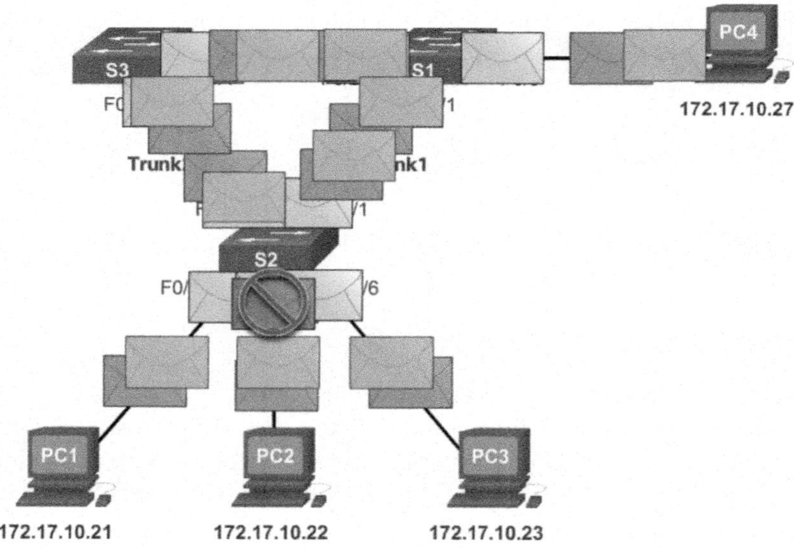

Figure 5-4 Broadcast Storm Example

A host caught in a Layer 2 loop is not accessible to other hosts on the network. Additionally, due to the constant changes in its MAC address table, the switch does not know out of which port to forward unicast frames. In the previous animation, the switches will have the incorrect ports listed for PC1. Any unknown unicast frame destined for PC1 loops around the network, just as the broadcast frames do. More and more frames looping around the network eventually creates a broadcast storm.

To prevent these issues from occurring in a redundant network, some type of spanning tree must be enabled on the switches. Spanning tree is enabled, by default, on Cisco switches to prevent Layer 2 loops from occurring.

The Spanning Tree Algorithm (5.1.7)

STP is based on an algorithm invented by Radia Perlman while working for Digital Equipment Corporation, and it was published in the 1985 paper "An Algorithm for Distributed Computation of a Spanning Tree in an Extended LAN." Her *spanning tree algorithm (STA)* creates a loop-free topology by selecting a single *root bridge* where all other switches determine a single least-cost path.

Without the loop prevention protocol, loops would occur, rendering a redundant switch network inoperable.

Figures 5-5 through 5-9 demonstrate how STA creates a loop-free topology.

This STA scenario uses an Ethernet LAN with redundant connections between multiple switches, as shown in Figure 5-5.

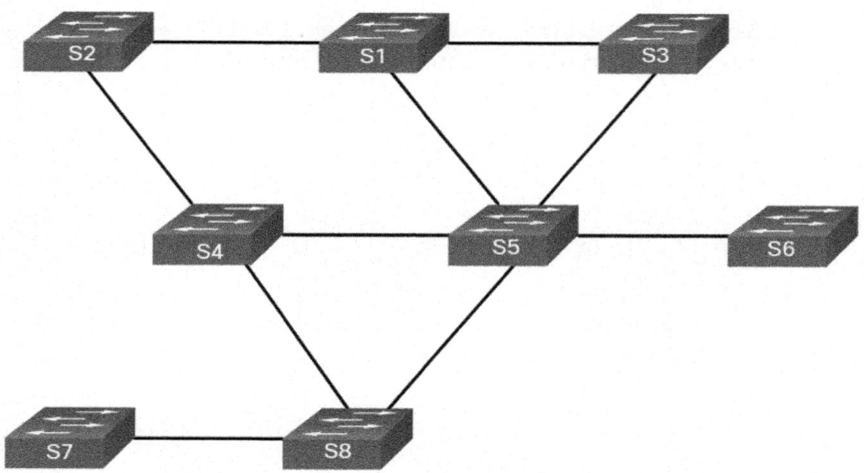

Figure 5-5 STA Scenario Topology

The spanning tree algorithm begins by selecting a single root bridge. Figure 5-6 shows that switch S1 has been selected as the root bridge. In this topology, all links are equal cost (same bandwidth). Each switch will determine a single, least cost path from itself to the root bridge.

Note

The STA and STP refer to switches as bridges. This is because in the early days of Ethernet, switches were referred to as bridges.

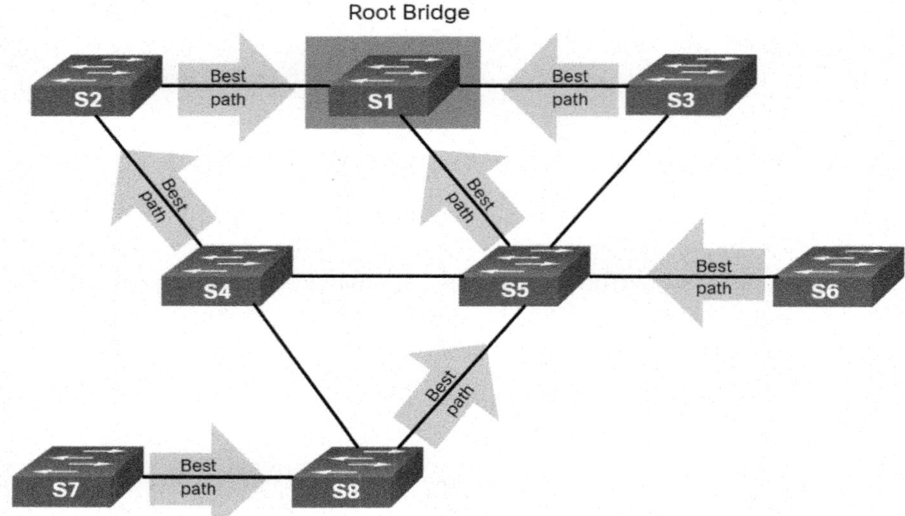

Figure 5-6 Select the Root Bridge

STP ensures that there is only one logical path between all destinations on the network by intentionally blocking redundant paths that could cause a loop, as shown in Figure 5-7. When a port is blocked, user data is prevented from entering or leaving that port. Blocking the redundant paths is critical to preventing loops on the network.

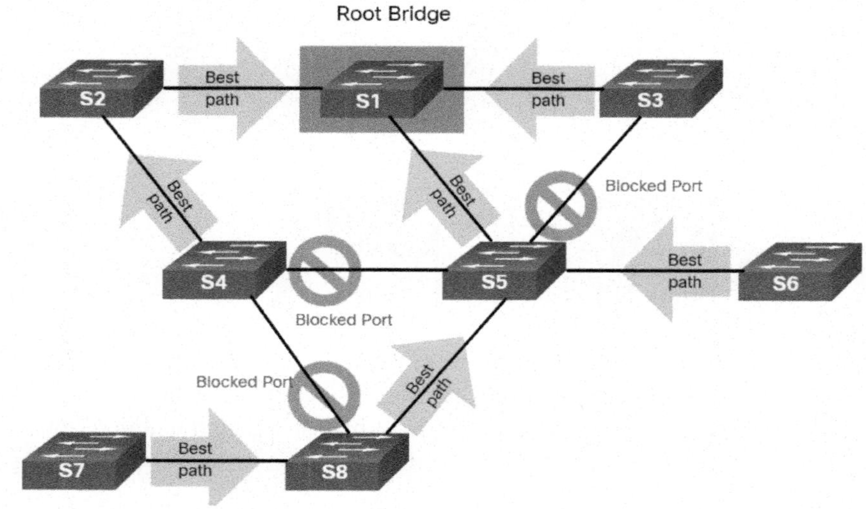

Figure 5-7 Block Redundant Paths

A blocked port has the effect of making that link a nonforwarding link between the two switches, as shown in Figure 5-8. Notice that this creates a topology where each switch has only a single path to the root bridge, similar to branches on a tree that connect to the root of the tree.

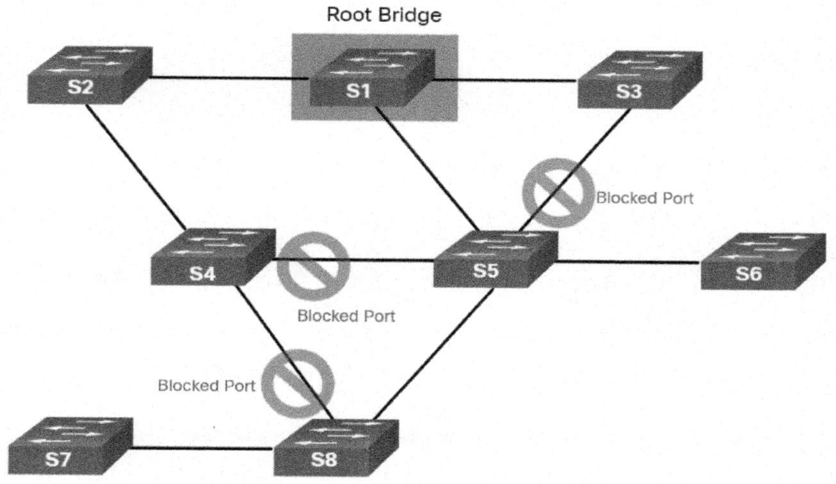

Each switch now has just one forwarding path to the root bridge.

Figure 5-8 Loop-Free Topology

The physical paths still exist to provide redundancy, but these paths are disabled to prevent the loops from occurring. If the path is ever needed to compensate for a network cable or switch failure, STP recalculates the paths and unblocks the necessary ports to allow the redundant path to become active. STP recalculations can also occur anytime a new switch or new inter-switch link is added to the network.

Figure 5-9 shows a link failure between switches S2 and S4, causing STP to recalculate. Notice that the previously redundant link between S4 and S5 is now forwarding to compensate for this failure. There is still only one path between every switch and the root bridge.

STP prevents loops from occurring by configuring a loop-free path through the network using strategically placed "blocking-state" ports. The switches running STP are able to compensate for failures by dynamically unblocking the previously blocked ports and permitting traffic to traverse the alternate paths.

Video

Video—Observe STP Operation (5.1.8)

Refer to the online course to view this video.

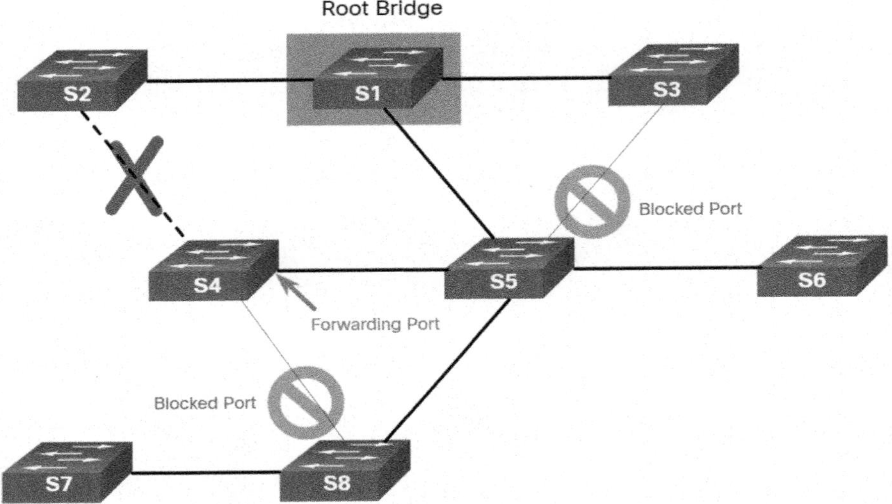

Figure 5-9 Link Failure Causes Recalculation

Packet Tracer Activity

Packet Tracer—Investigate STP Loop Prevention (5.1.9)

In this Packet Tracer activity, you complete the following objectives:

- Create and configure a simple three switch network with STP.
- View STP operation.
- Disable STP and view operation again.

Interactive Graphic

Check Your Understanding—Purpose of STP (5.1.10)

Refer to the online course to complete this activity.

STP Operations (5.2)

The focus of this section is to learn how to build a simple, switched network using STP.

Steps to a Loop-Free Topology (5.2.1)

Now you know how loops are created and the basics of using Spanning Tree Protocol to prevent them. This section will take you, step by step, through the operation of STP. Using the STA, STP builds a loop-free topology in a four-step process:

Step 1. Elect the root bridge.

Step 2. Elect the root ports.

Step 3. Elect designated ports.

Step 4. Elect alternate (blocked) ports.

During STA and STP functions, switches use *Bridge Protocol Data Units (BPDUs)* to share information about themselves and their connections. BPDUs are used to elect the root bridge, *root ports*, *designated ports*, and *alternate ports*. Each BPDU contains a *bridge ID (BID)* that identifies which switch sent the BPDU. The BID is involved in making many of the STA decisions, including root bridge and port roles. As shown in Figure 5-10, the BID contains a priority value, the MAC address of the switch, and an *extended system ID*. The lowest BID value is determined by the combination of these three fields.

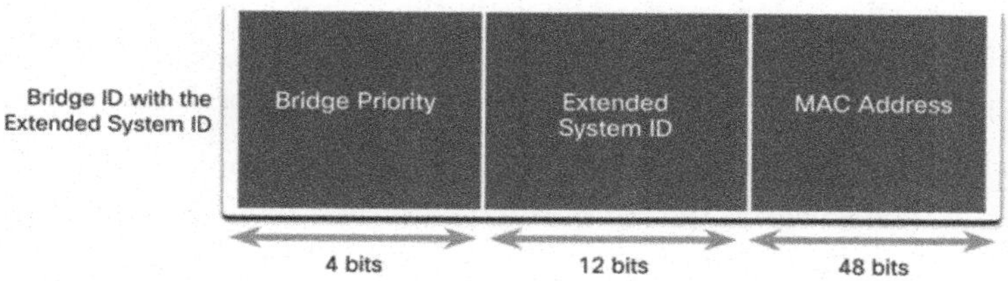

Figure 5-10 Fields in the Bridge ID

Bridge Priority

The default priority value for all Cisco switches is the decimal value 32768. The range is 0 to 61440 in increments of 4096. A lower *bridge priority* is preferable. A bridge priority of 0 takes precedence over all other bridge priorities.

Extended System ID

The extended system ID value is a decimal value added to the bridge priority value in the BID to identify the VLAN for this BPDU.

Early implementations of IEEE 802.1D were designed for networks that did not use VLANs. There was a single common spanning tree across all switches. For this reason, in older switches, the extended system ID was not included in the BPDUs. As VLANs became common for network infrastructure segmentation, 802.1D was enhanced to include support for VLANs, which required that the 12-bit VLAN ID be included in the BPDU frame. VLAN information is included in the BPDU frame through the use of the extended system ID.

The extended system ID allows later implementations of STP, such as *Rapid Spanning Tree Protocol (RSTP)*, to have different root bridges for different sets of VLANs. This can allow for redundant, nonforwarding links in an STP topology for one set of VLANs to be used by a different set of VLANs using a different root bridge.

MAC address

When two switches are configured with the same priority and have the same extended system ID, the switch having the MAC address with the lowest value, expressed in hexadecimal, will have the lower BID.

1. Elect the Root Bridge (5.2.2)

The STA designates a single switch as the root bridge and uses it as the reference point for all path calculations. Switches exchange BPDUs to build the loop-free topology beginning with selecting the root bridge.

An election process determines which switch becomes the root bridge. All switches in the broadcast domain participate in the election process. After a switch boots, it begins to send out BPDU frames every two seconds. These BPDU frames contain the BID of the sending switch and the BID of the root bridge, known as the Root ID.

The switch with the lowest BID will become the root bridge. At first, all switches declare themselves as the root bridge with their own BID set as the Root ID. Eventually, the switches learn through the exchange of BPDUs which switch has the lowest BID and will agree on one root bridge.

In Figure 5-11, S1 is elected the root bridge because it has the lowest BID.

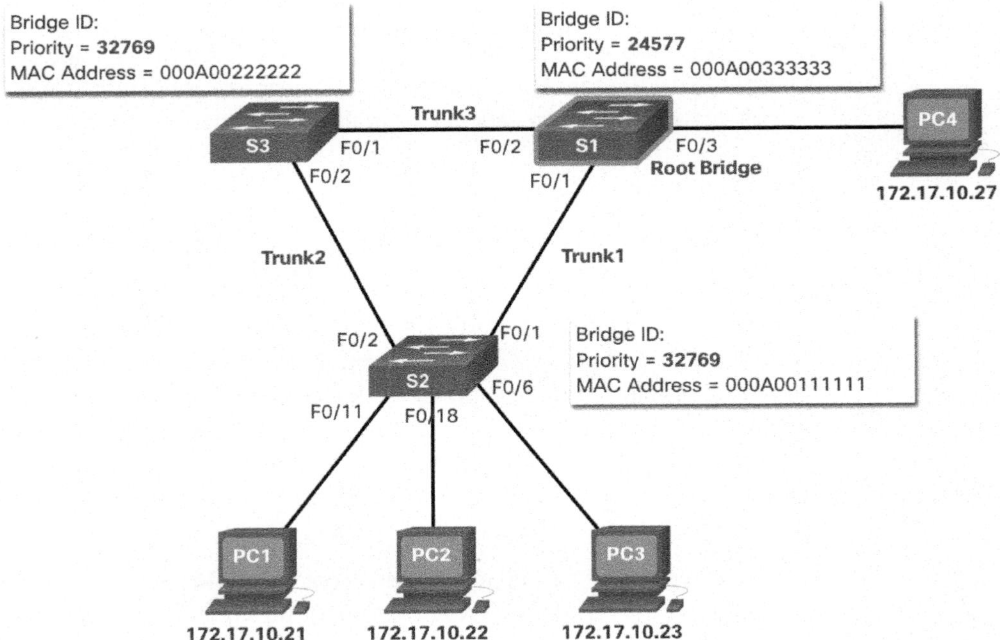

Figure 5-11 Root Bridge Selection

Impact of Default BIDs (5.2.3)

Because the default BID is 32768, it is possible for two or more switches to have the same priority. In this scenario, where the priorities are the same, the switch with the lowest MAC address will become the root bridge. To ensure that the root bridge decision best meets network requirements, it is recommended that the administrator configure the desired root bridge switch with a lower priority.

In Figure 5-12, all switches are configured with the same priority of 32769. Here the MAC address becomes the deciding factor as to which switch becomes the root bridge. The switch with the lowest hexadecimal MAC address value is the preferred root bridge. In this example, S2 has the lowest value for its MAC address and is elected as the root bridge for that *spanning tree instance*.

Note

In the example, the priority of all the switches is 32769. The value is based on the 32768 default bridge priority and the extended system ID (VLAN 1 assignment) associated with each switch (32768+1).

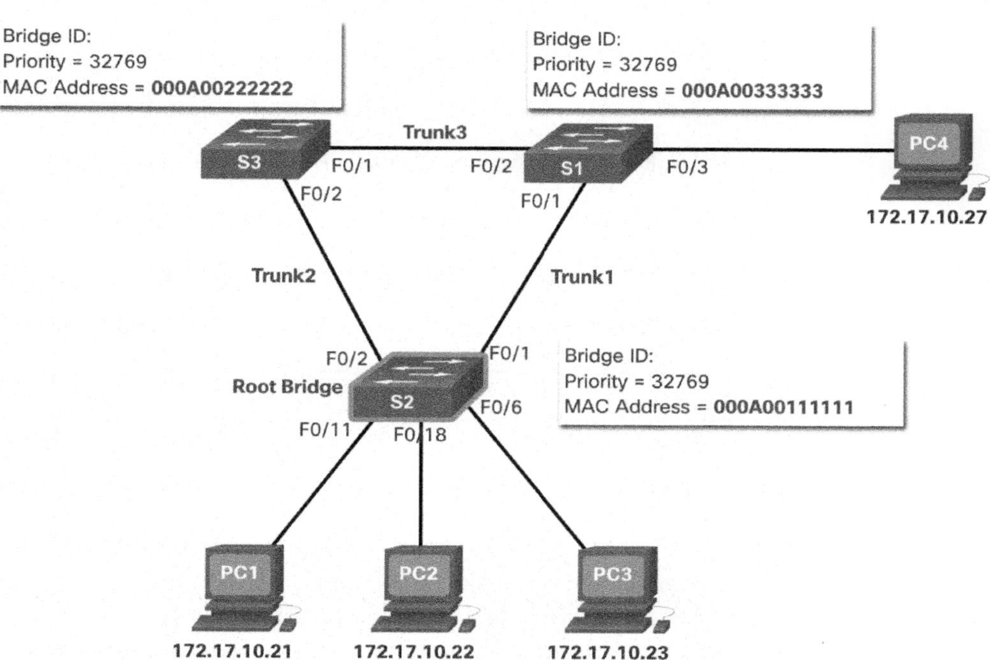

Figure 5-12 MAC Is Deciding Factor for Root Bridge Election

Determine the Root Path Cost (5.2.4)

When the root bridge has been elected for a given spanning tree instance, the STA starts the process of determining the best paths to the root bridge from all destinations in the broadcast domain. The path information, known as the internal *root path cost*, is determined by the sum of all the individual port costs along the path from the switch to the root bridge.

Note

The BPDU includes the root path cost. This is the cost of the path from the sending switch to the root bridge.

When a switch receives the BPDU, it adds the ingress port cost of the segment to determine its internal root path cost.

The *default port costs* are defined by the speed at which the port operates. Table 5-1 shows the default port costs suggested by IEEE. Cisco switches by default use the values as defined by the IEEE 802.1D standard, also known as the *short path cost*, for both STP and RSTP (that is, IEEE *802.1w*). However, the IEEE standard suggests using the values defined in the IEEE-802.1w, also known as *long path cost*, when using 10 Gbps links and faster.

Note

RSTP is discussed in more detail later in this module.

Table 5-1 Default Port Costs

Link Speed	STP Cost: IEEE 802.1D-1998	RSTP Cost: IEEE 802.1w-2004
10 Gbps	2	2,000
1 Gbps	4	20,000
100 Mbps	19	200,000
10 Mbps	100	2,000,000

Although switch ports have a default port cost associated with them, the port cost is configurable. The ability to configure individual port costs gives the administrator the flexibility to manually control the spanning tree paths to the root bridge.

2. Elect the Root Ports (5.2.5)

After the root bridge has been determined, the STA algorithm is used to select the root port. Every nonroot switch will select one root port. The root port is the port closest to the root bridge in terms of overall cost (best path) to the root bridge. This overall cost is known as the *internal root path cost*.

The internal root path cost is equal to the sum of all the port costs along the path to the root bridge, as shown in Figure 5-13. Paths with the lowest cost become preferred, and all other redundant paths are blocked. In the example, the internal root path cost from S2 to the root bridge S1 over path 1 is 19 (based on the IEEE-specified individual port cost) and the internal root path cost over path 2 is 38. Because path 1 has a lower overall path cost to the root bridge, it is the preferred path, and F0/1 becomes the root port on S2.

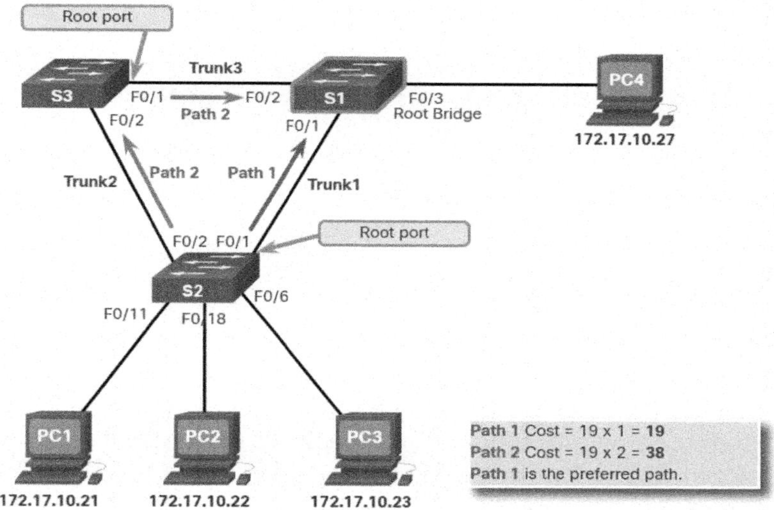

Figure 5-13 Path Cost Calculation Example

3. Elect Designated Ports (5.2.6)

The loop prevention part of spanning tree becomes evident during these next two steps. After each switch selects a root port, the switches will then select designated ports.

Every segment between two switches will have one designated port. The designated port is a port on the segment (with two switches) that has the internal root path cost to the root bridge. In other words, the designated port has the best path to receive traffic leading to the root bridge.

What is not a root port or a designated port becomes an alternate or blocked port. The end result is a single path from every switch to the root bridge.

All ports on the root bridge are designated ports, as shown in Figure 5-14. This is because the root bridge has the lowest cost to itself.

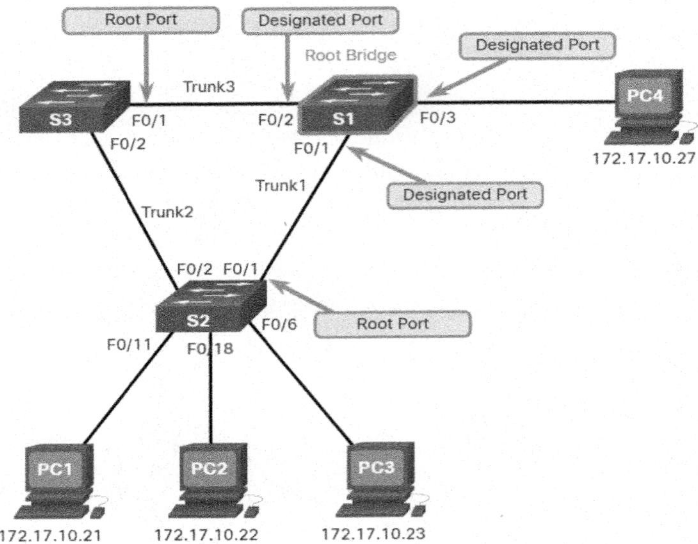

All the ports on the root bridge are designated ports.

Figure 5-14 Designated Ports on Root Bridge

If one end of a segment is a root port, then the other end is a designated port. To demonstrate this, Figure 5-15 shows that switch S4 is connected to S3. The Fa0/1 interface on S4 is its root port because it has the best and only path to the root bridge. The Fa0/3 interface on S3 at the other end of the segment would, therefore, be the designated port.

Note

All switch ports with end devices (hosts) attached are designated ports.

This leaves only segments between two switches where neither of the switches is the root bridge. In this case, the port on the switch with the least-cost path to the root bridge is the designated port for the segment. For example, in Figure 5-16, the last segment is the one between S2 and S3. Both S2 and S3 have the same path cost to the root bridge. The spanning tree algorithm will use the bridge ID as a tie breaker. Although not shown in the figure, S2 has a lower BID. Therefore, the F0/2 port of S2 will be chosen as the designated port. Designated ports are in forwarding state.

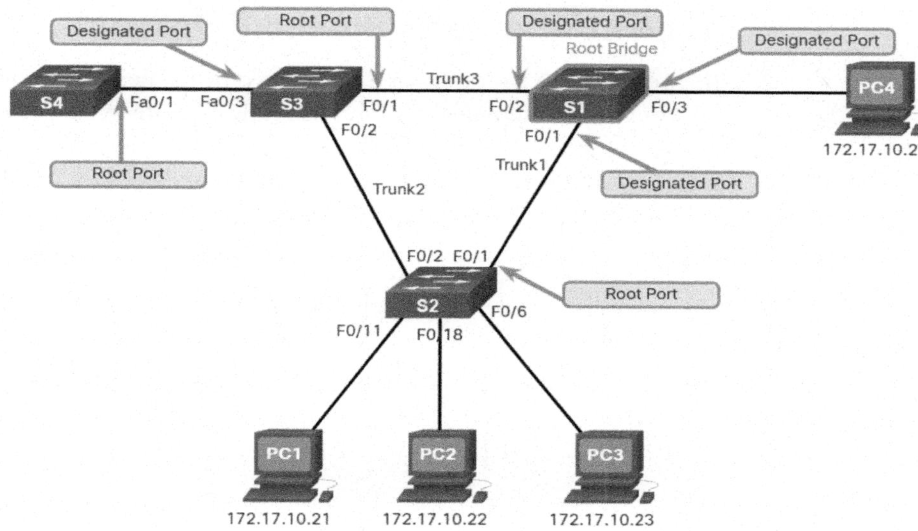

Fa0/1 interface on S4 is a designated port because the Fa0/3 interface of S3 is a root port.

Figure 5-15 Designated Port When There Is a Root Port

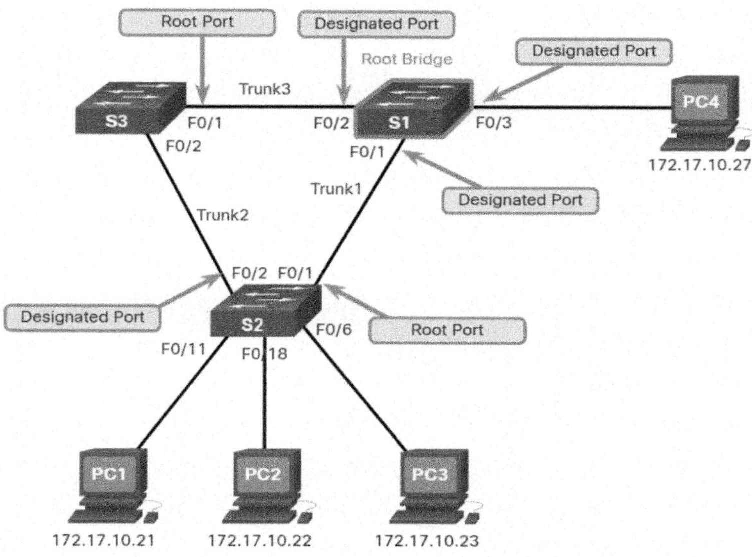

The Fa0/2 interface of S2 is the designated port on the segment with S3.

Figure 5-16 Designated Port Chosen Using Bridge ID

4. Elect Alternate (Blocked) Ports (5.2.7)

If a port is not a root port or a designated port, then it becomes an alternate (or backup) port. Alternate ports and backup ports are in discarding or blocking state to prevent loops. In Figure 5-17, the STA has configured port F0/2 on S3 in the alternate role. Port F0/2 on S3 is in the blocking state and will not forward Ethernet frames. All other inter-switch ports are in forwarding state. This is the loop-prevention part of STP.

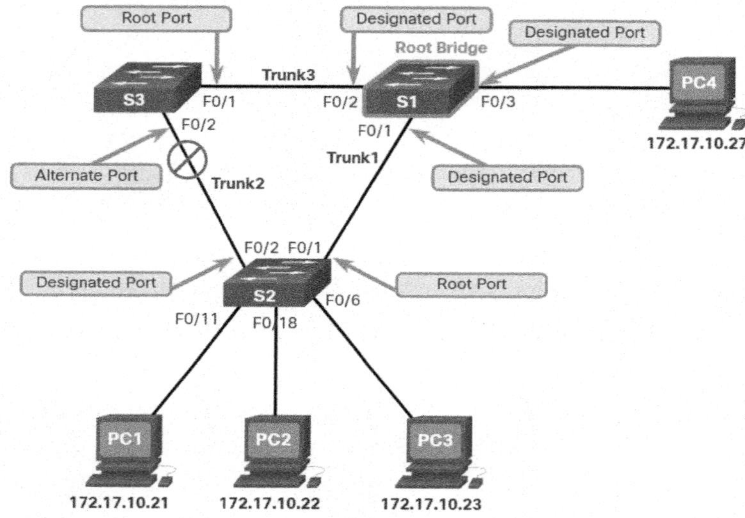

Figure 5-17 Elect Alternate Ports

Elect a Root Port from Multiple Equal-Cost Paths (5.2.8)

Root port and designated ports are based on the lowest path cost to the root bridge. But what happens if the switch has multiple equal-cost paths to the root bridge? How does a switch designate a root port?

When a switch has multiple equal-cost paths to the root bridge, the switch determines a port using the following criteria:

1. Lowest sender BID
2. Lowest sender port priority
3. Lowest sender port ID

1. Lowest Sender BID

Figure 5-18 shows a topology with four switches, including switch S1 as the root bridge.

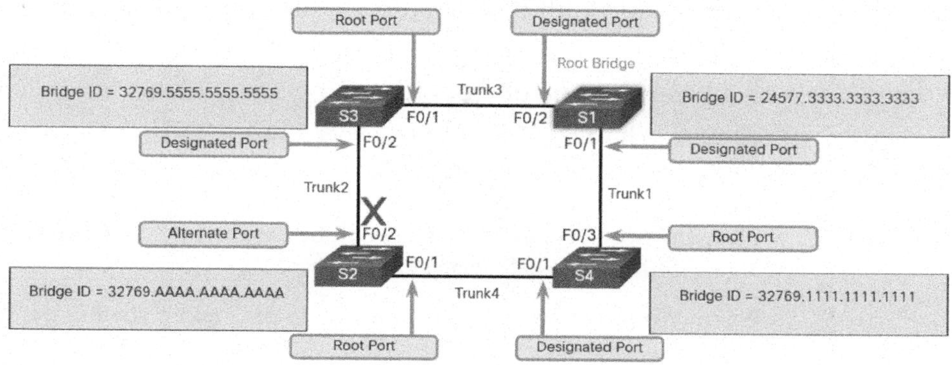

Figure 5-18 Four Switch Multiple Equal-Cost Paths Topology

When you examine the port roles, note that port F0/1 on switch S3 and port F0/3 on switch S4 have been selected as root ports because they have the lowest cost path (root path cost) to the root bridge for their respective switches. S2 has two ports, F0/1 and F0/2 with equal cost paths to the root bridge. In this case the bridge IDs of the neighboring switches, S3 and S4, will be used to break the tie. This is known as the sender's BID. S3 has a BID of 32769.5555.5555.5555 and S4 has a BID of 32769.1111.1111.1111. Because S4 has a lower BID, the F0/1 port of S2, which is the port connected to S4, will be the root port.

2. Lowest Sender Port Priority

To demonstrate these next two criteria, the topology is changed to one where two switches are connected with two equal-cost paths between them, as shown in Figure 5-19. S1 is the root bridge, so both of its ports are designated ports.

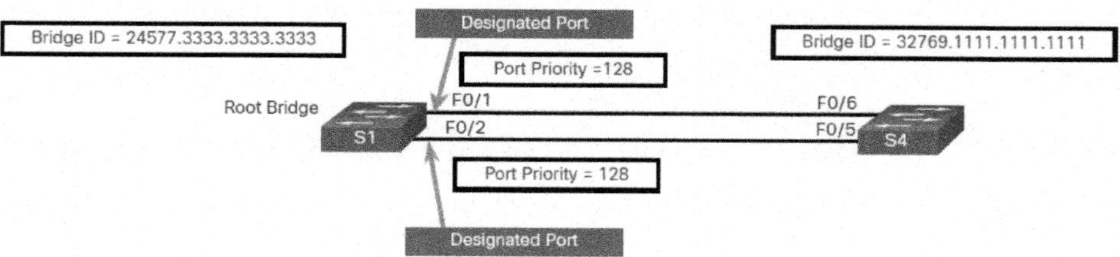

Figure 5-19 Selecting Designated Ports Based on Port Priority

S4 has two ports with equal-cost paths to the root bridge. Because both ports are connected to the same switch, the sender's BID (S1) is identical. Therefore, the first step results in a tie.

Next on the list is the sender's (S1) port priority. The default port priority is 128, so both ports on S1 have the same port priority. This is also a tie. However, if either port on S1 was configured with a lower port priority, S4 would put its adjacent port in forwarding state. The other port on S4 would be in a blocking state.

3. Lowest Sender Port ID

The last tiebreaker is the lowest sender's port ID. Switch S4 has received BPDUs from port F0/1 and port F0/2 on S1. Remember the decision is based on the sender's port ID, not the receiver's port ID. Because the port ID of F0/1 on S1 is lower than port F0/2, the port F0/6 on switch S4 will be the root port. This is the port on S4 that is connected to the F0/1 port on S1.

Port F0/5 on S4 will become an alternate port and placed in the blocking state, which is the loop-prevention part of STP, as shown in Figure 5-20.

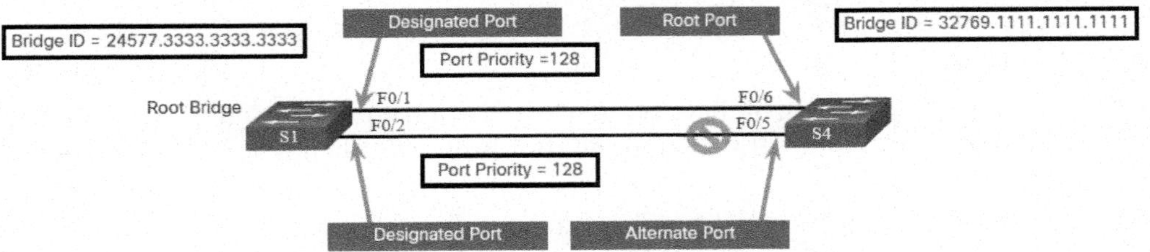

Figure 5-20 Selecting Root and Alternate Ports Based on Port ID

STP Timers and Port States (5.2.9)

STP convergence requires three timers, as follows:

- **Hello Timer:** The hello time is the interval between BPDUs. The default is 2 seconds but can be modified to between 1 and 10 seconds.
- **Forward Delay Timer:** The forward delay is the time that is spent in the *listening state* and *learning state*. The default is 15 seconds but can be modified to between 4 and 30 seconds.
- **Max Age Timer:** The max age is the maximum length of time that a switch waits before attempting to change the STP topology. The default is 20 seconds but can be modified to between 6 and 40 seconds.

Note

The default times can be changed on the root bridge, which dictates the value of these timers for the STP domain.

Note

The IEEE recommends a maximum *STP diameter* of seven switches for the default STP timers. The diameter is the maximum number of switches that data must cross to connect any two switches. The diameter can be increased, but the timers will need to be adjusted.

STP facilitates the logical loop-free path throughout the broadcast domain. The spanning tree is determined through the information learned by the exchange of the BPDU frames between the interconnected switches. If a switch port transitions directly from the blocking state to the *forwarding state* without information about the full topology during the transition, the port can temporarily create a data loop. For this reason, STP has five ports states, four of which are operational port states as shown in Figure 5-21. The *disabled state* is considered non-operational.

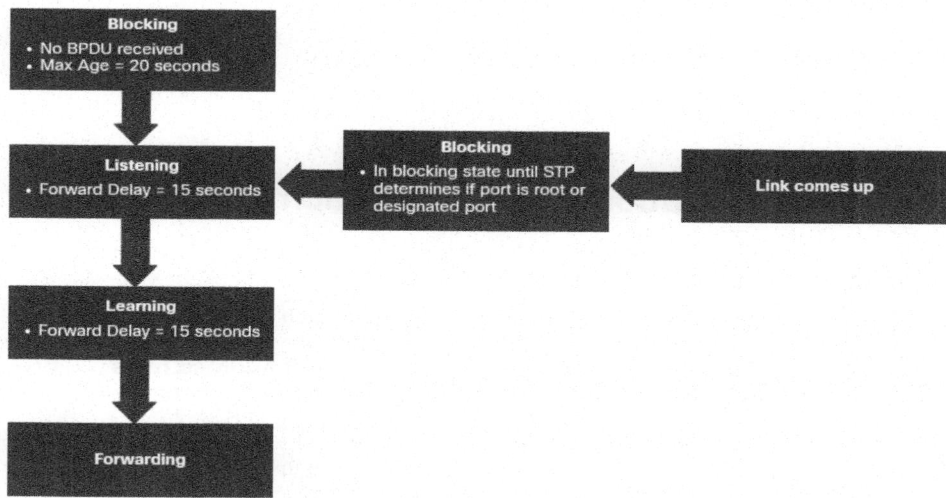

Figure 5-21 STP Operational States

The details of each port state are shown in the Table 5-2.

Table 5-2 STP Port States

Port State	Description
Blocking	The port is an alternate port and does not participate in frame forwarding. The port receives BPDU frames to determine the location and root ID of the root bridge. BPDU frames also determine which port roles each switch port should assume in the final active STP topology. With a Max Age timer of 20 seconds, a switch port that has not received an expected BPDU from a neighbor switch will go into the blocking state.

Port State	Description
Listening	After the blocking state, a port will move to the listening state. The port receives BPDUs to determine the path to the root. The switch port also transmits its own BPDU frames and informs adjacent switches that the switch port is preparing to participate in the active topology.
Learning	A switch port transitions to the learning state after the listening state. During the learning state, the switch port receives and processes BPDUs and prepares to participate in frame forwarding. It also begins to populate the MAC address table. However, in the learning state, user frames are not forwarded to the destination.
Forwarding	In the forwarding state, a switch port is considered part of the active topology. The switch port forwards user traffic and sends and receives BPDU frames.
Disabled	A switch port in the disabled state does not participate in spanning tree and does not forward frames. The disabled state is set when the switch port is administratively disabled.

Operational Details of Each Port State (5.2.10)

Table 5-3 summarizes the operational details of each port state.

Table 5-3 STP Operational Details of Each Port State

Port State	BPDU	MAC Address Table	Forwarding Data Frames
Blocking	Receive only	No update	No
Listening	Receive and send	No update	No
Learning	Receive and send	Updating table	No
Forwarding	Receive and send	Updating table	Yes
Disabled	None sent or received	No update	No

Per-VLAN Spanning Tree (5.2.11)

Up until now, we have discussed STP in an environment where there is only one VLAN. However, STP can be configured to operate in an environment with multiple VLANs.

In Per-VLAN Spanning Tree (PVST) versions of STP, there is a root bridge elected for each spanning tree instance. This makes it possible to have different root bridges for different sets of VLANs. STP operates a separate instance of STP for each individual VLAN. If all ports on all switches are members of VLAN 1, there is only one spanning tree instance.

Interactive Graphic

Check Your Understanding—STP Operations (5.2.12)

Refer to the online course to complete this activity.

Evolution of STP (5.3)

The focus of this section is to learn about the different spanning tree varieties.

Different Versions of STP (5.3.1)

This section details the many versions of STP and other options for preventing loops in your network.

Up to now, we have used the term Spanning Tree Protocol and the acronym STP, which can be misleading. Many professionals generically use these to refer to the various implementations of spanning tree, such as Rapid Spanning Tree Protocol (RSTP) and *Multiple Spanning Tree Protocol (MSTP)*. To communicate spanning tree concepts correctly, it is important to refer to the implementation or standard of spanning tree in context.

The latest standard for spanning tree is contained in IEEE-802-1D-2004, the IEEE standard for local and metropolitan area networks: Media Access Control (MAC) Bridges. This version of the standard states that switches and bridges that comply with the standard will use Rapid Spanning Tree Protocol (RSTP) instead of the older STP protocol specified in the original 802.1D standard.

In this curriculum, when the original Spanning Tree Protocol is the context of a discussion, the phrase "original 802.1D spanning tree" is used to avoid confusion. Because the two protocols share much of the same terminology and methods for the loop-free path, the primary focus will be on the current standard and the Cisco proprietary implementations of STP and RSTP.

Several varieties of spanning tree protocols have emerged since the original IEEE 802.1D specification, as shown in Table 5-4.

Table 5-4 Varieties of STP

STP Variety	Description
STP	This is the original STP (IEEE 802.1D-1998 and earlier) that provides a loop-free topology in a network with redundant links. Also called *Common Spanning Tree (CST)*, it assumes one spanning tree instance for the entire bridged network, regardless of the number of VLANs.

STP Variety	Description
PVST+	*Per-VLAN Spanning Tree (PVST+)* is a Cisco enhancement of STP that provides a separate 802.1D spanning tree instance for each VLAN configured in the network. PVST+ supports *PortFast*, UplinkFast, BackboneFast, *BPDU guard*, *BPDU filter*, *root guard*, and *loop guard*.
802.1D-2004	This is an updated version of the STP standard, incorporating RSTP.
RSTP	Rapid Spanning Tree Protocol (RSTP) (802.1w) is an evolution of STP that provides faster convergence than STP.
Rapid PVST+	This is a Cisco enhancement of RSTP that uses PVST+ and provides a separate instance of RSTP per VLAN. Each separate instance supports PortFast, BPDU guard, BPDU filter, root guard, and loop guard.
MSTP	Multiple Spanning Tree Protocol (MSTP) is an IEEE standard inspired by the earlier Cisco proprietary Multiple Instance STP (MISTP) implementation. MSTP maps multiple VLANs into the same spanning tree instance.
MST	*Multiple Spanning Tree (MST)* is the Cisco implementation of MSTP, which provides up to 16 instances of RSTP and combines many VLANs with the same physical and logical topology into a common RSTP instance. Each instance supports PortFast, BPDU guard, BPDU filter, root guard, and loop guard.

A network professional whose duties include switch administration may be required to decide which type of spanning tree protocol to implement.

Cisco switches running IOS 15.0 or later run PVST+ by default. This version incorporates many of the specifications of IEEE 802.1D-2004, such as alternate ports in place of the former nondesignated ports. Switches must be explicitly configured for rapid spanning tree mode in order to run the rapid spanning tree protocol.

RSTP Concepts (5.3.2)

RSTP (IEEE 802.1w) supersedes the original 802.1D while retaining backward compatibility. The 802.1w STP terminology remains primarily the same as the original IEEE 802.1D STP terminology. Most parameters have been left unchanged. Users who are familiar with the original STP standard can easily configure RSTP. The same spanning tree algorithm is used for both STP and RSTP to determine port roles and topology.

RSTP increases the speed of the recalculation of the spanning tree when the Layer 2 network topology changes. RSTP can achieve much faster convergence in a properly configured network, sometimes in as little as a few hundred milliseconds. If a port is configured to be an alternate port, it can immediately change to a forwarding state without waiting for the network to converge.

Note

Rapid PVST+ is the Cisco implementation of RSTP on a per-VLAN basis. With Rapid PVST+ an independent instance of RSTP runs for each VLAN.

RSTP Port States and Port Roles (5.3.3)

The port states and port roles between STP and RSTP are similar.

STP and RSTP Port States

As shown in Figure 5-22, only three port states in RSTP correspond to the three possible operational states in STP. The 802.1D disabled state, blocking state, and listening state are merged into a unique 802.1w *discarding state*.

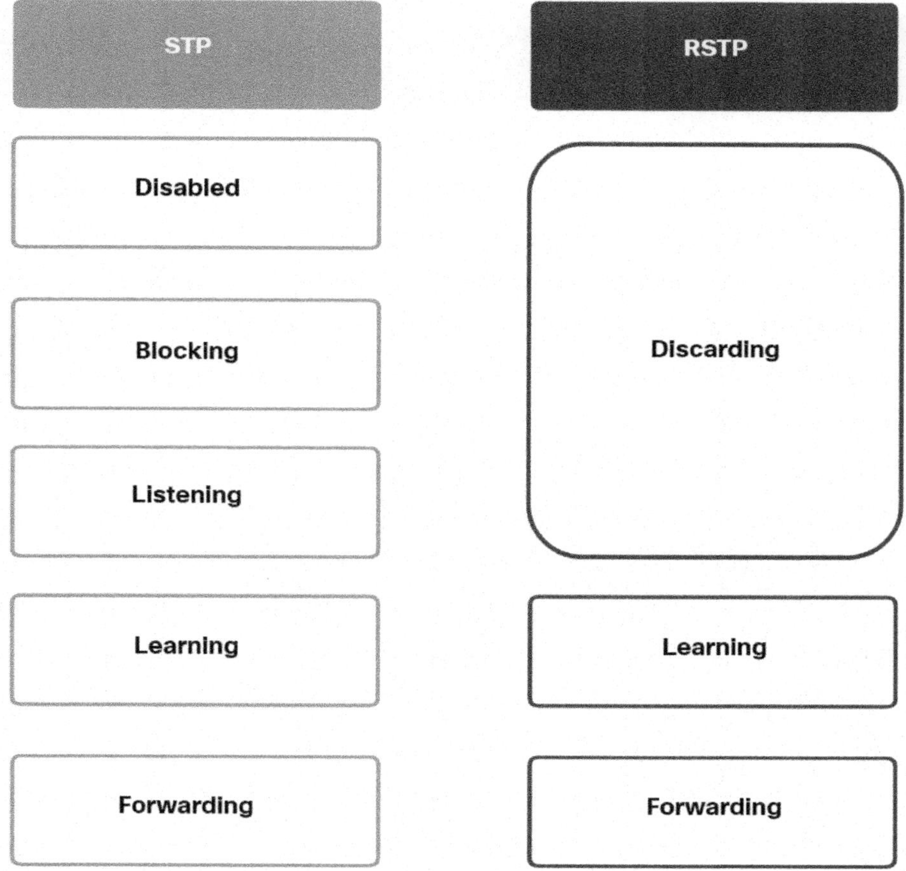

Figure 5-22 STP and RSTP Port States

As shown in Figure 5-23, root ports and designated ports are the same for both STP and RSTP. However, two RSTP port roles correspond to the blocking state of STP. In STP, a blocked port is defined as not being the designated or root port. RSTP has two port roles for this purpose.

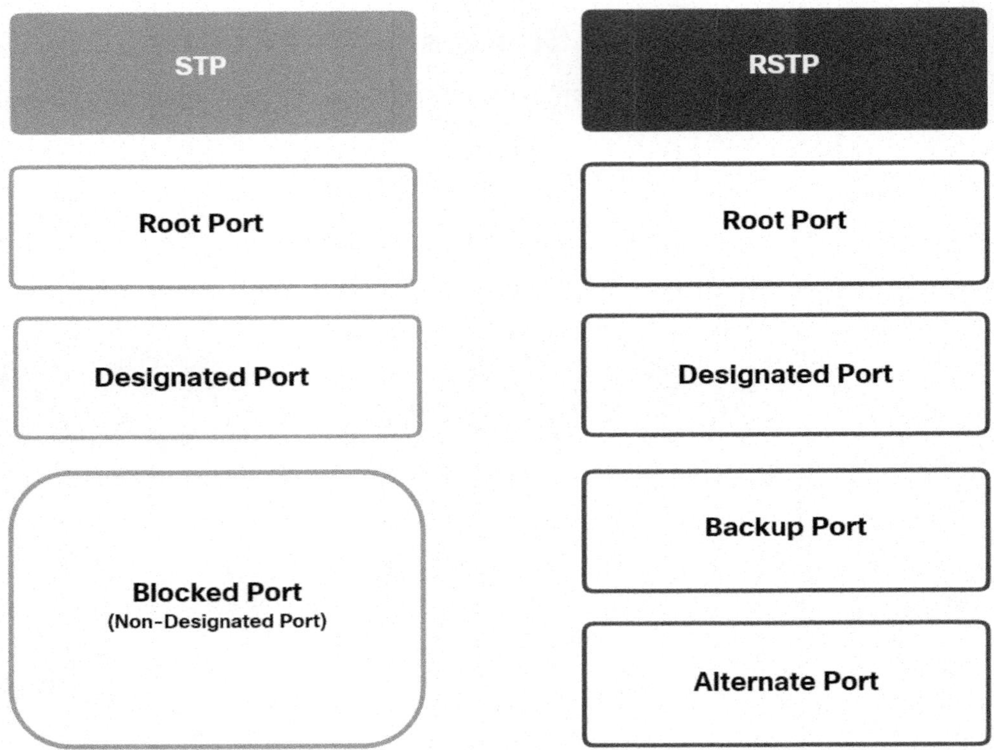

Figure 5-23 STP and RSTP Port Roles

As shown Figure 5-24, the alternate port has an alternate path to the root bridge. The backup port is a backup to a shared medium, such as a hub. A backup port is less common because hubs are now considered legacy devices.

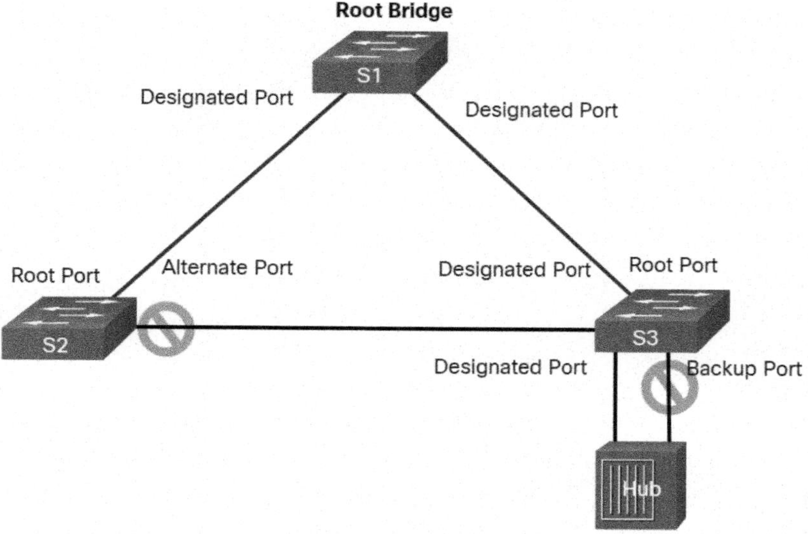

Figure 5-24 RSTP Alternate and Backup Ports

PortFast and BPDU Guard (5.3.4)

When a device is connected to a switch port, or when a switch powers up, the switch port goes through both the listening and learning states, each time waiting for the Forward Delay timer to expire. This delay is 15 seconds for each state, listening and learning, for a total of 30 seconds. This delay can present a problem for dynamic host configuration protocol (DHCP) clients trying to discover a DHCP server. DHCP messages from the connected host will not be forwarded for the 30 seconds of Forward Delay timers, and the DHCP process may time out. The result is that an IPv4 client will not receive a valid IPv4 address.

Note

Although this may occur with clients sending ICMPv6 Router Solicitation messages, the router will continue to send ICMPv6 Router Advertisement messages so the device will know how to obtain its address information.

When a switch port is configured with PortFast, that port transitions from blocking to forwarding state immediately, bypassing the usual 802.1D STP transition states (the listening and learning states) and avoiding a 30 second delay. You can use PortFast on access ports to allow devices connected to these ports, such as DHCP clients, to access the network immediately, rather than waiting for IEEE 802.1D STP to converge on each VLAN. Because the purpose of PortFast is to minimize the time that access ports must wait for spanning tree to converge, it should be used only on access ports. If you enable PortFast on a port connecting to another switch, you risk creating a spanning tree loop. PortFast is only for use on switch ports that connect to end devices, as shown in Figure 5-25.

In a valid PortFast configuration, BPDUs should never be received on PortFast-enabled switch ports because that would indicate that another bridge or switch is connected to the port. This potentially causes a spanning tree loop. To prevent this type of scenario from occurring, Cisco switches support a feature called BPDU guard. When enabled, BPDU guard immediately puts the switch port in an errdisabled (error-disabled) state on receipt of any BPDU. This protects against potential loops by effectively shutting down the port. The BPDU guard feature provides a secure response to invalid configurations because an administrator must manually put the interface back into service.

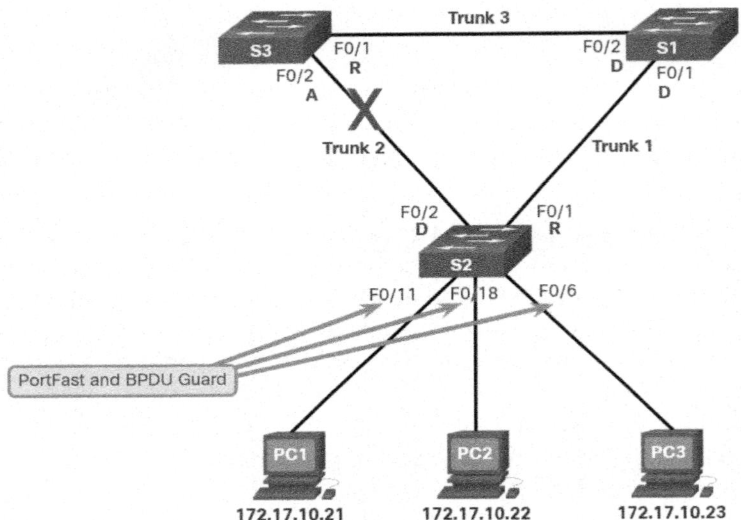

Figure 5-25 PortFast and BPDU Guard on Access Ports

Alternatives to STP (5.3.5)

STP was and still is an Ethernet loop-prevention protocol. Over the years, organizations required greater resiliency and availability in the LAN. Ethernet LANs went from a few interconnected switches connected to a single router, to a sophisticated hierarchical network design including access, distribution, and core layer switches, as shown in Figure 5-26.

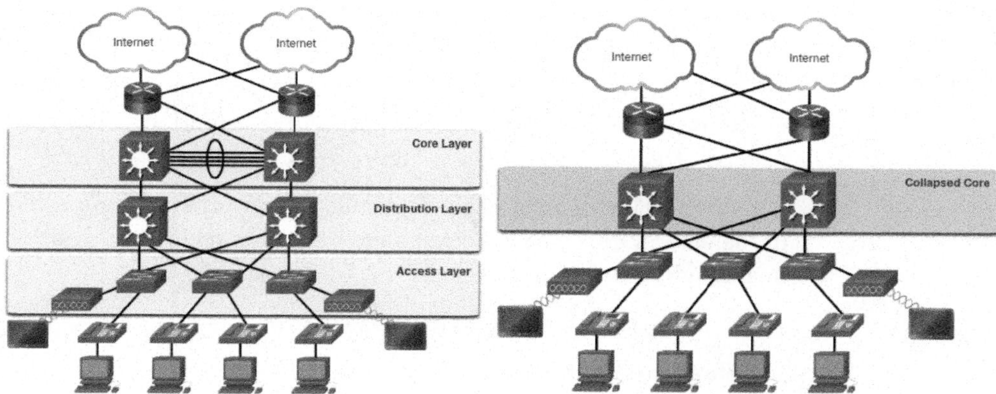

Figure 5-26 Hierarchical Network Designs

Depending on the implementation, Layer 2 may include not only the access layer, but also the distribution or even the core layers. These designs may include hundreds of switches, with hundreds or even thousands of VLANs. STP has adapted to the added redundancy and complexity with enhancements, as part of RSTP and MSTP.

An important aspect to network design is fast and predictable convergence when there is a failure or change in the topology. Spanning tree does not offer the same efficiencies and predictabilities provided by routing protocols at Layer 3. Figure 5-27 shows a traditional hierarchical network design with the distribution and core multilayer switches performing routing.

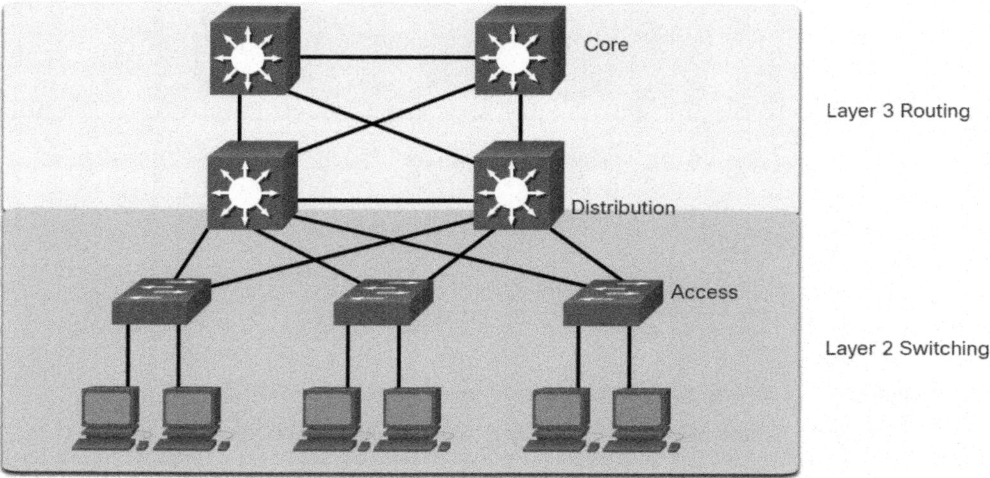

Figure 5-27 Traditional Hierarchical Network Design

Layer 3 routing allows for redundant paths and loops in the topology, without blocking ports. For this reason, some environments are transitioning to Layer 3 everywhere except where devices connect to the access layer switch. In other words, the connections between access layer switches and distribution switches would be Layer 3 instead of Layer 2, as shown in Figure 5-28.

Although STP will most likely continue to be used as a loop prevention mechanism in the enterprise, on access layer switches, other technologies are also being used, including the following:

- Multi System Link Aggregation (MLAG)
- Shortest Path Bridging (SPB)
- Transparent Interconnect of Lots of Links (TRILL)

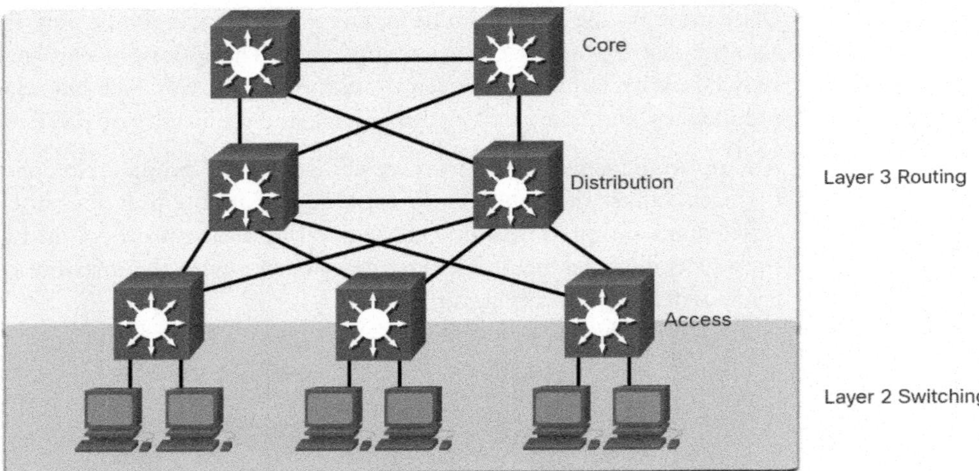

Figure 5-28 Hierarchical Network Design with Layer 3 Switches

Note

These technologies are beyond the scope of this course.

Interactive Graphic

Check Your Understanding—Evolution of STP (5.3.6)

Refer to the online course to complete this activity.

Summary (5.4)

The following is a summary of each section in the module:

Purpose of STP

Redundant paths in a switched Ethernet network may cause both physical and logical Layer 2 loops. A Layer 2 loop can result in MAC address table instability, link saturation, and high CPU utilization on switches and end devices. This results in the network becoming unusable. Unlike the Layer 3 protocols, IPv4 and IPv6, Layer 2 Ethernet does not include a mechanism to recognize and eliminate endlessly looping frames. Ethernet LANs require a loop-free topology with a single path between any two devices. STP is a loop-prevention network protocol that allows for redundancy while creating a loop-free Layer 2 topology. Without STP, Layer 2 loops can form, causing broadcast, multicast, and unknown unicast frames to loop endlessly, bringing down a network. A broadcast storm is an abnormally high number of broadcasts overwhelming the network during a specific amount of time. Broadcast storms can disable a network within seconds by overwhelming switches and end devices. STP is based on an algorithm invented by Radia Perlman. Her spanning tree algorithm (STA) creates a loop-free topology by selecting a single root bridge where all other switches determine a single least-cost path.

STP Operations

Using the STA, STP builds a loop-free topology in a four-step process: elect the root bridge, elect the root ports, elect designated ports, and elect alternate (blocked) ports. During STA and STP functions, switches use BPDUs to share information about themselves and their connections. BPDUs are used to elect the root bridge, root ports, designated ports, and alternate ports. Each BPDU contains a BID that identifies the switch that sent the BPDU. The BID is involved in making many of the STA decisions, including root bridge and port roles. The BID contains a priority value, the MAC address of the switch, and an extended system ID. The lowest BID value is determined by the combination of these three fields. The switch with the lowest BID will become the root bridge. Because the default BID is 32,768, it is possible for two or more switches to have the same priority. In this scenario, where the priorities are the same, the switch with the lowest MAC address will become the root bridge. When the root bridge has been elected for a given spanning tree instance, the STA determines the best paths to the root bridge from all destinations in the broadcast domain. The path information, known as the internal root path cost, is determined by the sum of all the individual port costs along the path from the switch to the root bridge. After the root bridge has been determined, the STA algorithm selects the root port. The root port is the port closest to the root bridge in terms of overall cost, which is called the internal root path cost. After each switch selects a root port, switches will select designated ports. The designated port is a port on the segment (with two switches) that has the internal root path cost to the root bridge. If a port is not a root port or a designated port, it becomes an alternate (or backup) port. Alternate ports and backup ports are in discarding or blocking state to prevent loops. When a switch has multiple equal-cost paths to the root bridge, the switch determines a port using the following criteria: lowest sender BID, then the lowest sender port priority, and finally, the lowest sender port ID. STP convergence requires three timers: the hello timer, the forward delay timer, and the max age timer. Port states are blocking, listening, learning, forwarding, and disabled. In PVST versions of STP, there is a root bridge elected for each spanning tree instance. This makes it possible to have different root bridges for different sets of VLANs.

Evolution of STP

The term Spanning Tree Protocol and the acronym STP can be misleading. STP is often used to refer to the various implementations of spanning tree, such as RSTP and MSTP. RSTP is an evolution of STP that provides faster convergence than STP. RSTP port states are learning, forwarding, and discarding. PVST+ is a Cisco enhancement of STP that provides a separate spanning tree instance for each VLAN configured in the network. PVST+ supports PortFast, UplinkFast, BackboneFast, BPDU guard, BPDU filter, root guard, and loop guard. Cisco switches running IOS 15.0 or later run PVST+ by default. Rapid PVST+ is a Cisco enhancement of RSTP that uses PVST+ and provides a separate instance of 802.1w per VLAN. When a switch port is configured with PortFast, that port transitions from blocking to forwarding state immediately, bypassing the STP listening and learning states and avoiding a 30 second delay. Use PortFast on access ports to allow devices connected to these ports, such as DHCP clients, to access the network immediately, rather than waiting for STP to converge on each VLAN. Cisco switches support a feature called BPDU guard, which immediately puts the switch port in an error-disabled state upon receipt of any BPDU to protect against potential loops. Over the years, Ethernet LANs went from a few interconnected switches that were connected to a single router to a sophisticated hierarchical network design. Depending on the implementation, Layer 2 may include not only the access layer, but also the distribution or even the core layers. These designs may include hundreds of switches, with hundreds or even thousands of VLANs. STP has adapted to the added redundancy and complexity with enhancements as part of RSTP and MSTP. Layer 3 routing allows for redundant paths and loops in the topology, without blocking ports. For this reason, some environments are transitioning to Layer 3 everywhere except where devices connect to the access layer switch.

Practice

The following activities provide practice with the topics introduced in this chapter. The Labs are available in the companion *Switching, Routing, and Wireless Essentials Labs and Study Guide (CCNAv7)* (ISBN 9780136634386). The Packet Tracer Activity instructions are also in the Labs & Study Guide. The PKA files are found in the online course.

Packet Tracer Activity

Packet Tracer Activity

Packet Tracer 5.1.9: Investigate STP Loop Prevention

Check Your Understanding Questions

Complete all the review questions listed here to test your understanding of the sections and concepts in this chapter. The appendix "Answers to 'Check Your Understanding' Questions" lists the answers.

1. Which three components are combined to form a bridge ID? (Choose three.)
   - A. Bridge priority
   - B. Cost
   - C. Extended system ID
   - D. IP address
   - E. MAC address
   - F. Port ID

2. Which STP port role is adopted by a switch port if there is no other port with a lower cost to the root bridge?
   - A. Alternate port
   - B. Designated port
   - C. Disabled port
   - D. Root port

3. Which is the default STP operation mode on Cisco Catalyst switches?
   - A. MST
   - B. MSTP
   - C. PVST+
   - D. Rapid PVST+
   - E. RSTP

4. What is an advantage of PVST+?

   A. PVST+ optimizes performance on the network through automatic selection of the root bridge.
   B. PVST+ optimizes performance on the network through load sharing using multiple root bridges.
   C. PVST+ reduces bandwidth consumption compared to traditional implementations of STP that use CST.
   D. PVST+ requires fewer CPU cycles for all the switches in the network.

5. In which two port states does a switch learn MAC addresses and process BPDUs in a PVST network? (Choose two.)

   A. Blocking
   B. Disabled
   C. Forwarding
   D. Learning
   E. Listening

6. What two features does Spanning Tree Protocol (STP) provide to ensure proper network operations? (Choose two.)

   A. Implementing VLANs to contain broadcasts
   B. Link-state dynamic routing that provides redundant routes
   C. Redundant links between Layer 2 switches
   D. Removing single points of failure with multiple Layer 2 switches
   E. Static default routes

7. What value determines the root bridge when all switches connected by trunk links have default STP configurations?

   A. Bridge priority
   B. Extended system ID
   C. Highest MAC address
   D. IP address
   E. Lowest MAC address

8. Which PVST+ feature ensures that configured switch edge ports do not cause Layer 2 loops if a port is mistakenly connected to another switch?

   A. BPDU guard

   B. Extended system ID

   C. PortFast

   D. PVST+

9. What is an advantage of using STP in a LAN?

   A. It combines multiple switch trunk links into a logical port channel link to increase bandwidth.

   B. It decreases the size of the failure domain.

   C. It provides firewall services to protect the LAN.

   D. It temporarily disables redundant paths to stop Layer 2 loops.

10. Which two statements regarding a PortFast enabled switch port are true? (Choose two.)

    A. The port immediately transitions from blocking to forwarding state.

    B. The port immediately transitions from listening to forwarding state.

    C. The port immediately processes any BPDUs before transitioning to the forwarding state.

    D. The port sends DHCP requests before transitioning to the forwarding state.

    E. The port should never receive BPDUs.

CHAPTER 6

EtherChannel

Objectives

Upon completion of this chapter, you will be able to answer the following questions:

- What is EtherChannel technology?
- How do you configure EtherChannel?
- How do you troubleshoot EtherChannel?

Key Terms

This chapter uses the following key terms. You can find the definitions in the Glossary.

EtherChannel *Page 176*

link aggregation *Page 176*

source MAC and destination MAC load balancing *Page 177*

source IP and destination IP load balancing *Page 177*

Port Aggregation Protocol (PAgP) *Page 179*

Link Aggregation Control Protocol (LACP) *Page 179*

802.3ad *Page 181*

port channel interface *Page 184*

Introduction (6.0)

Your network design includes redundant switches and links. You have some version of STP configured to prevent Layer 2 loops. But now, like most network administrators, you realize that you could use more bandwidth and redundancy in your network. Not to worry, EtherChannel is here to help! *EtherChannel* aggregates links between devices into bundles. These bundles include redundant links. STP may block one of those links, but it will not block all of them. With EtherChannel your network can have redundancy, loop prevention, and increased bandwidth!

There are two protocols, PAgP and LACP. This module explains them both and shows you how to configure, verify, and troubleshoot them! A Syntax Checker and two Packet Tracer activities help you to better understand these protocols. What are you waiting for?

EtherChannel Operation (6.1)

Link aggregation is commonly implemented between access layer and distribution layer switches to increase the uplink bandwidth. In this section you will learn about the link aggregation operation in a switched LAN environment.

This topic describes link aggregation.

Link Aggregation (6.1.1)

There are scenarios in which more bandwidth or redundancy between devices is needed than what can be provided by a single link. Multiple links could be connected between devices to increase bandwidth. However, Spanning Tree Protocol (STP), which is enabled on Layer 2 devices like Cisco switches by default, will block redundant links to prevent switching loops, as shown in Figure 6-1.

A link aggregation technology is needed that allows redundant links between devices that will not be blocked by STP. That technology is known as EtherChannel.

EtherChannel is a link aggregation technology that groups multiple physical Ethernet links together into one single logical link. It is used to provide fault tolerance, load sharing, increased bandwidth, and redundancy between switches, routers, and servers.

EtherChannel technology makes it possible to combine the number of physical links between the switches to increase the overall speed of switch-to-switch communication.

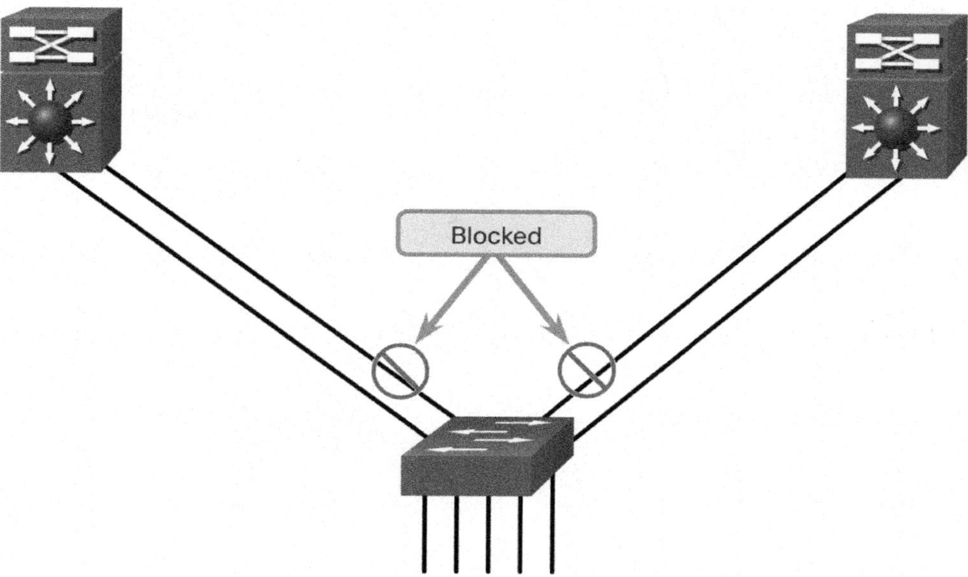

By default, STP will block redundant links.

Figure 6-1 Blocked Ports in STP

EtherChannel (6.1.2)

EtherChannel technology was originally developed by Cisco as a LAN switch-to-switch technique of grouping several Fast Ethernet or Gigabit Ethernet ports into one logical channel. When an EtherChannel is configured, the resulting virtual interface is called a port channel. The physical interfaces are bundled together into a port channel interface, as shown in Figure 6-2.

Advantages of EtherChannel (6.1.3)

EtherChannel technology has many advantages, including the following:

- Most configuration tasks can be done on the EtherChannel interface instead of on each individual port, ensuring configuration consistency throughout the links.
- EtherChannel relies on existing switch ports. There is no need to upgrade the link to a faster and more expensive connection to have more bandwidth.
- Load balancing takes place between links that are part of the same EtherChannel. Depending on the hardware platform, one or more load-balancing methods can be implemented. These methods include *source MAC and destination MAC load balancing*, or *source IP and destination IP load balancing*, across the physical links.

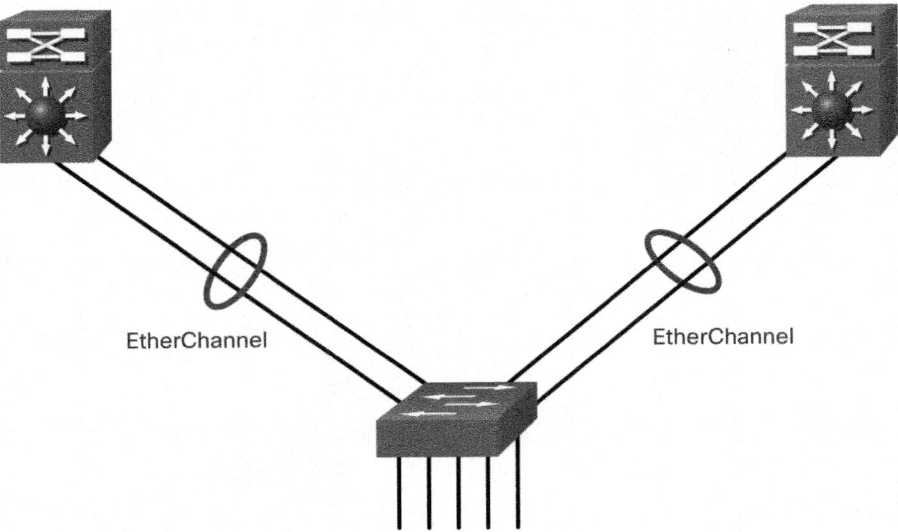

Figure 6-2 Bundling Links into an EtherChannel

- EtherChannel creates an aggregation that is seen as one logical link. When several EtherChannel bundles exist between two switches, STP may block one of the bundles to prevent switching loops. When STP blocks one of the redundant links, it blocks the entire EtherChannel. This blocks all the ports belonging to that EtherChannel link. Where there is only one EtherChannel link, all physical links in the EtherChannel are active because STP sees only one (logical) link.
- EtherChannel provides redundancy because the overall link is seen as one logical connection. Additionally, the loss of one physical link within the channel does not create a change in the topology. Therefore, a spanning tree recalculation is not required. Assuming at least one physical link is present, the EtherChannel remains functional, even if its overall throughput decreases because of a lost link within the EtherChannel.

Implementation Restrictions (6.1.4)

EtherChannel has certain implementation restrictions, including the following:

- Interface types cannot be mixed. For example, Fast Ethernet and Gigabit Ethernet cannot be mixed within a single EtherChannel.
- Currently each EtherChannel can consist of up to eight compatibly-configured Ethernet ports. EtherChannel provides full-duplex bandwidth up to 800 Mbps (Fast EtherChannel) or 8 Gbps (Gigabit EtherChannel) between one switch and another switch or host.

- The Cisco Catalyst 2960 Layer 2 switch currently supports up to six EtherChannels. However, as new IOSs are developed and platforms change, some cards and platforms may support increased numbers of ports within an EtherChannel link, as well as support an increased number of Gigabit EtherChannels.
- The individual EtherChannel group member port configuration must be consistent on both devices. If the physical ports of one side are configured as trunks, the physical ports of the other side must also be configured as trunks within the same native VLAN. Additionally, all ports in each EtherChannel link must be configured as Layer 2 ports.
- Each EtherChannel has a logical port channel interface, as shown in Figure 6-3. A configuration applied to the port channel interface affects all physical interfaces that are assigned to that interface.

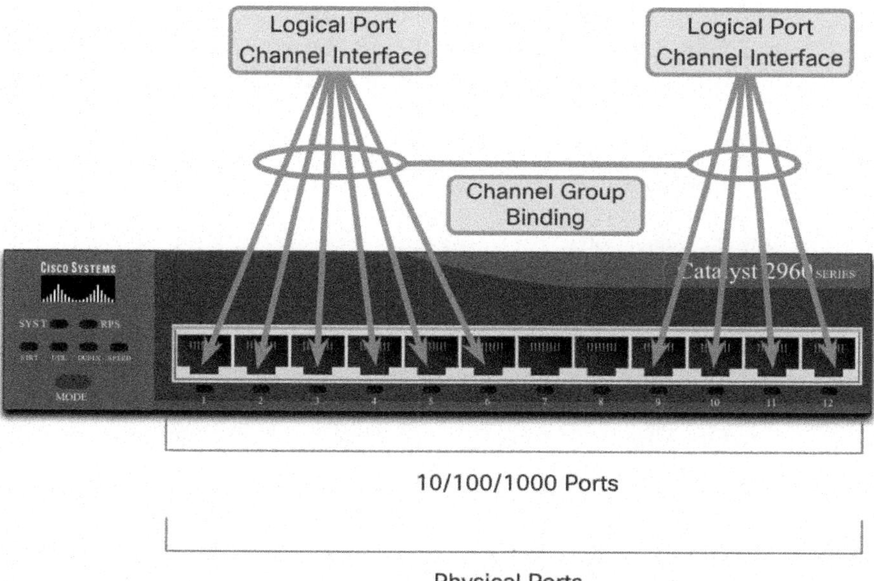

Figure 6-3 Physical and Logical Ports

AutoNegotiation Protocols (6.1.5)

EtherChannels can be formed through negotiation using one of two protocols, *Port Aggregation Protocol (PAgP)* or *Link Aggregation Control Protocol (LACP).* These protocols allow ports with similar characteristics to form a channel through dynamic negotiation with adjoining switches.

Note

It is also possible to configure a static or unconditional EtherChannel without PAgP or LACP.

PAgP Operation (6.1.6)

PAgP (pronounced "Pag - P") is a Cisco-proprietary protocol that aids in the automatic creation of EtherChannel links. When an EtherChannel link is configured using PAgP, PAgP packets are sent between EtherChannel-capable ports to negotiate the forming of a channel. When PAgP identifies matched Ethernet links, it groups the links into an EtherChannel. The EtherChannel is then added to the spanning tree as a single port.

When enabled, PAgP also manages the EtherChannel. PAgP packets are sent every 30 seconds. PAgP checks for configuration consistency and manages link additions and failures between two switches. It ensures that when an EtherChannel is created, all ports have the same type of configuration.

Note

In EtherChannel, it is mandatory that all ports have the same speed, duplex setting, and VLAN information. Any port modification after the creation of the channel also changes all other channel ports.

PAgP helps create the EtherChannel link by detecting the configuration of each side and ensuring that links are compatible so that the EtherChannel link can be enabled when needed. The modes for PAgP are as follows:

- **On:** This mode forces the interface to channel without PAgP. Interfaces configured in the on mode do not exchange PAgP packets.
- **PAgP desirable:** This PAgP mode places an interface in an active negotiating state in which the interface initiates negotiations with other interfaces by sending PAgP packets.
- **PAgP auto:** This PAgP mode places an interface in a passive negotiating state in which the interface responds to the PAgP packets that it receives but does not initiate PAgP negotiation.

The modes must be compatible on each side. If one side is configured to be in auto mode, it is placed in a passive state, waiting for the other side to initiate the EtherChannel negotiation. If the other side is also set to auto, the negotiation never starts, and the EtherChannel does not form. If all modes are disabled by using the **no** command, or if no mode is configured, the EtherChannel is disabled.

The on mode manually places the interface in an EtherChannel, without any negotiation. It works only if the other side is also set to on. If the other side is set to negotiate parameters through PAgP, no EtherChannel forms, because the side that is set to on mode does not negotiate.

No negotiation between the two switches means there is no checking to make sure that all the links in the EtherChannel are terminating on the other side, or that there is PAgP compatibility on the other switch.

PAgP Mode Settings Example (6.1.7)

Consider the two switches in Figure 6-4. Whether S1 and S2 establish an EtherChannel using PAgP depends on the mode settings on each side of the channel.

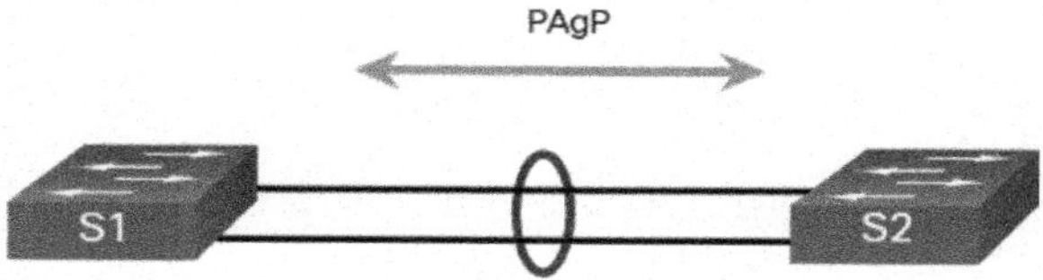

Figure 6-4 PAgP Topology

Table 6-1 shows the various combination of PAgP modes on S1 and S2 and the resulting channel establishment outcome.

Table 6-1 PAgP Modes

S1	S2	Channel Establishment
On	On	Yes
On	Desirable/Auto	No
Desirable	Desirable	Yes
Desirable	Auto	Yes
Auto	Desirable	Yes
Auto	Auto	No

LACP Operation (6.1.8)

LACP is specified in IEEE *802.3ad* and allows several physical ports to be bundled to form a single logical channel. LACP allows a switch to negotiate an automatic bundle by sending LACP packets to the other switch. It performs a function similar to PAgP with Cisco EtherChannel. Because LACP is an IEEE standard, it can be used to facilitate EtherChannels in multivendor environments. On Cisco devices, both protocols are supported.

Note

LACP was originally defined as IEEE 802.3ad. However, LACP is now defined in the newer IEEE 802.1AX standard for local and metropolitan area networks.

LACP provides the same negotiation benefits as PAgP. LACP helps create the EtherChannel link by detecting the configuration of each side and making sure that they are compatible so that the EtherChannel link can be enabled when needed. The modes for LACP are as follows:

- **On:** This mode forces the interface to channel without LACP. Interfaces configured in the on mode do not exchange LACP packets.
- **LACP active:** This LACP mode places a port in an active negotiating state. In this state, the port initiates negotiations with other ports by sending LACP packets.
- **LACP passive:** This LACP mode places a port in a passive negotiating state. In this state, the port responds to the LACP packets that it receives but does not initiate LACP packet negotiation.

Just as with PAgP, modes must be compatible on both sides for the EtherChannel link to form. The on mode is repeated because it creates the EtherChannel configuration unconditionally, without PAgP or LACP dynamic negotiation.

LACP allows for eight active links, and also eight standby links. A standby link will become active should one of the current active links fail.

LACP Mode Settings Example (6.1.9)

Consider the two switches in Figure 6-5. Whether S1 and S2 establish an EtherChannel using LACP depends on the mode settings on each side of the channel.

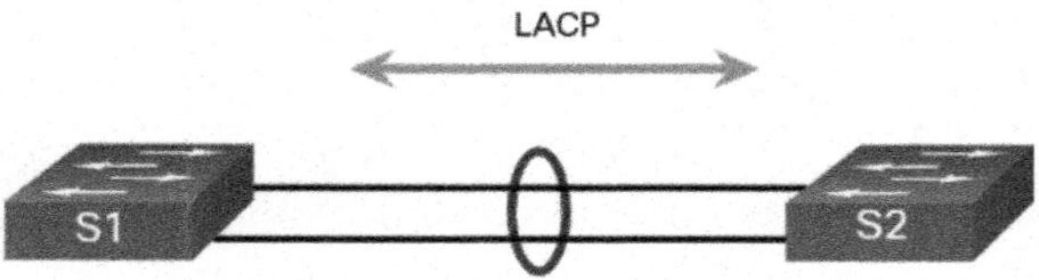

Figure 6-5 LACP Topology

Table 6-2 shows the various combination of LACP modes on S1 and S2 and the resulting channel establishment outcome.

Table 6-2 LACP Modes

S1	S2	Channel Establishment
On	On	Yes
On	Active/Passive	No
Active	Active	Yes
Active	Passive	Yes
Passive	Active	Yes
Passive	Passive	No

Interactive Graphic

Check Your Understanding—EtherChannel Operation (6.1.10)

Refer to the online course to complete this activity.

Configure EtherChannel (6.2)

In this topic you learn how to configure link aggregation.

Configuration Guidelines (6.2.1)

Now that you know what EtherChannel is, this topic explains how to configure it. The following guidelines and restrictions are useful for configuring EtherChannel:

- **EtherChannel support:** All Ethernet interfaces must support EtherChannel with no requirement that interfaces be physically contiguous.
- **Speed and duplex:** Configure all interfaces in an EtherChannel to operate at the same speed and in the same duplex mode.
- **VLAN match:** All interfaces in the EtherChannel bundle must be assigned to the same VLAN or be configured as a trunk (shown in the figure).
- **Range of VLANs:** An EtherChannel supports the same allowed range of VLANs on all the interfaces in a trunking EtherChannel. If the allowed range of VLANs is not the same, the interfaces do not form an EtherChannel, even when they are set to **auto** or **desirable** mode.

Figure 6-6 shows a configuration that would allow an EtherChannel to form between S1 and S2.

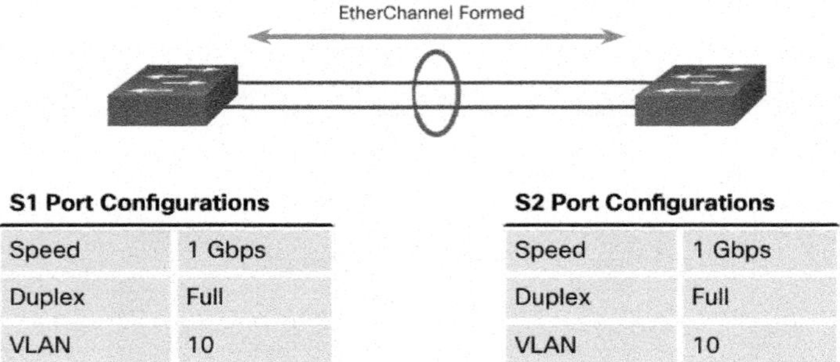

S1 Port Configurations

Speed	1 Gbps
Duplex	Full
VLAN	10

S2 Port Configurations

Speed	1 Gbps
Duplex	Full
VLAN	10

An EtherChannel is formed when configuration settings match on both switches.

Figure 6-6 EtherChannel Forms When Configuration Matches

In Figure 6-7, S1 ports are configured as half-duplex. Therefore, an EtherChannel will not form between S1 and S2.

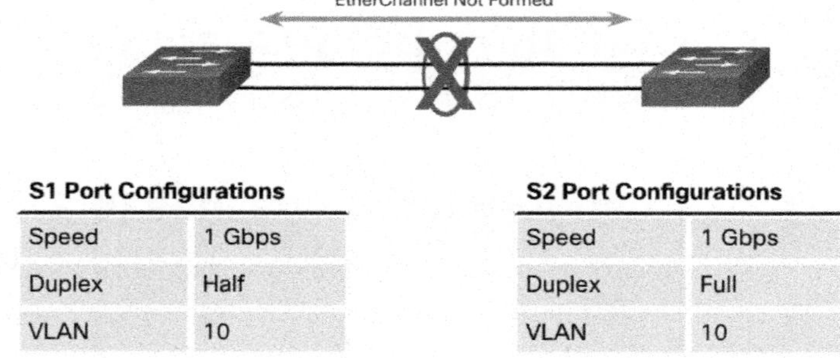

S1 Port Configurations

Speed	1 Gbps
Duplex	Half
VLAN	10

S2 Port Configurations

Speed	1 Gbps
Duplex	Full
VLAN	10

An EtherChannel is not formed when configuration settings are different on each switch.

Figure 6-7 Example of EtherChannel Not Forming Because of a Configuration Mismatch

If these settings must be changed, configure them in *port channel interface* configuration mode. Any configuration that is applied to the port channel interface also affects individual interfaces. However, configurations that are applied to the individual interfaces do not affect the port channel interface. Therefore, making configuration changes to an interface that is part of an EtherChannel link may cause interface compatibility issues.

The port channel can be configured in access mode, trunk mode (most common), or on a routed port.

LACP Configuration Example (6.2.2)

EtherChannel is disabled by default and must be configured. The topology in Figure 6-8 will be used to demonstrate an EtherChannel configuration example using LACP.

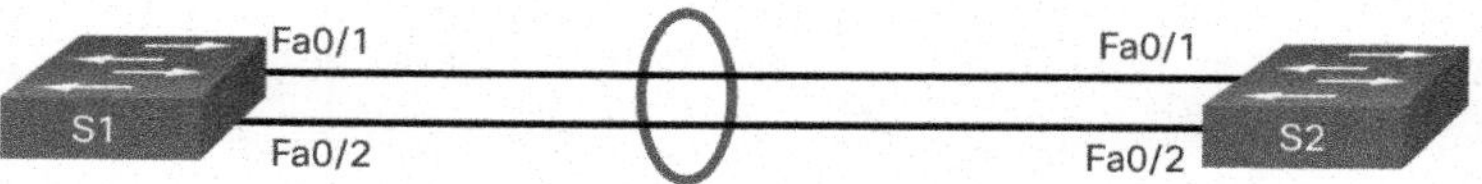

Figure 6-8 LACP Configuration Topology

Configuring EtherChannel with LACP requires the following three steps, as shown in Example 6-1:

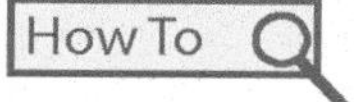

Step 1. Specify the interfaces that compose the EtherChannel group using the **interface range** interface global configuration mode command. The **range** keyword allows you to select several interfaces and configure them all together.

Step 2. Create the port channel interface with the **channel-group** *identifier* **mode active** command in interface range configuration mode. The identifier specifies a channel group number. The **mode active** keywords identify this as an LACP EtherChannel configuration.

Step 3. To change Layer 2 settings on the port channel interface, enter port channel interface configuration mode using the **interface port-channel** command, followed by the interface identifier. In the example, S1 is configured with an LACP EtherChannel. The port channel 1 is configured as a trunk interface with the allowed VLANs specified.

Example 6-1 LACP Configuration Example

```
S1(config)# interface range FastEthernet 0/1 - 2
S1(config-if-range)# channel-group 1 mode active
Creating a port-channel interface Port-channel 1
S1(config-if-range)# exit
S1(config)#
S1(config-if)# interface port-channel 1
S1(config-if)# switchport mode trunk
S1(config-if)# switchport trunk allowed vlan 1,2,20
S1(config-if)#
```

Interactive Graphic

Syntax Checker—Configure EtherChannel (6.2.3)

Configure the EtherChannel for S2 based on the specified requirements.

Packet Tracer—Configure EtherChannel (6.2.4)

Three switches have just been installed. There are redundant uplinks between the switches. As configured, only one of these links can be used; otherwise, a bridging loop might occur. However, using only one link utilizes only half of the available bandwidth. EtherChannel allows up to eight redundant links to be bundled together into one logical link. In this lab, you will configure Port Aggregation Protocol (PAgP), a Cisco EtherChannel protocol, and Link Aggregation Control Protocol (LACP), an IEEE 802.3ad open standard version of EtherChannel.

Verify and Troubleshoot EtherChannel (6.3)

In this topic you will learn how to troubleshoot a link aggregation implementation.

Verify EtherChannel (6.3.1)

As always, when you configure devices in your network, you must verify your configuration. If there are problems, you will also need to be able to troubleshoot and fix them. This topic gives you the commands to verify, as well as some common EtherChannel network problems and their solutions.

The verification command examples use the topology shown in Figure 6-9.

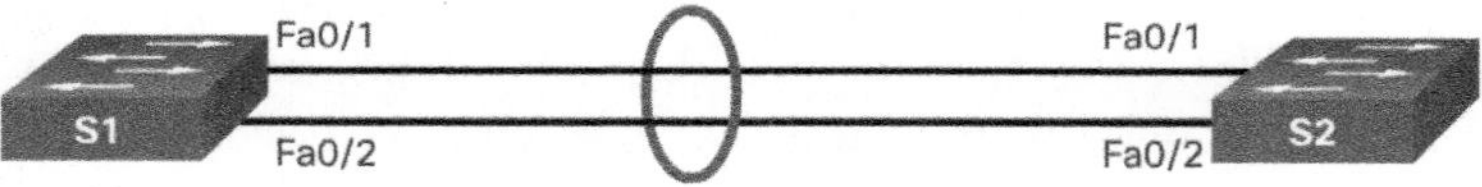

Figure 6-9 EtherChannel Topology

There are a number of commands to verify an EtherChannel configuration.

The **show interfaces port-channel** command displays the general status of the port channel interface. In Example 6-2, the Port Channel 1 interface is up.

Example 6-2 The **show interfaces port-channel** Command

```
S1# show interfaces port-channel 1
Port-channel1 is up, line protocol is up (connected)
  Hardware is EtherChannel, address is c07b.bcc4.a981 (bia c07b.bcc4.a981)
  MTU 1500 bytes, BW 200000 Kbit/sec, DLY 100 usec,
     reliability 255/255, txload 1/255, rxload 1/255
(output omitted)
```

When several port channel interfaces are configured on the same device, use the **show etherchannel summary** command to display one line of information per port channel. In Example 6-3, the switch has one EtherChannel configured; group 1 uses LACP.

The interface bundle consists of the FastEthernet0/1 and FastEthernet0/2 interfaces. The group is a Layer 2 EtherChannel, and it is in use, as indicated by the letters SU next to the port channel number.

Example 6-3 The **show etherchannel summary** Command

```
S1# show etherchannel summary
Flags:  D - down        P - bundled in port-channel
        I - stand-alone s - suspended
        H - Hot-standby (LACP only)
        R - Layer3      S - Layer2
        U - in use      N - not in use, no aggregation
        f - failed to allocate aggregator
        M - not in use, minimum links not met
        m - not in use, port not aggregated due to minimum links not met
        u - unsuitable for bundling
        w - waiting to be aggregated
        d - default port
        A - formed by Auto LAG
Number of channel-groups in use: 1
Number of aggregators:           1
Group  Port-channel  Protocol    Ports
------+-------------+-----------+-----------------------------------------------
1      Po1(SU)          LACP       Fa0/1(P)     Fa0/2(P)
```

Use the **show etherchannel port-channel** command to display information about a specific port channel interface. As shown in Example 6-4, the Port Channel 1 interface consists of two physical interfaces, FastEthernet0/1 and FastEthernet0/2. It uses LACP in active mode. It is properly connected to another switch with a compatible configuration, which is why the port channel is said to be in use.

Example 6-4 The **show etherchannel port-channel** Command

```
S1# show etherchannel port-channel
                Channel-group listing:
                ----------------------
Group: 1
----------
                Port-channels in the group:
                ---------------------------
Port-channel: Po1    (Primary Aggregator)
------------
```

```
Age of the Port-channel    = 0d:01h:02m:10s
Logical slot/port    = 2/1          Number of ports = 2
HotStandBy port = null
Port state           = Port-channel Ag-Inuse
Protocol             =    LACP
Port security        = Disabled
Load share deferral = Disabled
Ports in the Port-channel:
Index    Load    Port      EC state         No of bits
------+------+------+------------------+-----------
  0      00     Fa0/1     Active              0
  0      00     Fa0/2     Active              0
Time since last port bundled:    0d:00h:09m:30s    Fa0/2
```

On any physical interface member of an EtherChannel bundle, the **show interfaces** *interface* **etherchannel** command can provide information about the role of the interface in the EtherChannel, as shown in Example 6-5. The interface FastEthernet0/1 is part of the EtherChannel bundle 1. The protocol for this EtherChannel is LACP.

Example 6-5 The **show interfaces etherchannel** Command

```
S1# show interfaces f0/1 etherchannel
Port state      = Up Mstr Assoc In-Bndl
Channel group  = 1            Mode = Active             Gcchange = -
Port-channel   = Po1          GC    =   -               Pseudo port-channel = Po1
Port index     = 0            Load = 0x00               Protocol =    LACP
Flags:  S - Device is sending Slow LACPDUs   F - Device is sending fast LACPDUs.
        A - Device is in active mode.        P - Device is in passive mode.
Local information:
                             LACP port      Admin      Oper     Port
Port        Flags    State   Priority       Key        Number   State
Fa0/1       SA       bndl    32768          0x1        0x1      0x102        0x3D
Partner's information:
                    LACP port                           Admin   Oper    Port      Port
Port        Flags   Priority   Dev ID          Age      key     Key     Number    State
Fa0/1       SA      32768      c025.5cd7.ef00  12s      0x0     0x1     0x102     0x3Dof
  the port in the current state: 0d:00h:11m:51sllowed vlan 1,2,20
```

Common Issues with EtherChannel Configurations (6.3.2)

All interfaces within an EtherChannel must have the same configuration of speed and duplex mode, native and allowed VLANs on trunks, and access VLAN on access ports. Ensuring these configurations will significantly reduce network problems related to EtherChannel. Common EtherChannel issues include the following:

- Assigned ports in the EtherChannel are not part of the same VLAN, or not configured as trunks. Ports with different native VLANs cannot form an EtherChannel.
- Trunking was configured on some of the ports that make up the EtherChannel, but not all of them. It is not recommended that you configure trunking mode on individual ports that make up the EtherChannel. When configuring a trunk on an EtherChannel, verify the trunking mode on the EtherChannel.
- If the allowed range of VLANs is not the same, the ports do not form an EtherChannel even when PAgP is set to the **auto** or **desirable** mode.
- The dynamic negotiation options for PAgP and LACP are not compatibly configured on both ends of the EtherChannel.

Note

It is easy to confuse PAgP or LACP with DTP, because they are all protocols used to automate behavior on trunk links. PAgP and LACP are used for link aggregation (EtherChannel). DTP is used for automating the creation of trunk links. When an EtherChannel trunk is configured, typically EtherChannel (PAgP or LACP) is configured first and then DTP.

Troubleshoot EtherChannel Example (6.3.3)

In Figure 6-10, interfaces F0/1 and F0/2 on switches S1 and S2 are connected with an EtherChannel. However, the EtherChannel is not operational.

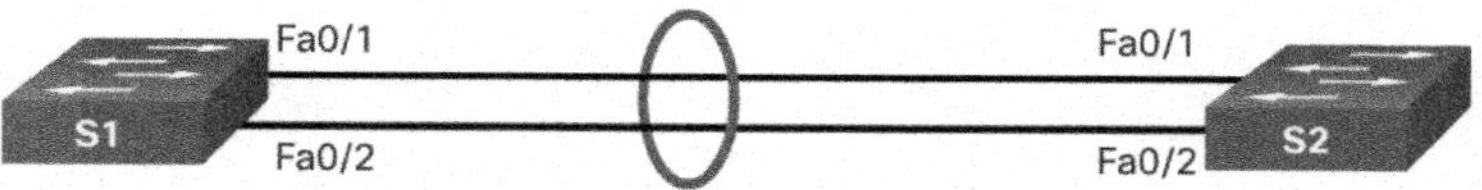

Figure 6-10 EtherChannel Topology

Use the following steps to troubleshoot the EtherChannel.

How To

Step 1. View the EtherChannel Summary Information

The output of the **show etherchannel summary** command in Example 6-6 indicates that the EtherChannel is down.

Example 6-6 Checking the EtherChannel Status

```
S1# show etherchannel summary
Flags:  D - down        P - bundled in port-channel
        I - stand-alone s - suspended
        H - Hot-standby (LACP only)
        R - Layer3      S - Layer2
        U - in use      N - not in use, no aggregation
        f - failed to allocate aggregator
        M - not in use, minimum links not met
        m - not in use, port not aggregated due to minimum links not met
        u - unsuitable for bundling
        w - waiting to be aggregated
        d - default port
        A - formed by Auto LAG
Number of channel-groups in use: 1
Number of aggregators:           1
Group  Port-channel  Protocol    Ports
------+-------------+-----------+-----------------------------------------------
1      Po1(SD)           -        Fa0/1(D)    Fa0/2(D)
```

Step 2. View Port Channel Configuration

In the **show run | begin interface port-channel** output in Example 6-7, more detailed output indicates that there are incompatible PAgP modes configured on S1 and S2.

Example 6-7 Piping **show run** to Check the EtherChannel Configuration

```
S1# show run | begin interface port-channel
interface Port-channel1
 switchport trunk allowed vlan 1,2,20
 switchport mode trunk
!
interface FastEthernet0/1
 switchport trunk allowed vlan 1,2,20
 switchport mode trunk
 channel-group 1 mode on
!
interface FastEthernet0/2
 switchport trunk allowed vlan 1,2,20
 switchport mode trunk
 channel-group 1 mode on
!=====================================
```

```
S2# show run | begin interface port-channel
interface Port-channel1
 switchport trunk allowed vlan 1,2,20
 switchport mode trunk
!
interface FastEthernet0/1
 switchport trunk allowed vlan 1,2,20
 switchport mode trunk
 channel-group 1 mode desirable
!
interface FastEthernet0/2
 switchport trunk allowed vlan 1,2,20
 switchport mode trunk
 channel-group 1 mode desirable
```

Step 3. Correct the Misconfiguration

To correct the issue, the PAgP mode on the EtherChannel is changed to desirable, as shown in Example 6-8.

Example 6-8 Correcting the PAgP Mode

```
S1(config)# no interface port-channel 1
S1(config)# interface range fa0/1 - 2
S1(config-if-range)# channel-group 1 mode desirable
Creating a port-channel interface Port-channel 1
S1(config-if-range)# no shutdown
S1(config-if-range)# exit
S1(config)# interface range fa0/1 - 2
S1(config-if-range)# channel-group 1 mode desirable
S1(config-if-range)# no shutdown
S1(config-if-range)# interface port-channel 1
S1(config-if)# switchport mode trunk
S1(config-if)# end
S1#
```

Note

EtherChannel and STP must interoperate. For this reason, the order in which EtherChannel-related commands are entered is important, which is why you see interface Port-Channel 1 removed and then re-added with the **channel-group** command, in contrast to directly changed. If one tries to change the configuration directly, STP errors cause the associated ports to go into blocking or errdisabled state.

Step 4. Verify EtherChannel Is Operational

The EtherChannel is now active, as verified by the output of the **show etherchannel summary** command in Example 6-9.

Example 6-9 Verifying the EtherChannel Is Now Operational

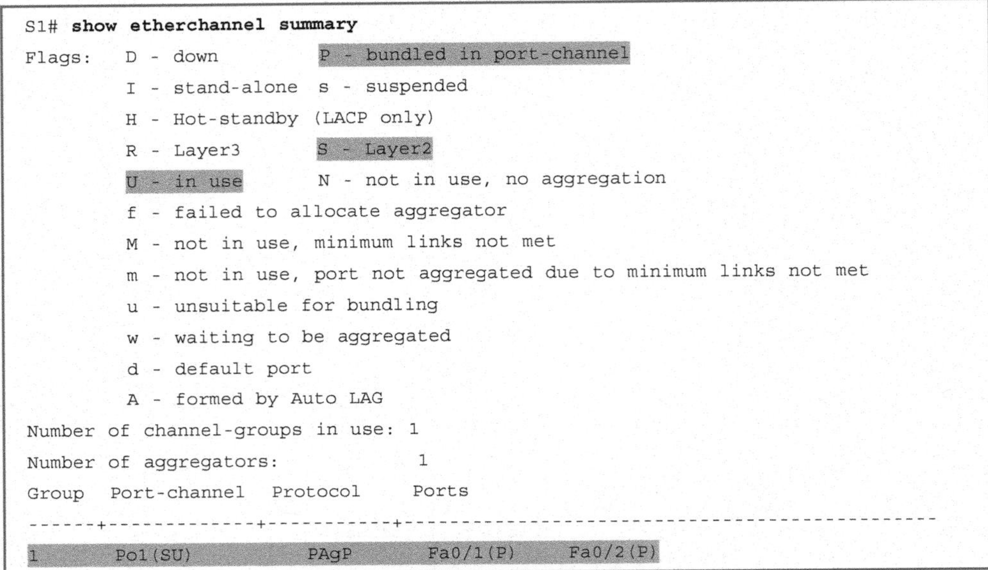

```
S1# show etherchannel summary
Flags:  D - down          P - bundled in port-channel
        I - stand-alone   s - suspended
        H - Hot-standby   (LACP only)
        R - Layer3        S - Layer2
        U - in use        N - not in use, no aggregation
        f - failed to allocate aggregator
        M - not in use, minimum links not met
        m - not in use, port not aggregated due to minimum links not met
        u - unsuitable for bundling
        w - waiting to be aggregated
        d - default port
        A - formed by Auto LAG
Number of channel-groups in use: 1
Number of aggregators:           1
Group  Port-channel  Protocol    Ports
------+-------------+-----------+-----------------------------------------------
1      Po1(SU)         PAgP      Fa0/1(P)     Fa0/2(P)
```

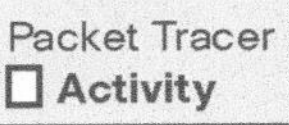

Packet Tracer—Troubleshoot EtherChannel (6.3.4)

Four switches were recently configured by a junior technician. Users are complaining that the network is running slowly and would like you to investigate.

Summary (6.4)

The following is a summary of each topic in the module:

EtherChannel Operation

To increase bandwidth or redundancy, multiple links could be connected between devices. However, STP will block redundant links to prevent switching loops. EtherChannel is a link aggregation technology that allows redundant links between devices that will not be blocked by STP. EtherChannel groups multiple physical Ethernet links together into one single logical link. It provides fault tolerance, load sharing, increased bandwidth, and redundancy between switches, routers, and servers. When an EtherChannel is configured, the resulting virtual interface is called a port channel. EtherChannel has several advantages, as well as some restrictions to implementation. EtherChannels can be formed through negotiation using one of two protocols, PAgP or LACP. These protocols allow ports with similar characteristics to form a channel through dynamic negotiation with adjoining switches. When an EtherChannel link is configured using Cisco-proprietary PAgP, PAgP packets are sent between EtherChannel-capable ports to negotiate the forming of a channel. Modes for PAgP are On, PAgP desirable, and PAgP auto. LACP performs a function similar to PAgP with Cisco EtherChannel. Because LACP is an IEEE standard, it can be used to facilitate EtherChannels in multivendor environments. Modes for LACP are On, LACP active, and LACP passive.

Configure EtherChannel

The following guidelines and restrictions are useful for configuring EtherChannel:

- **EtherChannel support:** All Ethernet interfaces on all modules must support EtherChannel with no requirement that interfaces be physically contiguous, or on the same module.
- **Speed and duplex:** Configure all interfaces in an EtherChannel to operate at the same speed and in the same duplex mode.
- **VLAN match:** All interfaces in the EtherChannel bundle must be assigned to the same VLAN or be configured as a trunk.
- **Range of VLANs:** An EtherChannel supports the same allowed range of VLANs on all the interfaces in a trunking EtherChannel.

Configuring EtherChannel with LACP requires three steps:

Step 1. Specify the interfaces that compose the EtherChannel group using the **interface range** *interface* global configuration mode command.

Step 2. Create the port channel interface with the **channel-group** *identifier* **mode active** command in interface range configuration mode.

Step 3. To change Layer 2 settings on the port channel interface, enter port channel interface configuration mode using the **interface port-channel** command, followed by the interface identifier.

Verify and Troubleshoot EtherChannel

There are a number of commands to verify an EtherChannel configuration, including **show interfaces port-channel**, **show etherchannel summary**, **show etherchannel port-channel**, and **show interfaces etherchannel**. Common EtherChannel issues include the following:

- Assigned ports in the EtherChannel are not part of the same VLAN, or not configured as trunks. Ports with different native VLANs cannot form an EtherChannel.
- Trunking was configured on some of the ports that make up the EtherChannel, but not all of them.
- If the allowed range of VLANs is not the same, the ports do not form an EtherChannel even when PAgP is set to the auto or desirable mode.
- The dynamic negotiation options for PAgP and LACP are not compatibly configured on both ends of the EtherChannel.

Packet Tracer—Implement EtherChannel (6.4.1)

You have been tasked with designing an EtherChannel implementation for a company that wants to improve the performance of the switch trunk links. You will try several ways of implementing the EtherChannel links to evaluate which is the best for the company. You will build the topology, configure trunk ports, and implement LACP and PAgP EtherChannels.

Lab—Implement EtherChannel (6.4.2)

In this lab, you complete the following objectives:

Part 1: Build the Network and Configure Basic Device Settings

Part 2: Create VLANs and Assign Switch Ports

Part 3: Configure 802.1Q Trunks Between the Switches

Part 4: Implement and Verify an EtherChannel Between the Switches

Practice

The following activities provide practice with the topics introduced in this chapter. The Labs are available in the companion *Switching, Routing, and Wireless Essentials Labs and Study Guide (CCNAv7)* (ISBN 9780136634386). The Packet Tracer Activity instructions are also in the Labs & Study Guide. The PKA files are found in the online course.

Lab

Lab 6.4.2: Implement EtherChannel

Packet Tracer Activities

Packet Tracer 6.2.4: Configure EtherChannel

Packet Tracer 6.3.4: Troubleshoot EtherChannel

Packet Tracer 6.4.1: Implement EtherChannel

Check Your Understanding Questions

Complete all the review questions listed here to test your understanding of the topics and concepts in this chapter. The appendix "Answers to the 'Check Your Understanding' Questions" lists the answers.

1. There has been an increase in network traffic between two Catalyst 2960 switches, and their FastEthernet trunk link has reached its capacity. How can traffic flow be improved?

   A. Add routers between the switches to create additional broadcast domains.

   B. Bundle physical ports using EtherChannel.

   C. Configure smaller VLANs to decrease the size of the collision domain.

   D. Increase the speed of the ports using the **bandwidth** command.

2. Which two load-balancing methods can be implemented with EtherChannel technology? (Choose two.)

   A. Destination IP to destination MAC
   B. Destination IP to source IP
   C. Destination MAC to destination IP
   D. Destination MAC to source MAC
   E. Source IP and destination IP
   F. Source MAC and destination MAC

3. Which statement is true regarding the use of PAgP to create EtherChannels?

   A. It increases the number of ports that are participating in spanning tree.
   B. It is Cisco proprietary.
   C. It mandates that an even number of ports (2, 4, 6, etc.) be used for aggregation.
   D. It requires full duplex.
   E. It requires more physical links than LACP does.

4. Which two protocols are link aggregation protocols? (Choose two.)

   A. 802.3ad
   B. EtherChannel
   C. PAgP
   D. RSTP
   E. STP

5. Which combination of channel-group modes will establish an EtherChannel?

   A. Switch 1 set to auto; switch 2 set to auto.
   B. Switch 1 set to auto; switch 2 set to on.
   C. Switch 1 set to desirable; switch 2 set to desirable.
   D. Switch 1 set to on; switch 2 set to desirable.

6. Which interface configuration command will enable the port to initiate an LACP EtherChannel?

   A. **channel-group mode active**
   B. **channel-group mode auto**
   C. **channel-group mode desirable**
   D. **channel-group mode on**
   E. **channel-group mode passive**

7. Which interface configuration command will enable the port to establish an EtherChannel only if it receives PAgP packets from the other switch?

   A. **channel-group mode active**

   B. **channel-group mode auto**

   C. **channel-group mode desirable**

   D. **channel-group mode on**

   E. **channel-group mode passive**

8. Which statement describes a characteristic of EtherChannel?

   A. It can combine up to a maximum of 4 physical links.

   B. It can bundle mixed types of 100 Mbps and 1 Gbps Ethernet links.

   C. It consists of multiple parallel links between a switch and a router.

   D. It is made by combining multiple physical links that are seen as one link between two switches.

9. What are two advantages of using LACP? (Choose two.)

   A. LACP allows automatic formation of EtherChannel links.

   B. LACP allows use of multivendor devices.

   C. LACP decreases the amount of configuration that is needed on a switch for EtherChannel.

   D. LACP eliminates the need for the Spanning Tree Protocol.

   E. LACP increases redundancy to Layer 3 devices.

   F. LACP provides a simulated environment for testing link aggregation.

10. Which three settings must match in order for switch ports to form an EtherChannel? (Choose three.)

   A. Non-trunk ports must belong to the same VLAN.

   B. Port security violation settings on interconnecting ports must match.

   C. The duplex settings on interconnecting ports must match.

   D. The port channel group number on interconnecting switches must match.

   E. The SNMP community strings must match.

   F. The speed settings on interconnecting ports must match.

CHAPTER 7

DHCPv4

Objectives

Upon completion of this chapter, you will be able to answer the following questions:

- How does DHCPv4 operate in a small- to medium-sized business network?
- How do you configure a router as a DHCPv4 server?
- How do you configure a router as a DHCPv4 client?

Key Terms

This chapter uses the following key terms. You can find the definitions in the Glossary.

Dynamic Host Configuration Protocol (DHCP) *Page 200*

DHCPv4 *Page 200*

DHCPv4 server *Page 200*

lease *Page 200*

DHCPv4 client *Page 201*

DHCP Discover (DHCPDISCOVER) *Page 201*

DHCP Offer (DHCPOFFER) *Page 201*

DHCP Request (DHCPREQUEST) *Page 201*

DHCP Acknowledgment (DHCPACK) *Page 201*

Cisco IOS helper address *Page 212*

DHCPv4 relay agent *Page 212*

Introduction (7.0)

The *Dynamic Host Configuration Protocol (DHCP)* dynamically assigns IPv4 addresses to devices. *DHCPv4* is for an IPv4 network. (Don't worry, you'll learn about DHCPv6 in another module.) This means that you, the network administrator, do not have to spend your day configuring IPv4 addresses for every device on your network. In a small home or office, that would not be very difficult, but any large network might have hundreds, or even thousands of devices.

In this module, you learn how to configure a Cisco IOS router to be a DHCPv4 server. Then you learn how to configure a Cisco IOS router as a client. This module includes a few Syntax Checkers and a Packet Tracer activity to help you try out your new knowledge. DHCPv4 configuration skills will significantly reduce your workload, and who doesn't want that?

DHCPv4 Concepts (7.1)

All hosts in a network require an IPv4 configuration. Although some devices will have their IPv4 configuration statically assigned, most devices will use DHCP to acquire a valid IPv4 configuration. Therefore, DHCP is a vital feature that must be managed and carefully implemented.

In this section, you learn how to implement DHCPv4 to operate across multiple LANs in a small- to medium-sized business network.

DHCPv4 Server and Client (7.1.1)

Dynamic Host Configuration Protocol v4 (DHCPv4) assigns IPv4 addresses and other network configuration information dynamically. Because desktop clients typically make up the bulk of network nodes, DHCPv4 is an extremely useful and timesaving tool for network administrators.

A dedicated *DHCPv4 server* is scalable and relatively easy to manage. However, in a small branch or small office or home office (SOHO) location, a Cisco router can be configured to provide DHCPv4 services without the need for a dedicated server. Cisco IOS software supports an optional, full-featured DHCPv4 server.

The DHCPv4 server dynamically assigns, or leases, an IPv4 address from a pool of addresses for a limited period of time chosen by the server, or until the client no longer needs the address.

Clients *lease* the information from the server for an administratively defined period. Administrators configure DHCPv4 servers to set the leases to time out at different intervals. The lease is typically anywhere from 24 hours to a week or more. When the lease expires, the client must ask for another address, although the client is typically reassigned the same address.

In Figure 7-1, the *DHCPv4 client* requests DHCPv4 services. The DHCPv4 server responds with network configuration information.

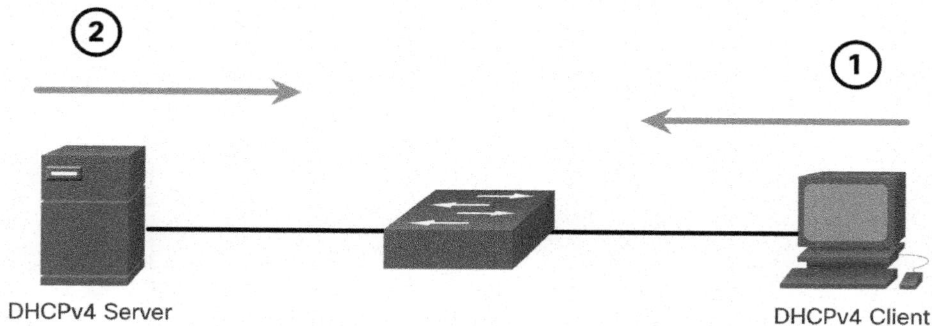

1. The DHCPv4 lease process begins with the client sending a message requesting the services of a DHCP server.
2. If there is a DHCPv4 server that receives the message, it will respond with an IPv4 address and possible other network configuration information.

Figure 7-1 DHCPv4 Server and Client Topology

DHCPv4 Operation (7.1.2)

DHCPv4 works in a client/server mode. When a client communicates with a DHCPv4 server, the server assigns or leases an IPv4 address to that client. The client connects to the network with that leased IPv4 address until the lease expires. The client must contact the DHCP server periodically to extend the lease. This lease mechanism ensures that clients that move or power off do not keep addresses that they no longer need. When a lease expires, the DHCP server returns the address to the pool, where it can be reallocated as necessary.

Steps to Obtain a Lease (7.1.3)

When the client boots (or otherwise wants to join a network), it begins a four-step process to obtain a lease:

Step 1. *DHCP Discover (DHCPDISCOVER)*

Step 2. *DHCP Offer (DHCPOFFER)*

Step 3. *DHCP Request (DHCPREQUEST)*

Step 4. *DHCP Acknowledgment (DHCPACK)*

This four-step process is summarized in Figure 7-2.

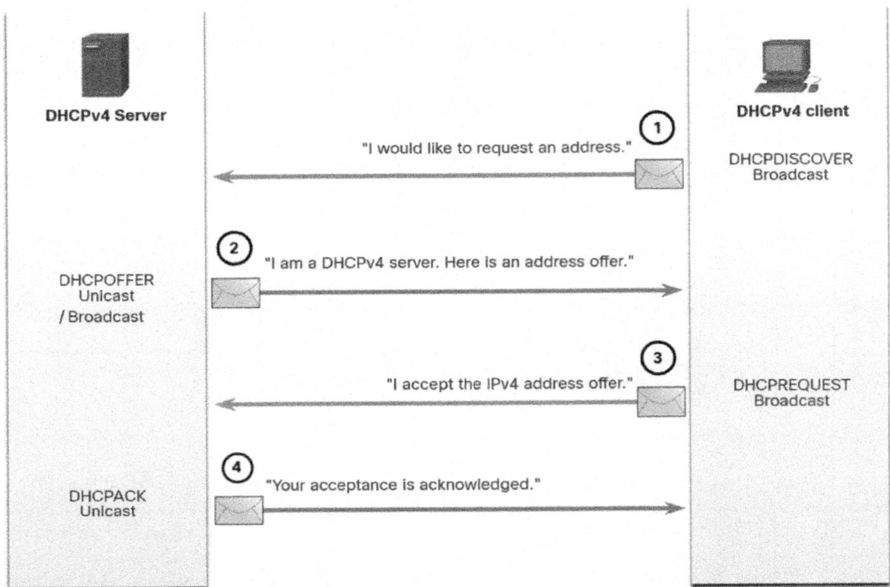

Figure 7-2 Four Steps to Obtain a Lease

Step 1. DHCP Discover (DHCPDISCOVER). The client starts the process using a broadcast DHCPDISCOVER message with its own MAC address to discover available DHCPv4 servers. Because the client has no valid IPv4 information at boot, it uses Layer 2 and Layer 3 broadcast addresses to communicate with the server. The purpose of the DHCPDISCOVER message is to find DHCPv4 servers on the network.

Step 2. DHCP Offer (DHCPOFFER). When the DHCPv4 server receives a DHCP-DISCOVER message, it reserves an available IPv4 address to lease to the client. The server also creates an ARP entry consisting of the MAC address of the requesting client and the leased IPv4 address of the client. The DHCPv4 server sends the binding DHCPOFFER message to the requesting client. The DHCPOFFER message can be sent as a broadcast or a unicast message.

Step 3. DHCP Request (DHCPREQUEST). When the client receives the DHCPOFFER from the server, it sends back a DHCPREQUEST message. This message is used for both lease origination and lease renewal. When used for lease origination, the DHCPREQUEST serves as a binding acceptance notice to the selected server for the parameters it has offered and an implicit decline to any other servers that may have provided the client a binding offer.

Many enterprise networks use multiple DHCPv4 servers. The DHCPREQUEST message is sent in the form of a broadcast to inform this DHCPv4 server and any other DHCPv4 servers about the accepted offer.

Step 4. DHCP Acknowledgment (DHCPACK). On receiving the DHCPREQUEST message, the server may verify the lease information with an Internet Control Message Protocol (ICMP) ping to that address to ensure it is not being used already, it will create a new address resolution protocol (ARP) entry for the client lease, and reply with a DHCPACK message. The DHCPACK message is a duplicate of the DHCPOFFER, except for a change in the message type field. When the client receives the DHCPACK message, it logs the configuration information and may perform an ARP lookup for the assigned address. If there is no reply to the ARP, the client knows that the IPv4 address is valid and starts using it as its own.

Steps to Renew a Lease (7.1.4)

Prior to lease expiration, the client begins a two-step process to renew the lease with the DHCPv4 server, as shown in Figure 7-3.

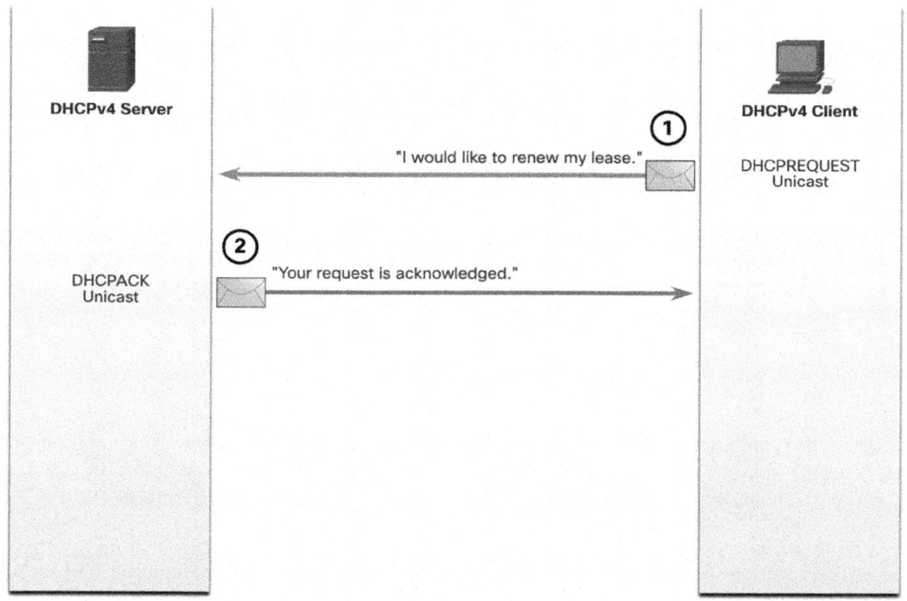

Figure 7-3 Two Steps to Renew a Lease

Step 1. DHCP Request (DHCPREQUEST)

Before the lease expires, the client sends a DHCPREQUEST message directly to the DHCPv4 server that originally offered the IPv4 address.

If a DHCPACK is not received within a specified amount of time, the client broadcasts another DHCPREQUEST so that one of the other DHCPv4 servers can extend the lease.

Step 2. DHCP Acknowledgment (DHCPACK)

On receiving the DHCPREQUEST message, the server verifies the lease information by returning a DHCPACK.

Note

These messages (primarily the DHCPOFFER and DHCPACK) can be sent as unicast or broadcast according to IETF RFC 2131.

Interactive Graphic

Check Your Understanding—DHCPv4 Concepts (7.1.5)

Refer to the online course to complete this activity.

Configure a Cisco IOS DHCPv4 Server (7.2)

In this section, you learn how to configure a router as a DHCPv4 server.

Cisco IOS DHCPv4 Server (7.2.1)

Now you have a basic understanding of how DHCPv4 works and how it can make your job a bit easier. If you do not have a separate DHCPv4 server, this section shows you how to configure a Cisco IOS router to act as one. A Cisco router running Cisco IOS software can be configured to act as a DHCPv4 server, as shown in Figure 7-4. The Cisco IOS DHCPv4 server assigns and manages IPv4 addresses from specified address pools within the router to DHCPv4 clients.

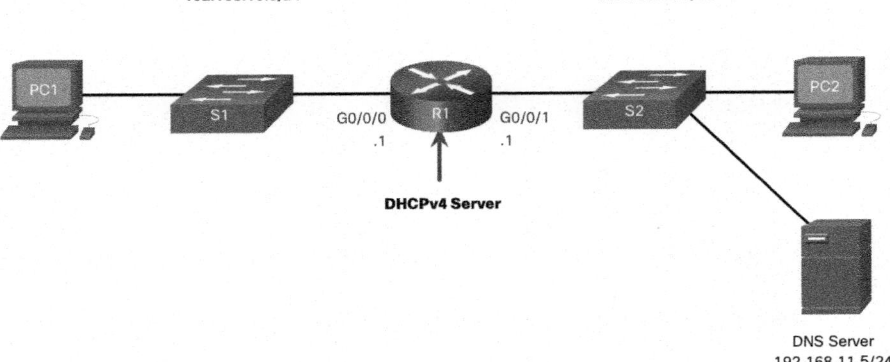

Figure 7-4 Cisco IOS DHCPv4 Server Topology

Steps to Configure a Cisco IOS DHCPv4 Server (7.2.2)

Use the following steps to configure a Cisco IOS DHCPv4 server.

Step 1. Exclude IPv4 Addresses

The router functioning as the DHCPv4 server assigns all IPv4 addresses in a DHCPv4 address pool unless it is configured to exclude specific addresses. Typically, some IPv4 addresses in a pool are assigned to network devices that require static address assignments. Therefore, these IPv4 addresses should not be assigned to other devices. The command syntax to exclude IPv4 addresses is the following:

```
Router(config)# ip dhcp excluded-address low-address [high-address]
```

A single address or a range of addresses can be excluded by specifying the *low-address* and *high-address* of the range. Excluded addresses should be those addresses that are assigned to routers, servers, printers, and other devices that have been, or will be, manually configured. You can also enter the command multiple times.

Step 2. Define a DHCPv4 Pool Name

Configuring a DHCPv4 server involves defining a pool of addresses to assign.

As shown in the example, the **ip dhcp pool** *pool-name* command creates a pool with the specified name and puts the router in DHCPv4 configuration mode, which is identified by the prompt Router(dhcp-config)#.

The command syntax to define the pool is the following:

```
Router(config)# ip dhcp pool pool-name
Router(dhcp-config)#
```

Step 3. Configure the DHCPv4 Pool

Table 7-1 lists the tasks to complete the DHCPv4 pool configuration.

Table 7-1 DHCPv4 Configuration Commands

Task	IOS Command
Define the address pool.	**network** *network-number* [*mask* | */prefix-length*]
Define the default router or gateway.	**default-router** *address* [*address2…address8*]
Define a DNS server.	**dns-server** *address* [*address2…address8*]
Define the domain name.	**domain-name** *domain*

Task	IOS Command
Define the duration of the DHCP lease.	**lease** {*days* [*hours* [*minutes*]] | **infinite**}
Define the NetBIOS WINS server.	**netbios-name-server** *address* [*address2*… *address8*]

The address pool and default gateway router must be configured. Use the **network** statement to define the range of available addresses. Use the **default-router** command to define the default gateway router. Typically, the gateway is the LAN interface of the router closest to the client devices. One gateway is required, but you can list up to eight addresses if there are multiple gateways.

Other DHCPv4 pool commands are optional. For example, the IPv4 address of the DNS server that is available to a DHCPv4 client is configured using the **dns-server** command. The **domain-name** command is used to define the domain name. The duration of the DHCPv4 lease can be changed using the **lease** command. The default lease value is one day. The **netbios-name-server** command is used to define the NetBIOS WINS server.

Note

Microsoft recommends using Domain Name System (DNS) for Windows name resolution instead of deploying WINS.

Configuration Example (7.2.3)

The topology for the example configuration is shown in Figure 7-5.

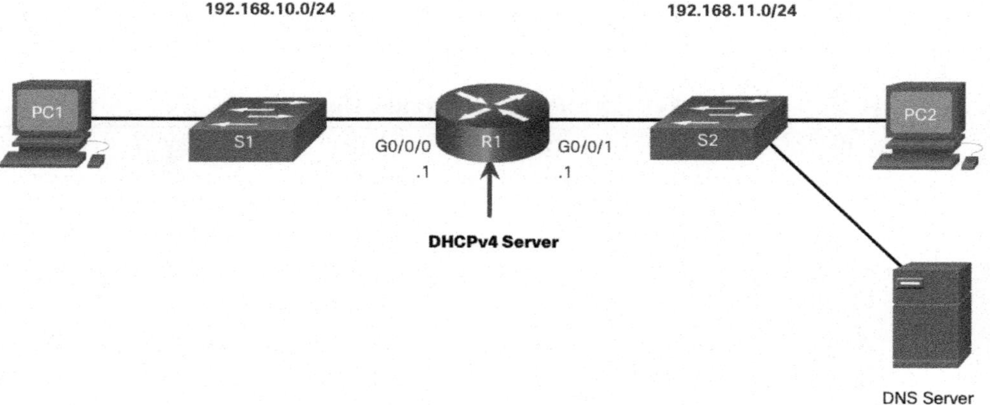

Figure 7-5 Cisco IOS DHCPv4 Server Topology

Example 7-1 shows the configuration to make R1 a DHCPv4 server for the 192.168.10.0/24 LAN.

Example 7-1 Cisco IOS DHCPv4 Server Configuration for R1

```
R1(config)# ip dhcp excluded-address 192.168.10.1 192.168.10.9
R1(config)# ip dhcp excluded-address 192.168.10.254
R1(config)# ip dhcp pool LAN-POOL-1
R1(dhcp-config)# network 192.168.10.0 255.255.255.0
R1(dhcp-config)# default-router 192.168.10.1
R1(dhcp-config)# dns-server 192.168.11.5
R1(dhcp-config)# domain-name example.com
R1(dhcp-config)# end
R1#
```

DHCPv4 Verification Commands (7.2.4)

Use the commands in Table 7-2 to verify that the Cisco IOS DHCPv4 server is operational.

Table 7-2 DHCPv4 Verification Commands

Command	Description
`show running-config | section dhcp`	Displays the DHCPv4 commands configured on the router.
`show ip dhcp binding`	Displays a list of all IPv4 address to MAC address bindings provided by the DHCPv4 service.
`show ip dhcp server statistics`	Displays count information regarding the number of DHCPv4 messages that have been sent and received.

Verify DHCPv4 is Operational (7.2.5)

R1 in Figure 7-5 has been configured to provide DHCPv4 services. PC1 has not been powered up and, therefore, does not have an IPv4 address.

The output for the following commands assumes PC1 has received its IPv4 addressing information from the DHCPv4 server. You may need to enter **ipconfig /renew** on a Windows PC to force it to send out a DHCPDISCOVER message.

You can use several commands to verify DHCPv4 is operational.

Verify the DHCPv4 Configuration

As shown in Example 7-2, the **show running-config | section dhcp** command output displays the DHCPv4 commands configured on R1. The **| section dhcp** parameter displays only the commands associated with DHCPv4 configuration.

Example 7-2 Verify the DHCPv4 Configuration

```
R1# show running-config | section dhcp
ip dhcp excluded-address 192.168.10.1 192.168.10.9
ip dhcp excluded-address 192.168.10.254
ip dhcp pool LAN-POOL-1
 network 192.168.10.0 255.255.255.0
 default-router 192.168.10.1
 dns-server 192.168.11.5
 domain-name example.com
```

Verify DHCPv4 Bindings

As shown in Example 7-3, the operation of DHCPv4 can be verified using the **show ip dhcp binding** command. This command displays a list of all IPv4 address to MAC address bindings that have been provided by the DHCPv4 service.

Example 7-3 Verify the DHCPv4 Bindings

```
R1# show ip dhcp binding
Bindings from all pools not associated with VRF:
IP address     Client-ID/          Lease expiration       Type        State   Interface
               Hardware address/
               User name
192.168.10.10  0100.5056.b3ed.d8   Sep 15 2019 8:42 AM  Automatic  Active
  GigabitEthernet0/0/0
```

Verify DHCPv4 Statistics

The output of the **show ip dhcp server statistics** command is used to verify that messages are being received or sent by the router, as shown in Example 7-4. This command displays count information regarding the number of DHCPv4 messages that have been sent and received.

Example 7-4 Verify DHCPv4 Statistics

```
R1# show ip dhcp server statistics
Memory usage          19465
Address pools         1
Database agents       0
Automatic bindings    2
Manual bindings       0
Expired bindings      0
Malformed messages    0
```

```
Secure arp entries    0
Renew messages        0
Workspace timeouts    0
Static routes         0
Relay bindings        0
Relay bindings active        0
Relay bindings terminated    0
Relay bindings selecting     0
Message               Received
BOOTREQUEST           0
DHCPDISCOVER          4
DHCPREQUEST           2
DHCPDECLINE           0
DHCPRELEASE           0
DHCPINFORM            0
```

Verify DHCPv4 Client Received IPv4 Addressing

The **ipconfig /all** command, when issued on PC1, displays the TCP/IP parameters, as shown in the example. Because PC1 was connected to the network segment 192.168.10.0/24, it automatically received a DNS suffix, IPv4 address, subnet mask, default gateway, and DNS server address from that pool. No DHCP-specific router interface configuration is required. If a PC is connected to a network segment that has a DHCPv4 pool available, the PC can obtain an IPv4 address from the appropriate pool automatically.

Example 7-5 Verify the DHCPv4 Client Configuration

```
C:\Users\Student> ipconfig /all
Windows IP Configuration
   Host Name . . . . . . . . . . . . . : ciscolab
   Primary Dns Suffix  . . . . . . . . :
   Node Type . . . . . . . . . . . . . : Hybrid
   IP Routing Enabled. . . . . . . . . : No
   WINS Proxy Enabled. . . . . . . . . : No
Ethernet adapter Ethernet0:
   Connection-specific DNS Suffix  . : example.com
   Description . . . . . . . . . . . : Realtek PCIe GBE Family Controller
   Physical Address. . . . . . . . . : 00-05-9A-3C-7A-00
   DHCP Enabled. . . . . . . . . . . : Yes
   Autoconfiguration Enabled . . . . : Yes
   IPv4 Address. . . . . . . . . . . : 192.168.10.10
   Subnet Mask . . . . . . . . . . . : 255.255.255.0
```

```
Lease Obtained . . . . . . . . . . : Saturday, September 14, 2019 8:42:22AM
Lease Expires  . . . . . . . . . . : Sunday, September 15, 2019 8:42:22AM
Default Gateway . . . . . . . . .  : 192.168.10.1
DHCP Server . . . . . . . . . . .  : 192.168.10.1
DNS Servers . . . . . . . . . . .  : 192.168.11.5
```

Interactive Graphic

Syntax Checker—Configure DHCPv4 (7.2.6)

In this Syntax Checker activity, you configure R1 to be the DHCPv4 server for the 192.168.11.0/24 network.

Refer to the online course to complete this activity.

Disable the Cisco IOS DHCPv4 Server (7.2.7)

The DHCPv4 service is enabled by default. To disable the service, use the **no service dhcp** global configuration mode command. Use the **service dhcp** global configuration mode command to reenable the DHCPv4 server process, as shown in Example 7-6. Enabling the service has no effect if the parameters are not configured.

Example 7-6 Disable and Reenable DHCPv4 Service

```
R1(config)# no service dhcp
R1(config)# service dhcp
R1(config)#
```

Note

Clearing the DHCP bindings or stopping and restarting the DHCP service may result in duplicate IPv4 addresses being temporarily assigned on the network.

DHCPv4 Relay (7.2.8)

In a complex hierarchical network, enterprise servers are usually located centrally. These servers may provide DHCP, DNS, Trivial File Transfer Protocol (TFTP), and File Transfer Protocol (FTP) services for the network. Network clients are not typically on the same subnet as those servers. To locate the servers and receive services, clients often use broadcast messages.

In Figure 7-6, PC1 is attempting to acquire an IPv4 address from a DHCPv4 server using a broadcast message. In this scenario, R1 is not configured as a DHCPv4 server and does not forward the broadcast. Because the DHCPv4 server is located on a different network, PC1 cannot receive an IPv4 address using DHCP. R1 must be configured to relay DHCPv4 messages to the DHCPv4 server.

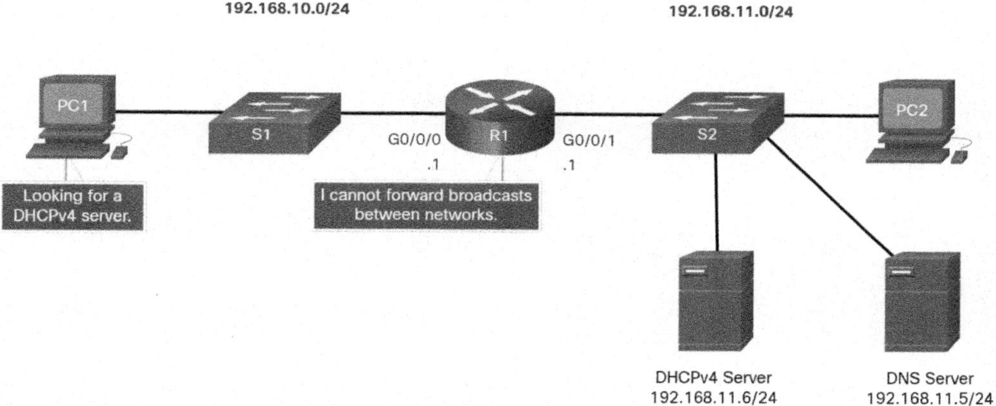

Figure 7-6 DHCPv4 Relay Topology

In this scenario, a network administrator is attempting to renew IPv4 addressing information for PC1. The following commands could be used by an administrator to resolve this issue.

The ipconfig /release Command

PC1 is a Windows computer. The network administrator releases all current IPv4 addressing information using the **ipconfig /release** command, as shown in Example 7-7. Notice that the IPv4 address is released and no address is shown.

Example 7-7 The **ipconfig /release** Command

```
C:\Users\Student> ipconfig /release
Windows IP Configuration
Ethernet adapter Ethernet0:
   Connection-specific DNS Suffix  . :
   Default Gateway . . . . . . . . . :
```

The ipconfig /renew Command

Next, the network administrator attempts to renew the IPv4 addressing information with the **ipconfig /renew** command, as shown in Example 7-8. This command causes PC1 to broadcast a DHCPDISCOVER message. The output shows that PC1 is unable to locate the DHCPv4 server. Because routers do not forward broadcasts, the request is not successful.

The network administrator could add DHCPv4 servers on R1 for all subnets. However, this would create additional cost and administrative overhead.

Example 7-8 The **ipconfig /renew** Command

```
C:\Users\Student> ipconfig /renew
Windows IP Configuration
An error occurred while renewing interface Ethernet0 : unable to connect to your
  DHCP server. Request has timed out.
```

The ip helper-address Command

A better solution is to configure R1 with the *Cisco IOS helper address* using the **ip helper-address** *address* interface configuration command. This will enable R1 to relay the DHCP client broadcast messages to the DHCPv4 server.

As shown in Example 7-9, the interface on R1 receiving the broadcast from PC1 is configured to relay DHCPv4 address to the DHCPv4 server at 192.168.11.6.

Example 7-9 The **ip helper-address** Command

```
R1(config)# interface g0/0/0
R1(config-if)# ip helper-address 192.168.11.6
R1(config-if)# end
R1#
```

The show ip interface Command

When R1 has been configured as a *DHCPv4 relay agent*, it accepts broadcast requests for the DHCPv4 service and then forwards those requests as a unicast to the IPv4 address 192.168.11.6. The network administrator can use the **show ip interface** command to verify the configuration, as shown in Example 7-10.

Example 7-10 The **show ip interface** Command

```
R1# show ip interface g0/0/0
GigabitEthernet0/0/0 is up, line protocol is up
  Internet address is 192.168.10.1/24
  Broadcast address is 255.255.255.255
  Address determined by setup command
  MTU is 1500 bytes
  Helper address is 192.168.11.6
(output omitted)
```

The ipconfig /all Command

As shown in Example 7-11, PC1 is now able to acquire an IPv4 address from the DHCPv4 server as verified with the **ipconfig /all** command.

Example 7-11 The **ipconfig /all** Command

```
C:\Users\Student> ipconfig /all
Windows IP Configuration

Ethernet adapter Ethernet0:
    Connection-specific DNS Suffix  . : example.com
    IPv4 Address. . . . . . . . . . . : 192.168.10.10
    Subnet Mask . . . . . . . . . . . : 255.255.255.0
    Default Gateway . . . . . . . . . : 192.168.10.1
(output omitted)
```

Other Service Broadcasts Relayed (7.2.9)

DHCPv4 is not the only service that the router can be configured to relay. By default, the **ip helper-address** command forwards the following eight UDP services:

- Port 37: Time
- Port 49: TACACS
- Port 53: DNS
- Port 67: DHCP/BOOTP server
- Port 68: DHCP/BOOTP client
- Port 69: TFTP
- Port 137: NetBIOS name service
- Port 138: NetBIOS datagram service

Packet Tracer—Configure DHCPv4 (7.2.10)

In this Packet Tracer Activity, you complete the following objectives:

Part 1: Configure a Router as a DHCP Server

Part 2: Configure DHCP Relay

Part 3: Configure a Router as a DHCP Client

Part 4: Verify DHCP and Connectivity

Configure a DHCPv4 Client (7.3)

In this section, you learn how to configure a router as a DHCPv4 client.

Cisco Router as a DHCPv4 Client (7.3.1)

There are scenarios where you might have access to a DHCP server through your ISP. In these instances, you can configure a Cisco IOS router as a DHCPv4 client. This section guides you through this process.

Sometimes, Cisco routers in a small office or home office (SOHO) and branch sites have to be configured as DHCPv4 clients in a similar manner to client computers. The method used depends on the ISP. However, in its simplest configuration, the Ethernet interface is used to connect to a cable or Digital Subscriber Line (DSL) modem.

To configure an Ethernet interface as a DHCP client, use the **ip address dhcp** interface configuration mode command.

In Figure 7-7, assume that an ISP has been configured to provide select customers with IPv4 addresses from the 209.165.201.0/27 network range after the G0/0/1 interface is configured with the **ip address dhcp** command.

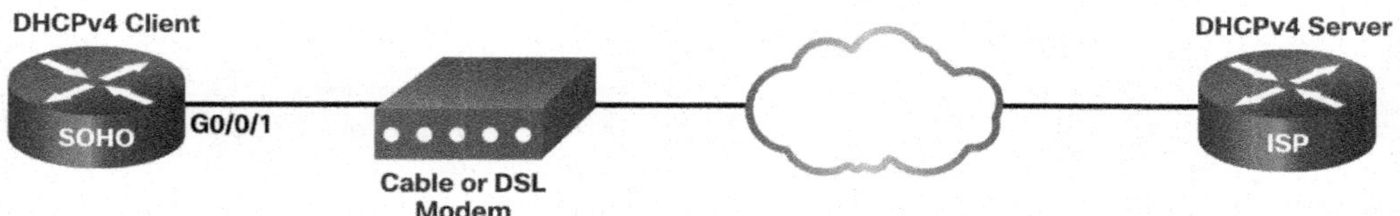

Figure 7-7 Topology with Router as a DHCPv4 Client

Configuration Example (7.3.2)

To configure an Ethernet interface as a DHCP client, use the **ip address dhcp** interface configuration mode command, as shown in Example 7-12. This configuration assumes that the ISP has been configured to provide select customers with IPv4 addressing information.

Example 7-12 Router as a DHCPv4 Client Configuration

```
SOHO(config)# interface G0/0/1
SOHO(config-if)# ip address dhcp
SOHO(config-if)# no shutdown
Sep 12 10:01:25.773: %DHCP-6-ADDRESS_ASSIGN: Interface GigabitEthernet0/0/1
  assigned DHCP address 209.165.201.12, mask 255.255.255.224, hostname SOHO
```

The **show ip interface g0/0/1** command confirms that the interface is up and that the address was allocated by a DHCPv4 server, as shown in Example 7-13.

Example 7-13 Verify the Router Received IPv4 Addressing Information

```
SOHO# show ip interface g0/0/1
GigabitEthernet0/0/1 is up, line protocol is up
  Internet address is 209.165.201.12/27
  Broadcast address is 255.255.255.255
  Address determined by DHCP
(output omitted)
```

Home Router as a DHCPv4 Client (7.3.3)

Home routers are typically already set to receive IPv4 addressing information automatically from the ISP. This is so that customers can easily set up the router and connect to the Internet.

For example, Figure 7-8 shows the default WAN setup page for a Packet Tracer wireless router. Notice that the Internet connection type is set to Automatic Configuration - DHCP. This selection is used when the router is connected to a DSL or cable modem and acts as a DHCPv4 client, requesting an IPv4 address from the ISP.

Various manufacturers of home routers will have a similar setup.

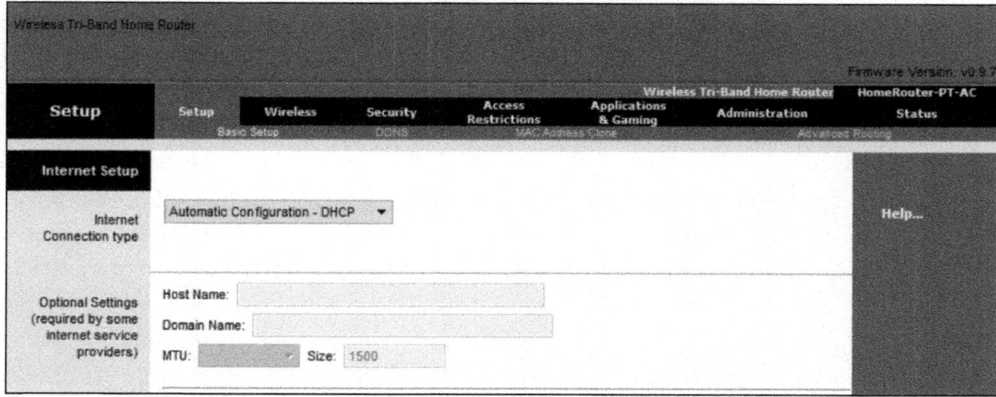

Figure 7-8 Configuring a Home Router as a DHCPv4 Client

Interactive Graphic

Syntax Checker—Configure a Cisco Router as DHCP Client (7.3.4)

In this Syntax Checker activity, you configure a Cisco router as DHCP client.

Refer to the online course to complete this activity.

Summary (7.4)

The following is a summary of each section in the module:

DHCPv4 Concepts

The DHCPv4 server dynamically assigns, or leases, an IPv4 address to a client from a pool of addresses for a limited period of time chosen by the server, or until the client no longer needs the address. The DHCPv4 lease process begins with the client sending a message requesting the services of a DHCP server. If there is a DHCPv4 server that receives the message, it will respond with an IPv4 address and possible other network configuration information. The client must contact the DHCP server periodically to extend the lease. This lease mechanism ensures that clients that move or power off do not keep addresses that they no longer need. When the client boots or otherwise wants to join the network, a four-step message exchange process is used between the client and the DHCPv4 server. Specifically, the DHCPDISCOVER, DHCPOFFER, DHCPREQUEST, and DHCPACK messages are used for the client to obtain its IP addressing information from the DHCPv4 server.

Configure a Cisco IOS DHCPv4 Server

A Cisco router running Cisco IOS software can be configured to act as a DHCPv4 server. Use the following steps to configure a Cisco IOS DHCPv4 server: exclude IPv4 addresses, define a DHCPv4 pool name, and configure the DHCPv4 pool. Verify your configuration using the **show running-config | section dhcp**, **show ip dhcp binding**, and **show ip dhcp server statistics** commands. The DHCPv4 service is enabled, by default. To disable the service, use the **no service dhcp** global configuration mode command. In a complex hierarchical network, enterprise servers are usually located centrally. These servers may provide DHCP, DNS, TFTP, and FTP services for the network. Network clients are not typically on the same subnet as those servers. To locate the servers and receive services, clients often use broadcast messages. A PC is attempting to acquire an IPv4 address from a DHCPv4 server using a broadcast message. If the router is not configured as a DHCPv4 server, it will not forward the broadcast. If the DHCPv4 server is located on a different network, the PC cannot receive an IPv4 address using DHCP. The router must be configured to relay DHCPv4 messages to the DHCPv4 server. The network administrator releases all current IPv4 addressing information using the **ipconfig /release** command. Next, the network administrator attempts to renew the IPv4 addressing information with the **ipconfig /renew** command. A better solution is to configure R1 with the **ip helper-address** *address* interface configuration command. The network administrator can use the **show ip interface** command to verify the configuration. The PC is now able to acquire an

IPv4 address from the DHCPv4 server as verified with the **ipconfig /all** command. By default, the **ip helper-address** command forwards the following eight UDP services:

- Port 37: Time
- Port 49: TACACS
- Port 53: DNS
- Port 67: DHCP/BOOTP server
- Port 68: DHCP/BOOTP client
- Port 69: TFTP
- Port 137: NetBIOS name service
- Port 138: NetBIOS datagram service

Configure a DHCPv4 Client

The Ethernet interface is used to connect to a cable or DSL modem. To configure an Ethernet interface as a DHCP client, use the **ip address dhcp** interface configuration mode command. Home routers are typically already set to receive IPv4 addressing information automatically from the ISP. The Internet connection type is set to Automatic Configuration - DHCP. This selection is used when the router is connected to a DSL or cable modem and acts as a DHCPv4 client, requesting an IPv4 address from the ISP.

Packet Tracer
☐ Activity

Packet Tracer—Implement DHCPv4 (7.4.1)

As the network technician for your company, you are tasked with configuring a Cisco router as a DHCP server to provide dynamic allocation of addresses to clients on the network. You are also required to configure the edge router as a DHCP client so that it receives an IP address from the ISP network. Because the server is centralized, you will need to configure the two LAN routers to relay DHCP traffic between the LANs and the router that is serving as the DHCP server.

Lab—Implement DHCPv4 (7.4.2)

In this lab, you complete the following objectives:

Part 1: Build the Network and Configure Basic Device Settings

Part 2: Configure and Verify Two DHCPv4 Servers on R1

Part 3: Configure and Verify a DHCP Relay on R2

Practice

The following activities provide practice with the topics introduced in this chapter. The Labs are available in the companion *Switching, Routing, and Wireless Essentials Labs and Study Guide (CCNAv7)* (ISBN 9780136634386). The Packet Tracer Activity instructions are also in the Labs & Study Guide. The PKA files are found in the online course.

Lab

Lab 7.4.2: Implement DHCPv4

Packet Tracer Activities

Packet Tracer 7.2.10: Configure DHCPv4

Packet Tracer 7.4.1: Implement DHCPv4

Check Your Understanding Questions

Complete all the review questions listed here to test your understanding of the sections and concepts in this chapter. The appendix "Answers to the 'Check Your Understanding' Questions" lists the answers.

1. Which DHCPv4 message will a client send to accept an IPv4 address that is offered by a DHCP server?

   A. Broadcast DHCPACK

   B. Broadcast DHCPREQUEST

   C. Unicast DHCPACK

   D. Unicast DHCPREQUEST

2. What is the reason why the DHCPREQUEST message is sent as a broadcast during the DHCPv4 process?

   A. For hosts on other subnets to receive the information

   B. For routers to fill their routing tables with this new information

   C. To notify other DHCP servers on the subnet that the IPv4 address was leased

   D. To notify other hosts not to request the same IPv4 address

3. Which address does a DHCPv4 server target when sending a DHCPOFFER message to a client that makes an address request?

   A. Broadcast MAC address
   B. Client MAC address
   C. Client IPv4 address
   D. Gateway IPv4 address

4. As a DHCPv4 client lease is about to expire, what is the message that the client sends the DHCP server?

   A. DHCPACK
   B. DHCPDISCOVER
   C. DHCPOFFER
   D. DHCPREQUEST

5. What is an advantage of configuring a Cisco router as a relay agent?

   A. It can forward both broadcast and multicast messages on behalf of clients.
   B. It can provide relay services for multiple UDP services.
   C. It reduces the response time from a DHCP server.
   D. It will allow DHCPDISCOVER messages to pass without alteration.

6. An administrator issues the **ip address dhcp** command on interface G0/0/1. What is the administrator trying to achieve?

   A. Configuring the router to act as a DHCPv4 server
   B. Configuring the router to act as a relay agent
   C. Configuring the router to obtain IPv4 parameters from a DHCPv4 server
   D. Configuring the router to resolve IPv4 address conflicts

7. Under which two circumstances would a router usually be configured as a DHCPv4 client? (Choose two.)

   A. The administrator needs the router to act as a relay agent.
   B. This is an ISP requirement.
   C. The router has a fixed IPv4 address.
   D. The router is intended to be used as a SOHO gateway.
   E. The router is meant to provide IPv4 addresses to the hosts.

8. A host on the 10.10.100.0/24 LAN is not being assigned an IPv4 address by an enterprise DHCP server with the address 10.10.200.10/24. What is the best way for the network engineer to resolve this problem?

   A. Issue the **default-router 10.10.200.10** command at the DHCP configuration prompt on the 10.10.100.0/24 LAN gateway router.

   B. Issue the **ip helper-address 10.10.100.0** command on the router interface that is the 10.10.200.0/24 gateway.

   C. Issue the **ip helper-address 10.10.200.10** command on the router interface that is the 10.10.100.0/24 gateway.

   D. Issue the **network 10.10.200.0 255.255.255.0** command at the DHCP configuration prompt on the 10.10.100.0/24 LAN gateway router.

9. What is accomplished by the **ip dhcp excluded-address 10.10.4.1 10.10.4.5** command?

   A. The DHCP server will ignore all traffic from clients with IPv4 addresses 10.10.4.1 to 10.10.4.5.

   B. The DHCP server will not issue IPv4 addresses ranging from 10.10.4.1 to 10.10.4.5.

   C. Traffic destined for 10.10.4.1 to 10.10.4.5 will be denied.

   D. Traffic from clients with IPv4 addresses 10.10.4.1 to 10.10.4.5 will be denied.

10. Which Windows command combination would enable a DHCPv4 client to reinstate its IPv4 configuration?

    A. Enter **ip config /release** and then **ip config /autonegotiate**

    B. Enter **ip config /release** and then **ip config /renew**

    C. Enter **ipconfig /release** and then **ipconfig /autonegotiate**

    D. Enter **ipconfig /release** and then **ipconfig /renew**

11. Which command issued on R1 can be used to verify the current IPv4 address and MAC address binding?

    A. R1# **show ip dhcp binding**

    B. R1# **show ip dhcp pool**

    C. R1# **show ip dhcp server statistics**

    D. R1# **show running-config | section dhcp**

12. Which DHCP operation statement is true?

    A. A DHCP client must wait for lease expiration before sending a new DHCPREQUEST message.

    B. If a DHCP client receives several DHCPOFFER messages from different servers, it sends a unicast DHCPACK message to the selected server.

    C. The DHCPDISCOVER message contains the IPv4 address and subnet mask to be assigned, the IPv4 address of the DNS server, and the IPv4 address of the default gateway.

    D. When a DHCP client boots, it broadcasts a DHCPDISCOVER message to identify an available DHCP server on the network.

CHAPTER 8

SLAAC and DHCPv6

Objectives

Upon completion of this chapter, you will be able to answer the following questions:

- How does an IPv6 host acquire its IPv6 configuration?
- How does SLAAC operate?
- How does DHCPv6 operate?
- How do you configure a stateful and stateless DHCPv6 server?

Key Terms

This chapter uses the following key terms. You can find the definitions in the Glossary.

Stateless Address Autoconfiguration (SLAAC) *Page 224*

DHCPv6 *Page 224*

DHCPv6 server *Page 224*

DHCPv6 client *Page 224*

DHCPv6 relay agent *Page 224*

global unicast addresses (GUA) *Page 224*

link-local addresses (LLAs) *Page 224*

ICMPv6 Router Advertisement (RA) message *Page 224*

Other Configuration flag (O flag) *Page 227*

stateless DHCPv6 server *Page 227*

Managed Address Configuration flag (M flag) *Page 227*

stateful DHCPv6 server *Page 227*

ICMPv6 Router Solicitation (RS) message *Page 229*

interface ID *Page 231*

Extended Unique Identifier method (EUI-64) *Page 231*

Duplicate Address Detection (DAD) *Page 234*

ICMPv6 Neighbor Solicitation (NS) message *Page 234*

solicited-node multicast address *Page 234*

ICMPv6 Neighbor Advertisement (NA) message *Page 234*

stateful DHCPv6 *Page 234*

DHCPv6 SOLICIT message *Page 235*

DHCPv6 ADVERTISE unicast message *Page 236*

Stateless DHCPv6 client *Page 236*

DHCPv6 INFORMATION-REQUEST message *Page 236*

Stateful DHCPv6 client *Page 236*

DHCPv6 REQUEST message *Page 236*

DHCPv6 REPLY unicast message *Page 236*

Introduction (8.0)

Stateless address autoconfiguration (SLAAC) and Dynamic Host Configuration Protocol version 6, *DHCPv6*, are dynamic addressing protocols for an IPv6 network. So, a little bit of configuring will make your day as a network administrator a lot easier. In this module, you learn how to use SLAAC to allow hosts to create their own IPv6 global unicast address, as well as configure a Cisco Internetwork Operating System (IOS) router to be a *DHCPv6 server*, a *DHCPv6 client*, or a *DHCPv6 relay agent*. This module includes a lab where you will configure DHCPv6 on real equipment!

IPv6 GUA Assignment (8.1)

In this section, you will learn about the operation of DHCPv6.

IPv6 Host Configuration (8.1.1)

First things first. To use either stateless address autoconfiguration (SLAAC) or DHCPv6, you should review IPv6 *global unicast addresses (GUAs)* and *link-local addresses (LLAs)*. This section covers both.

On a router, an IPv6 GUA is manually configured using the **ipv6 address** *ipv6-address/prefix-length* interface configuration command.

A Windows host can also be manually configured with an IPv6 GUA address configuration, as shown in Figure 8-1.

Manually entering an IPv6 GUA can be time consuming and somewhat error prone. Therefore, most Windows hosts are enabled to dynamically acquire an IPv6 GUA configuration, as shown in Figure 8-2.

IPv6 Host Link-Local Address (8.1.2)

When automatic IPv6 addressing is selected, the host will attempt to automatically obtain and configure IPv6 address information on the interface using Internet Control Message Protocol version 6 (ICMPv6) messages. Specifically, an IPv6 enabled router on the same link as the host sends out *ICMPv6 Router Advertisement (RA) messages* that suggest to hosts how they should obtain their IPv6 addressing information.

The IPv6 link-local address is automatically created by the host when it boots and the Ethernet interface is active. The Example 8-1 **ipconfig** output shows an automatically generated link-local address (LLA) on an interface.

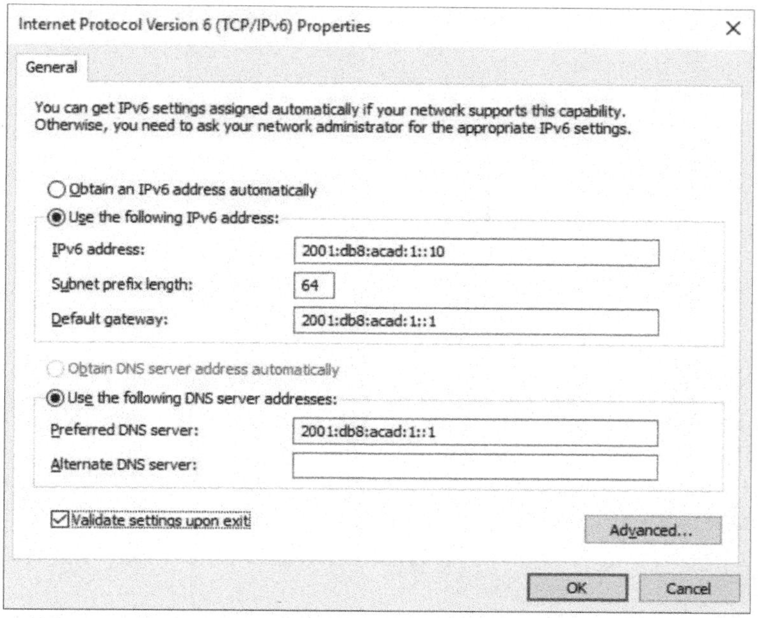

Figure 8-1 Manual Configuration of an IPv6 Windows Host

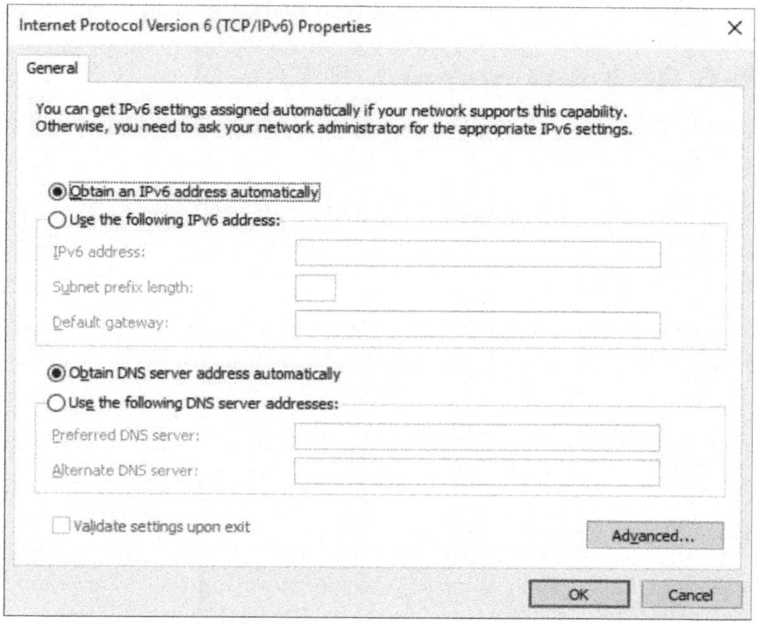

Figure 8-2 Automatic Configuration of an IPv6 Windows Host

Example 8-1 Viewing a Windows Host IPv6 Link-Local Address

```
C:\PC1> ipconfig
Windows IP Configuration
Ethernet adapter Ethernet0:
   Connection-specific DNS Suffix  . :
   IPv6 Address. . . . . . . . . . . :
   Link-local IPv6 Address . . . . . : fe80::fb:1d54:839f:f595%21
   IPv4 Address. . . . . . . . . . . : 169.254.202.140
   Subnet Mask . . . . . . . . . . . : 255.255.0.0
   Default Gateway . . . . . . . . . :
C:\PC1>
```

In Example 8-1, notice that the interface does not have an IPv6 GUA. The reason is because, in this example, the network segment does not have a router to provide network configuration instructions for the host, or the host has not been configured with a static IPv6 address.

Note

Host operating systems will at times show a link-local address appended with a "%" and a number. This is known as a Zone ID or Scope ID. It is used by the OS to associate the LLA with a specific interface.

IPv6 GUA Assignment (8.1.3)

IPv6 was designed to simplify how a host can acquire its IPv6 configuration. By default, an IPv6-enabled router advertises its IPv6 information. This allows a host to dynamically create or acquire its IPv6 configuration.

The IPv6 GUA can be assigned dynamically using stateless and stateful services, as shown Figure 8-3.

Three RA Message Flags (8.1.4)

The decision of how a client obtains an IPv6 GUA depends on the settings within the RA message.

An ICMPv6 RA message includes three flags to identify the dynamic options available to a host, as follows:

- **A flag:** This is the Address Autoconfiguration flag. Use Stateless Address Autoconfiguration (SLAAC) to create an IPv6 GUA.

- **O flag:** This is the *Other Configuration flag*. Other information is available from a *stateless DHCPv6 server*.
- **M flag:** This is the *Managed Address Configuration flag*. Use a *stateful DHCPv6 server* to obtain an IPv6 GUA.

All stateless and stateful methods in this module use ICMPv6 RA messages to suggest to the host how to create or acquire its IPv6 configuration. Although host operating systems follow the suggestion of the RA, the actual decision is ultimately up to the host.

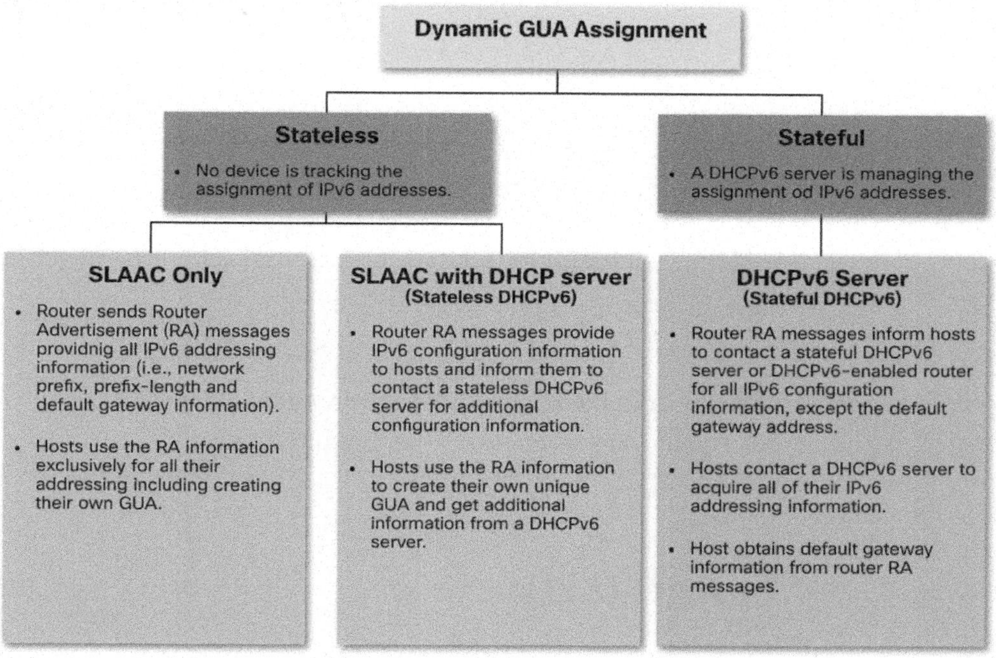

Figure 8-3 Methods for Dynamic IPv6 GUA Assignment

Using different combinations of the A, O, and M flags, RA messages inform the host about the dynamic options available.

Figure 8-4 illustrates these three methods.

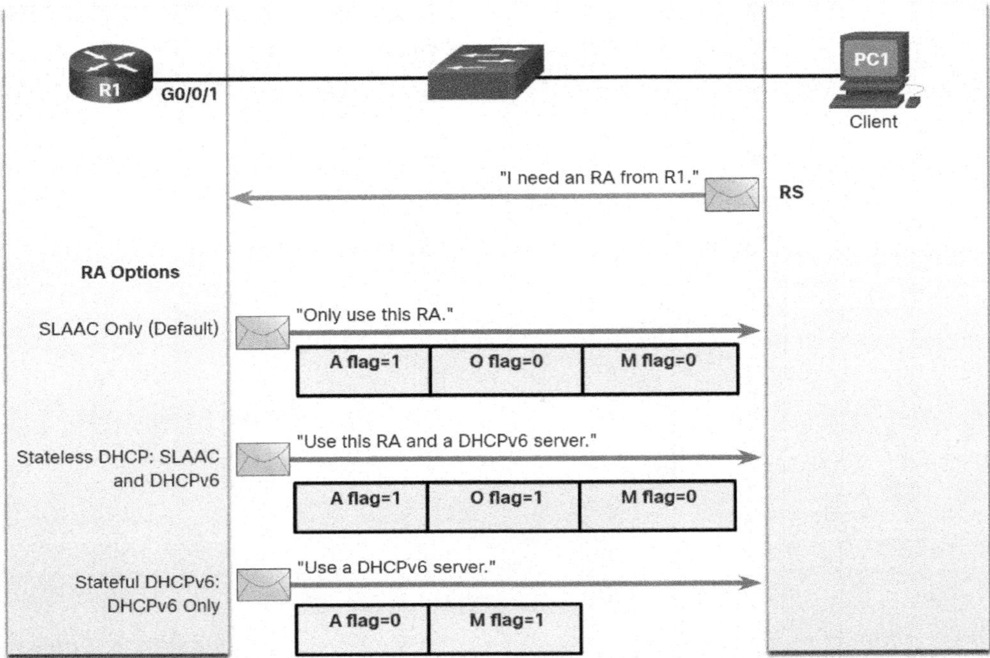

Figure 8-4 ICMPv6 RA Message Flags

Interactive Graphic

Check Your Understanding—IPv6 GUA Assignment (8.1.5)

Refer to the online course to complete this activity.

SLAAC (8.2)

In this section, you learn how to configure SLAAC.

SLAAC Overview (8.2.1)

Not every network has or needs access to a DHCPv6 server. But every device in an IPv6 network needs a GUA. The SLAAC method enables hosts to create their own unique IPv6 GUA without the services of a DHCPv6 server.

SLAAC is a stateless service. This means there is no server that maintains network address information to know which IPv6 addresses are being used and which ones are available.

SLAAC uses ICMPv6 RA messages to provide addressing and other configuration information that would normally be provided by a DHCP server. A host configures its IPv6 address based on the information that is sent in the RA. RA messages are sent by an IPv6 router every 200 seconds.

A host can also send an *ICMPv6 Router Solicitation (RS) message* requesting that an IPv6-enabled router send the host an RA.

SLAAC can be deployed as SLAAC only, or SLAAC with DHCPv6.

Enabling SLAAC (8.2.2)

Refer to the topology in Figure 8-5 to see how SLAAC is enabled to provide stateless dynamic GUA allocation.

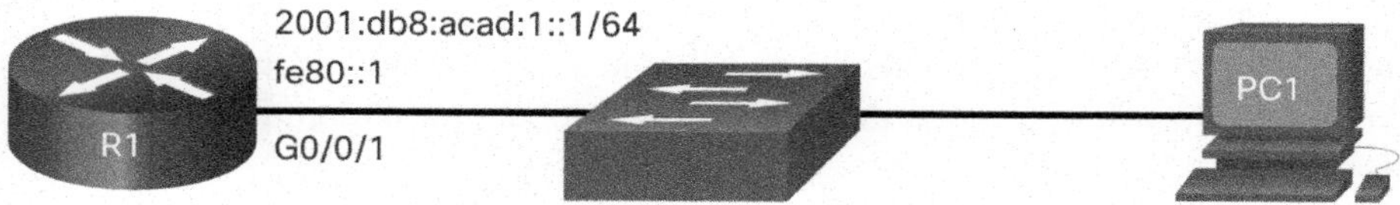

Figure 8-5 SLAAC Topology

Assume R1 GigabitEthernet 0/0/1 has been configured with the indicated IPv6 GUA and link-local addresses.

Verify IPv6 Addresses

The output of the **show ipv6 interface** command displays the current settings on the G0/0/1 interface.

As highlighted in Example 8-2, R1 has been assigned the following IPv6 addresses:

- **Link-local IPv6 address:** fe80::1
- **GUA and subnet:** 2001:db8:acad:1::1 and 2001:db8:acad:1::/64
- **IPv6 all-nodes group:** ff02::1

Example 8-2 Verifying IPv6 Addressing on an Interface

```
R1# show ipv6 interface G0/0/1
GigabitEthernet0/0/1 is up, line protocol is up
  IPv6 is enabled, link-local address is FE80::1
  No Virtual link-local address(es):
  Description: Link to LAN
  Global unicast address(es):
    2001:DB8:ACAD:1::1, subnet is 2001:DB8:ACAD:1::/64
  Joined group address(es):
    FF02::1
    FF02::1:FF00:1
(output omitted)
R1#
```

Enable IPv6 Routing

Although the router interface has an IPv6 configuration, it is still not yet enabled to send RAs containing address configuration information to hosts using SLAAC.

To enable the sending of RA messages, a router must join the IPv6 all-routers group using the **ipv6 unicast-routing** global config command, as shown in Example 8-3.

Example 8-3 Global Command to Enable IPv6 Routing

```
R1(config)# ipv6 unicast-routing
R1(config)# exit
R1#
```

Verify SLAAC Is Enabled

The IPv6 all-routers group responds to the IPv6 multicast address ff02::2. You can use the **show ipv6 interface** command to verify if a router is enabled, as shown in Example 8-4.

An IPv6-enabled Cisco router sends RA messages to the IPv6 all-nodes multicast address ff02::1 every 200 seconds.

Example 8-4 Verifying That SLAAC Is Enabled on an Interface

```
R1# show ipv6 interface G0/0/1 | section Joined
  Joined group address(es):
    FF02::1
    FF02::2
    FF02::1:FF00:1
R1#
```

SLAAC Only Method (8.2.3)

The SLAAC only method is enabled by default when the **ipv6 unicast-routing** command is configured. All enabled Ethernet interfaces with an IPv6 GUA configured will start sending RA messages with the A flag set to 1, and the O and M flags set to 0, as shown in Figure 8-6.

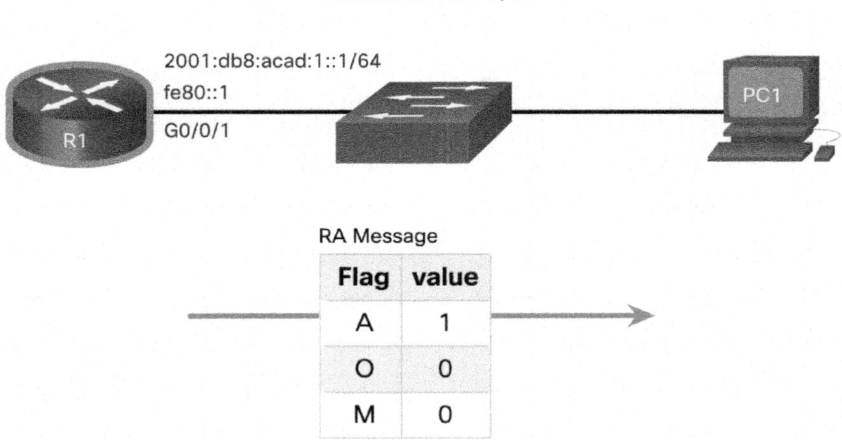

Figure 8-6 SLAAC Only Method Topology—A Flag Set to 1

The **A = 1** flag suggests to the client that it create its own IPv6 GUA using the prefix advertised in the RA. The client can create its own *Interface ID* using either *Extended Unique Identifier method (EUI-64)* or have it randomly generated.

The **O=0** and **M=0** flags instruct the client to use the information in the RA message exclusively. The RA includes the prefix, prefix-length, Domain Name System (DNS) server, maximum transmission unit (MTU), and default gateway information. There is no further information available from a DHCPv6 server.

In Example 8-5, PC1 is enabled to obtain its IPv6 addressing information automatically. Because of the settings of the A, O, and M flags, PC1 performs SLAAC only, using the information contained in the RA message sent by R1.

The default gateway address is the source IPv6 address of the RA message, which is the LLA for R1. The default gateway can only be obtained automatically from the RA message. A DHCPv6 server does not provide this information.

Example 8-5 Verifying a Windows Host Received IPv6 Addressing from SLAAC

```
C:\PC1> ipconfig
Windows IP Configuration
Ethernet adapter Ethernet0:
   Connection-specific DNS Suffix  . :
   IPv6 Address. . . . . . . . . . . : 2001:db8:acad:1:1de9:c69:73ee:ca8c
   Link-local IPv6 Address . . . . . : fe80::fb:1d54:839f:f595%21
   IPv4 Address. . . . . . . . . . . : 169.254.202.140
   Subnet Mask . . . . . . . . . . . : 255.255.0.0
   Default Gateway . . . . . . . . . : fe80::1%6
C:\PC1>
```

ICMPv6 RS Messages (8.2.4)

A router sends RA messages every 200 seconds. However, it will also send an RA message if it receives a router solicitation (RS) message from a host.

When a client is configured to obtain its addressing information automatically, it sends an RS message to the IPv6 all-routers multicast address of ff02::2.

Figure 8-7 illustrates how a host initiates the SLAAC method.

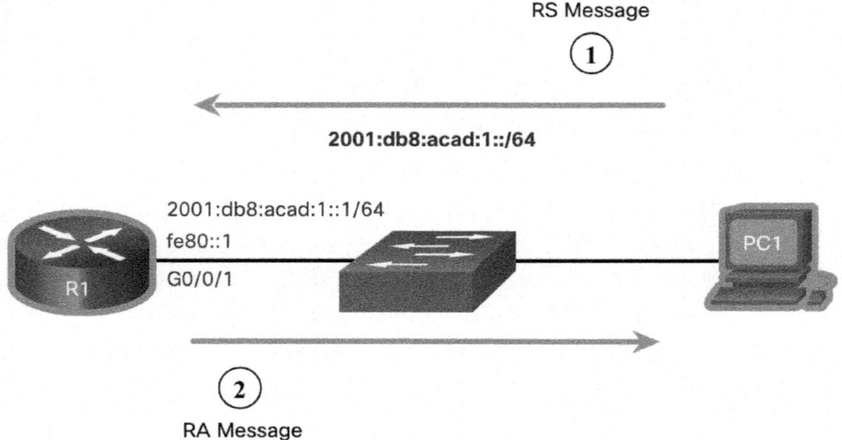

Figure 8-7 Host Initiates the SLAAC Method

1. In Figure 8-7, PC1 has just booted and has not yet received an RA message. Therefore, it sends an RS message to the IPv6 all-routers multicast address of ff02::2 requesting an RA.

2. R1 is part of the IPv6 all-routers group and has received the RS message. It generates an RA containing the local network prefix and prefix length (e.g., 2001:db8:acad:1::/64). It then sends the RA message to the IPv6 all-nodes multicast address of ff02::1. PC1 uses this information to create a unique IPv6 GUA.

Host Process to Generate Interface ID (8.2.5)

Using SLAAC, a host typically acquires its 64-bit IPv6 subnet information from the router RA. However, it must generate the remainder 64-bit interface identifier (ID) using one of two methods:

- **Randomly generated:** The 64-bit interface ID is randomly generated by the client operating system. This is the method now used by Windows 10 hosts.
- **EUI-64:** The host creates an interface ID using its 48-bit MAC address and inserts the hex value of **fffe** in the middle of the address. EUI-64 also flips the 7th bit, which modifies the second hex digit of the address. Some operating systems default to the randomly generated interface ID instead of the EUI-64 method, due to privacy concerns. This is because the Ethernet MAC address of the host is used by EUI-64 to create the interface ID.

Note

Windows, Linux, and Mac OS allow for the user to modify the generation of the interface ID to be either randomly generated or to use EUI-64.

For instance, in the following **ipconfig** output, the Windows 10 PC1 host used the IPv6 subnet information contained in the R1 RA and randomly generated a 64-bit interface ID, as highlighted in Example 8-6.

Example 8-6 Verifying a Windows Host Randomly Generated Its Interface ID

```
C:\PC1> ipconfig
Windows IP Configuration
Ethernet adapter Ethernet0:j
   Connection-specific DNS Suffix  . :
   IPv6 Address. . . . . . . . . . . : 2001:db8:acad:1:1de9:c69:73ee:ca8c
   Link-local IPv6 Address . . . . . : fe80::fb:1d54:839f:f595%21
   IPv4 Address. . . . . . . . . . . : 169.254.202.140
   Subnet Mask . . . . . . . . . . . : 255.255.0.0
   Default Gateway . . . . . . . . . : fe80::1%6
C:\PC1>
```

Duplicate Address Detection (8.2.6)

The process enables the host to create an IPv6 address. However, there is no guarantee that the address is unique on the network.

SLAAC is a stateless process; therefore, a host has the option to verify that a newly created IPv6 address is unique before it can be used. The *Duplicate Address Detection (DAD)* process is used by a host to ensure that the IPv6 GUA is unique.

DAD is implemented using ICMPv6. To perform DAD, the host sends an *ICMPv6 Neighbor Solicitation (NS) message* with a specially constructed multicast address, called a *solicited-node multicast address*. This address duplicates the last 24 bits of the IPv6 address of the host.

If no other devices respond with an *ICMPv6 Neighbor Advertisement (NA) message*, the address is virtually guaranteed to be unique and can be used by the host. If an NA is received by the host, then the address is not unique, and the operating system has to determine a new interface ID to use.

The Internet Engineering Task Force (IETF) recommends that DAD is used on all IPv6 unicast addresses regardless of whether it is created using SLAAC only, obtained using *stateful DHCPv6*, or manually configured. DAD is not mandatory because a 64-bit interface ID provides 18 quintillion possibilities, and the chance that there is a duplication is remote. However, most operating systems perform DAD on all IPv6 unicast addresses, regardless of how the address is configured.

Interactive Graphic

Check Your Understanding—SLAAC (8.2.7)

Refer to the online course to complete this activity.

DHCPv6 (8.3)

In this section, you learn about stateless DHCPv6 and stateful DHCPv6 for a small-to medium-sized business.

DHCPv6 Operation Steps (8.3.1)

This section explains stateless and stateful DHCPv6. Stateless DHCPv6 uses parts of SLAAC to ensure that all the necessary information is supplied to the host. Stateful DHCPv6 does not require SLAAC.

Although DHCPv6 is similar to DHCPv4 in what it provides, the two protocols are independent of each other.

Note

DHCPv6 is defined in RFC 3315.

The host begins the DHCPv6 client/server communications after stateless DHCPv6 or stateful DHCPv6 is indicated in the RA.

Server to client DHCPv6 messages use User Datagram Protocol (UDP) destination port 546 while client to server DHCPv6 messages use UDP destination port 547.

The steps shown in Figure 8-8 for DHCPv6 operations are as follows:

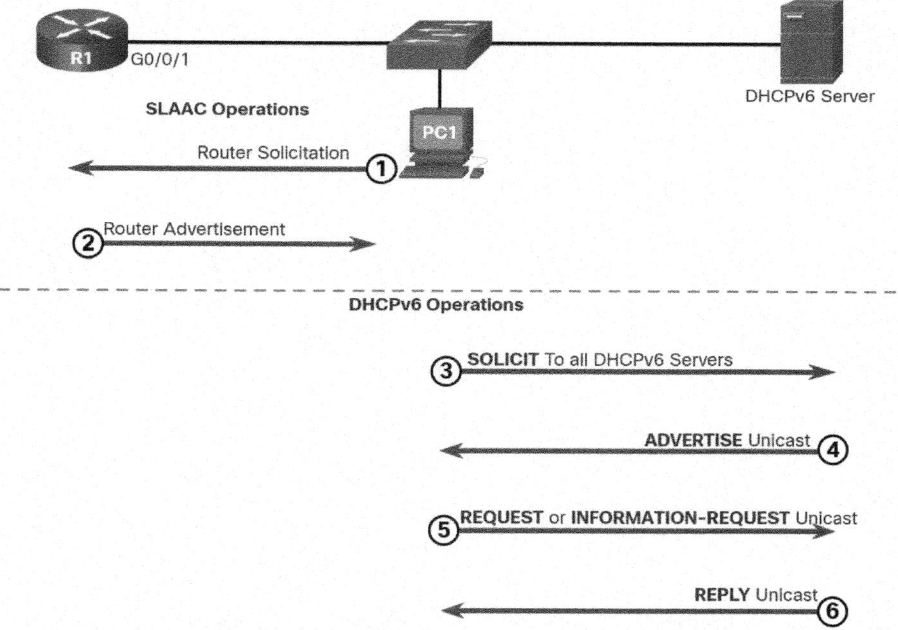

Figure 8-8 Steps in DHCPv6 Operation

Step 1. Host sends an RS message. PC1 sends an RS message to all IPv6-enabled routers.

Step 2. Router responds with an RA message. R1 receives the RS and responds with an RA indicating that the client is to initiate communication with a DHCPv6 server.

Step 3. Host sends a DHCPv6 SOLICIT message. The client, now a DHCPv6 client, needs to locate a DHCPv6 server and sends a *DHCPv6 SOLICIT message* to the reserved IPv6 multicast all-DHCPv6-servers address of ff02::1:2.

This multicast address has link-local scope, which means routers do not forward the messages to other networks.

Step 4. DHCPv6 server responds with an ADVERTISE message. One or more DHCPv6 servers respond with a *DHCPv6 ADVERTISE unicast message*. The ADVERTISE message informs the DHCPv6 client that the server is available for DHCPv6 service.

Step 5. Host responds to DHCPv6 server. The PC1 response depends on whether it is using stateful or stateless DHCPv6:

- *Stateless DHCPv6 client*: The client creates an IPv6 address using the prefix in the RA message and a self-generated Interface ID. The client then sends a *DHCPv6 INFORMATION-REQUEST message* to the DHCPv6 server requesting additional configuration parameters (for example, DNS server address).
- *Stateful DHCPv6 client*: The client sends a *DHCPv6 REQUEST message* to the DHCPv6 server to obtain all necessary IPv6 configuration parameters.

Step 6. DHCPv6 server sends a REPLY message. The server sends a *DHCPv6 REPLY unicast message* to the client. The content of the message varies, depending on if it is replying to a REQUEST or INFORMATION-REQUEST message.

Note

The client will use the source IPv6 link-local address of the RA as its default gateway address. A DHCPv6 server does not provide this information.

Stateless DHCPv6 Operation (8.3.2)

The stateless DHCPv6 option tells the client to use the information in the RA message for addressing, but additional configuration parameters are available from a DHCPv6 server.

This process is known as stateless DHCPv6 because the server is not maintaining any client state information (that is, a list of available and allocated IPv6 addresses). The stateless DHCPv6 server is only providing information that is identical for all devices on the network, such as the IPv6 address of a DNS server.

Figure 8-9 illustrates stateless DHCPv6 operation.

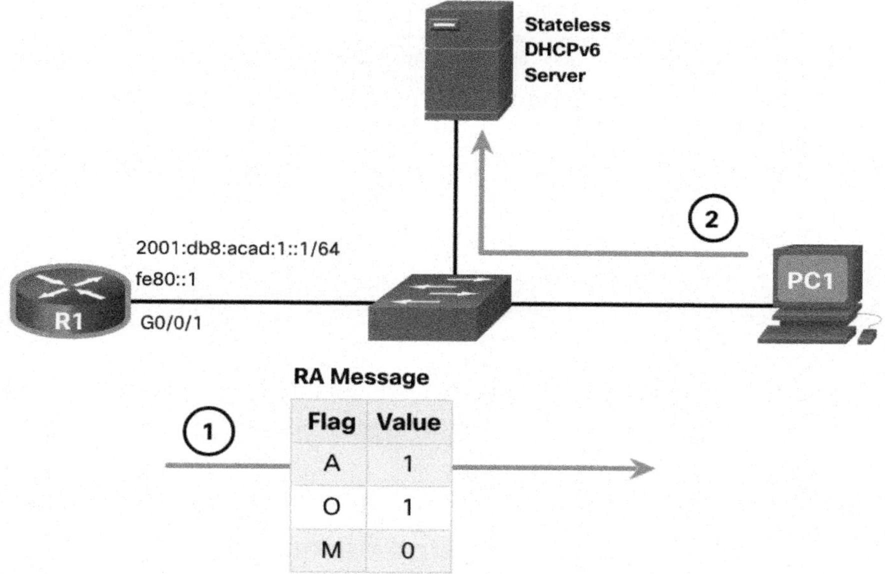

Figure 8-9 Stateless DHCPv6 Operation

1. In Figure 8-9, PC1 receives a stateless DHCP RA message. The RA message contains the network prefix and prefix length. The M flag for stateful DHCP is set to the default value 0. The A=1 flag tells the client to use SLAAC. The O=1 flag informs the client that additional configuration information is available from a stateless DHCPv6 server.
2. The client sends a DHCPv6 SOLICIT message looking for a stateless DHCPv6 server to obtain additional information (for example, DNS server addresses).

Enable Stateless DHCPv6 on an Interface (8.3.3)

Stateless DHCPv6 is enabled on a router interface using the **ipv6 nd other-config-flag** interface configuration command. This sets the O flag to 1.

The highlighted output in Example 8-7 confirms the RA will tell receiving hosts to use stateless autoconfigure (A flag = 1) and contact a DHCPv6 server to obtain another configuration information (O flag = 1).

Note

You can use the **no ipv6 nd other-config-flag** interface configuration command to reset the interface to the default SLAAC only option (O flag = 0).

Example 8-7 Enabling Stateless DHCPv6 on an Interface

```
R1(config-if)# ipv6 nd other-config-flag
R1(config-if)# end
R1#
R1# show ipv6 interface g0/0/1 | begin ND
  ND DAD is enabled, number of DAD attempts: 1
  ND reachable time is 30000 milliseconds (using 30000)
  ND advertised reachable time is 0 (unspecified)
  ND advertised retransmit interval is 0 (unspecified)
  ND router advertisements are sent every 200 seconds
  ND router advertisements live for 1800 seconds
  ND advertised default router preference is Medium
  Hosts use stateless autoconfig for addresses.
  Hosts use DHCP to obtain other configuration.
R1#
```

Stateful DHCPv6 Operation (8.3.4)

This option is most similar to DHCPv4. In this case, the RA message tells the client to obtain all addressing information from a stateful DHCPv6 server, except the default gateway address which is the source IPv6 link-local address of the RA.

This is known as stateful DHCPv6 because the DHCPv6 server maintains IPv6 state information. This is similar to a DHCPv4 server allocating addresses for IPv4.

Figure 8-10 illustrates stateful DHCPv6 operation.

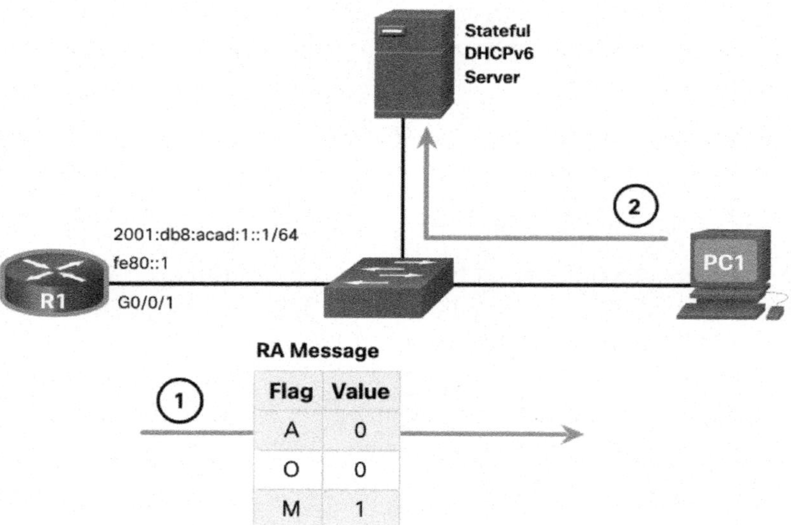

Figure 8-10 Stateful DHCPv6 Operation

1. In Figure 8-10, PC1 receives a DHCPv6 RA message with the O flag set to 0 and the M flag set to 1, indicating to PC1 that it will receive all its IPv6 addressing information from a stateful DHCPv6 server.
2. PC1 sends a DHCPv6 SOLICIT message looking for a stateful DHCPv6 server.

Note

If A=1 and M=1, some operating systems such as Windows will create an IPv6 address using SLAAC and obtain a different address from the stateful DHCPv6 server. In most cases it is recommended to manually set the A flag to 0.

Enable Stateful DHCPv6 on an Interface (8.3.5)

Stateful DHCPv6 is enabled on a router interface using the **ipv6 nd managed-config-flag** interface configuration command. This sets the M flag to 1. The **ipv6 nd prefix default no-autoconfig** interface command disables SLAAC by setting the A flag to 0.

The highlighted output in Example 8-8 confirms that the RA will tell the host to obtain all IPv6 configuration information from a DHCPv6 server (M flag = 1, A flag = 0).

Example 8-8 Configuring and Verifying Stateful DHCPv6 on an Interface

```
R1(config)# int g0/0/1
R1(config-if)# ipv6 nd managed-config-flag
R1(config-if)# ipv6 nd prefix default no-autoconfig
R1(config-if)# end
R1#
R1# show ipv6 interface g0/0/1 | begin ND
  ND DAD is enabled, number of DAD attempts: 1
  ND reachable time is 30000 milliseconds (using 30000)
  ND advertised reachable time is 0 (unspecified)
  ND advertised retransmit interval is 0 (unspecified)
  ND router advertisements are sent every 200 seconds
  ND router advertisements live for 1800 seconds
  ND advertised default router preference is Medium
  Hosts use DHCP to obtain routable addresses.
R1#
```

Interactive Graphic

Check Your Understanding—DHCPv6 (8.3.6)

Refer to the online course to complete this activity.

Configure DHCPv6 Server (8.4)

In this section, you learn how to configure a stateless DHCPv6 and a stateful DHCPv6 server for a small- to medium-sized business.

DHCPv6 Router Roles (8.4.1)

Cisco IOS routers are powerful devices. In smaller networks, you do not have to have separate devices to have a DHCPv6 server, client, or relay agent. A Cisco IOS router can be configured to provide DHCPv6 server services.

Specifically, it can be configured to be one of the following:

- **DHCPv6 Server:** Router provides stateless or stateful DHCPv6 services.
- **DHCPv6 Client:** Router interface acquires an IPv6 IP configuration from a DHCPv6 server.
- **DHCPv6 Relay Agent:** Router provides DHCPv6 forwarding services when the client and the server are located on different networks.

Configure a Stateless DHCPv6 Server (8.4.2)

The stateless DHCPv6 server option requires that the router advertise the IPv6 network addressing information in RA messages. However, the client must contact a DHCPv6 server for more information.

The topology in Figure 8-11 is used to demonstrate how to configure the stateless DHCPv6 server method.

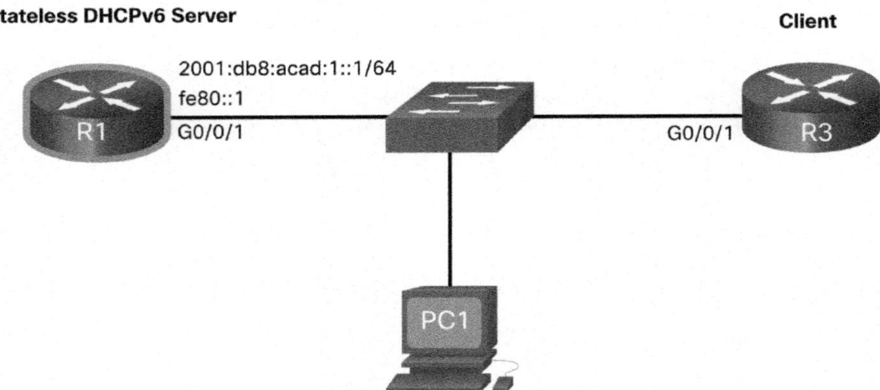

Figure 8-11 Topology for Stateless DHCPv6 Server

In this example, R1 will provide SLAAC services for the host IPv6 configuration and DHCPv6 services.

There are five steps to configure and verify a router as a stateless DHCPv6 server:

How To

Step 1. Enable IPv6 routing. The **ipv6 unicast-routing** command is required to enable IPv6 routing, as shown in Example 8-9. Although it is not necessary for the router to be a stateless DHCPv6 server, it is required for the router to source ICMPv6 RA messages.

Example 8-9 Enable IPv6 Routing

```
R1(config)# ipv6 unicast-routing
R1(config)#
```

Step 2. Define a DHCPv6 pool name. Create the DHCPv6 pool using the **ipv6 dhcp pool** *POOL-NAME* global config command. This enters DHCPv6 pool sub-configuration mode as identified by the Router(config-dhcpv6)# prompt, as shown in Example 8-10.

Note

The pool name does not have to be uppercase. However, using an uppercase name makes it easier to see in a configuration.

Example 8-10 Define a DHCPv6 Pool

```
R1(config)# ipv6 dhcp pool IPV6-STATELESS
R1(config-dhcpv6)#
```

Step 3. Configure the DHCPv6 pool. R1 will be configured to provide additional DHCP information, including DNS server address and domain name, as shown in Example 8-11.

Example 8-11 Configure Additional DHCP Information

```
R1(config-dhcpv6)# dns-server 2001:db8:acad:1::254
R1(config-dhcpv6)# domain-name example.com
R1(config-dhcpv6)# exit
R1(config)#
```

Step 4. Bind the DHCPv6 pool to an interface. The DHCPv6 pool has to be bound to the interface using the **ipv6 dhcp server** *POOL-NAME* interface config command, as shown in Example 8-12.

The router responds to stateless DHCPv6 requests on this interface with the information contained in the pool. The O flag needs to be manually changed from 0 to 1 using the interface command **ipv6 nd other-config-flag.** RA messages sent on this interface indicate that additional information is available from a stateless DHCPv6 server. The A flag is 1 by default, telling clients to use SLAAC to create their own GUA.

Example 8-12 Bind the DHCPv6 Pool to an Interface

```
R1(config)# interface GigabitEthernet0/0/1
R1(config-if)# description Link to LAN
R1(config-if)# ipv6 address fe80::1 link-local
R1(config-if)# ipv6 address 2001:db8:acad:1::1/64
R1(config-if)# ipv6 nd other-config-flag
R1(config-if)# ipv6 dhcp server IPV6-STATELESS
R1(config-if)# no shut
R1(config-if)# end
R1#
```

Step 5. Verify hosts received IPv6 addressing information. To verify stateless DHCP on a Windows host, use the **ipconfig /all** command. Example 8-13 displays the settings on PC1.

Notice in the output that PC1 created its IPv6 GUA using the 2001:db8:acad:1::/64 prefix. Also notice that the default gateway is the IPv6 link-local address of R1. This confirms that PC1 derived its IPv6 configuration from the RA of R1.

The highlighted output confirms that PC1 has learned the domain name and DNS server address information from the stateless DHCPv6 server.

Example 8-13 Verifying a Windows Host Received IPv6 Addressing Information from the DHCPv6 Server

```
C:\PC1> ipconfig /all
Windows IP Configuration
Ethernet adapter Ethernet0:
   Connection-specific DNS Suffix  . : example.com
   Description . . . . . . . . . . . : Intel(R) 82574L Gigabit Network Connection
   Physical Address. . . . . . . . . : 00-05-9A-3C-7A-00
   DHCP Enabled. . . . . . . . . . . : Yes
   Autoconfiguration Enabled . . . . : Yes
   IPv6 Address. . . . . . . . . . . : 2001:db8:acad:1:1dd:a2ea:66e7 Preferred)
   Link-local IPv6 Address . . . . . : fe80::fb:1d54:839f:f595%21(Preferred)
   IPv4 Address. . . . . . . . . . . : 169.254.102.23 (Preferred)
```

```
    Subnet Mask . . . . . . . . . . . . : 255.255.0.0
    Default Gateway . . . . . . . . . . : fe80::1%6
    DHCPv6 IAID . . . . . . . . . . . . : 318768538
    DHCPv6 Client DUID. . . . . . . . . : 00-01-00-01-21-F3-76-75-54-E1-AD-DE-DA-9A
    DNS Servers . . . . . . . . . . . . : 2001:db8:acad:1::1
    NetBIOS over Tcpip. . . . . . . . . : Enabled
C:\PC1>
```

Configure a Stateless DHCPv6 Client (8.4.3)

A router can also be a DHCPv6 client and get an IPv6 configuration from a DHCPv6 server, such as a router functioning as a DHCPv6 server. In Figure 8-12, R1 is a stateless DHCPv6 server.

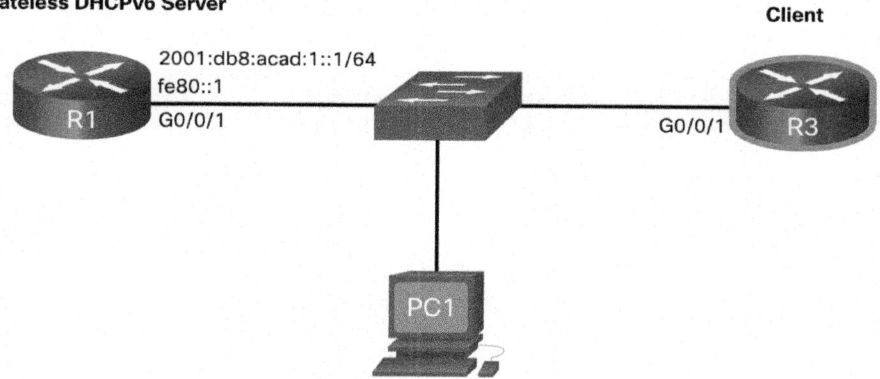

Figure 8-12 Topology with a Stateless DHCPv6 Client

There are five steps to configure and verify a router as a stateless DHCPv6 server.

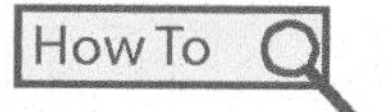

Step 1. Enable IPv6 routing. The DHCPv6 client router needs to have **ipv6 unicast-routing** enabled, as shown in Example 8-14.

Example 8-14 Enable IPv6 Routing

```
R3(config)# ipv6 unicast-routing
R3(config)#
```

Step 2. Configure client router to create an LLA. The client router needs to have a link-local address. An IPv6 link-local address is created on a router interface when a global unicast address is configured. It can also be created without a GUA using the **ipv6 enable** interface configuration command. Cisco IOS uses EUI-64 to create a randomized Interface ID.

In Example 8-15, the **ipv6 enable** command is configured on the Gigabit Ethernet 0/0/1 interface of the R3 client router.

Example 8-15 Configuring an Interface to Create an LLC

```
R3(config)# interface g0/0/1
R3(config-if)# ipv6 enable
R3(config-if)#
```

Step 3. Configure client router to use SLAAC. The client router needs to be configured to use SLAAC to create an IPv6 configuration. The **ipv6 address autoconfig** command enables the automatic configuration of IPv6 addressing using SLAAC, as shown in Example 8-16.

Example 8-16 Configuring an Interface to Use SLAAC

```
R3(config-if)# ipv6 address autoconfig
R3(config-if)# end
R3#
```

Step 4. Verify client router is assigned a GUA. Use the **show ipv6 interface brief** command to verify the host configuration, as shown in Example 8-17. The output confirms that the G0/0/1 interface on R3 was assigned a valid GUA.

Note

It may take the interface a few seconds to complete the process.

Example 8-17 Verifying the Client Router Received a GUA

```
R3# show ipv6 interface brief
GigabitEthernet0/0/0    [up/up]
    unassigned
GigabitEthernet0/0/1    [up/up]
    FE80::2FC:BAFF:FE94:29B1
    2001:DB8:ACAD:1:2FC:BAFF:FE94:29B1
Serial0/1/0             [up/up]
    unassigned
Serial0/1/1             [up/up]
    unassigned
R3#
```

Step 5. Verify client router received other DHCPv6 information. The **show ipv6 dhcp interface g0/0/1** command confirms that the DNS and domain names were also learned by R3, as shown in Example 8-18.

Example 8-18 Verifying the Client Router Received Other IPv6 Addressing

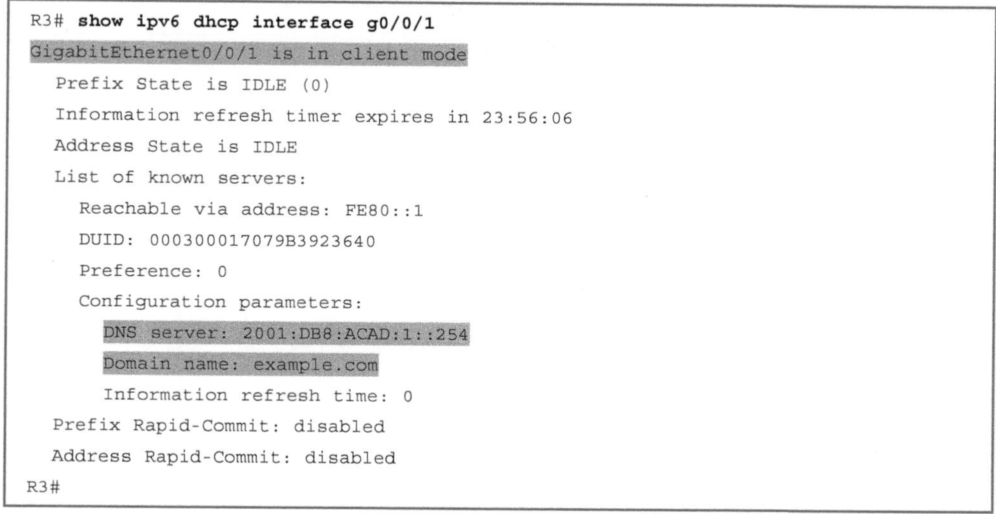

```
R3# show ipv6 dhcp interface g0/0/1
GigabitEthernet0/0/1 is in client mode
  Prefix State is IDLE (0)
  Information refresh timer expires in 23:56:06
  Address State is IDLE
  List of known servers:
    Reachable via address: FE80::1
    DUID: 000300017079B3923640
    Preference: 0
    Configuration parameters:
      DNS server: 2001:DB8:ACAD:1::254
      Domain name: example.com
      Information refresh time: 0
  Prefix Rapid-Commit: disabled
  Address Rapid-Commit: disabled
R3#
```

Configure a Stateful DHCPv6 Server (8.4.4)

The stateful DHCP server option requires that the IPv6 enabled router tells the host to contact a DHCPv6 server to obtain all necessary IPv6 network addressing information.

In Figure 8-13, R1 will provide stateful DHCPv6 services to all hosts on the local network. Configuring a stateful DHCPv6 server is similar to configuring a stateless server. The most significant difference is that a stateful DHCPv6 server also includes IPv6 addressing information similar to a DHCPv4 server.

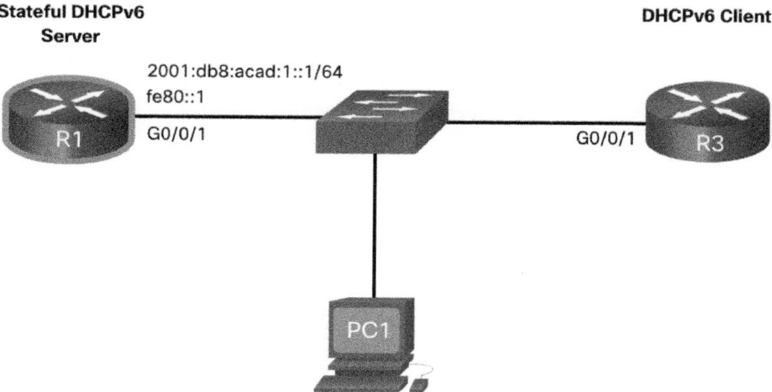

Figure 8-13 Topology for a Stateful DHCPv6 Server

There are five steps to configure and verify a router as a stateless DHCPv6 server:

How To

Step 1. Enable IPv6 routing. The **ipv6 unicast-routing** command is required to enable IPv6 routing, as shown in Example 8-19.

Example 8-19 Enabling IPv6 Routing

```
R1(config)# ipv6 unicast-routing
R1(config)#
```

Step 2. Define a DHCPv6 pool name. Create the DHCPv6 pool using the **ipv6 dhcp pool** *POOL-NAME* global config command, as shown in Example 8-20.

Example 8-20 Define a DHCPv6 Pool Name

```
R1(config)# ipv6 dhcp pool IPV6-STATEFUL
R1(config-dhcpv6)#
```

Step 3. Configure the DHCPv6 pool. R1 will be configured to provide IPv6 addressing, DNS server address, and domain name, as shown in Example 8-21. With stateful DHCPv6, all addressing and other configuration parameters must be assigned by the DHCPv6 server. The **address prefix** command is used to indicate the pool of addresses to be allocated by the server. Other information provided by the stateful DHCPv6 server typically includes the DNS server address and the domain name, as shown in the output.

Note

This example is setting the DNS server to Google's public DNS server.

Example 8-21 Configure the DHCPv6 Pool

```
R1(config-dhcpv6)# address prefix 2001:db8:acad:1::/64
R1(config-dhcpv6)# dns-server 2001:4860:4860::8888
R1(config-dhcpv6)# domain-name example.com
R1(config-dhcpv6)#
```

Step 4. Bind the DHCPv6 pool to an interface. Example 8-22 shows the full configuration of the GigabitEthernet 0/0/1 interface on R1.

Example 8-22 Bind the DHCPv6 Pool to an Interface

```
R1(config)# interface GigabitEthernet0/0/1
R1(config-if)# description Link to LAN
R1(config-if)# ipv6 address fe80::1 link-local
R1(config-if)# ipv6 address 2001:db8:acad:1::1/64
R1(config-if)# ipv6 nd managed-config-flag
R1(config-if)# ipv6 nd prefix default no-autoconfig
R1(config-if)# ipv6 dhcp server IPV6-STATEFUL
R1(config-if)# no shut
R1(config-if)# end
R1#
```

The DHCPv6 pool has to be bound to the interface using the **ipv6 dhcp server** *POOL-NAME* interface config command.

- The M flag is manually set from 0 to 1 using the **ipv6 nd managed-config-flag** interface command.
- The A flag is manually set from 1 to 0 using the **ipv6 nd prefix default no-autoconfig** interface command. The A flag can be left at 1, but some client operating systems such as Windows will create a GUA using SLAAC and get a GUA from the stateful DHCPv6 server. Setting the A flag to 0 tells the client not to use SLAAC to create a GUA.
- The **ipv6 dhcp server** command binds the DHCPv6 pool to the interface. R1 will now respond with the information contained in the pool when it receives stateful DHCPv6 requests on this interface.

Note

You can use the **no ipv6 nd managed-config-flag** command to set the M flag back to its default of 0. The **no ipv6 nd prefix default no-autoconfig** command sets the A flag back its default of 1.

Step 5. Verify hosts received IPv6 addressing information. To verify on a Windows host, use the **ipconfig /all** command to verify the stateless DHCP configuration method, as shown in Example 8-23. The output displays the settings on PC1. The highlighted output shows that PC1 has received its IPv6 GUA from a stateful DHCPv6 server.

Example 8-23 Verify the Host Received IPv6 Addressing Information

```
C:\PC1> ipconfig /all
Windows IP Configuration
Ethernet adapter Ethernet0:
   Connection-specific DNS Suffix  . : example.com
   Description . . . . . . . . . . . : IntelI 82574L Gigabit Network Connection
   Physical Address. . . . . . . . . : 00-05-9A-3C-7A-00
   DHCP Enabled. . . . . . . . . . . : Yes
   Autoconfiguration Enabled . . . . : Yes
   IPv6 Address. . . . . . . . . . . : 2001:db8:acad:1a43c:fd28:9d79:9e42
                                       (Preferred)
   Lease Obtained. . . . . . . . . . : Saturday, September 27, 2019, 10:45:30 AM
   Lease Expires . . . . . . . . . . : Monday, September 29, 2019 10:05:04 AM
   Link-local IPv6 Address . . . . . : fe80::192f:6fbc:9db:b749%6(Preferred)
   Autoconfiguration IPv4 Address. . : 169.254.102.73 (Preferred)
   Subnet Mask . . . . . . . . . . . : 255.255.0.0
   Default Gateway . . . . . . . . . : fe80::1%6
   DHCPv6 IAID . . . . . . . . . . . : 318768538
   DHCPv6 Client DUID. . . . . . . . : 00-01-00-01-21-F3-76-75-54-E1-AD-DE-DA-9A
   DNS Servers . . . . . . . . . . . : 2001:4860:4860::8888
   NetBIOS over Tcpip. . . . . . . . : Enabled
C:\PC1>
```

Configure a Stateful DHCPv6 Client (8.4.5)

A router can also be a DHCPv6 client. The client router needs to have **ipv6 unicast-routing** enabled and an IPv6 link-local address to send and receive IPv6 messages.

The topology in Figure 8-14 is used to demonstrate how to configure the stateful DHCPv6 client.

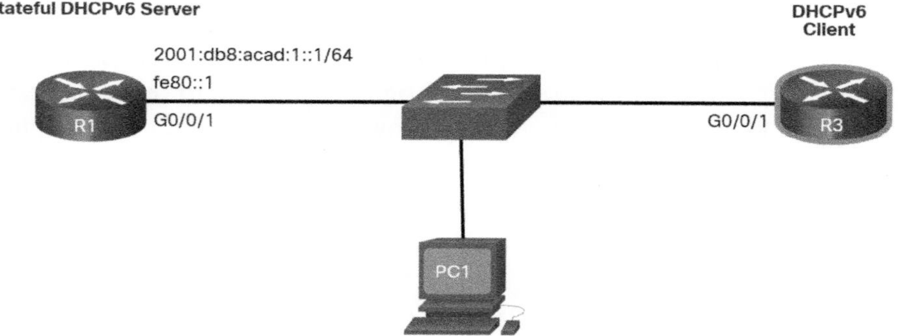

Figure 8-14 Topology with a Stateful DHCPv6 Client

There are five steps to configure and verify a router as a stateless DHCPv6 server:

Step 1. Enable IPv6 routing. The DHCPv6 client router needs to have **ipv6 unicast-routing** enabled, as shown in Example 8-24.

Example 8-24 Enable IPv6 Routing

```
R3(config)# ipv6 unicast-routing
R3(config)#
```

Step 2. Configure client router to create an LLA. In the output, the **ipv6 enable** command is configured on the R3 GigabitEthernet 0/0/1 interface as shown in Example 8-25. This enables the router to create an IPv6 LLA without needing a GUA.

Example 8-25 Configuring an Interface to Create an LLC

```
R3(config)# interface g0/0/1
R3(config-if)# ipv6 enable
R3(config-if)#
```

Step 3. Configure client router to use DHCPv6. The **ipv6 address dhcp** command configures R3 to solicit its IPv6 addressing information from a DHCPv6 server, as shown as Example 8-26.

Example 8-26 Configuring an Interface as a DHCPv6 Client

```
R3(config-if)# ipv6 address dhcp
R3(config-if)# end
R3#
```

Step 4. Verify client router is assigned a GUA. Use the **show ipv6 interface brief** command to verify the host configuration as shown in Example 8-27.

Example 8-27 Verifying the Client Router Received a GUA

```
R3# show ipv6 interface brief
GigabitEthernet0/0/0    [up/up]
    unassigned
GigabitEthernet0/0/1    [up/up]
    FE80::2FC:BAFF:FE94:29B1
    2001:DB8:ACAD:1:B4CB:25FA:3C9:747C
```

```
Serial0/1/0            [up/up]
    unassigned
Serial0/1/1            [up/up]
    unassigned
R3#
```

Step 5. Verify that the client router received other DHCPv6 information. The **show ipv6 dhcp interface g0/0/1** command confirms that the DNS and domain names were learned by R3, as shown in Example 8-28.

Example 8-28 Verifying the Client Router Received Other IPv6 Addressing

```
R3# show ipv6 dhcp interface g0/0/1
GigabitEthernet0/0/1 is in client mode
  Prefix State is IDLE
  Address State is OPEN
  Renew for address will be sent in 11:56:33
  List of known servers:
    Reachable via address: FE80::1
    DUID: 000300017079B3923640
    Preference: 0
    Configuration parameters:
      IA NA: IA ID 0x00060001, T1 43200, T2 69120
        Address: 2001:DB8:ACAD:1:B4CB:25FA:3C9:747C/128
                preferred lifetime 86400, valid lifetime 172800
                expires at Sep 29 2019 11:52 AM (172593 seconds)
      DNS server: 2001:4860:4860::8888
      Domain name: example.com
      Information refresh time: 0
  Prefix Rapid-Commit: disabled
  Address Rapid-Commit: disabled
R3#
```

DHCPv6 Server Verification Commands (8.4.6)

Use the **show ipv6 dhcp pool** and **show ipv6 dhcp binding** commands to verify DHCPv6 operation on a router.

The **show ipv6 dhcp pool** command verifies the name of the DHCPv6 pool and its parameters, as shown in Example 8-29. The command also identifies the number of active clients. In this example, the IPV6-STATEFUL pool currently has 2 clients, which reflects PC1 and R3 receiving their IPv6 global unicast address from this server.

When a router is providing stateful DHCPv6 services, it also maintains a database of assigned IPv6 addresses.

Example 8-29 Verify the DHCPv6 Pool Parameters

```
R1# show ipv6 dhcp pool
DHCPv6 pool: IPV6-STATEFUL
  Address allocation prefix: 2001:DB8:ACAD:1::/64 valid 172800 preferred 86400 (2
  in use,  0 conflicts)
  DNS server: 2001:4860:4860::8888
  Domain name: example.com
  Active clients: 2
R1#
```

Use the **show ipv6 dhcp binding** command output to display the IPv6 link-local address of the client and the global unicast address assigned by the server.

Example 8-30 displays the current stateful binding on R1. The first client in the output is PC1, and the second client is R3. This information is maintained by a stateful DHCPv6 server. A stateless DHCPv6 server would not maintain this information.

Example 8-30 Verify Clients Assigned DHCPv6 Addressing

```
R1# show ipv6 dhcp binding
Client: FE80::192F:6FBC:9DB:B749
  DUID: 0001000125148183005056B327D6
  Username : unassigned
  VRF : default
  IA NA: IA ID 0x03000C29, T1 43200, T2 69120
    Address: 2001:DB8:ACAD:1:A43C:FD28:9D79:9E42
            preferred lifetime 86400, valid lifetime 172800
            expires at Sep 27 2019 09:10 AM (171192 seconds)
Client: FE80::2FC:BAFF:FE94:29B1
  DUID: 0003000100FCBA9429B0
  Username : unassigned
  VRF : default
  IA NA: IA ID 0x00060001, T1 43200, T2 69120
    Address: 2001:DB8:ACAD:1:B4CB:25FA:3C9:747C
            preferred lifetime 86400, valid lifetime 172800
            expires at Sep 27 2019 09:29 AM (172339 seconds)
R1#
```

Configure a DHCPv6 Relay Agent (8.4.7)

If the DHCPv6 server is located on a different network from the client, the IPv6 router can be configured as a DHCPv6 relay agent. The configuration of a DHCPv6 relay agent is similar to the configuration of an IPv4 router as a DHCPv4 relay.

In Figure 8-15, R3 is configured as a stateful DHCPv6 server. PC1 is on the 2001:db8:acad:2::/64 network and requires the services of a stateful DHCPv6 server to acquire its IPv6 configuration. R1 needs to be configured as the DHCPv6 Relay Agent.

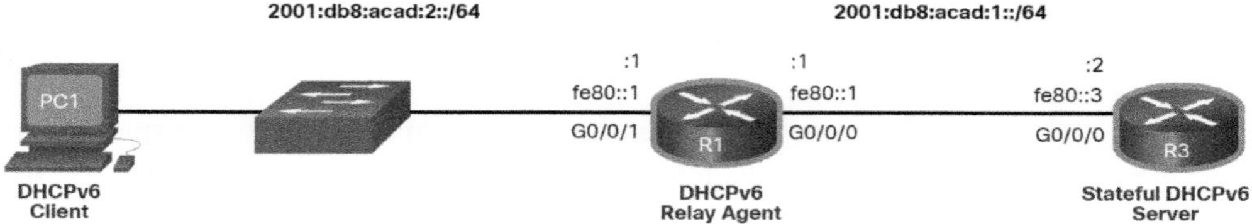

Figure 8-15 Topology with a DHCPv6 Relay Agent

The command syntax to configure a router as a DHCPv6 relay agent is as follows:

```
Router(config-if)# ipv6 dhcp relay destination ipv6-address [interface-type
   interface-number]
```

This command is configured on the interface facing the DHCPv6 clients and specifies the DHCPv6 server address and egress interface to reach the server, as shown in Example 8-31. The egress interface is required only when the next-hop address is an LLA.

Example 8-31 Configuring DHCPv6 Relay on the Interface

```
R1(config)# interface gigabitethernet 0/0/1
R1(config-if)# ipv6 dhcp relay destination 2001:db8:acad:1::2 G0/0/0
R1(config-if)# exit
R1(config)#
```

Verify the DHCPv6 Relay Agent (8.4.8)

Verify that the DHCPv6 relay agent is operational with the **show ipv6 dhcp interface** and **show ipv6 dhcp binding** commands. Verify Windows hosts received IPv6 addressing information with the **ipconfig /all** command.

The DHCPv6 relay agent can be verified using the **show ipv6 dhcp interface** command, as shown in Example 8-32. This will verify that the G0/0/1 interface is in relay mode.

Example 8-32 Verifying the Interface Is in DHCPv6 Relay Mode

```
R1# show ipv6 dhcp interface
GigabitEthernet0/0/1 is in relay mode
  Relay destinations:
    2001:DB8:ACAD:1::2
    2001:DB8:ACAD:1::2 via GigabitEthernet0/0/0
R1#
```

On R3, use the **show ipv6 dhcp binding** command to verify if any hosts have been assigned an IPv6 configuration, as shown in Example 8-33.

Notice that a client link-local address has been assigned an IPv6 GUA. We can assume that this is PC1.

Example 8-33 Verifying DHCPv6 Hosts Are Assigned an IPv6 GUA

```
R3# show ipv6 dhcp binding
Client: FE80::5C43:EE7C:2959:DA68
  DUID: 0001000124F5CEA2005056B3636D
  Username : unassigned
  VRF : default
  IA NA: IA ID 0x03000C29, T1 43200, T2 69120
    Address: 2001:DB8:ACAD:2:9C3C:64DE:AADA:7857
            preferred lifetime 86400, valid lifetime 172800
            expires at Sep 29 2019 08:26 PM (172710 seconds)
R3#
```

Finally, use **ipconfig /all** on PC1 to confirm that it has been assigned an IPv6 configuration. As you can see in Example 8-34, PC1 has indeed received its IPv6 configuration from the DHCPv6 server.

Example 8-34 Verifying the PC Received IPv6 Addressing Information

```
C:\PC1> ipconfig /all
Windows IP Configuration
Ethernet adapter Ethernet0:
   Connection-specific DNS Suffix  . : example.com
   Description . . . . . . . . . . . : Intel(R) 82574L Gigabit Network Connection
   Physical Address. . . . . . . . . : 00-05-9A-3C-7A-00
   DHCP Enabled. . . . . . . . . . . : Yes
   Autoconfiguration Enabled . . . . : Yes
   IPv6 Address. . . . . . . . . . . : 2001:db8:acad:2:9c3c:64de:aada:7857 (Preferred)
   Link-local IPv6 Address . . . . . : fe80::5c43:ee7c:2959:da68%6(Preferred)
```

```
    Lease Obtained . . . . . . . . . . : Saturday, September 27, 2019, 11:45:30 AM
    Lease Expires . . . . . . . . . . : Monday, September 29, 2019 11:05:04 AM
    IPv4 Address. . . . . . . . . . . : 169.254.102.73 (Preferred)
    Subnet Mask . . . . . . . . . . . : 255.255.0.0
    Default Gateway . . . . . . . . . : fe80::1%6
    DHCPv6 IAID . . . . . . . . . . . : 318768538
    DHCPv6 Client DUID. . . . . . . . : 00-01-00-01-21-F3-76-75-54-E1-AD-DE-DA-9A
    DNS Servers . . . . . . . . . . . : 2001:4860:4860::8888
    NetBIOS over Tcpip. . . . . . . . : Enabled
C:\PC1>
```

Interactive Graphic

Check Your Understanding—Configure DHCPv6 Server (8.4.9)

Refer to the online course to complete this activity.

Summary

The following is a summary of each section in the module:

IPv6 GUA Assignment

On a router, an IPv6 global unicast address (GUA) is manually configured using the **ipv6 address** *ipv6-address/prefix-length* interface configuration command. When automatic IPv6 addressing is selected, the host will attempt to automatically obtain and configure IPv6 address information on the interface. The IPv6 link-local address is automatically created by the host when it boots and the Ethernet interface is active. By default, an IPv6-enabled router advertises its IPv6 information enabling a host to dynamically create or acquire its IPv6 configuration. The IPv6 GUA can be assigned dynamically using stateless and stateful services. The decision of how a client will obtain an IPv6 GUA depends on the settings within the RA message. An ICMPv6 RA message includes three flags to identify the dynamic options available to a host:

- **A flag:** This is the Address Autoconfiguration flag. Use SLAAC to create an IPv6 GUA.
- **O flag:** This is the Other Configuration flag. Get Other information from a stateless DHCPv6 server.
- **M flag:** This is the Managed Address Configuration flag. Use a stateful DHCPv6 server to obtain an IPv6 GUA.

SLAAC

The SLAAC method enables hosts to create their own unique IPv6 global unicast address without the services of a DHCPv6 server. SLAAC, which is stateless, uses ICMPv6 RA messages to provide addressing and other configuration information that would normally be provided by a DHCP server. SLAAC can be deployed as SLAAC only, or SLAAC with DHCPv6. To enable the sending of RA messages, a router must join the IPv6 all-routers group using the **ipv6 unicast-routing** global config command. Use the **show ipv6 interface** command to verify if a router is enabled. The SLAAC only method is enabled by default when the **ipv6 unicast-routing** command is configured. All enabled Ethernet interfaces with an IPv6 GUA configured will start sending RA messages with the A flag set to 1, and the O and M flags set to 0. The A=1 flag suggests to the client to create its own IPv6 GUA using the prefix advertised in the RA. The O=0 and M=0 flag instructs the client to use the information in the RA message exclusively. A router sends RA messages every 200 seconds. However, it will also send an RA message if it receives an RS message from a host. Using SLAAC, a host typically acquires its 64-bit IPv6 subnet information from the router RA. However, it must generate the remainder 64-bit interface identifier (ID) using one of two methods: randomly generated, or EUI-64. The DAD process is used by a host to ensure that the IPv6 GUA is unique. DAD is implemented using ICMPv6. To perform DAD, the host sends an ICMPv6 NS message with a specially constructed multicast address, called a solicited-node multicast address. This address duplicates the last 24 bits of the IPv6 address of the host.

DHCPv6

The host begins the DHCPv6 client/server communications after stateless DHCPv6 or stateful DHCPv6 is indicated in the RA. Server to client DHCPv6 messages use UDP destination port 546, and client to server DHCPv6 messages use UDP destination port 547. The stateless DHCPv6 option informs the client to use the information in the RA message for addressing, but additional configuration parameters are available from a DHCPv6 server. This is called stateless DHCPv6 because the server is not maintaining any client state information. Stateless DHCPv6 is enabled on a router interface using the **ipv6 nd other-config-flag** interface configuration command. This sets the O flag to 1. In stateful DHCPv6, the RA message tells the client to obtain all addressing information from a stateful DHCPv6 server, except the default gateway address, which is the source IPv6 link-local address of the RA. It is called stateful because the DHCPv6 server maintains IPv6 state information. Stateful DHCPv6 is enabled on a router interface using the **ipv6 nd managed-config-flag** interface configuration command. This sets the M flag to 1.

Configure DHCPv6 Server

A Cisco IOS router can be configured to provide DHCPv6 server services as one of the following three types: DHCPv6 server, DHCPv6 client, or DHCPv6 relay agent. The stateless DHCPv6 server option requires that the router advertise the IPv6 network addressing information in RA messages. A router can also be a DHCPv6 client and get an IPv6 configuration from a DHCPv6 server. The stateful DHCP server option requires that the IPv6-enabled router tells the host to contact a DHCPv6 server to acquire all required IPv6 network addressing information. For a client router to be a DHCPv6 router, it needs to have **ipv6 unicast-routing** enabled and an IPv6 link-local address to send and receive IPv6 messages. Use the **show ipv6 dhcp pool** and **show ipv6 dhcp binding** commands to verify DHCPv6 operation on a router. If the DHCPv6 server is located on a different network from the client, the IPv6 router can be configured as a DHCPv6 relay agent using the **ipv6 dhcp relay destination** *ipv6-address* [*interface-type interface-number*] command. This command is configured on the interface facing the DHCPv6 clients and specifies the DHCPv6 server address and egress interface to reach the server. The egress interface is required only when the next-hop address is an LLA. Verify the DHCPv6 relay agent is operational with the **show ipv6 dhcp interface** and **show ipv6 dhcp binding** commands.

Lab—Configure DHCPv6 (8.5.1)

In this lab, you complete the following objectives:

- Part 1: Build the Network and Configure Basic Device Settings
- Part 2: Verify SLAAC Address Assignment from R1
- Part 3: Configure and Verify a Stateless DHCPv6 Server on R1
- Part 4: Configure and Verify a Stateful DHCPv6 Server on R1
- Part 5: Configure and Verify a DHCPv6 Relay on R2

Practice

The following activities provide practice with the topics introduced in this chapter. The Labs are available in the companion *Switching, Routing, and Wireless Essentials Labs and Study Guide (CCNAv7)* (ISBN 9780136634386).

Lab

Lab 8.5.1: Configure DHCPv6

Check Your Understanding Questions

Complete all the review questions listed here to test your understanding of the sections and concepts in this chapter. The appendix "Answers to the 'Check Your Understanding' Questions" lists the answers.

1. A company uses the SLAAC method to configure IPv6 addresses for the employee workstations. Which address will a client use as its default gateway?

    A. The all-routers multicast address

    B. The global unicast address of the router interface that is attached to the network

    C. The link-local address of the router interface that is attached to the network

    D. The unique local address of the router interface that is attached to the network

2. A network administrator configures a router to send RA messages with the A flag and O flag set to 1. The M flag is set to 0. Which statement describes the effect of this configuration when a PC tries to configure its IPv6 address?

   A. It should contact a DHCPv6 server for all the information that it needs.

   B. It should contact a DHCPv6 server for the prefix, the prefix-length information, and an interface ID that is both random and unique.

   C. It should use the information that is contained in the RA message and contact a DHCPv6 server for additional information.

   D. It should use the information that is contained in the RA message exclusively.

3. A company implements the stateless DHCPv6 method for configuring IPv6 addresses on employee workstations. After a workstation receives messages from multiple DHCPv6 servers to indicate their availability for DHCPv6 service, which message does it send to a server for configuration information?

   A. DHCPv6 ADVERTISE

   B. DHCPv6 INFORMATION-REQUEST

   C. DHCPv6 REQUEST

   D. DHCPv6 SOLICIT

4. An administrator wants to configure hosts to automatically assign IPv6 addresses to themselves by the use of Router Advertisement messages, but also to obtain the DNS server address from a DHCPv6 server. Which address assignment method should be configured?

   A. RA and EUI-64

   B. SLAAC

   C. Stateful DHCPv6

   D. SLAAC and stateless DHCPv6

5. How does an IPv6 client ensure that it has a unique address after it configures its IPv6 address using the SLAAC allocation method?

   A. It checks with the IPv6 address database that is hosted by the SLAAC server.

   B. It contacts the DHCPv6 server via a specially formed ICMPv6 message.

   C. It sends an ARP message with the IPv6 address as the destination IPv6 address.

   D. It sends an ICMPv6 Neighbor Solicitation message with the IPv6 address as the target IPv6 address.

6. What is used in the EUI-64 process to create an IPv6 interface ID on an IPv6 enabled interface?

   A. A randomly generated 64-bit hexadecimal address

   B. An IPv4 address that is configured on the interface

   C. An IPv6 address that is provided by a DHCPv6 server

   D. The MAC address of an Ethernet interface

7. A network administrator is implementing DHCPv6 for the company. The administrator configures a router to send RA messages with M flag as 1 by using the **ipv6 nd managed-config-flag** interface command, and the A flag is set to 0 using the **ipv6 nd prefix default no-autoconfig** command. What effect will this configuration have on the operation of the clients?

   A. Clients must use all configuration information that is provided by a DHCPv6 server.

   B. Clients must use the information that is contained in RA messages.

   C. Clients must use the prefix and prefix length that are provided by a DHCPv6 server and generate a random interface ID.

   D. Clients must use the prefix and prefix length that are provided by RA messages and obtain additional information from a DHCPv6 server.

8. An organization requires that LAN clients generate their IPv6 configuration using SLAAC. You have configured the IPv6 GUA on the router LAN interface and verified that the interface is UP. However, hosts are not generating an IPv6 GUA. Which other command should be configured to enable SLAAC?

   A. R1(config)# **ipv6 dhcp pool pool-name**

   B. R1(config)# **ipv6 unicast-routing**

   C. R1(config-if)# **ipv6 enable**

   D. R1(config-if)# **ipv6 nd other-config-flag**

9. A network administrator configures a router to send RA messages with M flag as 0 and O flag as 1. Which statement describes the effect of this configuration when a PC tries to configure its IPv6 address?

   A. It should contact a DHCPv6 server for all the information that it needs.

   B. It should contact a DHCPv6 server for the prefix, the prefix-length information, and an interface ID that is both random and unique.

   C. It should use the information that is contained in the RA message and contact a DHCPv6 server for additional information.

   D. It should use the information that is contained in the RA message exclusively.

10. When SLAAC is used, which address will a client use as its default gateway?

    A. The connecting router interface GUA
    B. The connecting router link-local address
    C. The IPv6 all-nodes group multicast IPv6 address FF02::1
    D. The IPv6 all-routers group multicast IPv6 address FF02::2

CHAPTER 9

FHRP Concepts

Objectives

Upon completion of this chapter, you will be able to answer the following questions:

- What is the purpose and operation of first hop redundancy protocols?
- How does HSRP operate?

Key Terms

This chapter uses the following key terms. You can find the definitions in the Glossary.

First Hop Redundancy Protocols (FHRP) *Page 262*

Hot Standby Router Protocol (HSRP) *Page 262*

active router *Page 264*

standby router *Page 265*

Virtual Router Redundancy Protocol (VRRP) *Page 266*

master router *Page 267*

backup router *Page 267*

Gateway Load Balancing Protocol (GLBP) *Page 267*

ICMP Router Discovery Protocol (IRDP) *Page 267*

Introduction (9.0)

Your network is up and running. You've conquered Layer 2 redundancy without any Layer 2 loops. All your devices get their addresses dynamically. You are *good* at network administration! But, wait. One of your routers, the default gateway router in fact, has gone down. None of your hosts can send any messages outside of the immediate network. It's going to take a while to get this default gateway router operating again. You've got a lot of angry people asking you how soon the network will be "back up."

You can avoid this problem easily. ***First Hop Redundancy Protocols (FHRPs)*** are the solution you need. This module discusses what FHRP does, and all the types of FHRPs that are available to you. One of these types is a Cisco-proprietary FHRP called ***Hot Standby Router Protocol (HSRP)***. You learn how HSRP works and then complete a Packet Tracer activity where you configure and verify HSRP. Don't wait, get started!

First Hop Redundancy Protocols (9.1)

If a router or router interface (that serves as a default gateway) fails, the hosts configured with that default gateway are isolated from outside networks. A mechanism is needed to provide alternate default gateways in switched networks where two or more routers are connected to the same VLANs.

In this section you will learn how FHRPs are used to provide default gateway redundancy.

Default Gateway Limitations (9.1.1)

If a router or router interface that serves as a default gateway fails, the hosts configured with that default gateway are isolated from outside networks. A mechanism is needed to provide alternate default gateways in switched networks where two or more routers are connected to the same network or VLANs. That mechanism is provided by first hop redundancy protocols (FHRPs).

In a switched network, each client receives only one default gateway. There is no way to use a secondary gateway, even if a second path exists to carry packets off the local segment.

In Figure 9-1, R1 is responsible for routing packets from PC1.

If R1 becomes unavailable, the routing protocols can dynamically converge. R2 now routes packets from outside networks that would have gone through R1. However, traffic from the inside network associated with R1, including traffic from workstations, servers, and printers configured with R1 as their default gateway, are still sent to R1 and dropped.

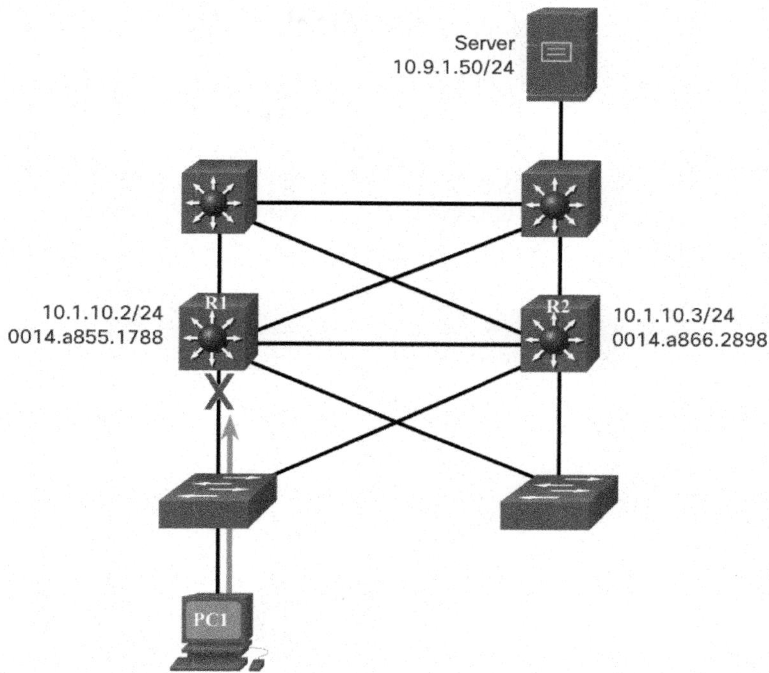

PC1 is unable to reach the default gateway.

Figure 9-1 Topology with Redundant Paths but Only One Default Gateway

Note

For the purposes of the discussion on router redundancy, there is no functional difference between a Layer 3 switch and a router at the distribution layer. In practice, it is common for a Layer 3 switch to act as the default gateway for each VLAN in a switched network. This discussion focuses on the functionality of routing, regardless of the physical device used.

End devices are typically configured with a single IPv4 address for a default gateway. This address does not change when the network topology changes. If that default gateway IPv4 address cannot be reached, the local device is unable to send packets off the local network segment, effectively disconnecting it from other networks. Even if a redundant router exists that could serve as a default gateway for that segment, there is no dynamic method by which these devices can determine the address of a new default gateway.

Note

IPv6 devices receive their default gateway address dynamically from the Internet Control Message Protocol version 6 (ICMPv6).

Router Redundancy (9.1.2)

One way to prevent a single point of failure at the default gateway is to implement a virtual router. To implement this type of router redundancy, multiple routers are configured to work together to present the illusion of a single router to the hosts on the LAN, as shown in Figure 9-2. By sharing an IP address and a MAC address of the virtual router, two or more routers can act as a single virtual router.

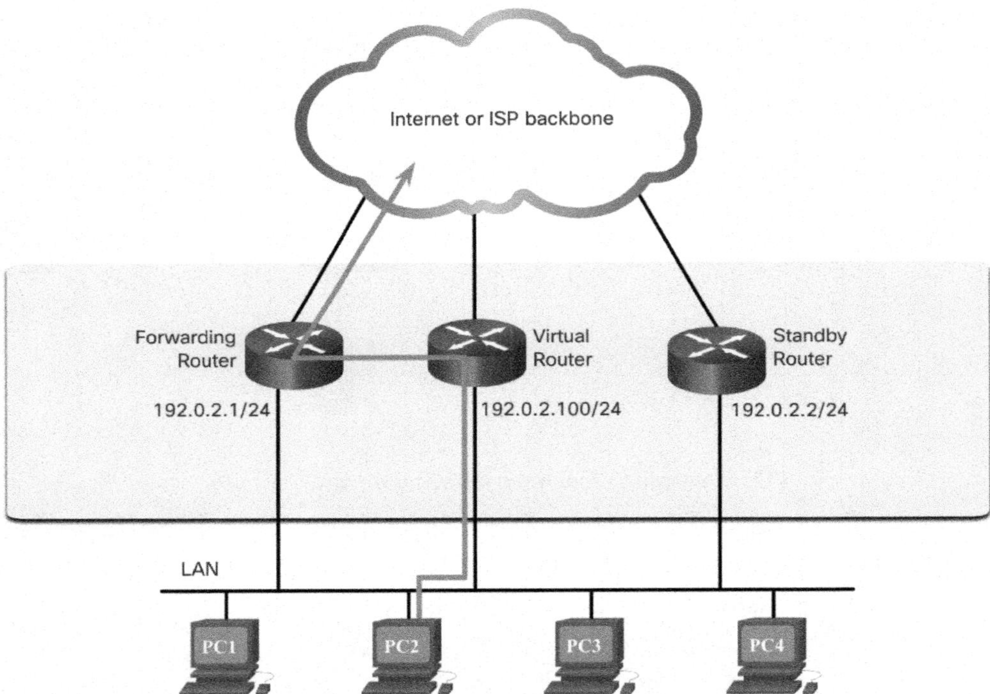

Figure 9-2 Redundant Routers Topology

The IPv4 address of the virtual router is configured as the default gateway for the workstations on a specific IPv4 segment. When frames are sent from host devices to the default gateway, the hosts use ARP to resolve the MAC address that is associated with the IPv4 address of the default gateway. The ARP resolution returns the MAC address of the virtual router. Frames that are sent to the MAC address of the virtual router can then be physically processed by the currently *active router* within the virtual router group. A protocol is used to identify two or more routers as the devices that are responsible for processing frames that are sent to the MAC or IP address of a single virtual router. Host devices send traffic to the address of the virtual router. The physical router that forwards this traffic is transparent to the host devices.

A redundancy protocol provides the mechanism for determining which router should take the active role in forwarding traffic. It also determines when the forwarding role must be taken over by a *standby router*. The transition from one forwarding router to another is transparent to the end devices.

The ability of a network to dynamically recover from the failure of a device acting as a default gateway is known as first-hop redundancy.

Steps for Router Failover (9.1.3)

When the active router fails, the redundancy protocol transitions the standby router to the new active router role, as shown in Figure 9-3.

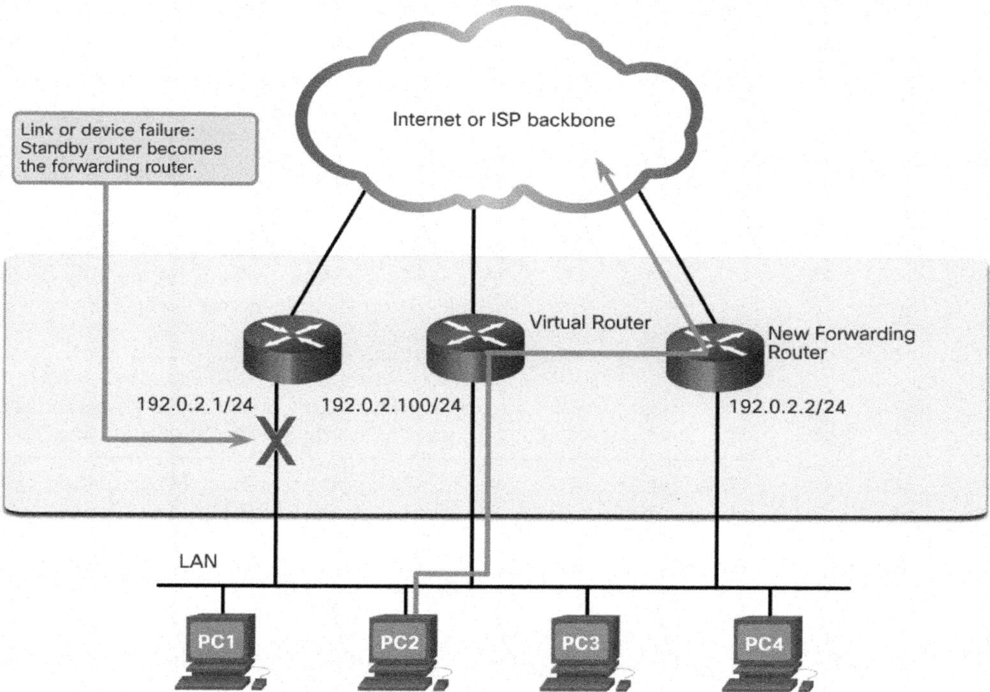

Figure 9-3 Standby Router Transitions to Active Router

These are the steps that take place when the active router fails:

Step 1. The standby router stops seeing Hello messages from the forwarding router.

Step 2. The standby router assumes the role of the forwarding router.

Step 3. Because the new forwarding router assumes both the IPv4 and MAC addresses of the virtual router, the host devices see no disruption in service.

FHRP Options (9.1.4)

The FHRP used in a production environment largely depends on the equipment and needs of the network. Table 9-1 lists all the options available for FHRPs.

Table 9-1 Options for Implementing an FHRP

FHRP Options	Description
Hot Standby Router Protocol (HSRP)	■ HSRP is a Cisco-proprietary FHRP that is designed to allow for transparent failover of a first-hop IPv4 device. ■ HSRP provides high network availability by providing first-hop routing redundancy for IPv4 hosts on networks configured with an IPv4 default gateway address. ■ HSRP is used in a group of routers for selecting an active device and a standby device. ■ In a group of device interfaces, the active device is the device that is used for routing packets; the standby device is the device that takes over when the active device fails, or when pre-set conditions are met. ■ The function of the HSRP standby router is to monitor the operational status of the HSRP group and to quickly assume packet-forwarding responsibility if the active router fails.
HSRP for IPv6	■ This is a Cisco-proprietary FHRP that provides the same functionality of HSRP, but in an IPv6 environment. ■ An HSRP IPv6 group has a virtual MAC address derived from the HSRP group number and a virtual IPv6 link-local address derived from the HSRP virtual MAC address. ■ Periodic router advertisements (RAs) are sent for the HSRP virtual IPv6 link-local address when the HSRP group is active. ■ When the group becomes inactive, these RAs stop after a final RA is sent.
Virtual Router Redundancy Protocol version 2 (VRRPv2)	■ This is a non-proprietary election protocol that dynamically assigns responsibility for one or more virtual routers to the VRRP routers on an IPv4 LAN. ■ This allows several routers on a multiaccess link to use the same virtual IPv4 address.

FHRP Options	Description
	■ A VRRP router is configured to run the VRRP protocol in conjunction with one or more other routers attached to a LAN. ■ In a VRRP configuration, one router is elected as the virtual router master (that is, *master router*), with the other routers acting as backups (that is, *backup router*), in case the virtual router master fails.
VRRPv3	■ This provides the capability to support IPv4 and IPv6 addresses. VRRPv3 works in multivendor environments and is more scalable than VRRPv2.
Gateway Load Balancing Protocol (GLBP)	■ This is a Cisco-proprietary FHRP that protects data traffic from a failed router or circuit, like HSRP and VRRP, while also allowing load balancing (also called load sharing) between a group of redundant routers.
GLBP for IPv6	■ This is a Cisco-proprietary FHRP that provides the same functionality of GLBP, but in an IPv6 environment. ■ GLBP for IPv6 provides automatic router backup for IPv6 hosts configured with a single default gateway on a LAN. ■ Multiple first-hop routers on the LAN combine to offer a single virtual first-hop IPv6 router while sharing the IPv6 packet forwarding load.
ICMP Router Discovery Protocol (IRDP)	■ Specified in RFC 1256, IRDP is a legacy FHRP solution. IRDP allows IPv4 hosts to locate routers that provide IPv4 connectivity to other (nonlocal) IP networks.

Interactive Graphic

Check Your Understanding—First Hop Redundancy Protocols (9.1.5)

Refer to the online course to complete this activity.

HSRP (9.2)

In this section you will learn how to implement HSRP.

HSRP Overview (9.2.1)

Cisco provides HSRP and HSRP for IPv6 as a way to avoid losing outside network access if your default router fails.

HSRP is a Cisco-proprietary FHRP that is designed to allow for transparent failover of a first-hop IP device.

HSRP ensures high network availability by providing first-hop routing redundancy for IP hosts on networks configured with an IP default gateway address. HSRP is used in a group of routers for selecting an active device and a standby device. In a group of device interfaces, the active device is the device that is used for routing packets; the standby device is the device that takes over when the active device fails, or when pre-set conditions are met. The function of the HSRP standby router is to monitor the operational status of the HSRP group and to quickly assume packet-forwarding responsibility if the active router fails.

HSRP Priority and Preemption (9.2.2)

The role of the active and standby routers is determined during the HSRP election process. By default, the router with the numerically highest IPv4 address is elected as the active router. However, it is always better to control how your network will operate under normal conditions rather than leaving it to chance.

HSRP Priority

HSRP priority can be used to determine the active router. The router with the highest HSRP priority will become the active router. By default, the HSRP priority is 100. If the priorities are equal, the router with the numerically highest IPv4 address is elected as the active router.

To configure a router to be the active router, use the **standby priority** interface command. The range of the HSRP priority is 0 to 255.

HSRP Preemption

By default, after a router becomes the active router, it will remain the active router even if another router comes online with a higher HSRP priority.

To force a new HSRP election process to take place when a higher priority router comes online, preemption must be enabled using the **standby preempt** interface command. Preemption is the ability of an HSRP router to trigger the reelection process. With preemption enabled, a router that comes online with a higher HSRP priority will assume the role of the active router.

Preemption only allows a router to become the active router if it has a higher priority. A router enabled for preemption, with equal priority but a higher IPv4 address, will not preempt an active router. Refer to the topology in Figure 9-4.

R1 has been configured with the HSRP priority of 150, and R2 has the default HSRP priority of 100. Preemption has been enabled on R1. With a higher priority, R1 is the active router and R2 is the standby router. Due to a power failure affecting only R1, the active router is no longer available and the standby router, R2, assumes the role of the active router. After power is restored, R1 comes back online. Because R1 has a higher priority and preemption is enabled, it will force a new election process. R1 will reassume the role of the active router, and R2 will fall back to the role of the standby router.

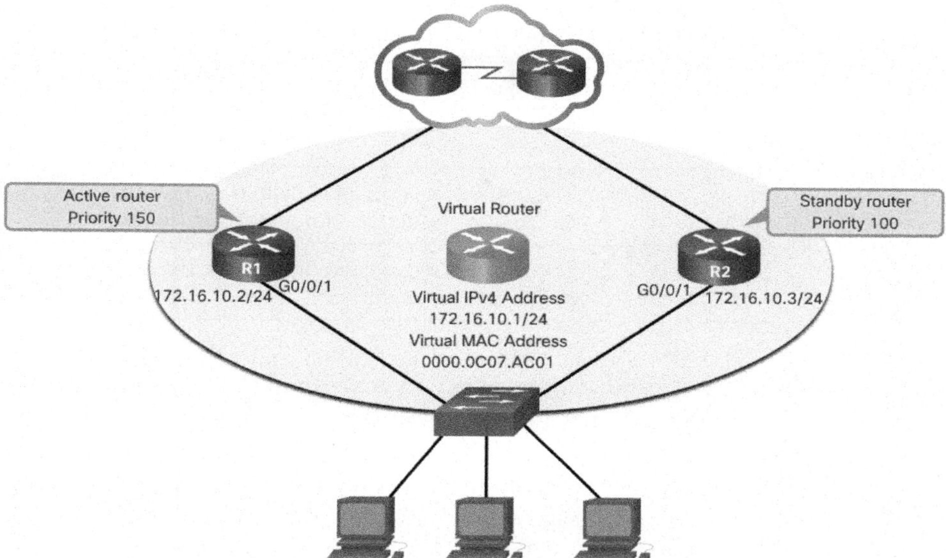

Figure 9-4 HSRP Topology with Active and Standby Routers

Note

With preemption disabled, the router that boots first will become the active router if there are no other routers online during the election process.

HSRP States and Timers (9.2.3)

A router can either be the active HSRP router responsible for forwarding traffic for the segment, or it can be a passive HSRP router on standby, ready to assume the active role if the active router fails. When an interface is configured with HSRP or is first activated with an existing HSRP configuration, the router sends and receives HSRP hello packets to begin the process of determining which state it will assume in the HSRP group.

Table 9-2 summarizes the HSRP states.

Table 9-2 HSRP States

HSRP State	Description
Initial	This state is entered through a configuration change or when an interface first becomes available.
Learn	The router has not determined the virtual IP address and has not yet seen a hello message from the active router. In this state, the router waits to hear from the active router.

HSRP State	Description
Listen	The router knows the virtual IP address, but the router is neither the active router nor the standby router. It listens for hello messages from those routers.
Speak	The router sends periodic hello messages and actively participates in the election of the active and/or standby router.
Standby	The router is a candidate to become the next active router and sends periodic hello messages.

The active and standby HSRP routers send hello packets to the HSRP group multicast address every 3 seconds by default. The standby router will become active if it does not receive a hello message from the active router after 10 seconds. You can lower these timer settings to speed up the failover or preemption. However, to avoid increased CPU usage and unnecessary standby state changes, do not set the hello timer below 1 second or the hold timer below 4 seconds.

Interactive Graphic

Check Your Understanding—HSRP (9.2.4)

Refer to the online course to complete this activity.

Summary (9.3)

The following is a summary of each section in the module.

First Hop Redundancy Protocols

If a router or router interface that serves as a default gateway fails, the hosts configured with that default gateway are isolated from outside networks. FHRP provides alternate default gateways in switched networks where two or more routers are connected to the same VLANs. One way to prevent a single point of failure at the default gateway is to implement a virtual router. With a virtual router, multiple routers are configured to work together to present the illusion of a single router to the hosts on the LAN. When the active router fails, the redundancy protocol transitions the standby router to the new active router role. These are the steps that take place when the active router fails:

Step 1. The standby router stops seeing hello messages from the forwarding router.

Step 2. The standby router assumes the role of the forwarding router.

Step 3. Because the new forwarding router assumes both the IPv4 and MAC addresses of the virtual router, the host devices see no disruption in service.

The FHRP used in a production environment largely depends on the equipment and needs of the network. These are the options available for FHRPs:

- HSRP and HSRP for IPv6
- VRRPv2 and VRRPv3
- GLBP and GLBP for IPv6
- IRDP

HSRP

HSRP is a Cisco-proprietary FHRP designed to allow for transparent failover of a first-hop IP device. HSRP is used in a group of routers for selecting an active device and a standby device. In a group of device interfaces, the active device is the device that is used for routing packets; the standby device is the device that takes over when the active device fails, or when pre-set conditions are met. The function of the HSRP standby router is to monitor the operational status of the HSRP group and to quickly assume packet-forwarding responsibility if the active router fails. The router with the highest HSRP priority will become the active router. Preemption is the ability of an HSRP router to trigger the reelection process. With preemption enabled, a router that comes online with a higher HSRP priority will assume the role of the active router. HSRP states include initial, learn, listen, speak, and standby.

Packet Tracer Activity

Packet Tracer—HSRP Configuration Guide (9.3.3)

Note: HSRP configuration is not a required skill for this module, course, or for the CCNA certification. However, we thought you might enjoy implementing HSRP in Packet Tracer. Completing this activity will help you better understand how FHRPs, and specifically HSRP, operate.

In this Packet Tracer activity, you learn how to configure Hot Standby Router Protocol (HSRP) to provide redundant default gateway devices to hosts on LANs. After configuring HSRP, you will test the configuration to verify that hosts are able to use the redundant default gateway if the current gateway device becomes unavailable.

Configure an HSRP active router.

Configure an HSRP standby router.

Verify HSRP operation.

Practice

The following activities provide practice with the topics introduced in this chapter. The Labs are available in the companion *Switching, Routing, and Wireless Essentials Labs and Study Guide (CCNAv7)* (ISBN 9780136634386). The Packet Tracer Activity instructions are also in the Labs & Study Guide. The PKA files are found in the online course.

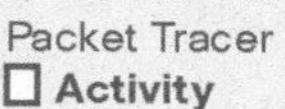

Packet Tracer Activity

Packet Tracer 9.3.3: HSRP Configuration Guide

Check Your Understanding Questions

Complete all the review questions listed here to test your understanding of the sections and concepts in this chapter. The appendix "Answers to the 'Check Your Understanding' Questions" lists the answers.

1. Which statement about HSRP operation is true?

   A. HSRP supports only clear-text authentication.

   B. The active router responds to requests for the virtual MAC and virtual IP address.

   C. The AVF responds to default gateway ARP requests.

   D. The HSRP virtual IP address must be the same as one of the router's interface addresses on the LAN.

2. Which statement about HSRP operation is true?

   A. HSRP supports only clear-text authentication.

   B. The active router responds to requests for the virtual MAC and virtual IP address.

   C. The AVF responds to default gateway ARP requests.

   D. The HSRP virtual IP address must be the same as one of the router's interface addresses on the LAN.

3. Which statement regarding VRRP is true?

   A. VRRP elects a master router and one or more other routers as backup routers.

   B. VRRP elects a master router, one backup router, and all other routers are standby routers.

   C. VRRP elects an active router and a standby router, and all other routers are backup routers.

   D. VRRP is a Cisco proprietary protocol.

4. A network administrator is overseeing the implementation of first hop redundancy protocols. Which protocol is a Cisco proprietary protocol?

   A. HSRP

   B. IRDP

   C. Proxy ARP

   D. VRRP

5. What is the purpose of HSRP?

   A. It enables an access port to immediately transition to the forwarding state.

   B. It prevents a rogue switch from becoming the STP root.

   C. It prevents malicious hosts from connecting to trunk ports.

   D. It provides a continuous network connection when a default gateway fails.

6. Which is a characteristic of the HSRP Learn state?

   A. The router actively participates in the active/standby election process.

   B. The router has not determined the virtual IP address.

   C. The router knows the virtual IP address.

   D. The router sends periodic hello messages.

7. A network administrator is analyzing the features that are supported by different first-hop router redundancy protocols. Which statement describes a feature that is associated with VRRP?

   A. VRRP assigns active and standby routers.

   B. VRRP assigns an IP address and default gateway to hosts.

   C. VRRP enables load balancing between a group of redundant routers.

   D. VRRP is a non-proprietary protocol.

8. When HSRP is used in a network, what destination MAC address is used in frames that are sent from the workstation to the default gateway?

   A. MAC address of the forwarding router

   B. MAC addresses of both the forwarding and standby routers

   C. MAC address of the standby router

   D. MAC address of the virtual router

9. What happens to a host in an HSRP network when the active router fails?

   A. The host initiates a new ARP request.

   B. The host stops seeing hello messages from the active router.

   C. The host uses the standby router IP and MAC addresses.

   D. The host will notice little or no disruption of service.

10. Which of the following correctly describes GLBP?

    A. It is a Cisco proprietary FHRP and provides redundancy and load sharing.

    B. It is an open standard FHRP.

    C. It uses virtual master routers and one or more backup routers.

    D. It is a legacy open standard FHRP that allows IPv4 hosts to discover gateway routers.

11. Which HSRP preemption statement is true?

    A. It enables a router that boots first to become the active router.

    B. It is enabled by default.

    C. It is enabled using the **standby preempt** interface command.

    D. It is enabled using the **standby priority** interface command.

CHAPTER 10

LAN Security Concepts

Objectives

Upon completion of this chapter, you will be able to answer the following questions:

- How do you use endpoint security to mitigate attacks?
- How are AAA and 802.1X used to authenticate LAN endpoints and devices?
- What are Layer 2 vulnerabilities?
- How does a MAC address table attack compromise LAN security?
- How do LAN attacks compromise LAN security?

Key Terms

This chapter uses the following key terms. You can find the definitions in the Glossary.

Distributed Denial of Service (DDoS) Page 277

zombies Page 277

data breach Page 277

malware Page 277

ransomware Page 277

WannaCry Page 277

virtual private network (VPN) Page 278

next-generation firewall (NGFW) Page 278

network access control (NAC) Page 278

VPN-enabled router Page 278

firewall Page 278

stateful packet inspection Page 278

next-generation intrusion prevention system (NGIPS) Page 278

advanced malware protection (AMP) Page 278

URL filtering Page 278

Cisco Identity Services Engine (ISE) Page 278

wireless LAN controllers (WLCs) Page 278

threat actor Page 278

host-based intrusion prevention systems (HIPSs) Page 278

email security appliance (ESA) Page 278

web security appliance (WSA) Page 278

Cisco Talos Intelligence Group Page 279

phishing Page 279

spear phishing Page 279

SANS Institute Page 279

blacklisting Page 280

authentication, authorization and accounting (AAA) Page 281

802.1X Page 281

Remote Authentication Dial-In User Service (RADIUS) Page 284

Terminal Access Controller Access Control System (TACACS+) Page 284

MAC table attacks Page 288

MAC address flooding Page 288

VLAN attacks Page 288

VLAN hopping Page 288

VLAN double-tagging Page 288

DHCP attacks Page 289

DHCP starvation Page 289

DHCP spoofing Page 289

ARP attacks Page 289

ARP spoofing Page 289

ARP poisoning Page 289

address spoofing attacks Page 289

MAC address spoofing Page 289

IP address spoofing Page 289

STP manipulation attacks Page 289

port security Page 289

DHCP snooping Page 289

Dynamic ARP Inspection (DAI) Page 289

IP Source Guard (IPSG) Page 289

Secure Copy Protocol (SCP) Page 289

Secure FTP (SFTP) Page 289

Secure Sockets Layer (SSL) Page 289

Transport Layer Security (TLS) Page 289

out-of-band management Page 289

man-in-the-middle attack Page 297

gratuitous ARP Page 300

Cisco Discovery Protocol (CDP) Page 305

Link Layer Discovery Protocol (LLDP) Page 306

Introduction (10.0)

If your career path is in IT, you won't just be building or maintaining networks. You will be responsible for the security of your network. For today's network architects and administrators, security is not an afterthought. It is a top priority for them! In fact, many people in IT now work exclusively in the area of network security.

Do you understand what makes a LAN secure? Do you know what threat actors can do to break network security? Do you know what you can do to stop them? This module is your introduction to the world of network security, so don't wait, click Next!

Endpoint Security (10.1)

This section explains how to use endpoint security to mitigate attacks.

Network Attacks Today (10.1.1)

The news media commonly covers attacks on enterprise networks. Simply search the Internet for "latest network attacks" to find up-to-date information on current attacks. Most likely, these attacks will involve one or more of the following:

- *Distributed Denial of Service (DDoS)*: This is a coordinated attack from many devices, called *zombies*, with the intention of degrading or halting public access to an organization's website and resources.
- *Data Breach*: This is an attack in which an organization's data servers or hosts are compromised to steal confidential information.
- *Malware*: This is an attack in which an organization's hosts are infected with malicious software that cause a variety of problems. For example, *ransomware* such as *WannaCry*, shown in Figure 10-1, encrypts the data on a host and locks access to it until a ransom is paid.

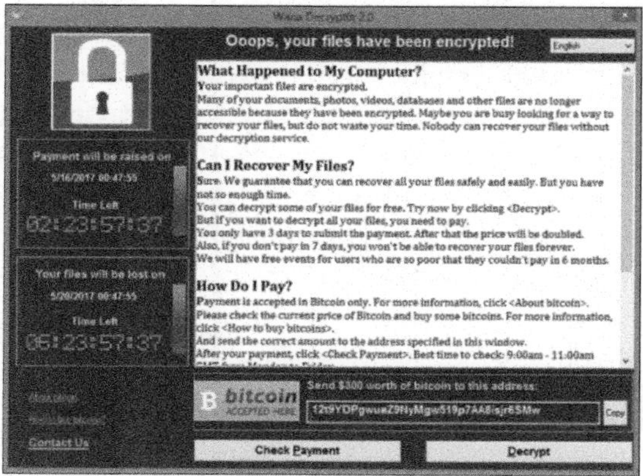

Figure 10-1 WannaCry Ransomware

Network Security Devices (10.1.2)

Various network security devices are required to protect the network perimeter from outside access. These devices could include a *virtual private network (VPN)* enabled router, a *next-generation firewall (NGFW)*, and a *network access control (NAC)* device.

- *VPN-Enabled Router*: A VPN-enabled router provides a secure connection to remote users across a public network and into the enterprise network. VPN services can be integrated into the *firewall*.
- **NGFW:** An NGFW provides *stateful packet inspection*, application visibility and control, a *next-generation intrusion prevention system (NGIPS)*, *advanced malware protection (AMP)*, and *URL filtering*.
- **NAC:** A NAC device includes authentication, authorization, and accounting (AAA) services. In larger enterprises, these services might be incorporated into an appliance that can manage access policies across a wide variety of users and device types. The *Cisco Identity Services Engine (ISE)* is an example of a NAC device.

Endpoint Protection (10.1.3)

LAN devices such as switches, *wireless LAN controllers (WLCs)*, and other access point (AP) devices interconnect endpoints. Most of these devices are susceptible to the LAN-related attacks that are covered in this module.

But many attacks can also originate from inside the network. If an internal host is infiltrated, it can become a starting point for a *threat actor* to gain access to critical system devices, such as servers and sensitive data.

Endpoints are hosts that commonly consist of laptops, desktops, servers, and IP phones, as well as employee-owned devices that are typically referred to as bring your own devices (BYODs). Endpoints are particularly susceptible to malware-related attacks that originate through email or web browsing. These endpoints have typically used traditional host-based security features, such as antivirus/antimalware, host-based firewalls, and *host-based intrusion prevention systems (HIPS)*. However, today endpoints are best protected by a combination of NAC, host-based AMP software, an *email security appliance (ESA)*, and a *web security appliance (WSA)*. Advanced Malware Protection (AMP) products include endpoint solutions, such as Cisco AMP for Endpoints.

Figure 10-2 is a simple topology representing all the network security devices and endpoint solutions discussed in this module.

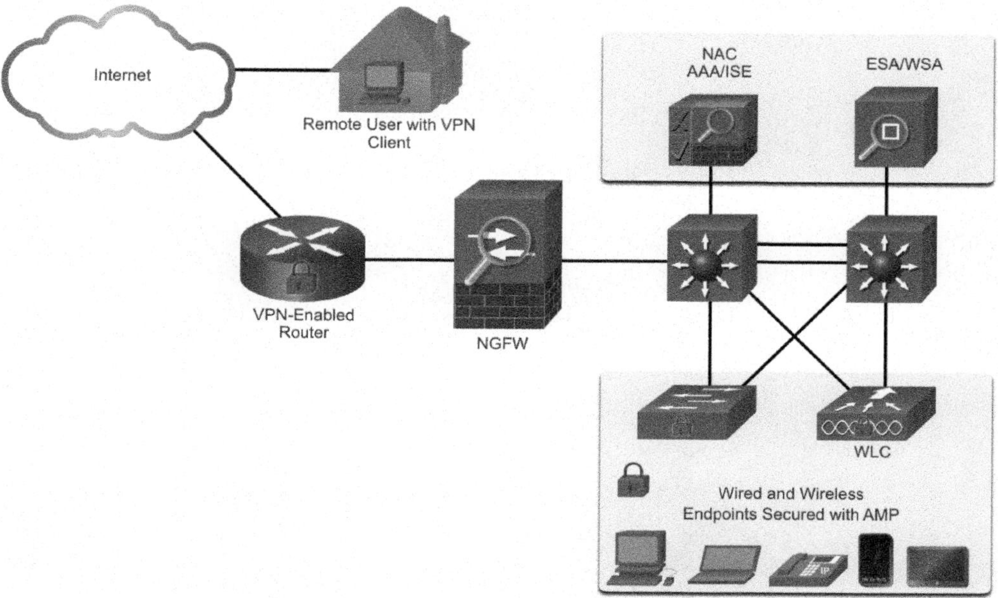

Figure 10-2 Topology with Secured Network Devices and Endpoints

Cisco Email Security Appliance (10.1.4)

Content security appliances include fine-grained control over email and web browsing for an organization's users.

According to the *Cisco Talos Intelligence Group*, in June 2019, 85% of all email sent was spam. *Phishing* attacks are a particularly virulent form of spam. Recall that a phishing attack entices the user to click a link or open an attachment. *Spear phishing* targets high-profile employees or executives who may have elevated login credentials. This is particularly crucial in today's environment where, according to the *SANS Institute*, 95% of all attacks on enterprise networks are the result of a successful spear phishing attack.

The Cisco ESA is a device that is designed to monitor Simple Mail Transfer Protocol (SMTP). The Cisco ESA is constantly updated by real-time feeds from the Cisco Talos, which detects and correlates threats and solutions by using a worldwide database monitoring system. This threat intelligence data is pulled by the Cisco ESA every three to five minutes. These are some of the functions of the Cisco ESA:

- Block known threats.
- Remediate against stealth malware that evaded initial detection.

- Discard emails with bad links (as shown in the figure).
- Block access to newly infected sites.
- Encrypt content in outgoing email to prevent data loss.

In Figure 10-3, the Cisco ESA discards the email with bad links.

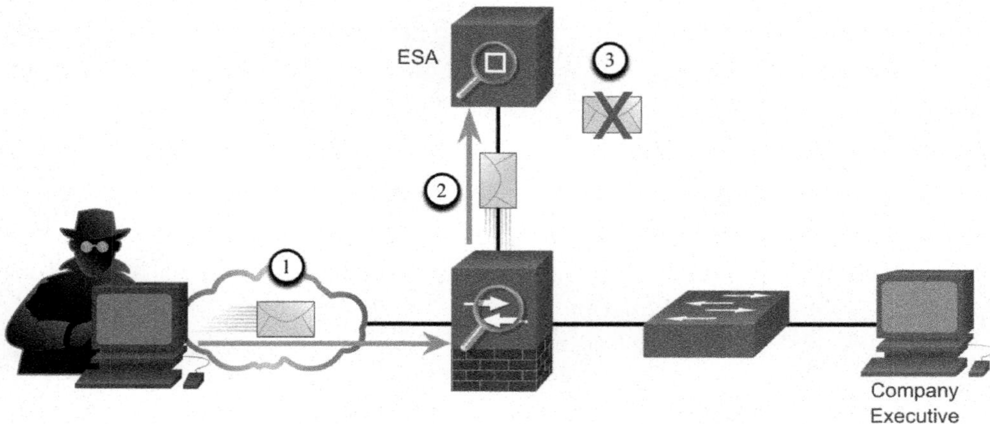

Figure 10-3 Example of Cisco ESA in Operation

In Figure 10-3, the following occurs:

1. Threat actor sends a phishing attack to an important host on the network.
2. The firewall forwards all email to the ESA.
3. The ESA analyzes the email, logs it, and if it is malware discards it.

Cisco Web Security Appliance (10.1.5)

The Cisco Web Security Appliance (WSA) is a mitigation technology for web-based threats. It helps organizations address the challenges of securing and controlling web traffic. The Cisco WSA combines advanced malware protection, application visibility and control, acceptable use policy controls, and reporting.

Cisco WSA provides complete control over how users access the Internet. Certain features and applications, such as chat, messaging, video, and audio, can be allowed, restricted with time and bandwidth limits, or blocked, according to the organization's requirements. The WSA can perform *blacklisting* of URLs, URL-filtering, malware scanning, URL categorization, Web application filtering, and encryption and decryption of web traffic.

In Figure 10-4, an internal corporate employee uses a smartphone to attempt to connect to a known blacklisted site.

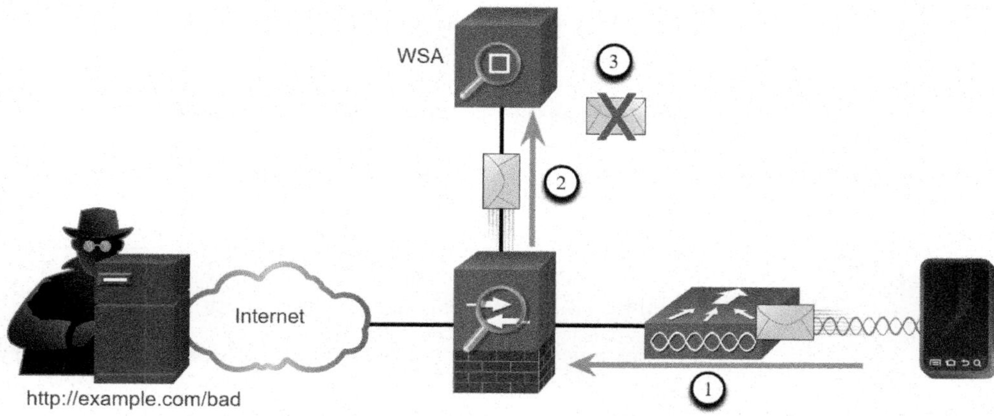

Figure 10-4 Example of Cisco WSA in Operation

In Figure 10-4, the following occurs:

1. A user attempts to connect to a website.
2. The firewall forwards the website request to the WSA.
3. The WSA evaluates the URL and determines it is a known blacklisted site. The WSA discards the packet and sends an access denied message to the user.

Interactive Graphic

Check Your Understanding—Endpoint Security (10.1.6)

Refer to the online course to complete this activity.

Access Control (10.2)

This section will explain how *authentication*, *authorization and accounting (AAA)* and *802.1X* are used to authenticate LAN endpoints and devices.

Authentication with a Local Password (10.2.1)

In the previous section, you learned that a NAC device provides AAA services. In this section, you learn more about AAA and the ways to control access.

Many types of authentication can be performed on networking devices, and each method offers varying levels of security. The simplest method of remote access authentication is to configure a login and password combination on console, vty lines, and aux ports, as shown in the vty lines in Example 10-1. This method is the easiest to implement, but it is also the weakest and least secure. This method provides no accountability, and the password is sent in plain text. Anyone with the password can gain entry to the device.

Example 10-1 Setting a Password on the VTY Lines

```
R1(config)# line vty 0 4
R1(config-line)# password ci5c0
R1(config-line)# login
```

Secure Shell (SSH) is a more secure form of remote access:

- It requires a username and a password, both of which are encrypted during transmission.
- The username and password can be authenticated by the local database method.
- It provides more accountability because the username is recorded when a user logs in.

Example 10-2 illustrates SSH and local database methods of remote access.

Example 10-2 Configuring SSH and Setting VTY Lines to Use SSH

```
R1(config)# ip domain-name example.com
R1(config)# crypto key generate rsa general-keys modulus 2048
R1(config)# username Admin secret Str0ng3rPa55w0rd
R1(config)# ssh version 2
R1(config)# line vty 0 4
R1(config-line)# transport input ssh
R1(config-line)# login local
```

The local database method has some limitations:

- **User accounts must be manually preconfigured:** User accounts must be configured locally on each device. In a large enterprise environment with multiple routers and switches to manage, it can take time to implement and change local databases on each device.
- **Local database method provides no fallback method:** The local database configuration provides no fallback authentication method. For example, what if the administrator forgets the username and password for that device? With no backup method available for authentication, password recovery becomes the only option.

A better solution is to have all devices refer to the same database of usernames and passwords from a central server.

AAA Components (10.2.2)

AAA stands for Authentication, Authorization, and Accounting. The AAA concept is similar to using a credit card, as shown Figure 10-5. The credit card identifies who can use it, how much that user can spend, and keeps an account of what items or services the user purchased.

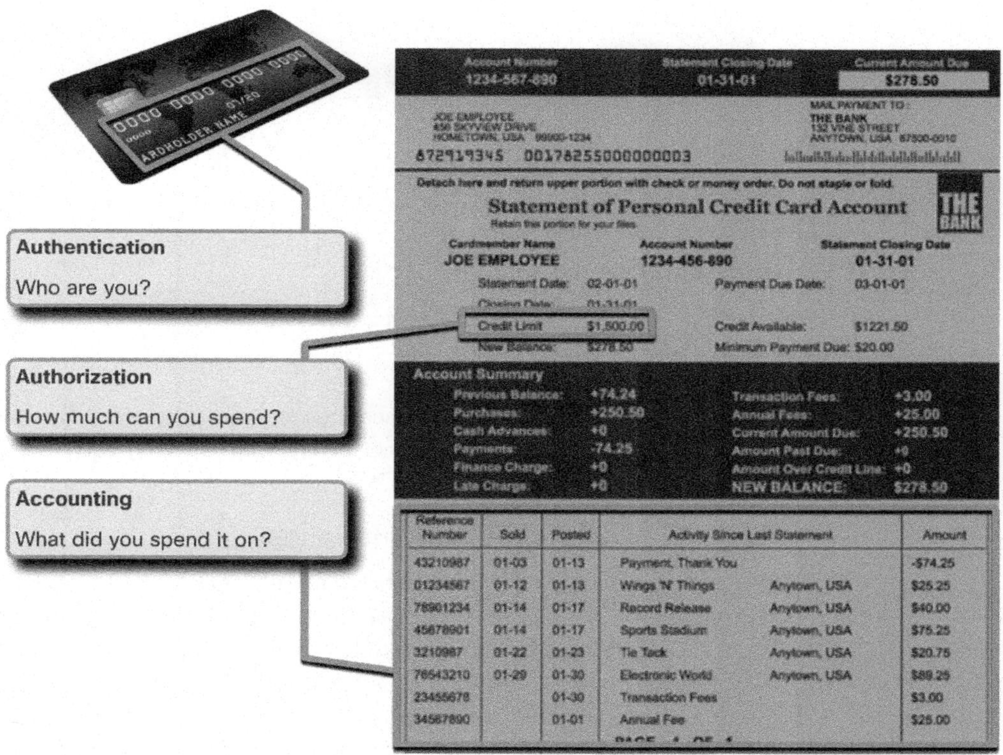

Figure 10-5 Analogy of AAA in a Credit Card Bill

AAA provides the primary framework to set up access control on a network device. AAA is a way to control who is permitted to access a network (authenticate), what they can do while they are there (authorize), and to audit what actions they performed while accessing the network (accounting).

Authentication (10.2.3)

Local and server-based are two common methods of implementing AAA authentication.

Local AAA Authentication

Local AAA stores usernames and passwords locally in a network device such as the Cisco router. Users authenticate against the local database, as shown Figure 10-6. Local AAA is ideal for small networks.

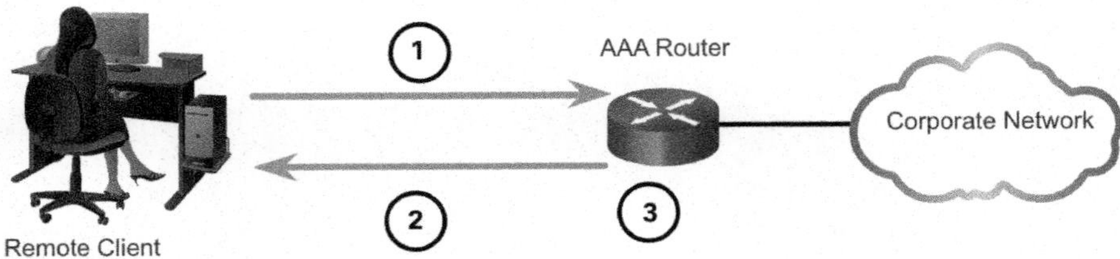

Figure 10-6 Example of Local AAA Authentication

In Figure 10-6, the following occurs:

1. The client establishes a connection with the router.
2. The AAA router prompts the user for a username and password.
3. The router authenticates the username and password using the local database, and the user is provided access to the network based on information in the local database.

Server-Based AAA Authentication

With the server-based method, the router accesses a central AAA server, as shown Figure 10-7. The AAA server contains the usernames and password for all users. The router uses either the *Remote Authentication Dial-In User Service (RADIUS)* or *Terminal Access Controller Access Control System (TACACS+)* protocols to communicate with the AAA server. When there are multiple routers and switches, server-based AAA is more appropriate.

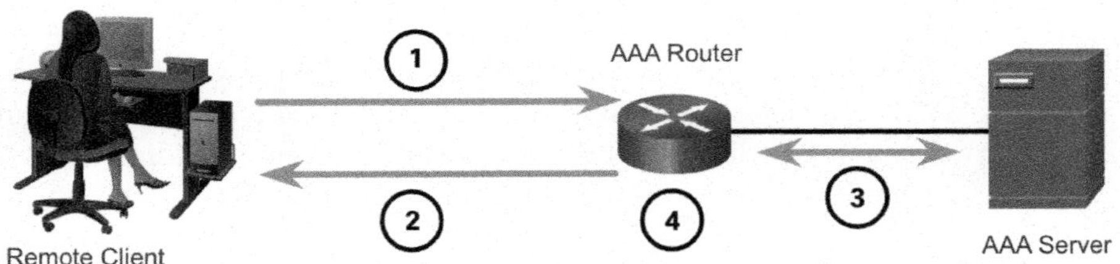

Figure 10-7 Example of Server-Based AAA Authentication

In Figure 10-7, the following occurs:

1. The client establishes a connection with the router.
2. The AAA router prompts the user for a username and password.

3. The router authenticates the username and password using a AAA server.
4. The user is provided access to the network based on information in the remote AAA server.

Authorization (10.2.4)

AAA authorization is automatic and does not require users to perform additional steps after authentication. Authorization governs what users can and cannot do on the network after they are authenticated.

Authorization uses a set of attributes that describes the user's access to the network. These attributes are used by the AAA server to determine privileges and restrictions for that user, as shown in Figure 10-8.

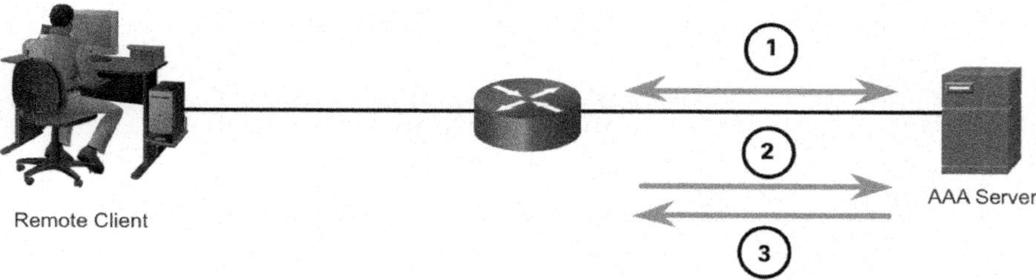

Figure 10-8 Example of AAA Authorization

In Figure 10-8, the following occurs:

1. When a user has been authenticated, a session is established between the router and the AAA server.
2. The router requests authorization from the AAA server for the client's requested service.
3. The AAA server returns a PASS/FAIL response for authorization.

Accounting (10.2.5)

AAA accounting collects and reports usage data. This data can be used for such purposes as auditing or billing. The collected data might include the start and stop connection times, executed commands, number of packets, and number of bytes.

A primary use of accounting is to combine it with AAA authentication. The AAA server keeps a detailed log of exactly what the authenticated user does on the device, as shown in Figure 10-9. This includes all EXEC and configuration commands issued by the user. The log contains numerous data fields, including the username, the date and time, and the actual command that was entered by the user. This information is useful when troubleshooting devices. It also provides evidence for when individuals perform malicious acts.

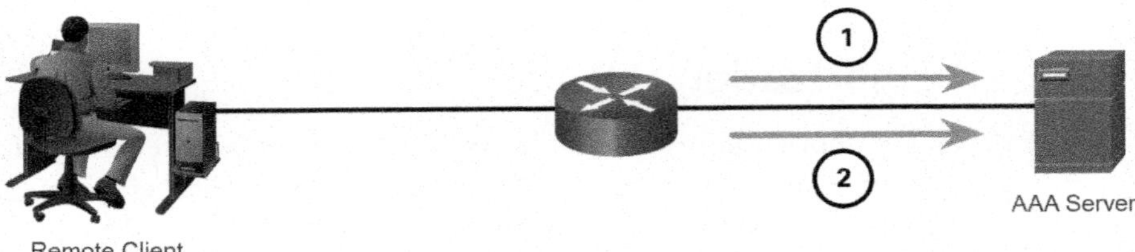

Figure 10-9 Example of AAA Accounting

In Figure 10-9, the following occurs:

1. When a user has been authenticated, a session is established between the router and the AAA server.
2. The router requests authorization from the AAA server for the client's requested service.

802.1X (10.2.6)

The IEEE 802.1X standard is a port-based access control and authentication protocol. This protocol restricts unauthorized workstations from connecting to a LAN through publicly accessible switch ports. The authentication server authenticates each workstation that is connected to a switch port before making available any services offered by the switch or the LAN.

With 802.1X port-based authentication, the devices in the network have specific roles, as shown in Figure 10-10.

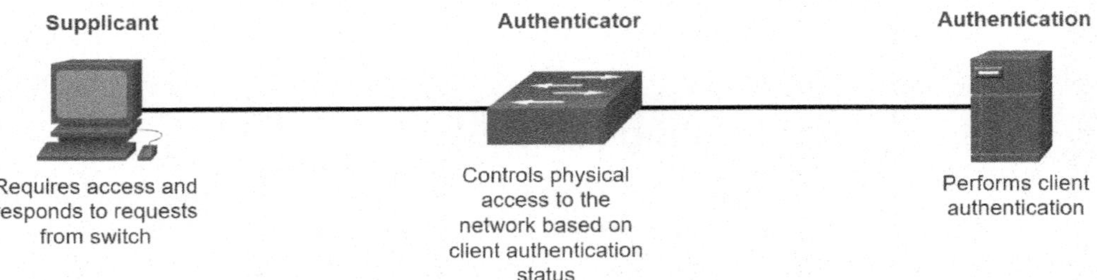

Figure 10-10 Example of 802.1X Port-Based Authentication

In Figure 10-10, the 802.1X roles are as follows:

- **Client (Supplicant):** This is a device running 802.1X-compliant client software, which is available for wired or wireless devices.
- **Switch (Authenticator):** The switch acts as an intermediary between the client and the authentication server. It requests identifying information from the client, verifies that information with the authentication server, and relays a response to the client. Another device that could act as authenticator is a wireless access point.
- **Authentication server:** The server validates the identity of the client and notifies the switch or wireless access point that the client is or is not authorized to access the LAN and switch services.

Interactive Graphic

Check Your Understanding—Access Control (10.2.7)

Refer to the online course to complete this activity.

Layer 2 Security Threats (10.3)

This section will identify Layer 2 vulnerabilities.

Layer 2 Vulnerabilities (10.3.1)

The previous two sections discussed securing endpoints. In this section, you continue to learn about ways to secure the LAN by focusing on the frames found in the data link layer (Layer 2) and the switch.

Recall that the OSI reference model is divided into seven layers that work independently of each other. Figure 10-11 shows the function of each layer and the core elements that can be exploited.

Network administrators routinely implement security solutions to protect the elements in Layer 3 up through Layer 7. They use VPNs, firewalls, and IPS devices to protect these elements. However, if Layer 2 is compromised, all the layers above it are also affected. For example, if a threat actor with access to the internal network captured Layer 2 frames, all the security implemented on the layers above would be useless. The threat actor could cause a lot of damage on the Layer 2 LAN networking infrastructure.

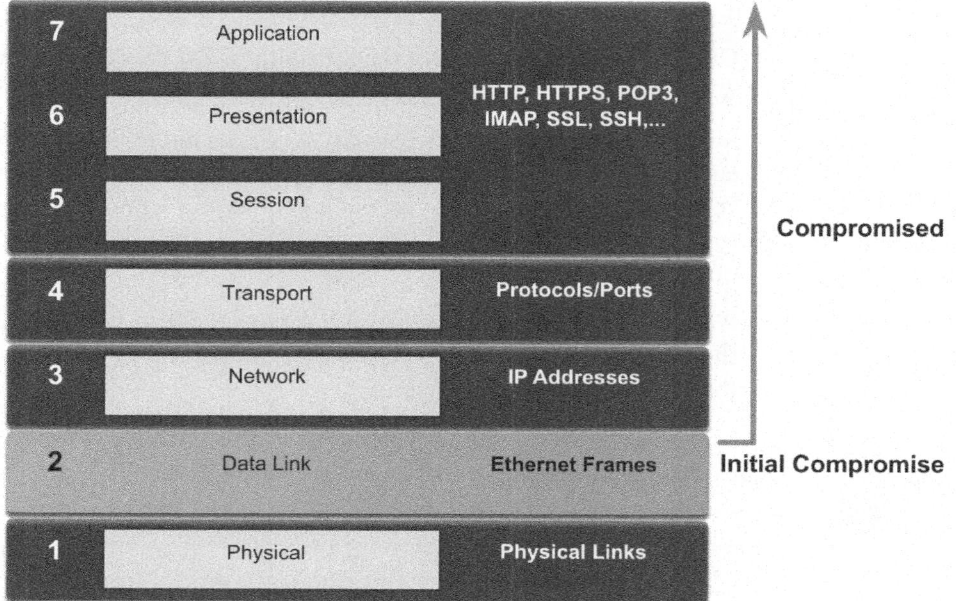

Figure 10-11 Compromising Layer 2 of the OSI Model

Switch Attack Categories (10.3.2)

Security is only as strong as the weakest link in the system, and Layer 2 is considered to be that weak link. This is because LANs were traditionally under the administrative control of a single organization. We inherently trusted all persons and devices connected to our LAN. Today, with BYOD and more sophisticated attacks, our LANs have become more vulnerable to penetration. Therefore, in addition to protecting Layer 3 to Layer 7, network security professionals must also mitigate attacks to the Layer 2 LAN infrastructure.

The first step in mitigating attacks on the Layer 2 infrastructure is to understand the underlying operation of Layer 2 and the threats posed by the Layer 2 infrastructure.

Attacks against the Layer 2 LAN infrastructure are described in Table 10-1 and are discussed in more detail later in this module.

Table 10-1 Layer 2 Attacks

Category	Examples
MAC Table Attacks	Includes *MAC address flooding* attacks.
VLAN Attacks	Includes *VLAN hopping* and *VLAN double-tagging* attacks. It also includes attacks between devices on a common VLAN.

Category	Examples
DHCP Attacks	Includes *DHCP starvation* and *DHCP spoofing* attacks.
ARP Attacks	Includes *ARP spoofing* and *ARP poisoning* attacks.
Address Spoofing Attacks	Includes *MAC address spoofing* and *IP address spoofing* attacks.
STP Attacks	Includes *STP manipulation attacks*.

Switch Attack Mitigation Techniques (10.3.3)

Table 10-2 provides an overview of Cisco solutions to help mitigate Layer 2 attacks.

Table 10-2 Layer 2 Attack Mitigation

Solution	Description
Port Security	Prevents many types of attacks including MAC address flooding attacks and DHCP starvation attacks.
DHCP Snooping	Prevents DHCP starvation and DHCP spoofing attacks.
Dynamic ARP Inspection (DAI)	Prevents ARP spoofing and ARP poisoning attacks.
IP Source Guard (IPSG)	Prevents MAC and IP address spoofing attacks.

These Layer 2 solutions will not be effective if the management protocols are not secured. For example, the management protocols Syslog, Simple Network Management Protocol (SNMP), Trivial File Transfer Protocol (TFTP), telnet, File Transfer Protocol (FTP), and most other common protocols are insecure; therefore, the following strategies are recommended:

- Always use secure variants of these protocols such as SSH, *Secure Copy Protocol (SCP)*, *Secure FTP (SFTP)*, *Secure Socket Layer (SSL) / Transport Layer Security (TLS)*.
- Consider using *out-of-band management* network to manage devices.
- Use a dedicated management VLAN where nothing but management traffic resides.
- Use ACLs to filter unwanted access.

Interactive Graphic

Check Your Understanding—Layer 2 Security Threats (10.3.4)

Refer to the online course to complete this activity.

MAC Address Table Attack (10.4)

This section will explain how a MAC address table attack compromises LAN security.

Switch Operation Review (10.4.1)

In this section, the focus is still on switches, specifically their MAC address tables and how these tables are vulnerable to attacks.

Recall that to make forwarding decisions, a Layer 2 LAN switch builds a table based on the source MAC addresses in received frames. Shown in Example 10-3, this is called a MAC address table. MAC address tables are stored in memory and are used to more efficiently forward frames.

Example 10-3 The MAC Address Table

```
S1# show mac address-table dynamic
          Mac Address Table
-------------------------------------------
Vlan    Mac Address       Type        Ports
----    -----------       --------    -----
   1    0001.9717.22e0    DYNAMIC     Fa0/4
   1    000a.f38e.74b3    DYNAMIC     Fa0/1
   1    0090.0c23.ceca    DYNAMIC     Fa0/3
   1    00d0.ba07.8499    DYNAMIC     Fa0/2
S1#
```

MAC Address Table Flooding (10.4.2)

All MAC tables have a fixed size and consequently, a switch can run out of resources in which to store MAC addresses. MAC address flooding attacks take advantage of this limitation by bombarding the switch with fake source MAC addresses until the switch MAC address table is full.

When this occurs, the switch treats the frame as an unknown unicast and begins to flood all incoming traffic out all ports on the same VLAN without referencing the

MAC table. This condition now allows a threat actor to capture all of the frames sent from one host to another on the local LAN or local VLAN.

Note

Traffic is flooded only within the local LAN or VLAN. The threat actor can capture traffic only within the local LAN or VLAN to which the threat actor is connected.

Figure 10-12 shows how a threat actor can easily use the network attack tool **macof** to overflow a MAC address table.

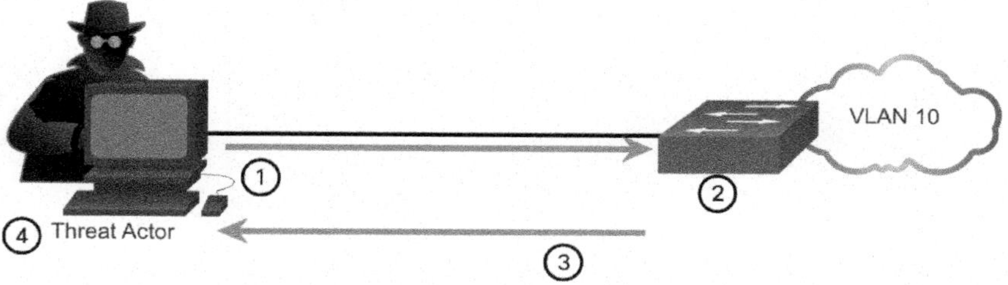

Figure 10-12 Flooding the MAC Address Table with an Attack Tool

In Figure 10-12, the following occurs:

1. The threat actor is connected to VLAN 10 and uses **macof** to rapidly generate many random source and destination MAC and IP addresses.
2. Over a short period of time, the switch's MAC table fills up.
3. When the MAC table is full, the switch begins to flood all frames that it receives. As long as **macof** continues to run, the MAC table remains full and the switch continues to flood all incoming frames out every port associated with VLAN 10.
4. The threat actor then uses packet sniffing software to capture frames from any and all devices connected to VLAN 10.

If the threat actor stops **macof** from running or is discovered and stopped, the switch eventually ages out the older MAC address entries from the table and begins to act like a switch again.

MAC Address Table Attack Mitigation (10.4.3)

What makes tools such as **macof** so dangerous is that an attacker can create a MAC table overflow attack very quickly. For instance, a Catalyst 6500 switch can store 132,000 MAC addresses in its MAC address table. A tool such as **macof** can flood a switch with up to 8,000 bogus frames per second, creating a MAC address table overflow attack in a matter of a few seconds. Example 10-4 shows a sample output of the **macof** command on a Linux host.

Example 10-4 Output from **macof** on a Linux Host

```
# macof -i eth1
36:a1:48:63:81:70 15:26:8d:4d:28:f8 0.0.0.0.26413 > 0.0.0.0.49492: S
  1094191437:1094191437(0) win 512
16:e8:8:0:4d:9c da:4d:bc:7c:ef:be 0.0.0.0.61376 > 0.0.0.0.47523: S
  446486755:446486755(0) win 512
18:2a:de:56:38:71 33:af:9b:5:a6:97 0.0.0.0.20086 > 0.0.0.0.6728: S
  105051945:105051945(0) win 512
e7:5c:97:42:ec:1 83:73:1a:32:20:93 0.0.0.0.45282 > 0.0.0.0.24898: S
  1838062028:1838062028(0) win 512
62:69:d3:1c:79:ef 80:13:35:4:cb:d0 0.0.0.0.11587 > 0.0.0.0.7723: S
  1792413296:1792413296(0) win 512
c5:a:b7:3e:3c:7a 3a:ee:c0:23:4a:fe 0.0.0.0.19784 > 0.0.0.0.57433: S
  1018924173:1018924173(0) win 512
88:43:ee:51:c7:68 b4:8d:ec:3e:14:bb 0.0.0.0.283 > 0.0.0.0.11466: S
  727776406:727776406(0) win 512
b8:7a:7a:2d:2c:ae c2:fa:2d:7d:e7:bf 0.0.0.0.32650 > 0.0.0.0.11324: S
  605528173:605528173(0) win 512
e0:d8:1e:74:1:e 57:98:b6:5a:fa:de 0.0.0.0.36346 > 0.0.0.0.55700: S
  2128143986:2128143986(0) win 512
```

Another reason why these attack tools are dangerous is because they not only affect the local switch, they can also affect other connected Layer 2 switches. When the MAC address table of a switch is full, it starts flooding out all ports, including those connected to other Layer 2 switches.

To mitigate MAC address table overflow attacks, network administrators must implement port security. Port security will allow only a specified number of source MAC addresses to be learned on the port. Port security is further discussed in another module.

Interactive Graphic

Check Your Understanding—MAC Address Table Attacks (10.4.4)

Refer to the online course to complete this activity.

LAN Attacks (10.5)

This section explains how LAN attacks compromise LAN security.

Video

Video—VLAN and DHCP Attacks (10.5.1)

Refer to the online course to view this video.

VLAN Hopping Attacks (10.5.2)

A VLAN hopping attack enables traffic from one VLAN to be seen by another VLAN without the aid of a router. In a basic VLAN hopping attack, the threat actor configures a host to act like a switch to take advantage of the automatic trunking port feature enabled by default on most switch ports.

The threat actor configures the host to spoof 802.1Q signaling and Cisco-proprietary Dynamic Trunking Protocol (DTP) signaling to trunk with the connecting switch. If successful, the switch establishes a trunk link with the host, as shown in Figure 10-13. Now the threat actor can access all the VLANs on the switch. The threat actor can send and receive traffic on any VLAN, effectively hopping between VLANs.

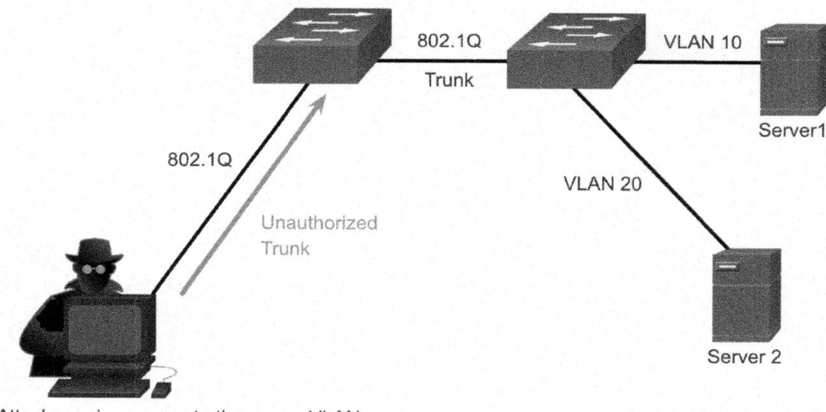

Figure 10-13 Example of a VLAN Hopping Attack

VLAN Double-Tagging Attack (10.5.3)

A threat actor in specific situations could embed a hidden 802.1Q tag inside the frame that already has an 802.1Q tag. This tag allows the frame to go to a VLAN that the original 802.1Q tag did not specify.

The following three steps provide an example and explanation of a double-tagging attack.

Step 1. **Threat actor sends a double-tagged 802.1Q frame to the switch.** In Figure 10-14, the outer header has the VLAN tag of the threat actor, which is the same as the native VLAN of the trunk port. For the purposes of this example, assume that this is VLAN 10. The inner tag is the victim VLAN, in this example, VLAN 20.

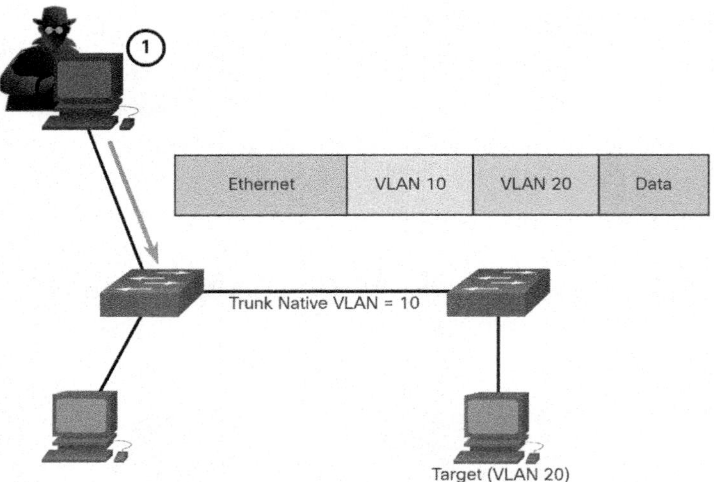

Figure 10-14 Step 1—Threat Actor Double-Tags a Frame

Step 2. **The frame arrives on the first switch.** In Figure 10-15, the switch sees that the frame is destined for VLAN 10, which is the native VLAN. The switch forwards the packet out all VLAN 10 ports after stripping the VLAN 10 tag. The frame is not retagged because it is part of the native VLAN. At this point, the VLAN 20 tag is still intact and has not been inspected by the first switch.

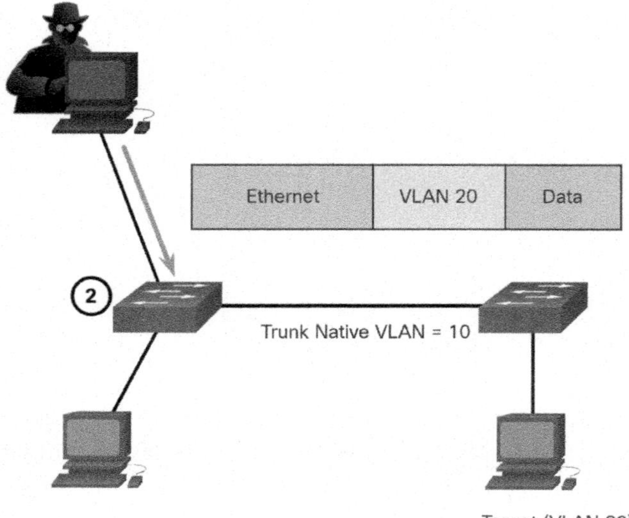

Figure 10-15 Step 2—Threat Actor's Tag Survives Tag Stripping

Step 3. **The frame arrives on the second switch.** In Figure 10-16, the second switch has no knowledge that the frame is supposed to be for VLAN 10. Native VLAN traffic is not tagged by the sending switch as specified in the 802.1Q specification. The second switch looks only at the inner 802.1Q tag that the threat actor inserted and sees that the frame is destined for VLAN 20, the target VLAN. The second switch sends the frame on to the target or floods it, depending on whether there is an existing MAC address table entry for the target.

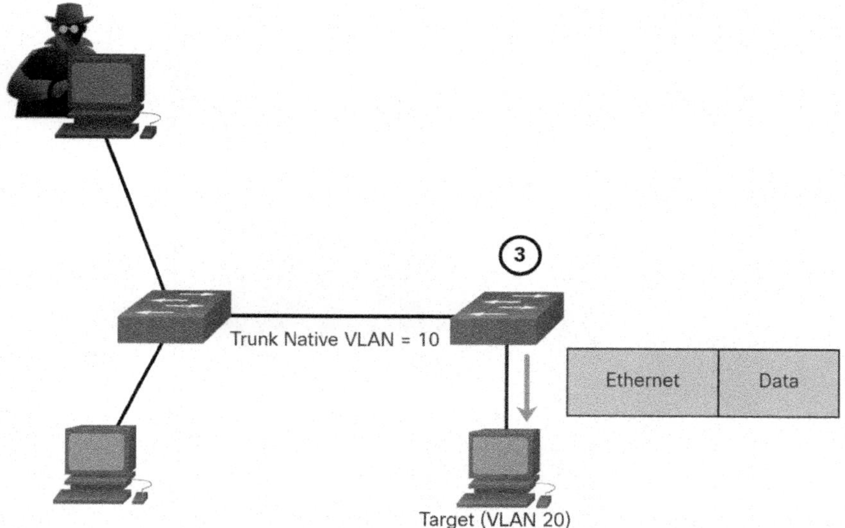

Figure 10-16 Step 3— Switch Forwards Threat Actor's Frame to the Victim

A VLAN double-tagging attack is unidirectional and works only when the attacker is connected to a port residing in the same VLAN as the native VLAN of the trunk port. The idea is that double tagging allows the attacker to send data to hosts or servers on a VLAN that otherwise would be blocked by some type of access control configuration. Presumably the return traffic will also be permitted, thus giving the attacker the ability to communicate with devices on the normally blocked VLAN.

VLAN Attack Mitigation

VLAN hopping and VLAN double-tagging attacks can be prevented by implementing the following trunk security guidelines, as discussed in a previous module:

- Disable trunking on all access ports.
- Disable auto trunking on trunk links so that trunks must be manually enabled.
- Be sure that the native VLAN is used only for trunk links.

DHCP Messages (10.5.4)

DHCP servers dynamically provide IP configuration information, including IP address, subnet mask, default gateway, DNS servers, and more to clients. A review of the sequence of the DHCP message exchange between client and server is shown in Figure 10-17.

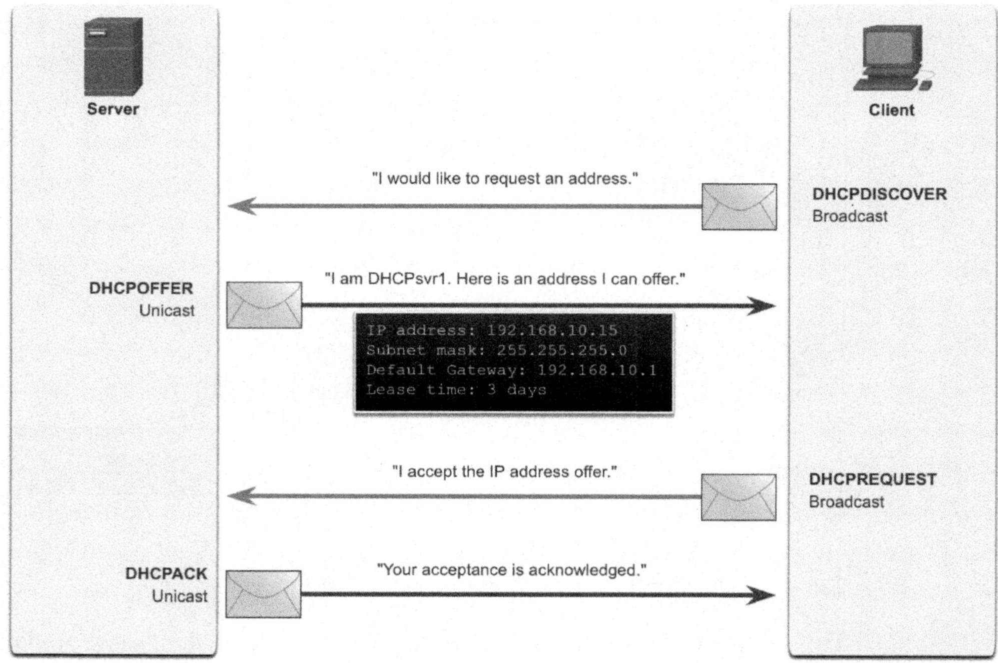

Figure 10-17 DHCP Operation and Messages

DHCP Attacks (10.5.5)

Two types of DHCP attacks are DHCP starvation and DHCP spoofing. Both attacks are mitigated by implementing DHCP snooping.

DHCP Starvation Attack

The goal of the DHCP Starvation attack is to create a DoS for connecting clients. DHCP starvation attacks require an attack tool such as Gobbler.

Gobbler has the ability to look at the entire scope of leasable IP addresses and tries to lease them all. Specifically, it creates DHCP discovery messages with bogus MAC addresses.

DHCP Spoofing Attack

A DHCP spoofing attack occurs when a rogue DHCP server is connected to the network and provides false IP configuration parameters to legitimate clients. A rogue server can provide a variety of misleading information:

- **Wrong default gateway:** The rogue server provides an invalid gateway or the IP address of its host to create a *man-in-the-middle attack*. This may go entirely undetected as the intruder intercepts the data flow through the network.
- **Wrong DNS server:** The rogue server provides an incorrect DNS server address pointing the user to a nefarious website.
- **Wrong IP address:** The rogue server provides an invalid IP address effectively creating a DoS attack on the DHCP client.

The following steps provide an example and explanation of a DHCP spoofing attack.

Step 1. **Threat Actor Connects Rogue DHCP Server.** In Figure 10-18, a threat actor successfully connects a rogue DHCP server to a switch port on the same subnet and VLANs as the target clients. The goal of the rogue server is to provide clients with false IP configuration information.

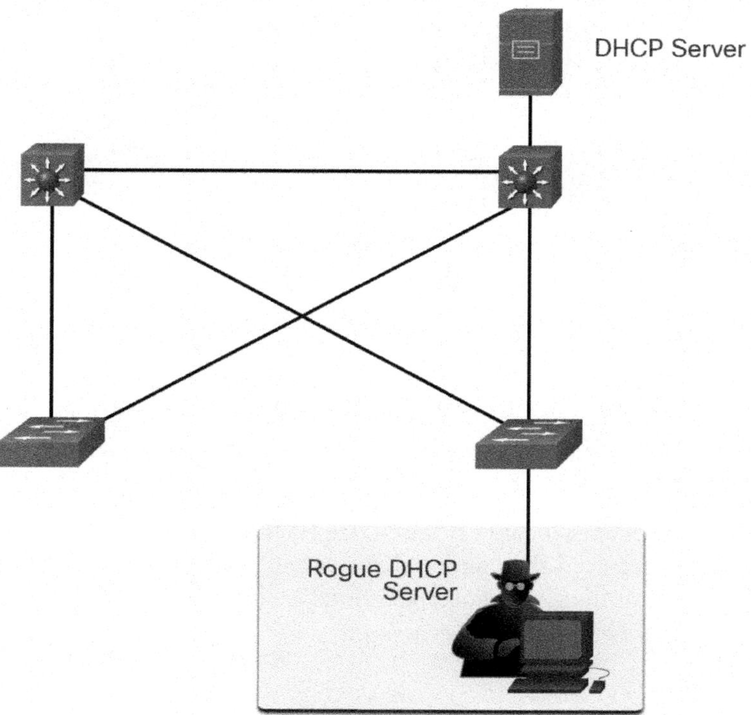

Figure 10-18 Threat Actor Connects Rogue DHCP Server

Step 2. **Client Broadcasts DHCP Discovery Messages.** In Figure 10-19, a legitimate client connects to the network and requires IP configuration parameters. Therefore, the client broadcasts a DHCP Discovery request looking for a response from a DHCP server. Both servers will receive the message and respond.

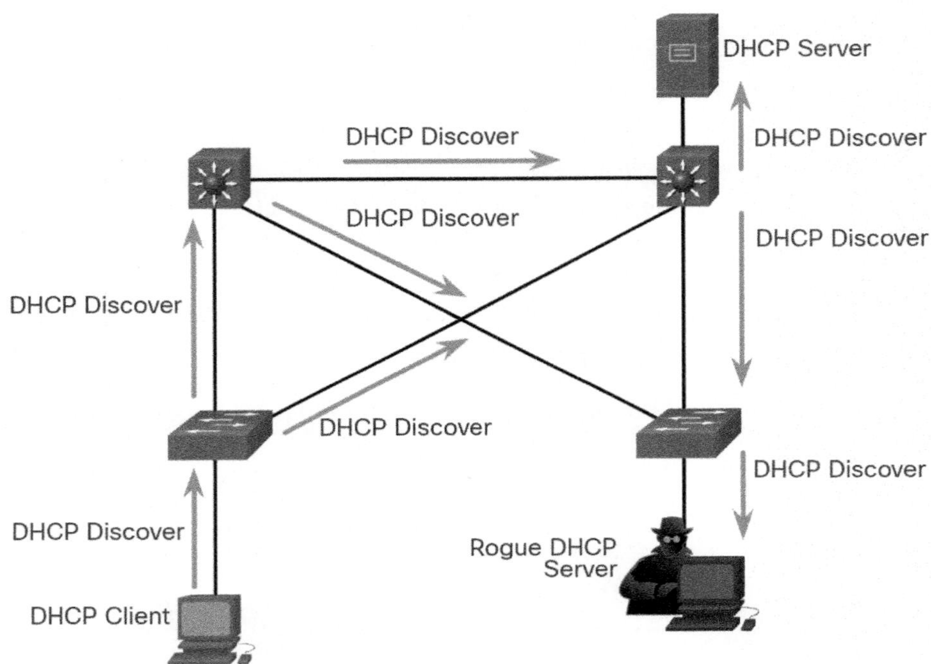

Figure 10-19 Client Broadcasts DHCP Discovery Messages

Step 3. **Legitimate and Rogue DHCP Reply.** In Figure 10-20, the legitimate DHCP server responds with valid IP configuration parameters. However, the rogue server also responds with a DHCP offer containing IP configuration parameters defined by the threat actor. The client will reply to the first offer received.

Step 4. **Client Accepts Rogue DHCP Offer.** In Figure 10-21, the rogue offer was received first, and therefore, the client broadcasts a DHCP request accepting the IP parameters defined by the threat actor. The legitimate and rogue server will receive the request.

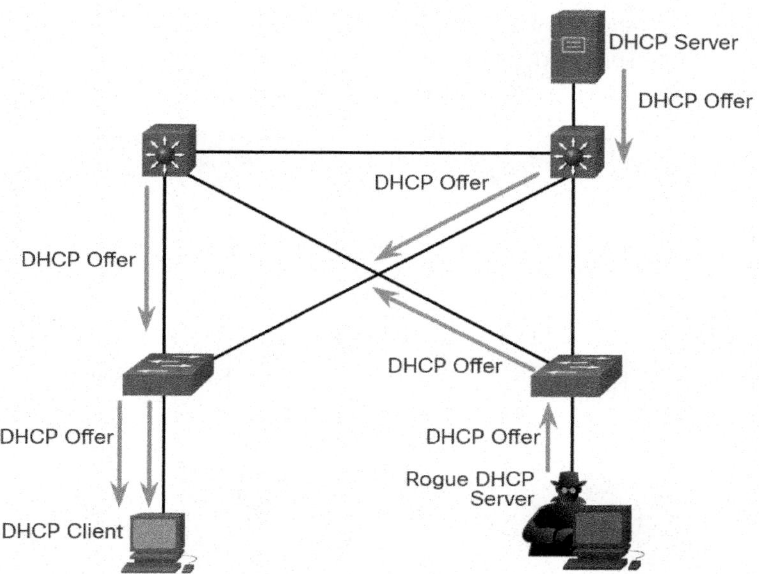

Figure 10-20 Legitimate and Rogue Send DHCPOFFER Message

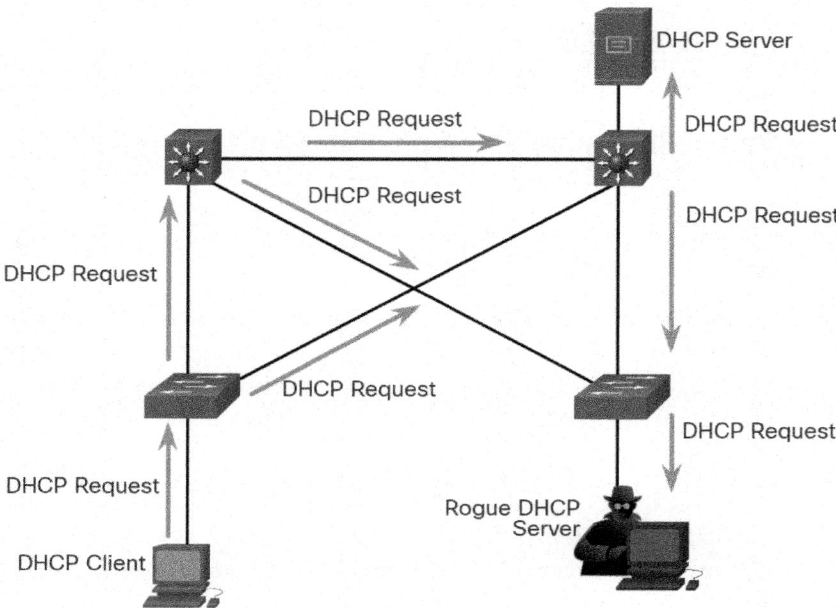

Figure 10-21 Client Accepts Rogue DHCP Offer

Step 5. **Rogue Server Acknowledges.** In Figure 10-22, the rogue server unicasts a reply to the client to acknowledge its request. The legitimate server will cease communicating with the client.

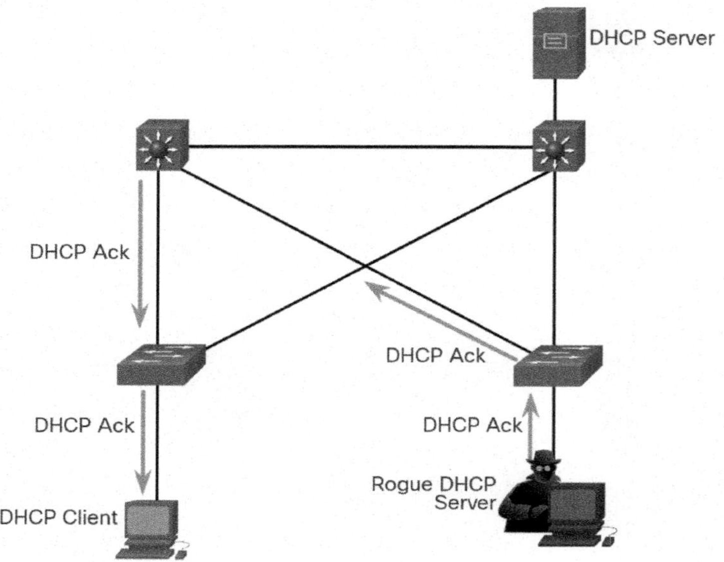

Figure 10-22 Rogue Server Acknowledges

Video

Video—ARP Attacks, STP Attacks, and CDP Reconnaissance (10.5.6)

Refer to the online course to view this video.

ARP Attacks (10.5.7)

Recall that hosts broadcast address resolution protocol (ARP) Requests to determine the MAC address of a host with a particular IPv4 address. This is typically done to discover the MAC address of the default gateway. All hosts on the subnet receive and process the ARP Request. The host with the matching IPv4 address in the ARP Request sends an ARP Reply.

According to the ARP Request for Comments (RFC), a client is allowed to send an unsolicited ARP Reply called a "*gratuitous ARP*." When a host sends a gratuitous ARP, other hosts on the subnet store the MAC address and IPv4 address contained in the gratuitous ARP in their ARP tables.

The problem is that an attacker can send a gratuitous ARP message containing a spoofed MAC address to a switch, and the switch would update its MAC table accordingly. Therefore, any host can claim to be the owner of any IPv4 and MAC address combination they choose. In a typical attack, a threat actor can send unsolicited ARP Replies to other hosts on the subnet with the MAC Address of the threat actor and the IPv4 address of the default gateway.

There are many tools available on the Internet to create ARP man-in-the-middle attacks, including dsniff, Cain & Abel, ettercap, Yersinia, and others. IPv6 uses ICMPv6 Neighbor Discovery Protocol for Layer 2 address resolution. IPv6 includes strategies to mitigate Neighbor Advertisement spoofing, similar to the way IPv6 prevents a spoofed ARP Reply.

ARP spoofing and ARP poisoning are mitigated by implementing DAI.

The following steps provide an example and explanation of ARP spoofing and ARP poisoning.

Step 1. **Normal State with Converged MAC Tables.** In Figure 10-23, each device has an accurate MAC table with the correct IPv4 and MAC addresses for the other devices on the LAN.

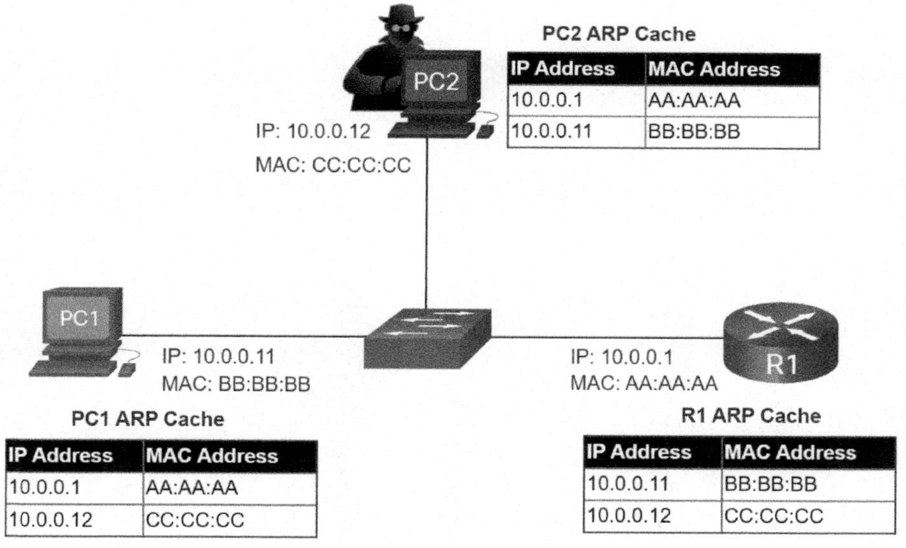

Figure 10-23 Normal State with Converged MAC Tables

Step 2. **ARP Spoofing Attack.** In Figure 10-24, the threat actor sends two spoofed gratuitous ARP Replies in an attempt to replace R1 as the default gateway:

- The first one informs all devices on the LAN that the threat actor's MAC address (CC:CC:CC) maps to R1's IPv4 address, 10.0.0.1.
- The second one informs all devices on the LAN that the threat actor's MAC address (CC:CC:CC) maps to PC1's IPv4 address, 10.0.0.11.

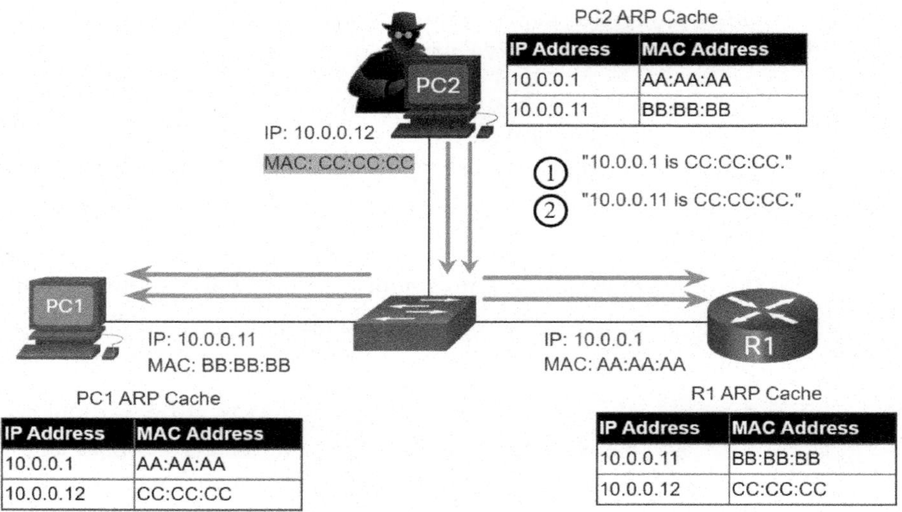

Figure 10-24 ARP Spoofing Attack

Step 3. **ARP Poisoning Attack with Man-in-the-Middle Attack.** In Figure 10-25, R1 and PC1 remove the correct entry for each other's MAC address and replace it with PC2's MAC address. The threat actor has now poisoned the ARP caches of all devices on the subnet. ARP poisoning leads to various man-in-the-middle attacks, posing a serious security threat to the network.

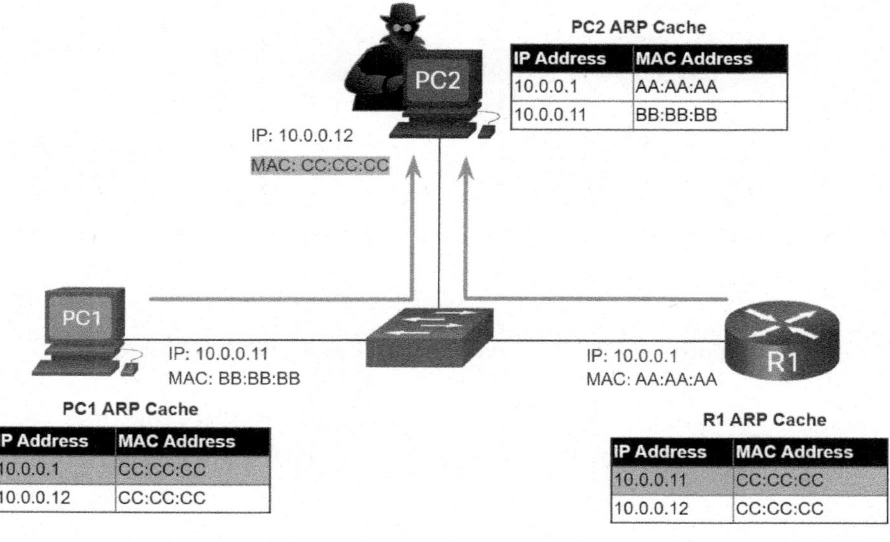

Figure 10-25 ARP Poisoning Attack with Man-in-the-Middle Attack

Address Spoofing Attack (10.5.8)

IPv4 addresses and MAC addresses can be spoofed for a variety of reasons. IPv4 address spoofing is when a threat actor hijacks a valid IPv4 address of another device on the subnet, or uses a random IPv4 address. IPv4 address spoofing is difficult to mitigate, especially when it is used inside a subnet in which the IPv4 belongs.

MAC address spoofing attacks occur when the threat actors alter the MAC address of their host to match another known MAC address of a target host. The attacking host then sends a frame throughout the network with the newly configured MAC address. When the switch receives the frame, it examines the source MAC address. The switch overwrites the current MAC table entry and assigns the MAC address to the new port, as shown in Figure 10-26. It then inadvertently forwards frames destined for the target host to the attacking host.

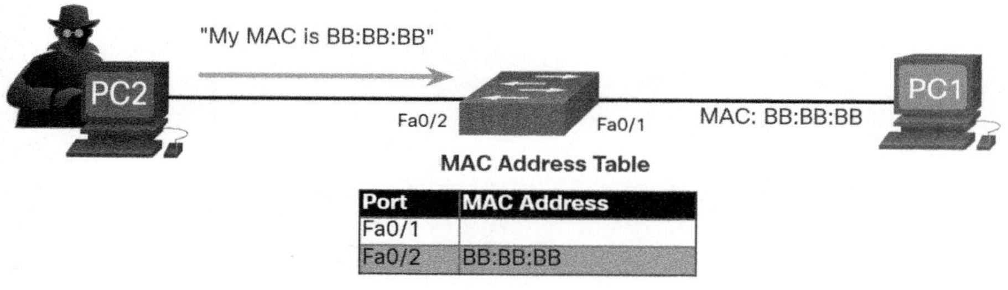

Figure 10-26 Example of MAC Address Spoofing

When the target host sends traffic, the switch will correct the error, realigning the MAC address to the original port. To stop the switch from returning the port assignment to its correct state, the threat actor can create a program or script that will constantly send frames to the switch so that the switch maintains the incorrect or spoofed information. There is no security mechanism at Layer 2 that allows a switch to verify the source of MAC addresses, which is what makes it so vulnerable to spoofing.

IPv4 and MAC address spoofing can be mitigated by implementing IPSG.

STP Attack (10.5.9)

Network attackers can manipulate the Spanning Tree Protocol (STP) to conduct an attack by spoofing the root bridge and changing the topology of a network. Attackers can make their hosts appear as root bridges, and thereby, capture all traffic for the immediate switched domain.

Note

An STP attack can also occur without malicious intent, such as when users want to add a switch to the network in their office space.

To conduct an STP manipulation attack, the attacking host broadcasts STP bridge protocol data units (BPDUs) containing configuration and topology changes that will force spanning-tree recalculations, as shown in Figure 10-27. The BPDUs sent by the attacking host announce a lower bridge priority in an attempt to be elected as the root bridge.

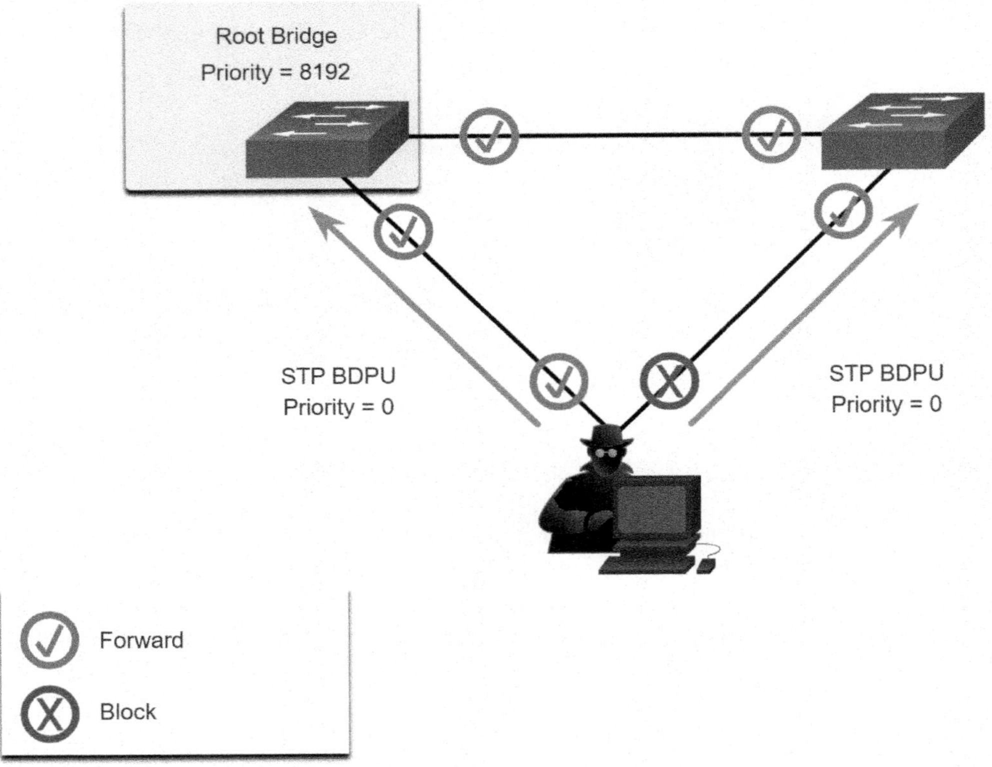

Figure 10-27 Threat Actor Sends Out Priority 0 BPDUs

If successful, the attacking host becomes the root bridge, as shown in Figure 10-28, and can now capture a variety of frames that would otherwise not be accessible.

This STP attack is mitigated by implementing BPDU Guard on all access ports. BPDU Guard is discussed in more detail later in the course.

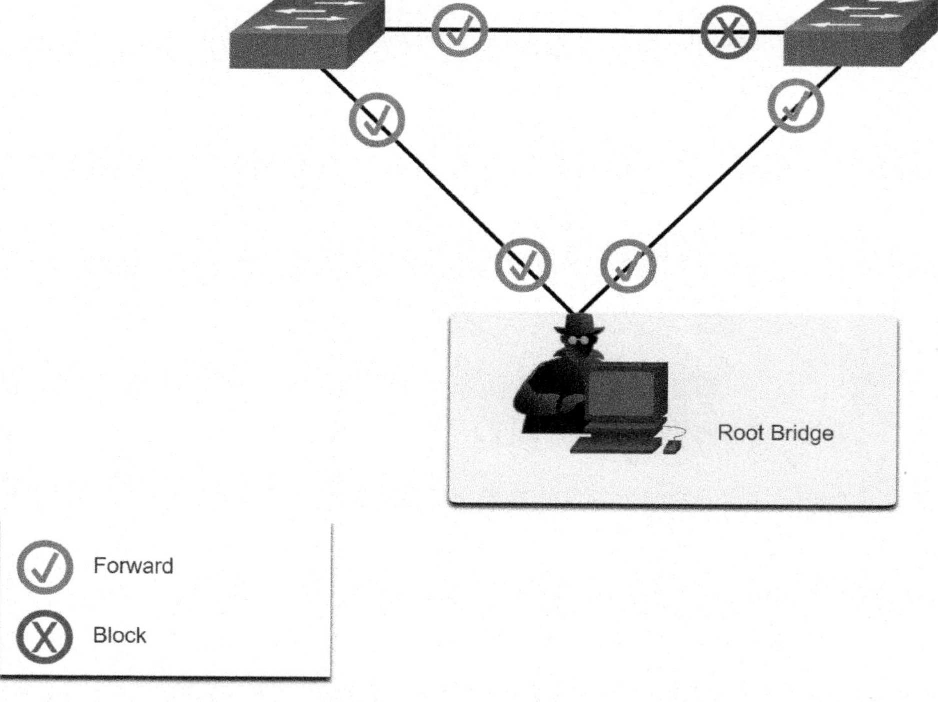

Figure 10-28 Threat Actor Becomes Root Bridge

CDP Reconnaissance (10.5.10)

The *Cisco Discovery Protocol (CDP)* is a proprietary Layer 2 link discovery protocol. It is enabled on all Cisco devices by default. CDP can automatically discover other CDP-enabled devices and help autoconfigure their connection. Network administrators also use CDP to help configure and troubleshoot network devices.

CDP information is sent out CDP-enabled ports in periodic, unencrypted broadcasts. CDP information includes the IPv4 address of the device, IOS software version, platform, capabilities, and the native VLAN. The device receiving the CDP message updates its CDP database.

CDP information is extremely useful in network troubleshooting. For example, CDP can be used to verify Layer 1 and 2 connectivity. If an administrator cannot ping a directly connected interface, but still receives CDP information, the problem is most likely related to the Layer 3 configuration.

However, the information provided by CDP can also be used by a threat actor to discover network infrastructure vulnerabilities.

In Figure 10-29, a sample Wireshark capture displays the contents of a CDP packet. The attacker is able to identify the Cisco IOS software version used by the device.

This allows the attacker to determine whether there were any security vulnerabilities specific to that particular version of IOS.

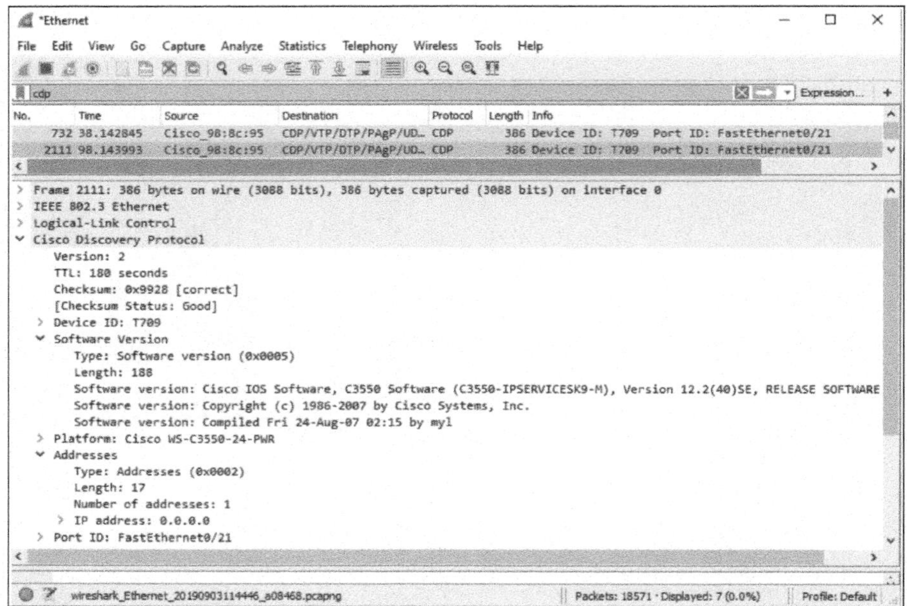

Figure 10-29 Wireshark Capture of a CDP Frame

CDP broadcasts are sent unencrypted and unauthenticated. Therefore, an attacker could interfere with the network infrastructure by sending crafted CDP frames containing bogus device information to directly connected Cisco devices.

To mitigate the exploitation of CDP, limit the use of CDP on devices or ports. For example, disable CDP on edge ports that connect to untrusted devices.

To disable CDP globally on a device, use the **no cdp run** global configuration mode command. To enable CDP globally, use the **cdp run** global configuration command.

To disable CDP on a port, use the **no cdp enable** interface configuration command. To enable CDP on a port, use the **cdp enable** interface configuration command.

Note

Link Layer Discovery Protocol (LLDP) is also vulnerable to reconnaissance attacks. Configure **no lldp run** to disable LLDP globally. To disable LLDP on the interface, configure **no lldp transmit** and **no lldp receive**.

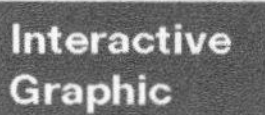

Check Your Understanding—LAN Attacks (10.5.11)

Refer to the online course to complete this activity.

Summary (10.6)

Endpoints are particularly susceptible to malware-related attacks that originate through email or web browsing, such as DDoS, data breaches, and malware. These endpoints have typically used traditional host-based security features, such as antivirus/antimalware, host-based firewalls, and host-based intrusion prevention systems (HIPSs). Endpoints are best protected by a combination of NAC, host-based AMP software, an email security appliance (ESA), and a web security appliance (WSA). Cisco WSA can perform blacklisting of URLs, URL-filtering, malware scanning, URL categorization, Web application filtering, and encryption and decryption of web traffic.

AAA controls who is permitted to access a network (authenticate), what they can do while they are there (authorize), and audits what actions they performed while accessing the network (accounting). Authorization uses a set of attributes that describes the user's access to the network. Accounting is combined with AAA authentication. The AAA server keeps a detailed log of exactly what the authenticated user does on the device. The IEEE 802.1X standard is a port-based access control and authentication protocol that restricts unauthorized workstations from connecting to a LAN through publicly accessible switch ports.

If Layer 2 is compromised, all layers above it are also affected. The first step in mitigating attacks on the Layer 2 infrastructure is to understand the underlying operation of Layer 2 and the Layer 2 solutions: Port Security, DHCP Snooping, DAI, and IPSG. These won't work unless management protocols are secured.

MAC address flooding attacks bombard the switch with fake source MAC addresses until the switch MAC address table is full. At this point, the switch treats the frame as an unknown unicast and begins to flood all incoming traffic out all ports on the same VLAN without referencing the MAC table. The threat actor can now capture all the frames sent from one host to another on the local LAN or local VLAN. The threat actor uses **macof** to rapidly generate many random source and destination MAC and IPv4 addresses. To mitigate MAC table overflow attacks, network administrators must implement port security.

A VLAN hopping attack enables traffic from one VLAN to be seen by another VLAN without the aid of a router. The threat actor configures a host to act like a switch to take advantage of the automatic trunking port feature enabled by default on most switch ports.

A VLAN double-tagging attack is unidirectional and works only when the threat actor is connected to a port residing in the same VLAN as the native VLAN of the trunk port. Double tagging allows the threat actor to send data to hosts or servers on a VLAN that otherwise would be blocked by some type of access control configuration. Return traffic will also be permitted, letting the threat actor communicate with devices on the normally blocked VLAN.

VLAN hopping and VLAN double-tagging attacks can be prevented by implementing the following trunk security guidelines:

- Disable trunking on all access ports.
- Disable auto trunking on trunk links so that trunks must be manually enabled.
- Be sure that the native VLAN is used only for trunk links.

DHCP Attack: DHCP servers dynamically provide IPv4 configuration information including IPv4 address, subnet mask, default gateway, DNS servers, and more to clients. Two types of DHCP attacks are DHCP starvation and DHCP spoofing. Both attacks are mitigated by implementing DHCP snooping.

ARP Attack: A threat actor sends a gratuitous ARP message containing a spoofed MAC address to a switch, and the switch updates its MAC table accordingly. Now the threat actor sends unsolicited ARP Replies to other hosts on the subnet with the MAC Address of the threat actor and the IPv4 address of the default gateway. ARP spoofing and ARP poisoning are mitigated by implementing DAI.

Address Spoofing Attack: IPv4 address spoofing is when a threat actor hijacks a valid IPv4 address of another device on the subnet or uses a random IPv4 address. MAC address spoofing attacks occur when the threat actors alter the MAC address of their host to match another known MAC address of a target host. IPv4 and MAC address spoofing can be mitigated by implementing IPSG.

STP Attack: Threat actors manipulate STP to conduct an attack by spoofing the root bridge and changing the topology of a network. Threat actors make their hosts appear as root bridges, thereby capturing all traffic for the immediate switched domain. This STP attack is mitigated by implementing BPDU Guard on all access ports.

CDP Reconnaissance: CDP information is sent out CDP-enabled ports in periodic, unencrypted broadcasts. CDP information includes the IPv4 address of the device, IOS software version, platform, capabilities, and the native VLAN. The device receiving the CDP message updates its CDP database. The information provided by CDP can also be used by a threat actor to discover network infrastructure vulnerabilities. To mitigate the exploitation of CDP, limit the use of CDP on devices or ports.

Practice

There are no labs or Packet Tracer activities in this module.

Check Your Understanding Questions

Complete all the review questions listed here to test your understanding of the sections and concepts in this chapter. The appendix "Answers to the 'Check Your Understanding' Questions" lists the answers.

1. Which of the following encrypts the data on end-devices, which can be decrypted only if a payment is made?

   A. DDoS

   B. Ransomware

   D. Virus

   E. Worm

2. Which network security device monitors and encrypts SMTP traffic to block threats and prevent data loss?

   A. ESA

   B. NAC

   C. NGFW

   D. WSA

3. Which AAA component is responsible for determining what access is permitted?

   A. Accounting

   B. Administration

   C. Authentication

   D. Authorization

4. Which small network router authentication method authenticates device access by referring to local usernames and passwords?

   A. Local AAA authentication

   B. Local AAA over RADIUS or TACACS+

   C. Server-based AAA

   D. Server-based AAA over RADIUS or TACACS+

5. Which 802.1X term is used to describe the device that is responsible for relaying 802.1X responses?

   A. Authenticator

   B. Authentication server

   C. Client

   D. Supplicant

6. Which 802.1X term is used to describe the device that is requesting authentication?

   A. Authenticator
   B. Authentication server
   C. Client
   D. Supplicant

7. Which mitigation technique prevents MAC address table overflow attacks?

   A. DAI
   B. Firewalls
   C. Port security
   D. VPNs

8. Which mitigation technique prevents ARP spoofing and ARP poisoning attacks?

   A. DAI
   B. Firewalls
   C. Port security
   D. VPNs

9. Which type of attack does IPSG mitigate?

   A. It prevents ARP spoofing and ARP poisoning attacks.
   B. It prevents DHCP starvation and DHCP spoofing attacks.
   C. It prevents MAC address table overflow attacks.
   D. It prevents MAC and IP address spoofing.

10. What happens to a compromised switch during a MAC address table attack?

    A. The switch interfaces will transition to error-disabled state.
    B. The switch will drop all received frames.
    C. The switch will flood all incoming frames to all other ports in the VLAN.
    D. The switch will shut down.

11. Why would a threat actor launch a MAC address overflow attack on a small network?

    A. To capture frames destined for other LAN devices
    B. To ensure legitimate hosts cannot forward traffic
    C. To launch a DoS attack
    D. To overwhelm the switch and drop frames

12. Which is an example of a DHCP starvation attack?

    A. A threat actor changes the MAC address of the threat actor's device to the MAC address of the default gateway.

    B. A threat actor configures a host with the 802.1Q protocol and forms a trunk with the connected switch.

    C. A threat actor discovers the IOS version and IP addresses of the local switch.

    D. A threat actor leases all the available IP addresses on a subnet to deny legitimate clients DHCP resources.

    D. A threat actor sends a BPDU message with priority 0.

    E. A threat actor sends a message that causes all other devices to believe the MAC address of the threat actor's device is the default gateway.

13. Which is an example of an STP attack?

    A. A threat actor changes the MAC address of the threat actor's device to the MAC address of the default gateway.

    B. A threat actor configures a host with the 802.1Q protocol and forms a trunk with the connected switch.

    C. A threat actor discovers the IOS version and IP addresses of the local switch.

    D. A threat actor leases all the available IP addresses on a subnet to deny legitimate clients DHCP resources.

    E. A threat actor sends a BPDU message with priority 0.

    F. A threat actor sends a message that causes all other devices to believe the MAC address of the threat actor's device is the default gateway.

14. Which is an example of an address spoofing attack?

    A. A threat actor changes the MAC address of the threat actor's device to the MAC address of the default gateway.

    B. A threat actor configures a host with the 802.1Q protocol and forms a trunk with the connected switch.

    C. A threat actor discovers the IOS version and IP addresses of the local switch.

    D. A threat actor leases all the available IP addresses on a subnet to deny legitimate clients DHCP resources.

    E. A threat actor sends a BPDU message with priority 0.

    F. A threat actor sends a message that causes all other devices to believe the MAC address of the threat actor's device is the default gateway.

15. Which is an example of an ARP spoofing attack?

    A. A threat actor changes the MAC address of the threat actor's device to the MAC address of the default gateway.
    B. A threat actor configures a host with the 802.1Q protocol and forms a trunk with the connected switch.
    C. A threat actor discovers the IOS version and IP addresses of the local switch.
    D. A threat actor leases all the available IP addresses on a subnet to deny legitimate clients DHCP resources.
    E. A threat actor sends a BPDU message with priority 0.
    F. A threat actor sends a message that causes all other devices to believe the MAC address of the threat actor's device is the default gateway.

16. Which is an example of a CDP reconnaissance attack?

    A. A threat actor changes the MAC address of the threat actor's device to the MAC address of the default gateway.
    B. A threat actor configures a host with the 802.1Q protocol and forms a trunk with the connected switch.
    C. A threat actor discovers the IOS version and IP addresses of the local switch.
    D. A threat actor leases all the available IP addresses on a subnet to deny legitimate clients DHCP resources.
    E. A threat actor sends a BPDU message with priority 0.
    F. A threat actor sends a message that causes all other devices to believe the MAC address of the threat actor's device is the default gateway.

CHAPTER 11

Switch Security Configuration

Objectives

Upon completion of this chapter, you will be able to answer the following questions:

- How do you implement port security to mitigate MAC address table attacks?
- How do you configure DTP and native VLAN to mitigate VLAN attacks?
- How do you configure DHCP snooping to mitigate DHCP attacks?
- How do you configure ARP inspection to mitigate ARP attacks?
- How do you configure PortFast and BPDU Guard to mitigate STP attacks?

Key Terms

This chapter uses the following key terms. You can find the definitions in the Glossary.

MAC address table overflow *Page 315*

error-disabled state *Page 317*

DHCP snooping binding table *Page 329*

Introduction (11.0)

An important part of your responsibility as a network professional is to keep the network secure. Most of the time we only think about security attacks coming from outside the network, but threats can come from within the network as well. These threats can range anywhere from an employee innocently adding an Ethernet switch to the corporate network so they can have more ports, to malicious attacks caused by a disgruntled employee. It is your job to keep the network safe and ensure that business operations continue uncompromised.

How do we keep the network safe and stable? How do we protect it from malicious attacks from within the network? How do we make sure employees are not adding switches, servers, and other devices to the network that might compromise network operations?

This module is your introduction to keeping your network secure from within!

Implement Port Security (11.1)

In this section, you learn how to configure the port security feature to restrict network access and mitigate MAC address table attacks.

Secure Unused Ports (11.1.1)

Layer 2 devices are considered to be the weakest link in a company's security infrastructure. Layer 2 attacks are some of the easiest for hackers to deploy, but these threats can also be mitigated with some common Layer 2 solutions.

All switch ports (interfaces) should be secured before the switch is deployed for production use. How a port is secured depends on its function.

A simple method that many administrators use to help secure the network from unauthorized access is to disable all unused ports on a switch. For example, if a Catalyst 2960 switch has 24 ports and there are three Fast Ethernet connections in use, it is good practice to disable the 21 unused ports. Navigate to each unused port and issue the **shutdown** command. If a port must be reactivated at a later time, it can be enabled with the **no shutdown** command.

To configure a range of ports, use the **interface range** command.

```
Switch(config)# interface range type module/first-number - last-number
```

For example, to shut down ports for Fa0/8 through Fa0/24 on S1, you enter the command shown in Example 11-1.

Example 11-1 Shutting Down Unused Ports

```
S1(config)# interface range fa0/8 - 24
S1(config-if-range)# shutdown
%LINK-5-CHANGED: Interface FastEthernet0/8, changed state to administratively down
(output omitted)
%LINK-5-CHANGED: Interface FastEthernet0/24, changed state to administratively
  down
S1(config-if-range)#
```

Mitigate MAC Address Table Attacks (11.1.2)

The simplest and most effective method to prevent *MAC address table overflow* attacks is to enable port security.

Port security limits the number of valid MAC addresses allowed on a port. It allows an administrator to manually configure MAC addresses for a port or to permit the switch to dynamically learn a limited number of MAC addresses. When a port configured with port security receives a frame, the source MAC address of the frame is compared to the list of secure source MAC addresses that were manually configured or dynamically learned on the port.

By limiting the number of permitted MAC addresses on a port to one, port security can be used to control unauthorized access to the network, as shown in Figure 11-1.

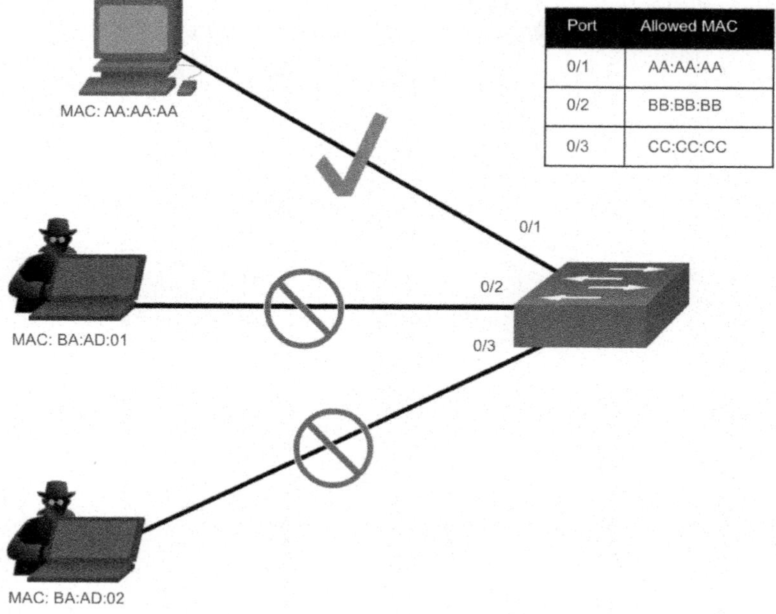

Figure 11-1 Port Security Limits the Number of MAC Addresses on a Port

Enable Port Security (11.1.3)

Notice in Example 11-2, the **switchport port-security** interface configuration command was rejected.

Example 11-2 Enabling Port Security on an Access Port

```
S1(config)# interface f0/1
S1(config-if)# switchport port-security
Command rejected: FastEthernet0/1 is a dynamic port.
S1(config-if)# switchport mode access
S1(config-if)# switchport port-security
S1(config-if)# end
S1#
```

This is because port security can be configured only on manually configured access ports or manually configured trunk ports. By default, Layer 2 switch ports are set to dynamic auto (trunking on). Therefore, the port is configured with the **switchport mode access** interface configuration command.

Note

Trunk port security is beyond the scope of this course.

Use the **show port-security interface** command to display the current port security settings for FastEthernet 0/1, as shown in Example 11-3.

Example 11-3 Displaying Current Port Security

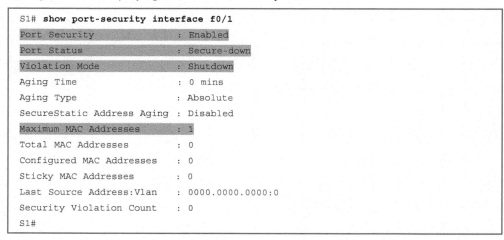

```
S1# show port-security interface f0/1
Port Security              : Enabled
Port Status                : Secure-down
Violation Mode             : Shutdown
Aging Time                 : 0 mins
Aging Type                 : Absolute
SecureStatic Address Aging : Disabled
Maximum MAC Addresses      : 1
Total MAC Addresses        : 0
Configured MAC Addresses   : 0
Sticky MAC Addresses       : 0
Last Source Address:Vlan   : 0000.0000.0000:0
Security Violation Count   : 0
S1#
```

Notice how port security is enabled; port status is Secure-down, which means there are no devices attached and no violation has occurred; the violation mode is Shutdown; and how the maximum number of MAC addresses is 1. If a device is connected to the port, the switch will automatically add the device's MAC address as a secure MAC. Currently, in our example, no device is connected to the port.

Note

If an active port is configured with the **switchport port-security** interface configuration command and more than one device is connected to that port, the port will transition to the *error-disabled state*. This condition is discussed later in this section.

After port security is enabled, other port security specifics can be configured, as shown in Example 11-4.

Example 11-4 Port Security Command Options

```
S1(config-if)# switchport port-security ?
  aging         Port-security aging commands
  mac-address   Secure mac address
  maximum       Max secure addresses
  violation     Security violation mode
  <cr>
S1(config-if)# switchport port-security
```

Limit and Learn MAC Addresses (11.1.4)

To set the maximum number of MAC addresses allowed on a port, use the following command:

```
Switch(config-if)# switchport port-security maximum value
```

The default port security value is 1. The maximum number of secure MAC addresses that can be configured depends on the switch and the IOS. In Example 11-5, the maximum is 8192.

Example 11-5 Verifying Maximum Number of MAC Addresses Allowed on a Port

```
S1(config)# interface f0/1
S1(config-if)# switchport port-security maximum ?
  <1-8192>  Maximum addresses
S1(config-if)# switchport port-security maximum
```

The switch can be configured to learn about MAC addresses on a secure port in one of three ways:

1. **Manually Configured.** The administrator manually configures a static MAC address(es) by using the following command for each secure MAC address on the port:

   ```
   Switch(config-if)# switchport port-security mac-address mac-address
   ```

2. **Dynamically Learned.** When the **switchport port-security** interface configuration command is entered, the current source MAC for the device connected to the port is automatically secured but is not added to the startup configuration. If the switch is rebooted, the port will have to relearn the device's MAC address.

3. **Dynamically Learned—Sticky.** The administrator can enable the switch to dynamically learn the MAC address and "stick" them to the running configuration by using the following command:

   ```
   Switch(config-if)# switchport port-security mac-address sticky
   ```

   Saving the running configuration will commit the dynamically learned MAC address to NVRAM.

Example 11-6 demonstrates a complete port security configuration for FastEthernet 0/1 with a host connected to port Fa0/1. The administrator specifies a maximum of two MAC addresses, manually configures one secure MAC address, and then configures the port to dynamically learn additional secure MAC addresses up to the two secure MAC address maximum. Use the **show port-security interface** and the **show port-security address** commands to verify the configuration.

Example 11-6 Configuring and Verifying Port Security

```
S1(config)# interface fa0/1
S1(config-if)# switchport mode access
S1(config-if)# switchport port-security
S1(config-if)# switchport port-security maximum 2
S1(config-if)# switchport port-security mac-address aaaa.bbbb.1234
S1(config-if)# switchport port-security mac-address sticky
S1(config-if)# end
S1#
S1# show port-security interface fa0/1
Port Security                : Enabled
Port Status                  : Secure-up
Violation Mode               : Shutdown
Aging Time                   : 0 mins
Aging Type                   : Absolute
SecureStatic Address Aging   : Disabled
Maximum MAC Addresses        : 2
Total MAC Addresses          : 2
```

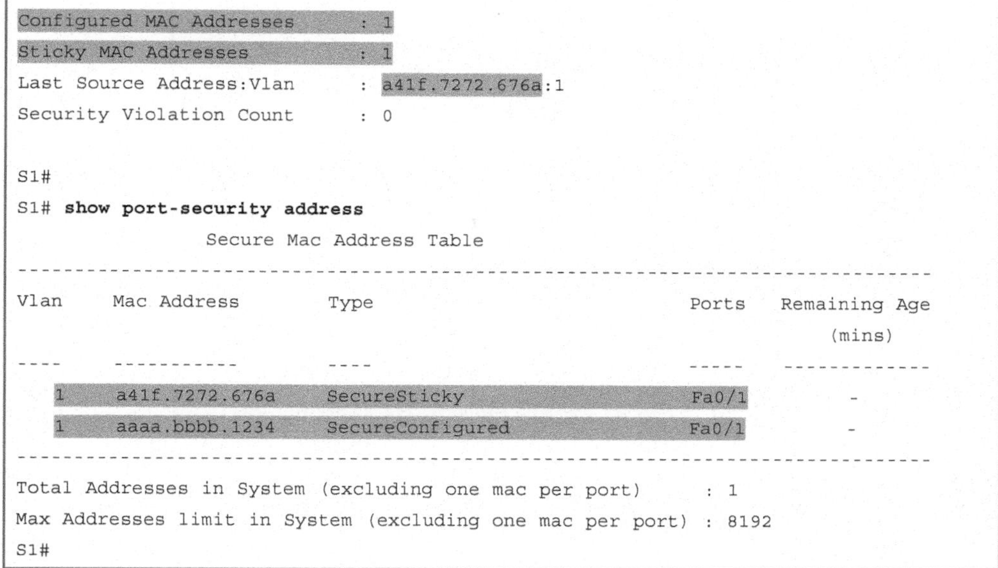

```
Configured MAC Addresses        : 1
Sticky MAC Addresses            : 1
Last Source Address:Vlan        : a41f.7272.676a:1
Security Violation Count        : 0

S1#
S1# show port-security address
                Secure Mac Address Table
-------------------------------------------------------------------------------
Vlan    Mac Address       Type                          Ports   Remaining Age
                                                                    (mins)
----    -----------       ----                          -----   -------------
   1    a41f.7272.676a    SecureSticky                  Fa0/1          -
   1    aaaa.bbbb.1234    SecureConfigured              Fa0/1          -
-------------------------------------------------------------------------------
Total Addresses in System (excluding one mac per port)     : 1
Max Addresses limit in System (excluding one mac per port) : 8192
S1#
```

The output of the **show port-security interface** command verifies that port security is enabled, there is a host connected to the port (that is, Secure-up), a total of 2 MAC addresses will be allowed, and S1 has learned one MAC address statically and one MAC address dynamically (that is, sticky).

The output of the **show port-security address** command lists the two learned MAC addresses.

Port Security Aging (11.1.5)

Port security aging can be used to set the aging time for static and dynamic secure addresses on a port. Two types of aging are supported per port:

- **Absolute:** The secure addresses on the port are deleted after the specified aging time.
- **Inactivity:** The secure addresses on the port are deleted only if they are inactive for the specified aging time.

Use aging to remove secure MAC addresses on a secure port without manually deleting the existing secure MAC addresses. Aging time limits can also be increased to ensure past secure MAC addresses remain, even while new MAC addresses are added. Aging of statically configured secure addresses can be enabled or disabled on a per-port basis.

Use the **switchport port-security aging** interface configuration command to enable or disable static aging for the secure port, or to set the aging time or type.

```
Switch(config-if)# switchport port-security aging {static | time time | type
   {absolute | inactivity}}
```

The parameters for the command are described in Table 11-1.

Table 11-1 Parameters for the Port Security Aging

Parameter	Description
`static`	■ Enable aging for statically configured secure addresses on this port.
`time` *`time`*	■ Specify the aging time for this port. ■ The range is 0 to 1440 minutes. ■ If the time is 0, aging is disabled for this port.
`type absolute`	■ Set the absolute aging time. ■ All the secure addresses on this port age out exactly after the time (in minutes) specified and are removed from the secure address list.
`type inactivity`	■ Set the inactivity aging type. ■ The secure addresses on this port age out only if there is no data traffic from the secure source address for the specified time period.

Note

MAC addresses are shown as 24 bits for simplicity.

Example 11-7 shows an administrator configuring the aging type to 10 minutes of inactivity and then using the **show port-security interface** command to verify the configuration.

Example 11-7 Configuring and Verifying Port Security Aging

```
S1(config)# interface fa0/1
S1(config-if)# switchport port-security aging time 10
S1(config-if)# switchport port-security aging type inactivity
S1(config-if)# end
S1#
S1# show port-security interface fa0/1
Port Security              : Enabled
Port Status                : Secure-up
Violation Mode             : Shutdown
Aging Time                 : 10 mins
Aging Type                 : Inactivity
SecureStatic Address Aging : Disabled
```

```
Maximum MAC Addresses          : 2
Total MAC Addresses            : 2
Configured MAC Addresses       : 1
Sticky MAC Addresses           : 1
Last Source Address:Vlan       : a41f.7272.676a:1
Security Violation Count       : 0

S1#
```

Port Security Violation Modes (11.1.6)

If the MAC address of a device attached to the port differs from the list of secure addresses, a port violation occurs. By default, the port enters the error-disabled state.

To set the port security violation mode, use the following interface configuration command:

```
Switch(config-if)# switchport port-security violation {shutdown | restrict |
  protect}
```

Table 11-2 describes the command syntax and Table 11-3 summarizes how a switch reacts based on the configured violation mode.

Table 11-2 Security Violation Mode Descriptions

Mode	Description
shutdowan *(default)*	■ The port transitions to the error-disabled state immediately, turns off the port LED, and sends a syslog message. ■ It increments the violation counter. ■ When a secure port is in the error-disabled state, an administrator must reenable it by entering the **shutdown** and **no shutdown** commands.
restrict	■ The port drops packets with unknown source addresses until you remove a sufficient number of secure MAC addresses to drop below the maximum value or increase the maximum value. ■ This mode causes the Security Violation counter to increment and generates a syslog message.
protect	■ This is the least secure of the security violation modes. ■ The port drops packets with unknown MAC source addresses until you remove a sufficient number of secure MAC addresses to drop below the maximum value or increase the maximum value. ■ No syslog message is sent.

Table 11-3 Security Violation Mode Comparison

Violation Mode	Discards Offending Traffic	Sends Syslog Message	Increase Violation Counter	Shuts Down Port
Shutdown	Yes	Yes	Yes	Yes
Restrict	Yes	Yes	Yes	No
Protect	Yes	No	No	No

Example 11-8 shows an administrator changing the security violation to "restrict". The output of the **show port-security interface** command confirms that the change has been made.

Example 11-8 Modifying Port Security to Restrict

```
S1(config)# interface f0/1
S1(config-if)# switchport port-security violation restrict
S1(config-if)# end
S1#
S1# show port-security interface f0/1
Port Security              : Enabled
Port Status                : Secure-up
Violation Mode             : Restrict
Aging Time                 : 10 mins
Aging Type                 : Inactivity
SecureStatic Address Aging : Disabled
Maximum MAC Addresses      : 2
Total MAC Addresses        : 2
Configured MAC Addresses   : 1
Sticky MAC Addresses       : 1
Last Source Address:Vlan   : a41f.7272.676a:1
Security Violation Count   : 0

S1#
```

Ports in error-disabled State (11.1.7)

What happens when the port security violation is shut down and a port violation occurs? The port is physically shut down and placed in the error-disabled state, and no traffic is sent or received on that port.

In Example 11-9, the port security violation is changed back to the default shutdown setting. Then the host with MAC address a41f.7272.676a is disconnected, and a new host is plugged into Fa0/1.

Notice how a series of port security related messages are generated on the console, as shown in Example 11-9.

Example 11-9 Log Messages Showing a Port Security Violation

```
S1(config)# int fa0/1
S1(config-if)# switchport port-security violation shutdown
S1(config-if)# end
S1#
*Mar  1 00:24:15.599: %LINEPROTO-5-UPDOWN: Line protocol on Interface
  FastEthernet0/1, changed state to down
*Mar  1 00:24:16.606: %LINK-3-UPDOWN: Interface FastEthernet0/1, changed state to
  down
*Mar  1 00:24:19.114: %LINK-3-UPDOWN: Interface FastEthernet0/1, changed state to
  up
*Mar  1 00:24:20.121: %LINEPROTO-5-UPDOWN: Line protocol on Interface
  FastEthernet0/1, changed state to up
S1#
*Mar  1 00:24:32.829: %PM-4-ERR_DISABLE: psecure-violation error detected on
  Fa0/1, putting Fa0/1 in err-disable state
*Mar  1 00:24:32.838: %PORT_SECURITY-2-PSECURE_VIOLATION: Security violation
  occurred, caused by MAC address a41f.7273.018c on port FastEthernet0/1.
*Mar  1 00:24:33.836: %LINEPROTO-5-UPDOWN: Line protocol on Interface
  FastEthernet0/1, changed state to down
*Mar  1 00:24:34.843: %LINK-3-UPDOWN: Interface FastEthernet0/1, changed state to
  down
S1#
```

Note

The port protocol and link status are changed to down, and the port LED is turned off.

In Example 11-10, the **show interface** command identifies the port status as **err-disabled** (that is, error-disabled state). The output of the **show port-security interface** command now shows the port status as Secure-shutdown instead of Secure-up. The Security Violation counter increments by 1.

Example 11-10 Verifying a Port Is Disabled

```
S1# show interface fa0/1 | include down
FastEthernet0/1 is down, line protocol is down (err-disabled)
S1#
S1# show port-security interface fa0/1
Port Security              : Enabled
Port Status                : Secure-shutdown
Violation Mode             : Shutdown
Aging Time                 : 10 mins
```

```
Aging Type                   : Inactivity
SecureStatic Address Aging   : Disabled
Maximum MAC Addresses        : 2
Total MAC Addresses          : 2
Configured MAC Addresses     : 1
Sticky MAC Addresses         : 1
Last Source Address:Vlan     : a41f.7273.018c:1
Security Violation Count     : 1

S1#
```

The administrator should determine what caused the security violation. If an unauthorized device is connected to a secure port, the security threat is eliminated before reenabling the port.

In the next example, the first host is reconnected to Fa0/1. To reenable the port, first use the **shutdown** command, then use the **no shutdown** command to make the port operational, as shown Example 11-11.

Example 11-11 Reenabling a Disabled Port

```
S1(config)# interface fastEthernet 0/1
S1(config-if)# shutdown
S1(config-if)#
*Mar  1 00:39:54.981: %LINK-5-CHANGED: Interface FastEthernet0/1, changed state to
  administratively
S1(config-if)# no shutdown
S1(config-if)#
*Mar  1 00:40:04.275: %LINK-3-UPDOWN: Interface FastEthernet0/1, changed state to
  up
*Mar  1 00:40:05.282: %LINEPROTO-5-UPDOWN: Line protocol on Interface
  FastEthernet0/1, changed state to up
S1(config-if)#
```

Verify Port Security (11.1.8)

After configuring port security on a switch, check each interface to verify that the port security is set correctly, and check to ensure that the static MAC addresses have been configured correctly.

Port Security for All Interfaces

To display port security settings for the switch, use the **show port-security** command. Example 11-12 indicates that only one port is configured with the **switchport port-security** command.

Example 11-12 Verifying Port Security on All Interfaces

```
S1# show port-security
Secure Port  MaxSecureAddr  CurrentAddr  SecurityViolation  Security Action
                (Count)        (Count)          (Count)
---------------------------------------------------------------------------
      Fa0/1              2            2                  0        Shutdown
---------------------------------------------------------------------------
Total Addresses in System (excluding one mac per port)      : 1
Max Addresses limit in System (excluding one mac per port) : 8192
S1#
```

Port Security for a Specific Interface

Use the **show port-security interface** command to view details for a specific interface, as shown previously and in Example 11-13.

Example 11-13 Verifying Port Security on a Specific Interface

```
S1# show port-security interface fastEthernet 0/1
Port Security              : Enabled
Port Status                : Secure-up
Violation Mode             : Shutdown
Aging Time                 : 10 mins
Aging Type                 : Inactivity
SecureStatic Address Aging : Disabled
Maximum MAC Addresses      : 2
Total MAC Addresses        : 2
Configured MAC Addresses   : 1
Sticky MAC Addresses       : 1
Last Source Address:Vlan   : a41f.7273.018c:1
Security Violation Count   : 0

S1#
```

Verify Learned MAC Addresses

To verify that MAC addresses are "sticking" to the configuration, use the **show run** command, as shown in Example 11-14 for FastEthernet 0/1.

Example 11-14 Verifying the MAC Addresses Learned on an Interface

```
S1# show run interface fa0/1
Building configuration...

Current configuration : 365 bytes
!
interface FastEthernet0/1
 switchport mode access
 switchport port-security maximum 2
 switchport port-security mac-address sticky
 switchport port-security mac-address sticky a41f.7272.676a
 switchport port-security mac-address aaaa.bbbb.1234
 switchport port-security aging time 10
 switchport port-security aging type inactivity
 switchport port-security
end

S1#
```

Verify Secure MAC Addresses

To display all secure MAC addresses that are manually configured or dynamically learned on all switch interfaces, use the **show port-security address** command, as shown in Example 11-15.

Example 11-15 Verifying All Secure MAC Addresses

```
S1# show port-security address
                Secure Mac Address Table
-------------------------------------------------------------------------------
Vlan    Mac Address       Type                          Ports   Remaining Age
                                                                     (mins)
----    -----------       ----                          -----   -------------
   1    a41f.7272.676a    SecureSticky                  Fa0/1         -
   1    aaaa.bbbb.1234    SecureConfigured              Fa0/1         -
-------------------------------------------------------------------------------
Total Addresses in System (excluding one mac per port)      : 1
Max Addresses limit in System (excluding one mac per port)  : 8192
S1#
```

Interactive Graphic

Syntax Checker—Implement Port Security (11.1.9)

Refer to the online course to complete this activity.

Packet Tracer—Implement Port Security (11.1.10)

In this activity, you configure and verify port security on a switch. Port security allows you to restrict a port's ingress traffic by limiting the MAC addresses that are allowed to send traffic into the port.

Mitigate VLAN Attacks (11.2)

In this section, you learn how to configure dynamic trunking protocol (DTP) and native VLAN to mitigate VLAN attacks.

VLAN Attacks Review (11.2.1)

As a quick review, a VLAN hopping attack can be launched in one of three ways:

- Spoofing DTP messages from the attacking host to cause the switch to enter trunking mode. From here, the attacker can send traffic tagged with the target VLAN, and the switch then delivers the packets to the destination.
- Introducing a rogue switch and enabling trunking. The attacker can then access all the VLANs on the victim switch from the rogue switch.
- Another type of VLAN hopping attack is a double-tagging (or double-encapsulated) attack. This attack takes advantage of the way hardware on most switches operate.

Steps to Mitigate VLAN Hopping Attacks (11.2.2)

Use the following steps to mitigate VLAN hopping attacks:

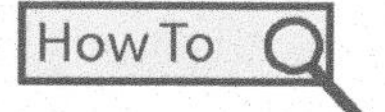

Step 1. Disable DTP (auto trunking) negotiations on non-trunking ports by using the **switchport mode access** interface configuration command.

Step 2. Disable unused ports using the **shutdown** interface configuration command and assign them to an unused VLAN.

Step 3. Manually enable the trunk link on a trunking port by using the **switchport mode trunk** interface configuration command.

Step 4. Disable DTP (auto trunking) negotiations on trunking ports by using the **switchport nonegotiate** interface configuration command.

Step 5. Set the native VLAN to a VLAN other than VLAN 1 by using the **switchport trunk native vlan** *vlan_number* interface configuration command.

For example, assume the following:

- FastEthernet ports 0/1 through fa0/16 are active access ports.
- FastEthernet ports 0/17 through 0/20 are not currently in use.
- FastEthernet ports 0/21 through 0/24 are trunk ports.

VLAN hopping can be mitigated by implementing the configuration in Example 11-16.

Example 11-16 Configuration to Mitigate VLAN Hopping Attacks

```
S1(config)# interface range fa0/1 - 16
S1(config-if-range)# switchport mode access
S1(config-if-range)# exit
S1(config)#
S1(config)# interface range fa0/17 - 20
S1(config-if-range)# switchport mode access
S1(config-if-range)# switchport access vlan 1000
S1(config-if-range)# shutdown
S1(config-if-range)# exit
S1(config)#
S1(config)# interface range fa0/21 - 24
S1(config-if-range)# switchport mode trunk
S1(config-if-range)# switchport nonegotiate
S1(config-if-range)# switchport trunk native vlan 999
S1(config-if-range)# end
S1#
```

- FastEthernet ports 0/1 to 0/16 are access ports, and therefore trunking is disabled by explicitly making them access ports.
- FastEthernet ports 0/17 to 0/20 are unused ports and are disabled (that is, shut down) and assigned to an unused VLAN.
- FastEthernet ports 0/21 to 0/24 are trunk links and are manually enabled as trunks with DTP disabled. The native VLAN is also changed from the default VLAN 1 to an unused VLAN 999.

Interactive Graphic

Syntax Checker—Mitigate VLAN Hopping Attacks (11.2.3)

Refer to the online course to complete this activity.

Mitigate DHCP Attacks (11.3)

In this section, you learn how to configure DHCP snooping to mitigate DHCP attacks.

DHCP Attack Review (11.3.1)

The goal of a DHCP starvation attack is to create a Denial of Service (DoS) for connecting clients. DHCP starvation attacks require an attack tool such as Gobbler. Recall that DHCP starvation attacks can be effectively mitigated by using port security because Gobbler uses a unique source MAC address for each DHCP request sent.

However, mitigating DHCP spoofing attacks requires more protection. Gobbler could be configured to use the actual interface MAC address as the source Ethernet address, but specify a different Ethernet address in the DHCP payload. This would render port security ineffective because the source MAC address would be legitimate.

DHCP spoofing attacks can be mitigated by using DHCP snooping on trusted ports.

DHCP Snooping (11.3.2)

DHCP snooping does not rely on source MAC addresses. Instead, DHCP snooping determines whether DHCP messages are from an administratively configured trusted or untrusted source. It then filters DHCP messages and rate-limits DHCP traffic from untrusted sources.

Devices under your administrative control, such as switches, routers, and servers, are trusted sources. Any device beyond the firewall or outside your network is an untrusted source. In addition, all access ports are generally treated as untrusted sources. Figure 11-2 shows an example of trusted and untrusted ports.

Notice that the rogue DHCP server would be on an untrusted port after enabling DHCP snooping. All interfaces are treated as untrusted by default. Trusted interfaces are typically trunk links and ports directly connected to a legitimate DHCP server. These interfaces must be explicitly configured as trusted.

A DHCP table is built that includes the source MAC address of a device on an untrusted port and the IP address assigned by the DHCP server to that device. The MAC address and IP address are bound together. Therefore, this table is called the *DHCP snooping binding table*.

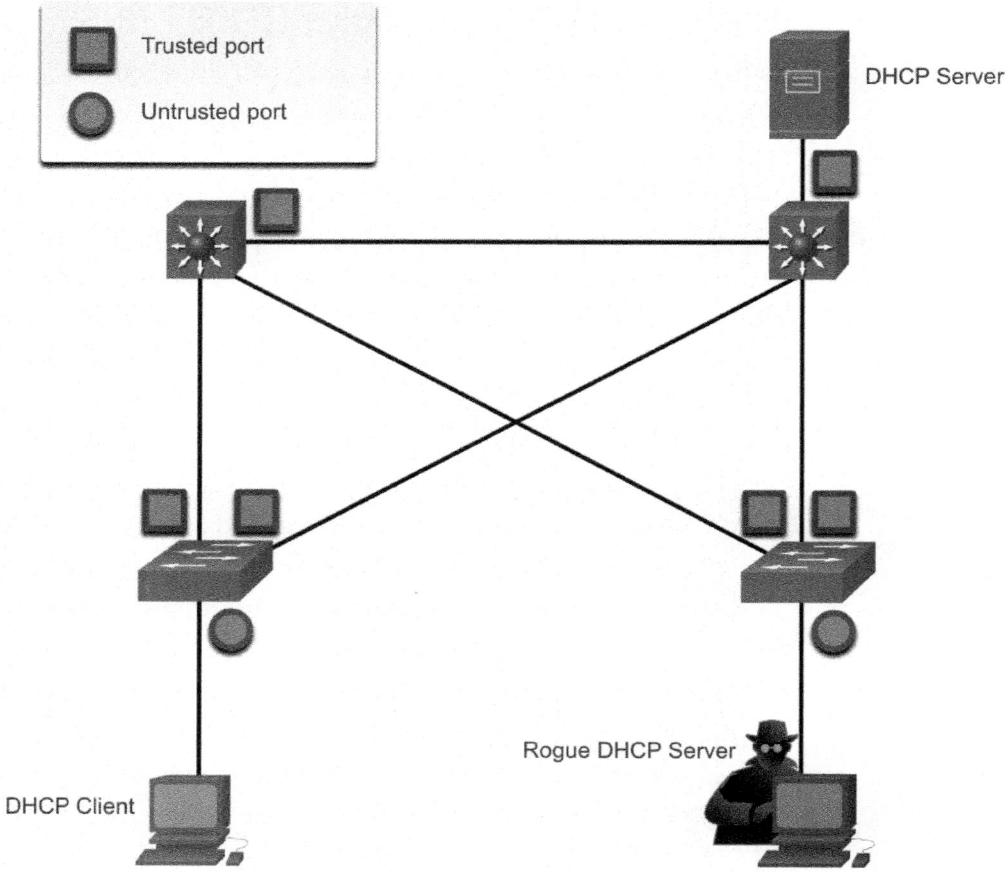

Figure 11-2 Trusted and Untrusted Ports in a DHCP Snooping Implementation

Steps to Implement DHCP Snooping (11.3.3)

Use the following steps to enable DHCP snooping:

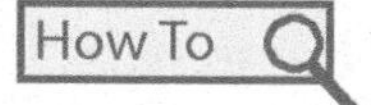

Step 1. Enable DHCP snooping by using the **ip dhcp snooping** global configuration command.

Step 2. On trusted ports, use the **ip dhcp snooping trust** interface configuration command.

Step 3. On untrusted ports, limit the number of DHCP discovery messages that can be received per second by using the **ip dhcp snooping limit rate** *packets* interface configuration command.

Step 4. Enable DHCP snooping by VLAN, or by a range of VLANs, by using the **ip dhcp snooping** *vlan* global configuration command.

DHCP Snooping Configuration Example (11.3.4)

The reference topology for this DHCP snooping example is shown in Figure 11-3. Notice that F0/5 is an untrusted port because it connects to a PC. F0/1 is a trusted port because it connects to the DHCP server.

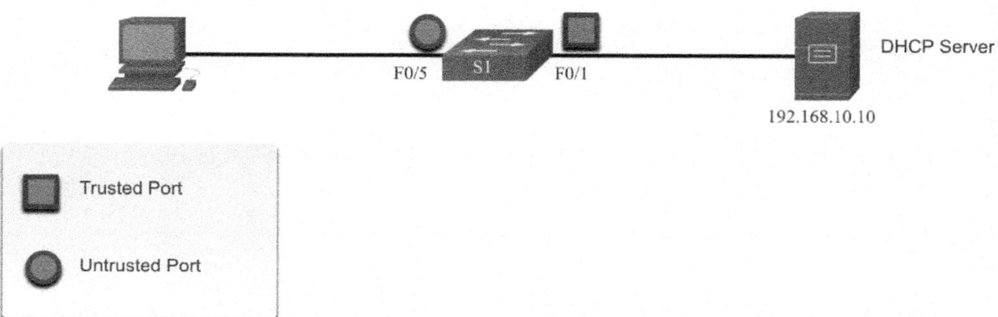

Figure 11-3 DHCP Snooping Reference Topology

Example 11-17 demonstrates how to configure DHCP snooping on S1.

Example 11-17 Configuring DHCP Snooping

```
S1(config)# ip dhcp snooping
S1(config)# interface f0/1
S1(config-if)# ip dhcp snooping trust
S1(config-if)# exit
S1(config)#
S1(config)# interface range f0/5 - 24
S1(config-if-range)# ip dhcp snooping limit rate 6
S1(config-if-range)# exit
S1(config)#
S1(config)# ip dhcp snooping vlan 5,10,50-52
S1(config)# end
S1#
```

Notice how DHCP snooping is first enabled. Then the upstream interface to the DHCP server is explicitly trusted. Next, the range of FastEthernet ports from F0/5 to F0/24 are untrusted by default, so a rate limit is set to six packets per second. Finally, DHCP snooping is enabled on VLANS 5, 10, 50, 51, and 52.

Use the **show ip dhcp snooping** privileged EXEC command to verify DHCP snooping and **show ip dhcp snooping binding** to view the clients that have received DHCP information, as shown Example 11-18.

Note

DHCP snooping is also required by Dynamic ARP Inspection (DAI), which is the next section.

Example 11-18 Verifying DHCP Snooping

```
S1# show ip dhcp snooping
Switch DHCP snooping is enabled
DHCP snooping is configured on following VLANs:
5,10,50-52
DHCP snooping is operational on following VLANs:
none
DHCP snooping is configured on the following L3 Interfaces:
Insertion of option 82 is enabled
   circuit-id default format: vlan-mod-port
   remote-id: 0cd9.96d2.3f80 (MAC)
Option 82 on untrusted port is not allowed
Verification of hwaddr field is enabled
Verification of giaddr field is enabled
DHCP snooping trust/rate is configured on the following Interfaces:
Interface                  Trusted     Allow option     Rate limit (pps)
-----------------------    -------     ------------     ----------------
FastEthernet0/1            yes         yes              unlimited
  Custom circuit-ids:
FastEthernet0/5            no          no               6
  Custom circuit-ids:
FastEthernet0/6            no          no               6
  Custom circuit-ids:
S1#
S1# show ip dhcp snooping binding
MacAddress          IpAddress        Lease(sec)  Type           VLAN  Interface
------------------  ---------------  ----------  -------------  ----  ------------------
00:03:47:B5:9F:AD   192.168.10.10    193185      dhcp-snooping  5     FastEthernet0/5
```

Interactive Graphic

Syntax Checker—Mitigate DHCP Attacks (11.3.5)

Refer to the online course to complete this activity.

Mitigate ARP Attacks (11.4)

In this section, you learn how to configure dynamic ARP inspection (DAI) to mitigate ARP attacks.

Dynamic ARP Inspection (11.4.1)

In a typical address resolution protocol (ARP) attack, a threat actor can send unsolicited ARP replies to other hosts on the subnet with the MAC address of the threat actor and the IP address of the default gateway. To prevent ARP spoofing and the resulting ARP poisoning, a switch must ensure that only valid ARP Requests and Replies are relayed.

Dynamic ARP inspection (DAI) requires DHCP snooping and helps prevent ARP attacks by

- Not relaying invalid or gratuitous ARP Replies out to other ports in the same VLAN.
- Intercepting all ARP Requests and Replies on untrusted ports.
- Verifying each intercepted packet for a valid IP-to-MAC binding.
- Dropping and logging invalid ARP Replies to prevent ARP poisoning.
- Error-disabling the interface if the configured DAI number of ARP packets is exceeded.

DAI Implementation Guidelines (11.4.2)

To mitigate the chances of ARP spoofing and ARP poisoning, follow these DAI implementation guidelines:

- Enable DHCP snooping globally.
- Enable DHCP snooping on selected VLANs.
- Enable DAI on selected VLANs.
- Configure trusted interfaces for DHCP snooping and ARP inspection.

It is generally advisable to configure all access switch ports as untrusted and to configure all uplink ports that are connected to other switches as trusted.

The topology in Figure 11-4 identifies trusted and untrusted ports.

DAI Configuration Example (11.4.3)

In the previous topology, S1 is connecting two users on VLAN 10. DAI will be configured to mitigate against ARP spoofing and ARP poisoning attacks.

As shown in Example 11-19, DHCP snooping is enabled because DAI requires the DHCP snooping binding table to operate. Next, DHCP snooping and ARP inspection are enabled for the PCs on VLAN 10. The uplink port to the router is trusted, and therefore is configured as trusted for DHCP snooping and ARP inspection.

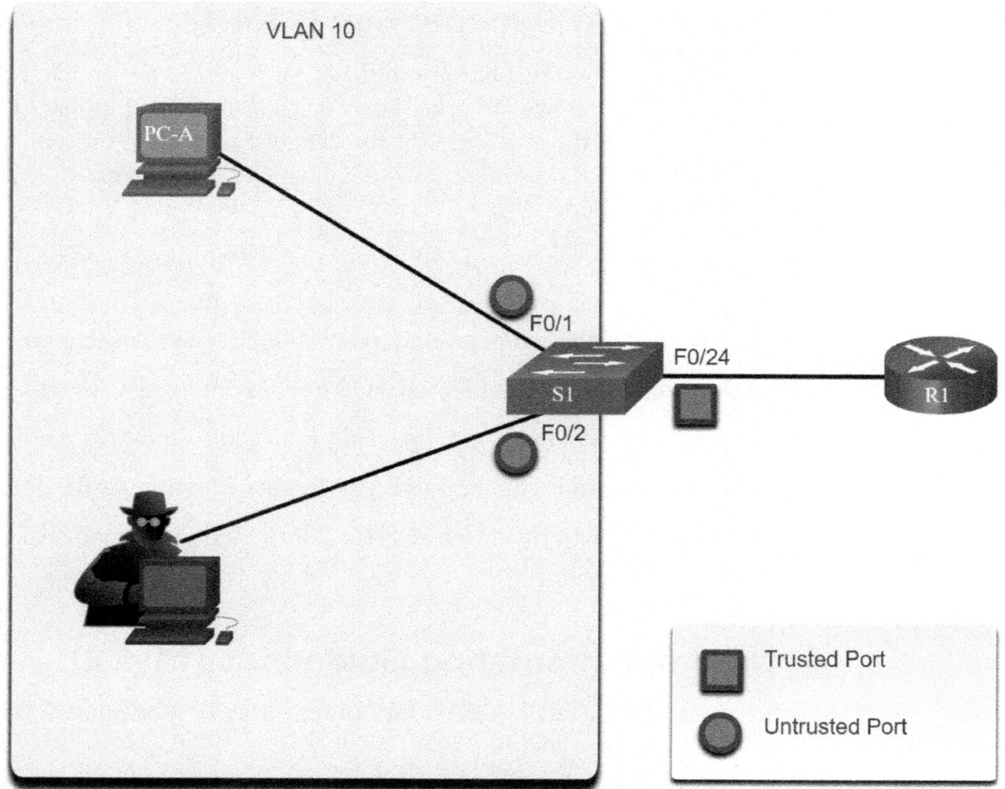

Figure 11-4 DAI Implementation Topology

Example 11-19 Configuring DAI

```
S1(config)# ip dhcp snooping
S1(config)# ip dhcp snooping vlan 10
S1(config)# ip arp inspection vlan 10
S1(config)# interface fa0/24
S1(config-if)# ip dhcp snooping trust
S1(config-if)# ip arp inspection trust
```

DAI can also be configured to check for both destination or source MAC and IP addresses:

- **Destination MAC:** Checks the destination MAC address in the Ethernet header against the target MAC address in ARP body.
- **Source MAC:** Checks the source MAC address in the Ethernet header against the sender MAC address in the ARP body.

- **IP address:** Checks the ARP body for invalid and unexpected IP addresses, including addresses 0.0.0.0, 255.255.255.255, and all IP multicast addresses.

The **ip arp inspection validate {[src-mac] [dst-mac] [ip]}** global configuration command is used to configure DAI to drop ARP packets when the IPv4 addresses are invalid. It can be used when the MAC addresses in the body of the ARP packets do not match the addresses that are specified in the Ethernet header. Notice in the following example how only one command can be configured. Therefore, entering multiple **ip arp inspection validate** commands overwrites the previous command. To include more than one validation method, enter them on the same command line, as shown and verified in Example 11-20.

Example 11-20 Configuring DAI to Inspect and Drop Invalid Packets

```
S1(config)# ip arp inspection validate ?
  dst-mac  Validate destination MAC address
  ip       Validate IP addresses
  src-mac  Validate source MAC address
S1(config)# ip arp inspection validate src-mac
S1(config)# ip arp inspection validate dst-mac
S1(config)# ip arp inspection validate ip
S1(config)# do show run | include validate
ip arp inspection validate ip
S1(config)# ip arp inspection validate src-mac dst-mac ip
S1(config)# do show run | include validate
ip arp inspection validate src-mac dst-mac ip
S1(config)#
```

Interactive Graphic

Syntax Checker—Mitigate ARP Attacks (11.4.4)

Refer to the online course to complete this activity.

Mitigate STP Attacks (11.5)

In this section, you learn how to configure PortFast and BPDU Guard to mitigate STP attacks.

PortFast and BPDU Guard (11.5.1)

Recall that network attackers can manipulate the Spanning Tree Protocol (STP) to conduct an attack by spoofing the root bridge and changing the topology of a network. To mitigate STP manipulation attacks, use PortFast and Bridge Protocol Data Unit (BPDU) Guard:

- **PortFast:** PortFast immediately brings an interface configured as an access or trunk port to the forwarding state from a blocking state, bypassing the listening and learning states. Apply to all end-user ports. PortFast should be configured only on ports attached to end devices.
- **BPDU Guard:** BPDU guard immediately error-disables a port that receives a BPDU. Like PortFast, BPDU guard should be configured only on interfaces attached to end devices.

In Figure 11-5, the access ports for S1 should be configured with PortFast and BPDU Guard.

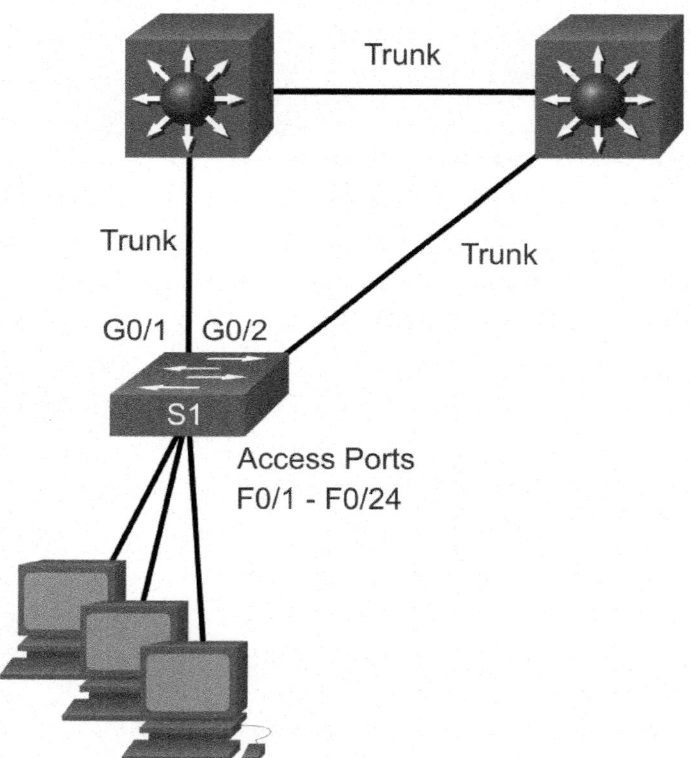

Figure 11-5 PortFast and BPDU Reference Topology

Configure PortFast (11.5.2)

PortFast bypasses the STP listening and learning states to minimize the time that access ports must wait for STP to converge. If PortFast is enabled on a port connecting to another switch, there is a risk of creating a spanning-tree loop.

PortFast can be enabled on an interface by using the **spanning-tree portfast** interface configuration command. Alternatively, Portfast can be configured globally on all access ports by using the **spanning-tree portfast default** global configuration command.

To verify whether PortFast is enabled globally, you can use either the **show running-config | begin span** command or the **show spanning-tree summary** command. To verify if PortFast is enabled an interface, use the **show running-config** command, as shown in Example 11-21. The **show spanning-tree interface** *type/number* **detail** command can also be used for verification.

Notice that when PortFast is enabled, warning messages are displayed.

Example 11-21 Configuring and Verifying PortFast

```
S1(config)# interface fa0/1
S1(config-if)# switchport mode access
S1(config-if)# spanning-tree portfast
%Warning: portfast should only be enabled on ports connected to a single
 host. Connecting hubs, concentrators, switches, bridges, etc... to this
 interface when portfast is enabled, can cause temporary bridging loops.
 Use with CAUTION
%Portfast has been configured on FastEthernet0/1 but will only
 have effect when the interface is in a non-trunking mode.
S1(config-if)# exit
S1(config)#
S1(config)# spanning-tree portfast default
%Warning: this command enables portfast by default on all interfaces. You
 should now disable portfast explicitly on switched ports leading to hubs,
 switches and bridges as they may create temporary bridging loops.
S1(config)# exit
S1#
S1# show running-config | begin span
spanning-tree mode pvst
spanning-tree portfast default
spanning-tree extend system-id
!
interface FastEthernet0/1
 switchport mode access
 spanning-tree portfast
!
interface FastEthernet0/2
!
interface FastEthernet0/3
!
```

```
interface FastEthernet0/4
!
interface FastEthernet0/5
!
(output omitted)
S1#
```

Configure BPDU Guard (11.5.3)

Even though PortFast is enabled, the interface will still listen for BPDUs. Unexpected BPDUs might be accidental, or part of an unauthorized attempt to add a switch to the network.

If any BPDUs are received on a BPDU Guard enabled port, that port is put into error-disabled state. This means the port is shut down and must be manually reenabled or automatically recovered through the **errdisable recovery cause psecure_violation** global command.

BPDU Guard can be enabled on a port by using the **spanning-tree bpduguard enable** interface configuration command. Alternatively, use the **spanning-tree portfast bpduguard default** global configuration command to globally enable BPDU guard on all PortFast-enabled ports.

To display information about the state of spanning tree, use the **show spanning-tree summary** command. In Example 11-22, PortFast default and BPDU Guard are both enabled as the default state for ports configured as access mode.

Note

Always enable BPDU Guard on all PortFast-enabled ports.

Example 11-22 Configuring and Verifying BPDU Guard

```
S1(config)# interface fa0/1
S1(config-if)# spanning-tree bpduguard enable
S1(config-if)# exit
S1(config)#
S1(config)# spanning-tree portfast bpduguard default
S1(config)# end
S1#
S1# show spanning-tree summary
Switch is in pvst mode
Root bridge for: none
Extended system ID           is enabled
Portfast Default             is enabled
PortFast BPDU Guard Default  is enabled
```

```
Portfast BPDU Filter Default is disabled
Loopguard Default            is disabled
EtherChannel misconfig guard is enabled
UplinkFast                   is disabled
BackboneFast                 is disabled
Configured Pathcost method used is short
(output omitted)
S1#
```

Interactive Graphic

Syntax Checker—Mitigate STP Attacks (11.5.4)

Refer to the online course to complete this activity.

Summary (11.6)

All switch ports (interfaces) should be secured before the switch is deployed for production use. The simplest and most effective method to prevent MAC address table overflow attacks is to enable port security. By default, Layer 2 switch ports are set to dynamic auto (trunking on). The switch can be configured to learn about MAC addresses on a secure port in one of three ways: manually configured, dynamically learned, and dynamically learned—sticky. Port security aging can be used to set the aging time for static and dynamic secure addresses on a port. Two types of aging are supported per port: absolute and inactivity. If the MAC address of a device attached to the port differs from the list of secure addresses, a port violation occurs. By default, the port enters the error-disabled state. When a port is shut down and placed in the error-disabled state, no traffic is sent or received on that port. To display port security settings for the switch, use the **show port-security** command.

To mitigate VLAN hopping attacks:

Step 1. Disable DTP negotiations on non-trunking ports.

Step 2. Disable unused ports.

Step 3. Manually enable the trunk link on a trunking port.

Step 4. Disable DTP negotiations on trunking ports.

Step 5. Set the native VLAN to a VLAN other than VLAN 1.

The goal of a DHCP starvation attack is to create a Denial of Service (DoS) for connecting clients. DHCP spoofing attacks can be mitigated by using DHCP snooping on trusted ports. DHCP snooping determines whether DHCP messages are from an administratively configured trusted or untrusted source. It then filters DHCP messages and rate-limits DHCP traffic from untrusted sources. Use the following steps to enable DHCP snooping:

Step 1. Enable DHCP snooping.

Step 2. On trusted ports, use the **ip dhcp snooping trust** interface configuration command.

Step 3. Limit the number of DHCP discovery messages that can be received per second on untrusted ports.

Step 4. Enable DHCP snooping by VLAN, or by a range of VLANs.

Dynamic ARP inspection (DAI) requires DHCP snooping and helps prevent ARP attacks by

- Not relaying invalid or gratuitous ARP Replies out to other ports in the same VLAN.

- Intercepting all ARP Requests and Replies on untrusted ports.
- Verifying each intercepted packet for a valid IP-to-MAC binding.
- Dropping and logging ARP Replies coming from invalid addresses to prevent ARP poisoning.
- Error-disabling the interface if the configured DAI number of ARP packets is exceeded.

To mitigate the chances of ARP spoofing and ARP poisoning, follow these DAI implementation guidelines:

- Enable DHCP snooping globally.
- Enable DHCP snooping on selected VLANs.
- Enable DAI on selected VLANs.
- Configure trusted interfaces for DHCP snooping and ARP inspection.

As a general guideline, configure all access switch ports as untrusted and all uplink ports that are connected to other switches as trusted.

DAI can also be configured to check for both destination or source MAC and IP addresses:

- **Destination MAC:** Checks the destination MAC address in the Ethernet header against the target MAC address in ARP body.
- **Source MAC:** Checks the source MAC address in the Ethernet header against the sender MAC address in the ARP body.
- **IP address:** Checks the ARP body for invalid and unexpected IP addresses, including addresses 0.0.0.0, 255.255.255.255, and all IP multicast addresses.

To mitigate Spanning Tree Protocol (STP) manipulation attacks, use PortFast and Bridge Protocol Data Unit (BPDU) Guard:

- **PortFast:** PortFast immediately brings an interface configured as an access or trunk port to the forwarding state from a blocking state, bypassing the listening and learning states. Apply to all end-user ports. PortFast should be configured only on ports attached to end devices. PortFast bypasses the STP listening and learning states to minimize the time that access ports must wait for STP to converge. If PortFast is enabled on a port connecting to another switch, there is a risk of creating a spanning-tree loop.

- **BPDU Guard:** BPDU guard immediately error-disables a port that receives a BPDU. Like PortFast, BPDU guard should be configured only on interfaces attached to end devices. BPDU Guard can be enabled on a port by using the **spanning-tree bpduguard enable** interface configuration command. Alternatively, use the **spanning-tree portfast bpduguard default** global configuration command to globally enable BPDU guard on all PortFast-enabled ports.

Packet Tracer
Activity

Packet Tracer—Switch Security Configuration (11.6.1)

In this Packet Tracer activity, you will do the following:

- Secure unused ports.
- Implement port security.
- Mitigate VLAN hopping attacks.
- Mitigate DHCP attacks.
- Mitigate ARP attacks.
- Mitigate STP attacks.
- Verify the switch security configuration.

Lab—Switch Security Configuration (11.6.2)

In this lab, you will do the following:

- Secure unused ports.
- Implement port security.
- Mitigate VLAN hopping attacks.
- Mitigate DHCP attacks.
- Mitigate ARP attacks.
- Mitigate STP attacks.
- Verify the switch security configuration.

Practice

The following activities provide practice with the topics introduced in this chapter. The Labs are available in the companion *Switching, Routing, and Wireless Essentials Labs and Study Guide (CCNAv7)* (ISBN 9780136634386). The Packet Tracer

Activity instructions are also in the Labs & Study Guide. The PKA files are found in the online course.

Lab

Lab 11.6.2: Switch Security Configuration

Packet Tracer Activities

Packet Tracer 11.1.10: Implement Port Security

Packet Tracer 11.6.1: Switch Security Configuration

Check Your Understanding Questions

Complete all the review questions listed here to test your understanding of the sections and concepts in this chapter. The appendix "Answers to the 'Check Your Understanding' Questions" lists the answers.

1. Which method would mitigate a MAC address flooding attack?

   A. Configuring port security

   B. Increasing the size of the CAM table

   C. Increasing the speed of switch ports

   D. Using ACLs to filter broadcast traffic on the switch

2. Which action will bring an error-disabled switch port back to an operational state?

   A. Clear the MAC address table on the switch.

   B. Issue the **shutdown** and **no shutdown** interface config commands.

   C. Issue the **switchport mode access** interface config command.

   D. Remove and reconfigure port security on the interface.

3. Which two statements are true regarding switch port security? (Choose two.)

   A. After entering the **sticky** parameter, only MAC addresses subsequently learned are converted to secure MAC addresses.

   B. Dynamically learned secure MAC addresses are lost when the switch reboots.

C. If fewer than the maximum number of MAC addresses for a port are configured statically, dynamically learned addresses are added to CAM until the maximum number is reached.

D. The three configurable violation modes all log violations via SNMP.

E. The three configurable violation modes all require user intervention to reenable ports.

4. Port security has been enabled on access ports to allow a maximum of two MAC addresses. Which port security violation would drop the frame and send a notification to the syslog server if the maximum number of MAC addresses is exceeded?

   A. Protect

   B. Restrict

   C. Shutdown

   D. Warning

5. Which feature should be configured on PortFast enabled switches to prevent rogue switches from being added to a network?

   A. BPDU guard

   B. DAI

   C. DHCP snooping

   D. Port security

6. Which port security feature enables switches to automatically learn and retain MAC addresses for each port?

   A. Auto secure MAC addresses

   B. Dynamic secure MAC addresses

   C. Static secure MAC addresses

   D. Sticky secure MAC addresses

7. Assume that BPDU Guard has been enabled globally on all access ports. However, one port must not be configured with the feature. Which command would explicitly disable BPDU Guard on that switch port?

   A. S1(config)# **no spanning-tree bpduguard default**

   B. S1(config)# **no spanning-tree portfast bpduguard default**

   C. S1(config-if)# **no enable spanning-tree bpduguard**

   D. S1(config-if)# **no spanning-tree bpduguard enable**

   E. S1(config-if)# **no spanning-tree portfast bpduguard**

8. Which DAI command checks the source MAC address in the Ethernet header against the target MAC address in the ARP body?

   A. **ip arp inspection validate dst-mac**

   B. **ip arp inspection validate dst-mac ip**

   C. **ip arp inspection validate ip**

   D. **ip arp inspection validate src-mac**

9. What is the result of entering the **ip dhcp snooping limit rate 4 interface** configuration command?

   A. The port can receive up to 4 DHCP discovery messages per second.

   B. The port can receive up to 4 DHCP offer messages per second.

   C. The port can send up to 4 DHCP messages per second.

   D. The port can send up to 4 DHCP offer discovery messages per second.

10. Port security has been enabled on a switch port. What is the default violation mode in use by default?

    A. Restrict

    B. Disabled

    C. Protect

    D. Shutdown

11. What techniques should be done to mitigate VLAN attacks? (Choose three.)

    A. Disable DTP.

    B. Enable BPDU guard.

    C. Enable Source Guard.

    D. Enable trunking manually.

    E. Set the native VLAN to an unused VLAN.

    F. Use private VLANs.

12. Port security has been enabled on interface Fa0/1 and the **show port-security interface fa0/1** command has been entered. What does the Port Status "Secure-up" message indicate?

    A. The Fa0/1 port is currently error-disabled.

    B. The Fa0/1 port violation mode is "protect".

    C. There are no hosts connected to the secured Fa0/1 port.

    D. There is a host connected to the secured Fa0/1 port.

CHAPTER 12

WLAN Concepts

Objectives

Upon completion of this chapter, you will be able to answer the following questions:

- What is WLAN technology and standards?
- What are the components of a WLAN infrastructure?
- How does wireless technology enable WLAN operation?
- How does a WLC use CAPWAP to manage multiple APs?
- What is channel management in a WLAN?
- What are threats to WLANs?
- What are WLAN security mechanisms?

Key Terms

This chapter uses the following key terms. You can find the definitions in the Glossary.

Wireless LAN (WLAN) Page 349

Wireless Personal-Area Networks (WPAN) Page 349

Wireless MANs (WMAN) Page 350

Wireless Wide-Area Networks (WWANs) Page 350

WiMAX (Worldwide Interoperability for Microwave Access) Page 352

4G/5G Page 352

wireless access points (APs) Page 353

multiple-input and multiple-output (MIMO) Page 353

802.11 Page 353

802.11a Page 353

802.11b Page 354

802.11g Page 354

802.11n Page 354

802.11ac Page 354

802.11ax Page 354

service set identifier (SSID) Page 357

Wi-Fi range extenders Page 358

autonomous APs Page 359

lightweight APs (LAPs) Page 359

Lightweight Access Point Protocol (LWAPP) Page 359

WLAN controller (WLC) Page 359

controller-based APs Page 359

link aggregation group (LAG) Page 360

omnidirectional antennas Page 360

directional antennas Page 360

Yagi antenna Page 360

parabolic dish antenna Page 360

ad hoc mode Page 362

independent basic service set (IBSS) Page 363

infrastructure mode Page 363

tethering Page 363

hotspot Page 363

Basic Service Set (BSS) Page 364

Extended Service Set (ESS) Page 364

Basic Service Area (BSA) Page 364

Basic Service Set Identifier (BSSID) Page 365

distribution system (DS) Page 365

Extended Service Area (ESA) Page 365

carrier sense multiple access with collision avoidance (CSMA/CA) Page 367

CAPWAP Page 370

Datagram Transport Layer Security (DTLS) Page 370

split MAC Page 371

FlexConnect Page 372

Direct-Sequence Spread Spectrum (DSSS) Page 373

Frequency-Hopping Spread Spectrum (FHSS) Page 374

Orthogonal Frequency-Division Multiplexing (OFDM) Page 375

SSID cloaking Page 383

MAC address filtering Page 383

Wired Equivalent Privacy (WEP) Page 386

Wi-Fi Protected Access (WPA) Page 386

Temporal Key Integrity Protocol (TKIP) Page 386

WPA2 Page 386

Advanced Encryption Standard (AES) Page 386

WPA3 Page 386

Protected Management Frames (PMF) Page 386

pre-shared key (PSK) Page 386

Extensible Authentication Protocol (EAP) Page 386

Message Integrity Check (MIC) Page 387

Simultaneous Authentication of Equals (SAE) Page 389

Wi-Fi Protected Setup (WPS) Page 390

Device Provisioning Protocol (DPP) Page 390

Introduction (12.0)

Do you use a wireless connection at home, work, or school? Ever wonder how it works?

There are many ways to connect wirelessly. Like everything else involving networks, these connection types are best used in particular situations. They require specific devices and are also prone to certain types of attacks. And of course, there are solutions to mitigate these attacks. Want to learn more? The WLAN Concepts module gives you the foundational knowledge you need to understand what Wireless LANs are, what they can do, and how to protect them.

If you are curious, don't wait, get started today!

Introduction to Wireless (12.1)

This section explains WLAN technology and standards.

Benefits of Wireless (12.1.1)

A *Wireless LAN (WLAN)* is a type of wireless network that is commonly used in homes, offices, and campus environments. Networks must support people who are on the move. People connect using computers, laptops, tablets, and smart phones. There are many network infrastructures that provide network access, such as wired LANs, service provider networks, and cell phone networks. But it's the WLAN that makes mobility possible within the home and business environments.

In businesses with a wireless infrastructure in place, there can be a cost savings anytime equipment changes, or when relocating an employee within a building, reorganizing equipment or a lab, or moving to temporary locations or project sites. A wireless infrastructure can adapt to rapidly changing needs and technologies.

Types of Wireless Networks (12.1.2)

Wireless networks are based on the Institute of Electrical and Electronics Engineers (IEEE) standards and can be classified broadly into four main types: WPAN, WLAN, WMAN, and WWAN.

- *Wireless Personal-Area Networks (WPAN)*: Uses low-powered transmitters for a short-range network, usually 20 to 30 ft. (6 to 9 meters), as shown in Figure 12-1. Bluetooth and ZigBee based devices are commonly used in WPANs. WPANs are based on the 802.15 standard and a 2.4-GHz radio frequency.

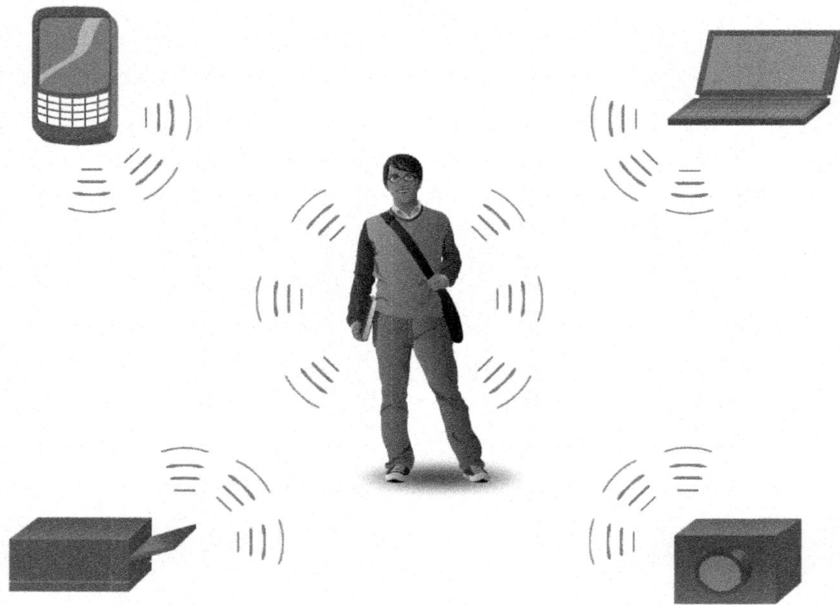

Figure 12-1 Example of WPANs

- **Wireless LANs (WLAN):** Uses transmitters to cover a medium-sized network (Figure 12-2), usually up to 300 feet. WLANs are suitable for use in a home, office, and even a campus environment. WLANs are based on the 802.11 standard and a 2.4-GHz or 5-GHz radio frequency.
- *Wireless MANs (WMAN)*: Uses transmitters to provide wireless service over a larger geographic area, as shown in Figure 12-3. WMANs are suitable for providing wireless access to a metropolitan city or specific district. WMANs use specific licensed frequencies.
- *Wireless Wide-Area Networks (WWANs)*: Uses transmitters to provide coverage over an extensive geographic area, as shown in Figure 12-4. WWANs are suitable for national and global communications. WWANs also use specific licensed frequencies.

Wireless Technologies (12.1.3)

Wireless technology uses the unlicensed radio spectrum to send and receive data. The unlicensed spectrum is accessible to anyone who has a wireless router and wireless technology in the device they are using.

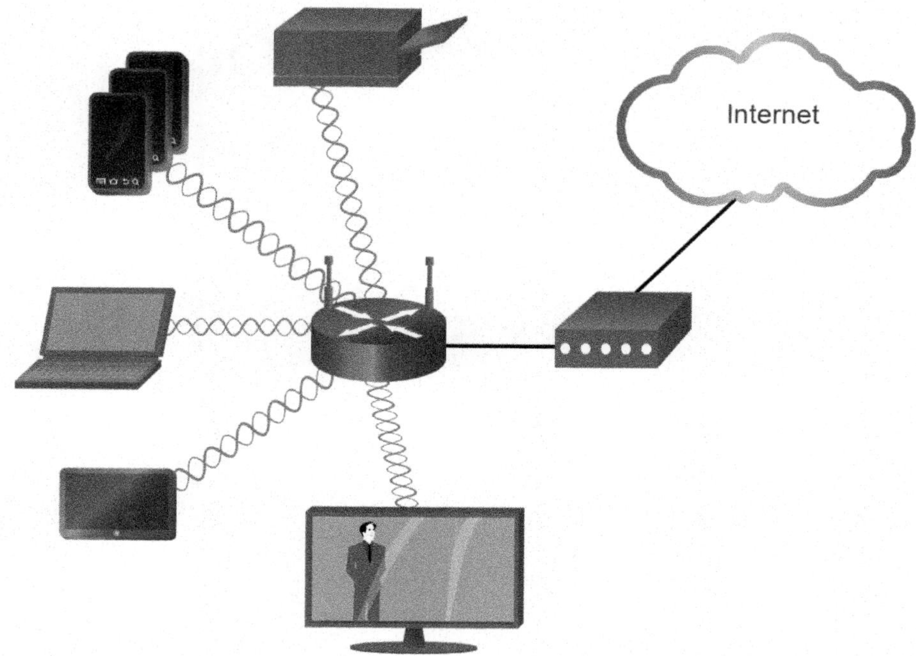

Figure 12-2 Example of a WLAN

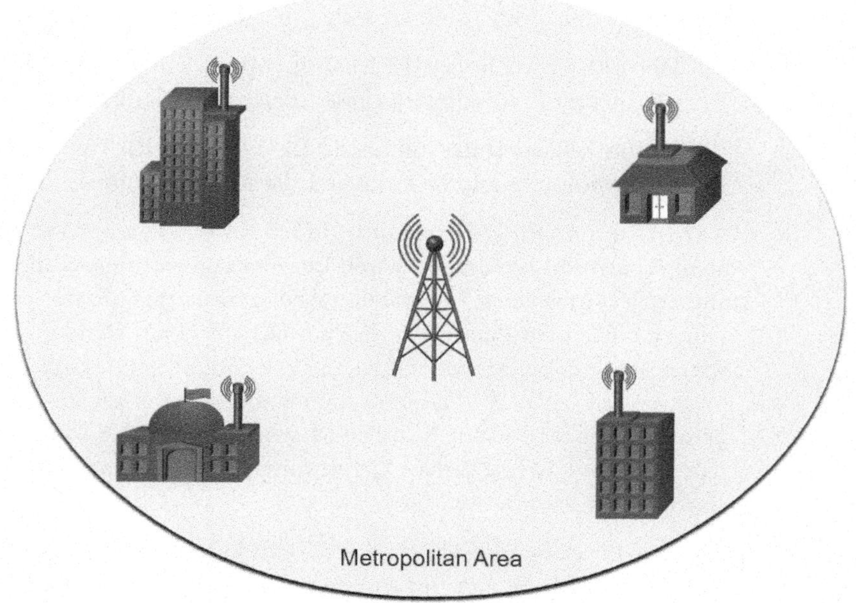

Figure 12-3 Example of a WMAN

Figure 12-4 Example of a WWAN

The following describes some common wireless technologies:

- **Bluetooth:** An IEEE 802.15 WPAN standard that uses a device-pairing process to communicate over distances up to 300 ft. (100m). It can be found in smart home devices, audio connections, automobiles, and other devices that require a short distance connection. There are two types of Bluetooth radios:
  - **Bluetooth Low Energy (BLE):** This supports multiple network technologies, including mesh topology to large scale network devices.
  - **Bluetooth Basic Rate/Enhanced Rate (BR/EDR):** This supports point-to-point topologies and is optimized for audio streaming.
- *WiMAX (Worldwide Interoperability for Microwave Access)*: WiMAX is an alternative to broadband wired Internet connections, competing with DSL and cable. However, it is typically used in areas that are not yet connected to a DSL or cable provider. It is an IEEE 802.16 WWAN standard that provides high-speed wireless broadband access of up to 30 miles (50 km). WiMAX operates in a similar way to Wi-Fi, but at higher speeds, over greater distances, and for a greater number of users. It uses a network of WiMAX towers that are similar to cell phone towers. WiMAX transmitters and cellular transmitters may share space on the same tower.
- **Cellular Broadband:** Cellular *4G/5G* are wireless mobile networks primarily used by cellular phones but can be used in automobiles, tablets, and laptops. Cellular networks are multi-access networks carrying both data and voice

communications. A cell site is created by a cellular tower transmitting signals in a given area. Interconnecting cell sites form the cellular network. The two types of cellular networks are Global System for Mobile (GSM) and Code Division Multiple Access (CDMA). GSM is internationally recognized, whereas CDMA is primarily used in the United States.

- The 4th Generation GSM network (4G) is the current mobile network. 4G delivers speeds that are 10 times the previous 3G networks. The new 5G holds the promise of delivering 100 times faster speeds than 4G and connecting more devices to the network than ever before.
- **Satellite Broadband:** Provides network access to remote sites through the use of a directional satellite dish that is aligned with a specific geostationary Earth orbit satellite. It is usually more expensive and requires a clear line of sight. Typically, it is used by rural homeowners and businesses where cable and DSL are not available.

802.11 Standards (12.1.4)

The world of wireless communications is vast. However, for particular job-related skills, we want to focus on specific aspects of Wi-Fi. The best place to start is with the IEEE 802.11 WLAN standards. These standards define how radio frequencies are used for wireless links. Most of the standards specify that wireless devices have one antenna to transmit and receive wireless signals on the specified radio frequency (2.4 GHz or 5 GHz). Some of the newer standards that transmit and receive at higher speeds require *wireless access points (APs)* and wireless clients to have multiple antennas using the *multiple-input and multiple-output (MIMO)* technology. MIMO uses multiple antennas as both the transmitter and receiver to improve communication performance. Up to eight transmit and receive antennas can be used to increase throughput.

Various implementations of the IEEE 802.11 standard have been developed over the years. Table 12-1 highlights these standards.

Table 12-1 802.11 Standards

IEEE WLAN Standard	Radio Frequency	Description
802.11	2.4 GHz	■ Speeds of up to 2 Mbps.
802.11a	5 GHz	■ Speeds of up to 54 Mbps. ■ Small coverage area. ■ Less effective at penetrating building structures. ■ Not interoperable with the 802.11b and 802.11g.

IEEE WLAN Standard	Radio Frequency	Description
802.11b	2.4 GHz	■ Speeds of up to 11 Mbps. ■ Longer range than 802.11a. ■ Better able to penetrate building structures.
802.11g	2.4 GHz	■ Speeds of up to 54 Mbps. ■ Backward compatible with 802.11b with reduced bandwidth capacity.
802.11n	2.4 GHz 5 GHz	■ Data rates range from 150 Mbps to 600 Mbps with a distance range of up to 70 m (230 feet). ■ APs and wireless clients require multiple antennas using MIMO technology. ■ Backward compatible with 802.11a/b/g devices with limiting data rates.
802.11ac	5 GHz	■ Provides data rates ranging from 450 Mbps to 1.3 Gbps (1300 Mbps) using MIMO technology. ■ Up to eight antennas can be supported. ■ Backward compatible with 802.11a/n devices with limiting data rates.
802.11ax	2.4 GHz 5 GHz	■ Latest standard released in 2019. ■ Also known as Wi-Fi 6 or High-Efficiency Wireless (HEW). ■ Provides improved power efficiency, higher data rates, increased capacity, and handles many connected devices. ■ Currently operates using 2.4 GHz and 5 GHz but will use 1 GHz and 7 GHz when those frequencies become available. ■ Search the Internet for Wi-Fi Generation 6 for more information.

Radio Frequencies (12.1.5)

All wireless devices operate in the radio waves range of the electromagnetic spectrum. WLAN networks operate in the 2.4 GHz frequency band and the 5 GHz band. Wireless LAN devices have transmitters and receivers tuned to specific frequencies of the radio waves range, as shown in Figure 12-5. Specifically, the following frequency bands are allocated to 802.11 wireless LANs:

- 2.4 GHz (UHF) - 802.11b/g/n/ax
- 5 GHz (SHF) - 802.11a/n/ac/ax

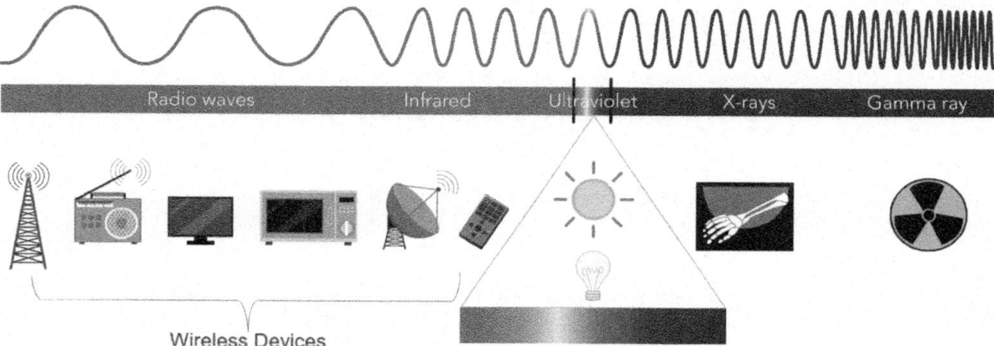

Figure 12-5 The Electromagnetic Spectrum

Wireless Standards Organizations (12.1.6)

Standards ensure interoperability between devices that are made by different manufacturers. Internationally, the three organizations influencing WLAN standards are the ITU-R, the IEEE, and the Wi-Fi Alliance.

- The *International Telecommunication Union (ITU)* regulates the allocation of the radio frequency spectrum and satellite orbits through the ITU-R. ITU-R stands for the ITU Radiocommunication Sector.
- The *IEEE* specifies how a radio frequency is modulated to carry information. It maintains the standards for local and metropolitan area networks (MAN) with the IEEE 802 LAN/MAN family of standards. The dominant standards in the IEEE 802 family are 802.3 Ethernet and 802.11 WLAN.
- The *Wi-Fi Alliance* is a global, nonprofit, industry trade association devoted to promoting the growth and acceptance of WLANs. It is an association of vendors whose objective is to improve the interoperability of products that are based on the 802.11 standard by certifying vendors for conformance to industry norms and adherence to standards.

Interactive Graphic

Check Your Understanding—Introduction to Wireless (12.1.7)

Refer to the online course to complete this activity.

WLAN Components (12.2)

The section identifies the components of a WLAN infrastructure.

Video

Video—WLAN Components (12.2.1)

Refer to the online course to view this video.

Wireless NICs (12.2.2)

Wireless deployments require a minimum of two devices that have a radio transmitter and a radio receiver tuned to the same radio frequencies:

- End devices with wireless network interface cards (NICs)
- A network device, such as a wireless router or wireless AP

To communicate wirelessly, laptops, tablets, smart phones, and even the latest automobiles include integrated wireless NICs that incorporate a radio transmitter/receiver. However, if a device does not have an integrated wireless NIC, a USB wireless adapter can be used, as shown in Figure 12-6.

Note

Many wireless devices you are familiar with do not have visible antennas. They are embedded inside smart phones, laptops, and wireless home routers.

Figure 12-6 USB Wireless Adapter

Wireless Home Router (12.2.3)

The type of infrastructure device that an end device associates and authenticates with varies based on the size and requirement of the WLAN.

For example, a home user typically interconnects wireless devices using a small, wireless router, such as the one in Figure 12-7.

Figure 12-7 Example of a Wireless Home Router

The wireless router serves as the following:

- **Access point:** This provides 802.11a/b/g/n/ac wireless access.
- **Switch:** This provides a four-port, full-duplex, 10/100/1000 Ethernet switch to interconnect wired devices.
- **Router:** This provides a default gateway for connecting to other network infrastructures, such as the Internet.

A wireless router is commonly implemented as a small business or residential wireless access device. The wireless router advertises its wireless services by sending beacons containing its shared *service set identifier (SSID)*. Devices wirelessly discover the SSID and attempt to associate and authenticate with it to access the local network and Internet.

Most wireless routers also provide advanced features, such as high-speed access, support for video streaming, IPv6 addressing, quality of service (QoS), configuration utilities, and USB ports to connect printers or portable drives.

Additionally, home users who want to extend their network services can implement *Wi-Fi range extenders*. A device can connect wirelessly to the extender, which boosts its communications to be repeated to the wireless router.

Wireless Access Points (12.2.4)

Although range extenders are easy to set up and configure, the best solution would be to install another wireless access point to provide dedicated wireless access to the user devices. Wireless clients use their wireless NIC to discover nearby APs advertising their SSID. Clients then attempt to associate and authenticate with an AP. After being authenticated, wireless users have access to network resources. The Cisco Meraki Go APs are shown in the Figure 12-8.

Figure 12-8 Cisco Meraki Go APs

AP Categories (12.2.5)

APs can be categorized as either autonomous APs or controller-based APs.

Autonomous APs

These are standalone devices configured using a command line interface (CLI) or a graphical user interface (GUI), as shown in Figure 12-9. *Autonomous APs* are useful in situations where only a couple of APs are required in the organization. A home router is an example of an autonomous AP, because the entire AP configuration resides on the device. If the wireless demands increase, more APs would be required. Each AP would operate independently of other APs, and each AP would require manual configuration and management. This would become overwhelming if many APs were needed.

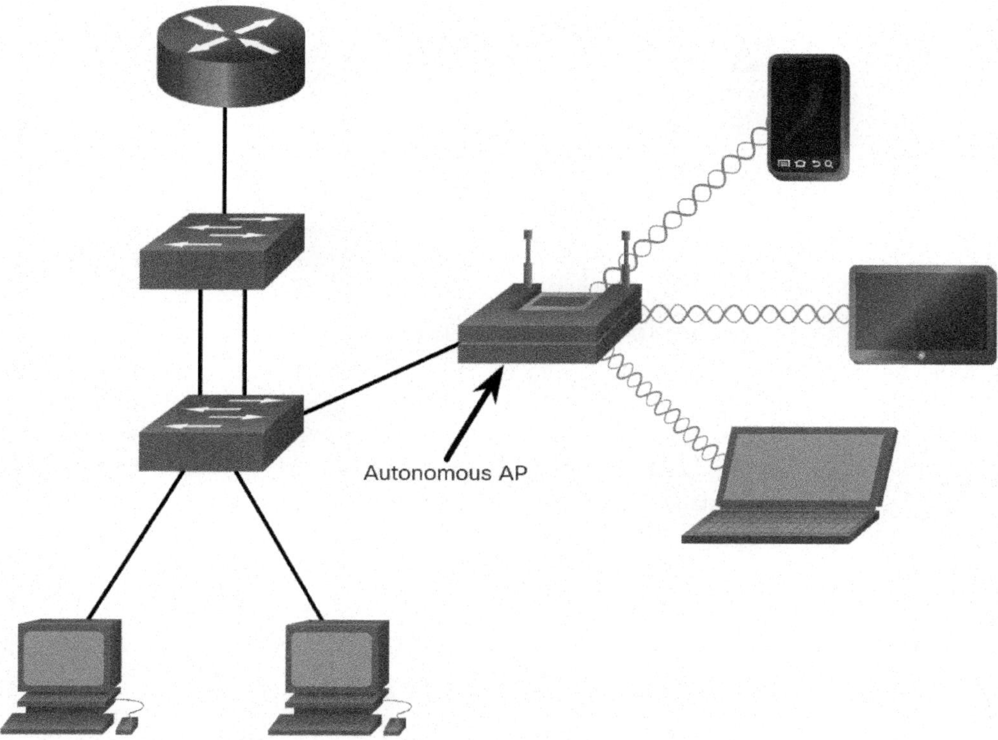

Figure 12-9 Example Topology with an Autonomous AP

Controller-Based APs

These devices require no initial configuration and are often called *lightweight APs (LAPs)*. LAPs use the *Lightweight Access Point Protocol (LWAPP)* to communicate with a *WLAN controller (WLC)*, as shown in Figure 12-10. *Controller-based APs* are useful in situations where many APs are required in the network. As more APs are added, each AP is automatically configured and managed by the WLC.

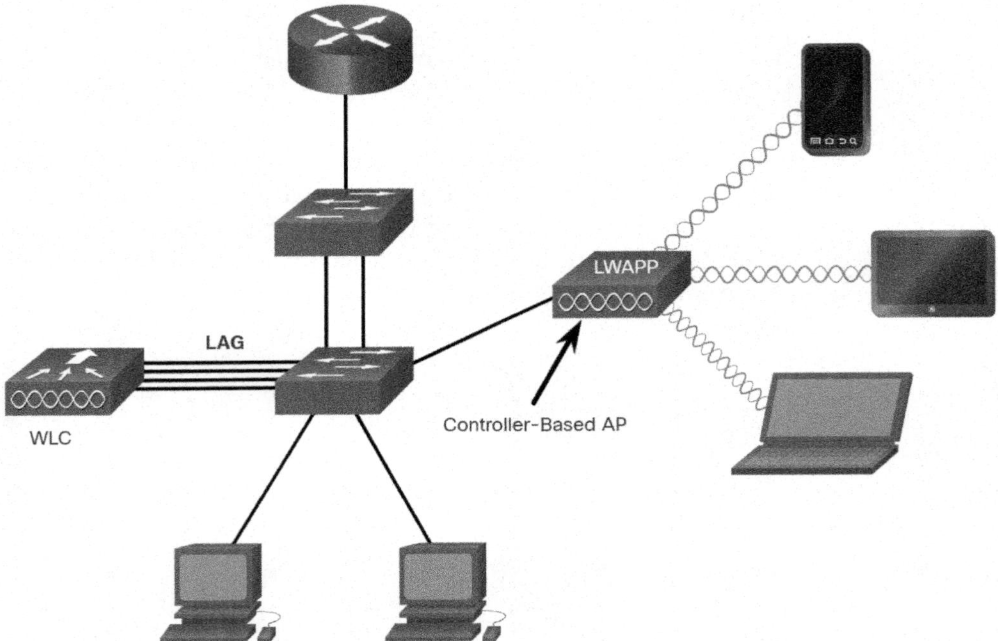

Figure 12-10 Example Topology with a Controller-Based AP

Notice in the figure that the WLC has four ports connected to the switching infrastructure. These four ports are configured as a *link aggregation group (LAG)* to bundle them together. Much like how EtherChannel operates, LAG provides redundancy and load-balancing. All the ports on the switch that are connected to the WLC need to be trunking and configured with EtherChannel on. However, LAG does not operate exactly like EtherChannel. The WLC does not support Port Aggregation Protocol (PaGP) or Link Aggregation Control Protocol (LACP).

Wireless Antennas (12.2.6)

Most business class APs require external antennas to make them fully functioning units.

Omnidirectional antennas, such as the one shown in Figure 12-11, provide 360-degree coverage and are ideal in houses, open office areas, conference rooms, and outside areas.

Directional antennas focus the radio signal in a given direction, such as the one shown in Figure 12-12. This enhances the signal to and from the AP in the direction the antenna is pointing This provides a stronger signal strength in one direction and reduced signal strength in all other directions. Examples of directional Wi-Fi antennas include *Yagi antenna* and *parabolic dish antenna*.

Figure 12-11 Omnidirectional Antenna

Figure 12-12 Directional Antenna

Multiple Input Multiple Output (MIMO) uses multiple antennas to increase available bandwidth for IEEE 802.11n/ac/ax wireless networks, such as the one in Figure 12-13. Up to eight transmit and receive antennas can be used to increase throughput.

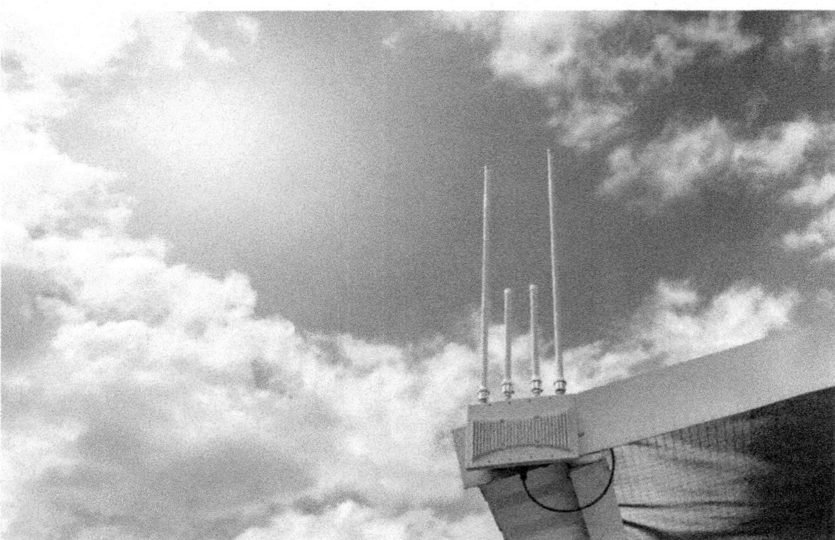

Figure 12-13 MIMO Antennas

Interactive Graphic

Check Your Understanding—WLAN Components (12.2.7)

Refer to the online course to complete this activity.

WLAN Operation (12.3)

This section explains how wireless technology enables WLAN operation.

Video

Video—WLAN Operation (12.3.1)

Refer to the online course to view this video.

802.11 Wireless Topology Modes (12.3.2)

Wireless LANs can accommodate various network topologies. The 802.11 standard identifies two main wireless topology modes: Ad hoc mode and Infrastructure mode. Tethering is also a mode sometimes used to provide quick wireless access.

- *Ad hoc mode*: This is when two devices connect wirelessly in a peer-to-peer (P2P) manner without using APs or wireless routers, as shown in Figure 12-14. Examples include wireless clients connecting directly to each other using

Bluetooth or Wi-Fi Direct. The IEEE 802.11 standard refers to an ad hoc network as an *independent basic service set (IBSS)*.

Figure 12-14 Example of an Ad Hoc Wireless Network

- *Infrastructure mode*: This is when wireless clients interconnect via a wireless router or AP, such as in WLANs. APs connect to the network infrastructure using the wired distribution system, such as Ethernet shown in Figure 12-15.

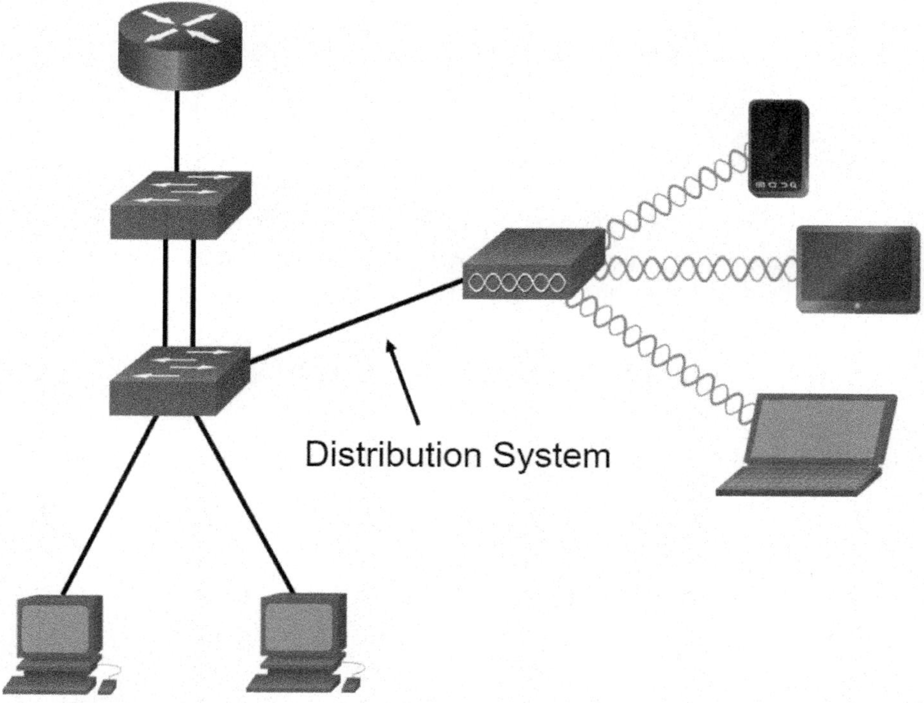

Figure 12-15 Example of Infrastructure Mode Wireless Network

- *Tethering*: A variation of the ad hoc topology is when a smart phone or tablet with cellular data access is enabled to create a personal *hotspot*, as shown in Figure 12-16. This feature is sometimes referred to as tethering. A hotspot is usually a temporary, quick solution that enables a smart phone to provide the wireless services of a Wi-Fi router. Other devices can associate and authenticate with the smart phone to use the Internet connection.

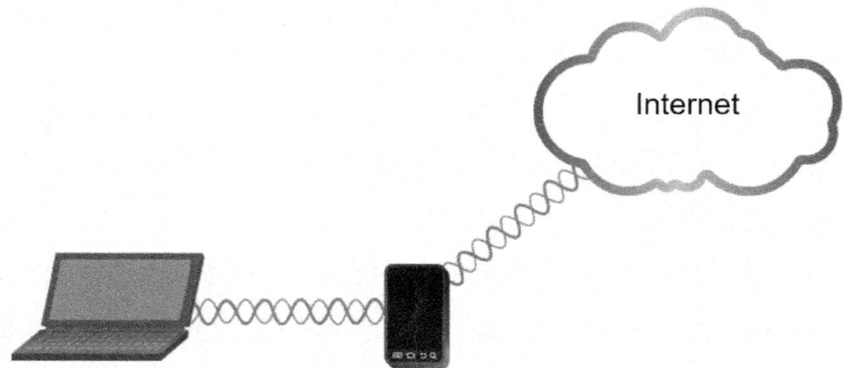

Figure 12-16 Example of Tethering

BSS and ESS (12.3.3)

Infrastructure mode defines two topology building blocks: A *Basic Service Set (BSS)* and an *Extended Service Set (ESS)*.

Basic Service Set

A BSS consists of a single AP interconnecting all associated wireless clients. Two BSSs are shown in Figure 12-17. The circles depict the coverage area for the BSS, which is called the *Basic Service Area (BSA)*. If a wireless client moves out of its BSA, it can no longer directly communicate with other wireless clients within the BSA.

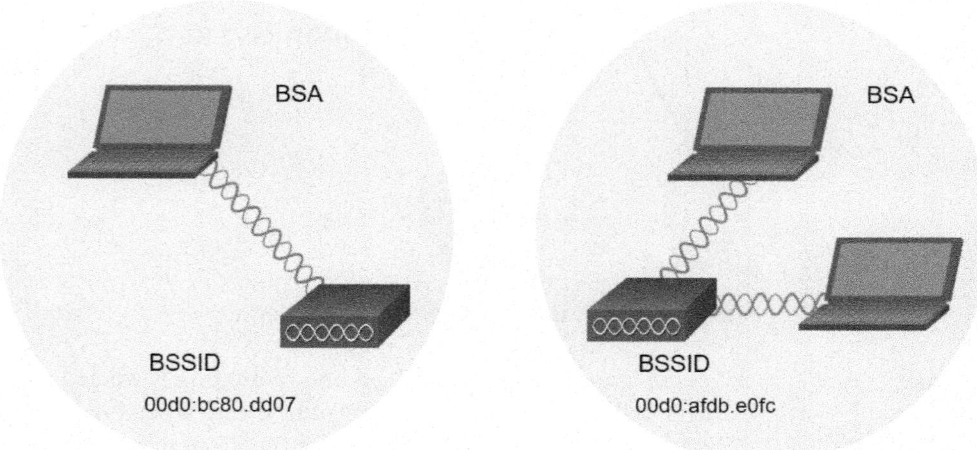

Figure 12-17 Example of BSSs

The Layer 2 MAC address of the AP is used to uniquely identify each BSS, which is called the *Basic Service Set Identifier (BSSID)*. Therefore, the BSSID is the formal name of the BSS and is always associated with only one AP.

Extended Service Set

When a single BSS provides insufficient coverage, two or more BSSs can be joined through a common *distribution system (DS)* into an ESS. An ESS is the union of two or more BSSs interconnected by a wired DS. Each ESS is identified by a SSID, and each BSS is identified by its BSSID.

Wireless clients in one BSA can now communicate with wireless clients in another BSA within the same ESS. Roaming mobile wireless clients may move from one BSA to another (within the same ESS) and seamlessly connect.

The rectangular area in Figure 12-18 depicts the coverage area within which members of an ESS may communicate. This area is called the *Extended Service Area (ESA)*.

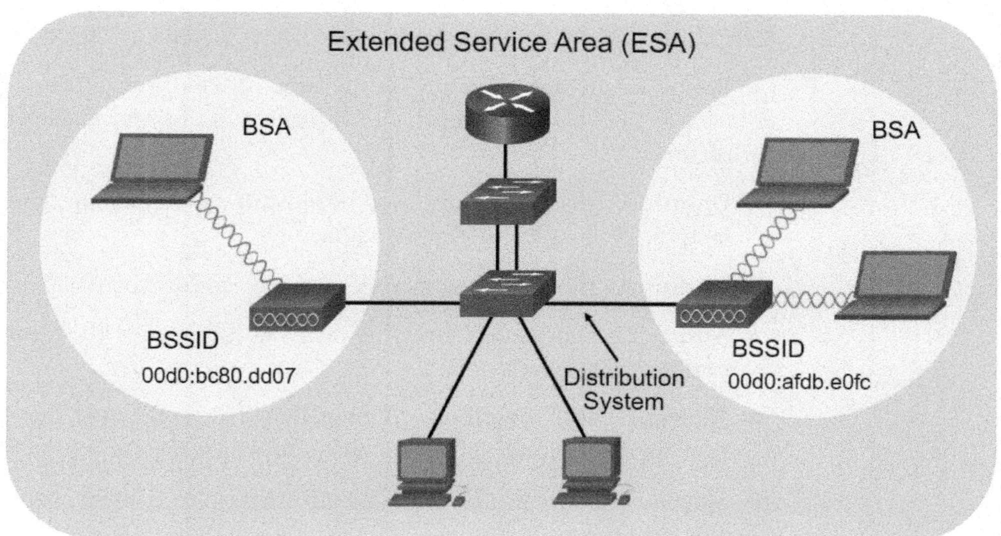

Figure 12-18 Example of an ESS

802.11 Frame Structure (12.3.4)

Recall that all Layer 2 frames consist of a header, payload, and Frame Check Sequence (FCS) section. The 802.11 frame format is similar to the Ethernet frame format, except that it contains more fields, as shown in Figure 12-19.

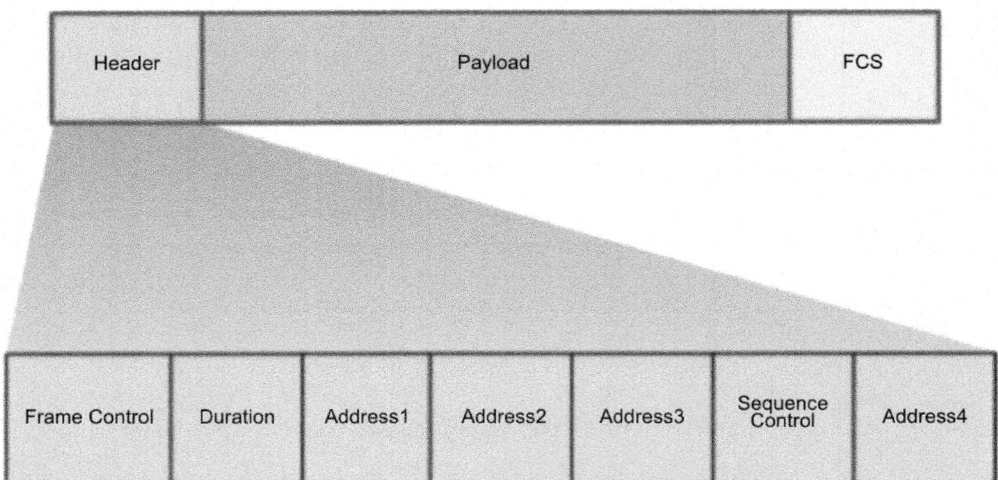

Figure 12-19 802.11 Frame Structure

All 802.11 wireless frame from a wireless device contains the following fields:

- **Frame Control:** This identifies the type of wireless frame and contains subfields for protocol version, frame type, address type, power management, and security settings.
- **Duration:** This is typically used to indicate the remaining duration needed to receive the next frame transmission.
- **Address1:** This usually contains the MAC address of AP.
- **Address2:** This usually contains the MAC address of the transmitting (that is, sending) wireless device.
- **Address3:** This sometimes contains the MAC address of the destination, such as the router interface (default gateway) to which the AP is attached.
- **Sequence Control:** This contains information to control sequencing and fragmented frames.
- **Address4:** This is usually missing because it is used only in ad hoc mode.
- **Payload:** This contains the data for transmission.
- **FCS:** This is used for Layer 2 error control.

Note

Addresses 1 through 3 would be different if this was a frame from an AP replying to the wireless device.

CSMA/CA (12.3.5)

WLANs are half-duplex, shared media configurations. Half-duplex means that only one client can transmit or receive at any given moment. Shared media means that wireless clients can all transmit and receive on the same radio channel. This creates a problem because a wireless client cannot hear while it is sending, which makes it impossible to detect a collision.

To resolve this problem, WLANs use *carrier sense multiple access with collision avoidance (CSMA/CA)* as the method to determine how and when to send data on the network. A wireless client does the following:

1. Listens to the channel to see if it is idle, which means that it senses no other traffic is currently on the channel. The channel is also called the carrier.
2. Sends a ready to send (RTS) message to the AP to request dedicated access to the network.
3. Receives a clear to send (CTS) message from the AP granting access to send.
4. If the wireless client does not receive a CTS message, it waits a random amount of time before restarting the process.
5. After it receives the CTS, it transmits the data.
6. All transmissions are acknowledged. If a wireless client does not receive an acknowledgment, it assumes a collision occurred and restarts the process.

Wireless Client and AP Association (12.3.6)

For wireless devices to communicate over a network, they must first associate with an AP or wireless router. An important part of the 802.11 process is discovering a WLAN and subsequently connecting to it. Wireless devices complete the following three-stage process, as shown in Figure 12-20:

1. Discover a wireless AP.
2. Authenticate with AP.
3. Associate with AP.

To have a successful association, a wireless client and an AP must agree on specific parameters. Parameters must then be configured on the AP and subsequently on the client to enable the negotiation of a successful association.

- **SSID:** The SSID name appears in the list of available wireless networks on a client. In larger organizations that use multiple VLANs to segment traffic, each SSID is mapped to one VLAN. Depending on the network configuration, several APs on a network can share a common SSID.

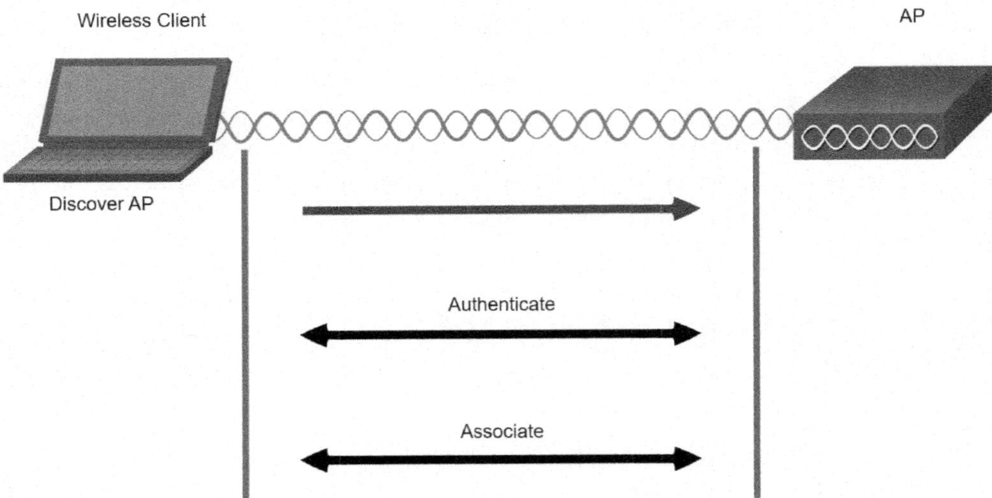

Figure 12-20 Discover, Authenticate, and Associate with an AP

- **Password:** This is required from the wireless client to authenticate to the AP.
- **Network mode:** This refers to the 802.11a/b/g/n/ac/ad WLAN standards. APs and wireless routers can operate in a Mixed mode, meaning that they can simultaneously support clients connecting via multiple standards.
- **Security mode:** This refers to the security parameter settings, such as Wired Equivalent Privacy (WEP), Wi-Fi Protected Access (WPA), WPA2, or WPA3. Always enable the highest security level supported.
- **Channel settings:** This refers to the frequency bands used to transmit wireless data. Wireless routers and APs can scan the radio frequency channels and automatically select an appropriate channel setting. The channel can also be set manually if there is interference with another AP or wireless device.

Passive and Active Discover Mode (12.3.7)

Wireless devices must discover and connect to an AP or wireless router. Wireless clients connect to the AP using a scanning (probing) process. This process can be passive or active.

Passive Mode

In passive mode, the AP openly advertises its service by periodically sending broadcast beacon frames containing the SSID, supported standards, and security settings, as shown in Figure 12-21. The primary purpose of the beacon is to allow wireless clients to learn which networks and APs are available in a given area. This allows the wireless clients to choose which network and AP to use.

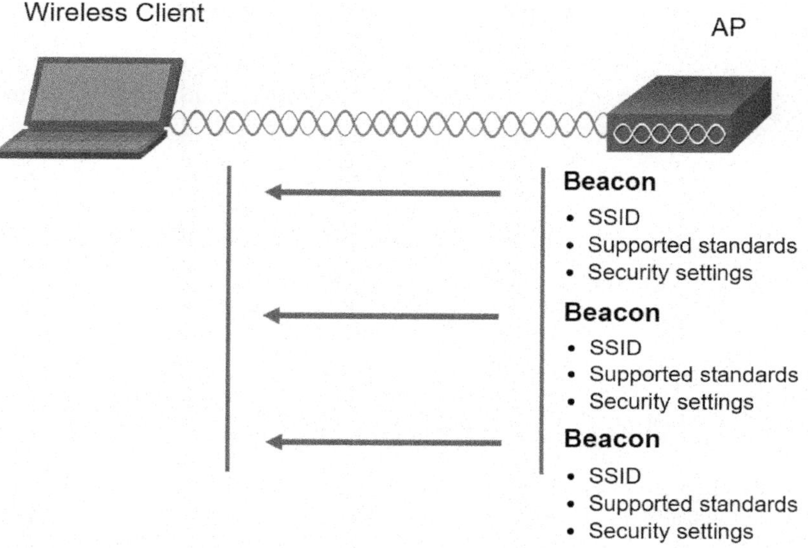

Figure 12-21 Passive Discover Mode

Active Mode

In active mode, wireless clients must know the name of the SSID. The wireless client initiates the process by broadcasting a probe request frame on multiple channels, as shown in Figure 12-22.

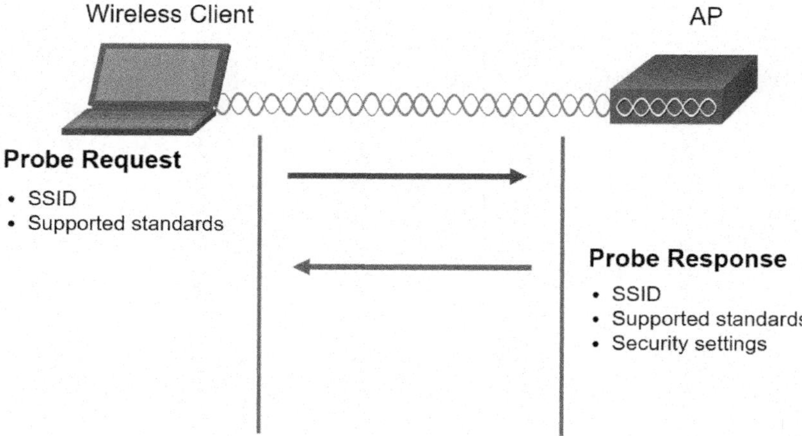

Figure 12-22 Active Discover Mode

The probe request includes the SSID name and standards supported. APs configured with the SSID will send a probe response that includes the SSID, supported standards, and security settings. Active mode may be required if an AP or wireless router is configured to not broadcast beacon frames.

A wireless client could also send a probe request without a SSID name to discover nearby WLAN networks. APs configured to broadcast beacon frames would respond to the wireless client with a probe response and provide the SSID name. APs with the broadcast SSID feature disabled do not respond.

Interactive Graphic

Check Your Understanding—WLAN Operation (12.3.8)

Refer to the online course to complete this activity.

CAPWAP Operation (12.4)

This section explains how Wireless LAN Controller (WLC) uses Control and Provisioning of Wireless Access Points (CAPWAP) protocol to manage multiple APs.

Video

Video—CAPWAP (12.4.1)

Refer to the online course to view this video.

Introduction to CAPWAP (12.4.2)

CAPWAP is an IEEE standard protocol that enables a WLC to manage multiple APs and WLANs. CAPWAP is also responsible for the encapsulation and forwarding of WLAN client traffic between an AP and a WLC.

CAPWAP is based on Lightweight Access Point Protocol (LWAPP) but adds additional security with *Datagram Transport Layer Security (DTLS)*. CAPWAP establishes tunnels on User Datagram Protocol (UDP) ports. CAPWAP can operate either over IPv4 or IPv6, as shown in Figure 12-23, but uses IPv4 by default.

IPv4 and IPv6 can use UDP ports 5246 and 5247. However, CAPWAP tunnels use different IP protocols in the frame header. IPv4 uses IP protocol 17 and IPv6 uses IP protocol 136.

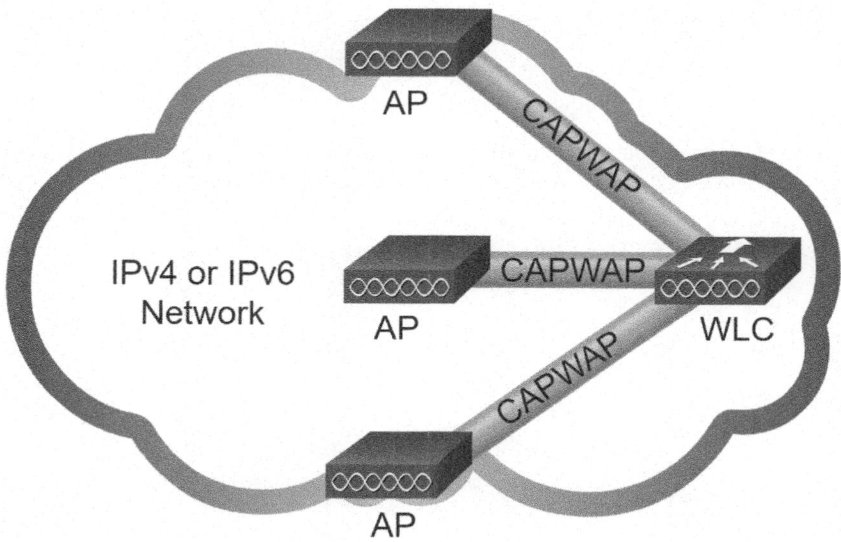

Figure 12-23 CAPWAP Tunnels

Split MAC Architecture (12.4.3)

A key component of CAPWAP is the concept of a split media access control (MAC). The CAPWAP *split MAC* concept does all the functions normally performed by individual APs and distributes them between two functional components:

- AP MAC Functions
- WLC MAC Functions

Table 12-2 shows some of the MAC functions performed by each.

Table 12-2 AP and WLC MAC Functions

AP MAC Functions	WLC MAC Functions
Beacons and probe responses	Authentication
Packet acknowledgments and retransmissions	Association and reassociation of roaming clients
Frame queuing and packet prioritization	Frame translation to other protocols
MAC layer data encryption and decryption	Termination of 802.11 traffic on a wired interface

DTLS Encryption (12.4.4)

DTLS is a protocol that provides security between the AP and the WLC. It allows them to communicate using encryption and prevents eavesdropping or tampering.

DTLS is enabled by default to secure the CAPWAP control channel but is disabled by default for the data channel, as shown in Figure 12-24. All CAPWAP management and control traffic exchanged between an AP and WLC is encrypted and secured by default to provide control plane privacy and prevent man-in-the-middle (MITM) attacks.

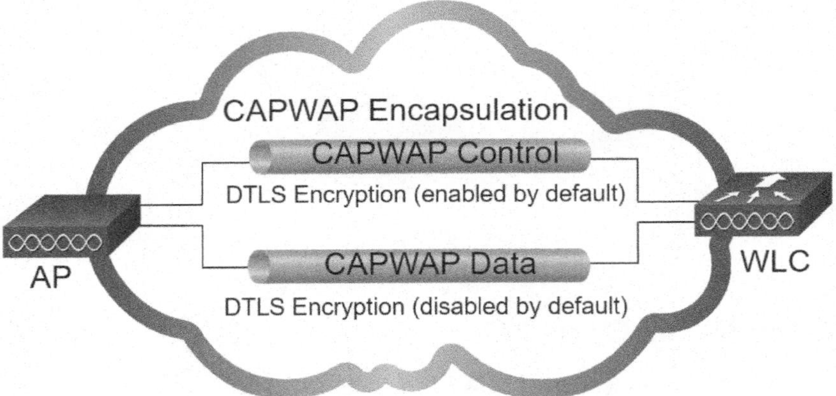

Figure 12-24 DTLS Encryption on CAPWAP Control and Data Tunnels

CAPWAP data encryption is optional and is enabled per AP. Data encryption requires a DTLS license to be installed on the WLC prior to being enabled on an AP. When enabled, all WLAN client traffic is encrypted at the AP before being forwarded to the WLC, and vice versa.

FlexConnect APs (12.4.5)

FlexConnect is a wireless solution for branch office and remote office deployments. It lets you configure and control access points in a branch office from the corporate office through a WAN link, without deploying a controller in each office.

There are two modes of operation for the FlexConnect AP.

- **Connected mode:** The WLC is reachable. In this mode the FlexConnect AP has CAPWAP connectivity with its WLC and can send traffic through the CAPWAP tunnel, as shown in Figure 12-25. The WLC performs all its CAPWAP functions.
- **Standalone mode:** The WLC is unreachable. The FlexConnect has lost or failed to establish CAPWAP connectivity with its WLC. In this mode, a FlexConnect

AP can assume some of the WLC functions, such as switching client data traffic locally and performing client authentication locally.

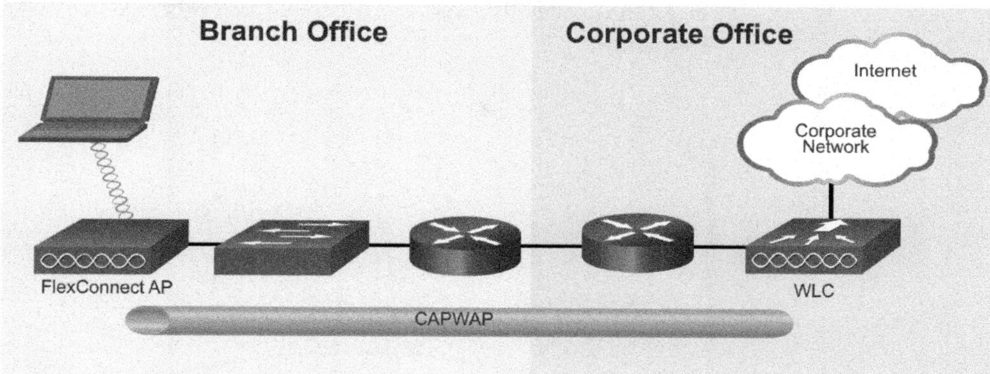

Figure 12-25 Topology Example of FlexConnect

Interactive Graphic

Check Your Understanding—CAPWAP Operation (12.4.6)

Refer to the online course to complete this activity.

Channel Management (12.5)

This section explains how channel management is used on a WLAN.

Frequency Channel Saturation (12.5.1)

Wireless LAN devices have transmitters and receivers tuned to specific frequencies of radio waves to communicate. A common practice is for frequencies to be allocated as ranges. Such ranges are then split into smaller ranges called channels.

If the demand for a specific channel is too high, that channel is likely to become oversaturated. The saturation of the wireless medium degrades the quality of the communication. Over the years, a number of techniques have been created to improve wireless communication and alleviate saturation. These techniques mitigate channel saturation by using the channels in a more efficient way.

- *Direct-Sequence Spread Spectrum (DSSS)*: This is a modulation technique designed to spread a signal over a larger frequency band. Spread spectrum techniques were developed during wartime to make it more difficult for enemies to intercept or jam a communication signal. It does this by spreading the signal over a wider frequency, which effectively hides the discernable peak of the signal, as shown in Figure 12-26. A properly configured receiver can reverse the DSSS modulation and reconstruct the original signal. DSSS is used by 802.11b devices to avoid interference from other devices using the same 2.4 GHz frequency.

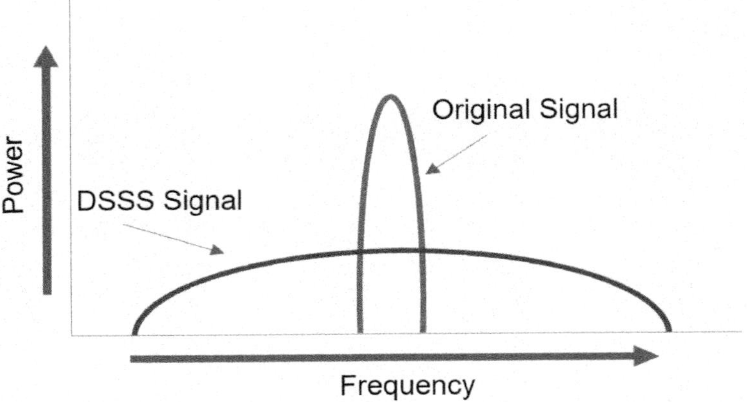

Figure 12-26 DSSS

- *Frequency-Hopping Spread Spectrum (FHSS)*: This relies on spread spectrum methods to communicate. It transmits radio signals by rapidly switching a carrier signal among many frequency channels, as shown in Figure 12-27. With FHSS, the sender and receiver must be synchronized to "know" which channel to jump to. This channel-hopping process allows for a more efficient usage of the channels, decreasing channel congestion. FHSS was used by the original 802.11 standard. Walkie-talkies and 900 MHz cordless phones also use FHSS, and Bluetooth uses a variation of FHSS.

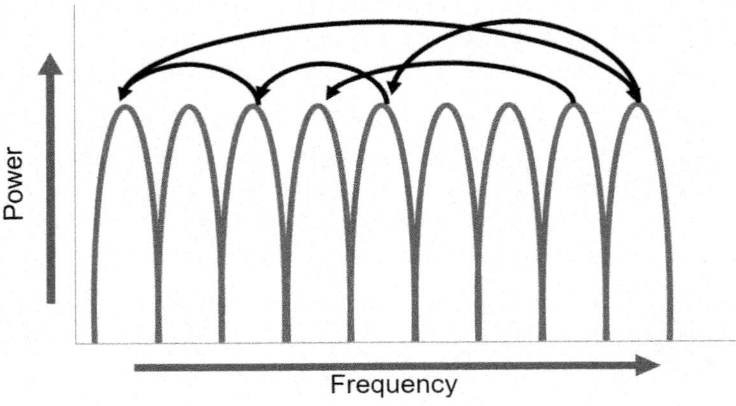

Figure 12-27 FHSS

- *Orthogonal Frequency-Division Multiplexing (OFDM)*: This is a subset of frequency division multiplexing in which a single channel uses multiple subchannels on adjacent frequencies, as shown in Figure 12-28. Subchannels in an OFDM system are precisely orthogonal to one another, which allows the subchannels to overlap without interfering. OFDM is used by a number of communication systems, including 802.11a/g/n/ac. The new 802.11ax uses a variation of OFDM called orthogonal frequency-division multiaccess (OFDMA).

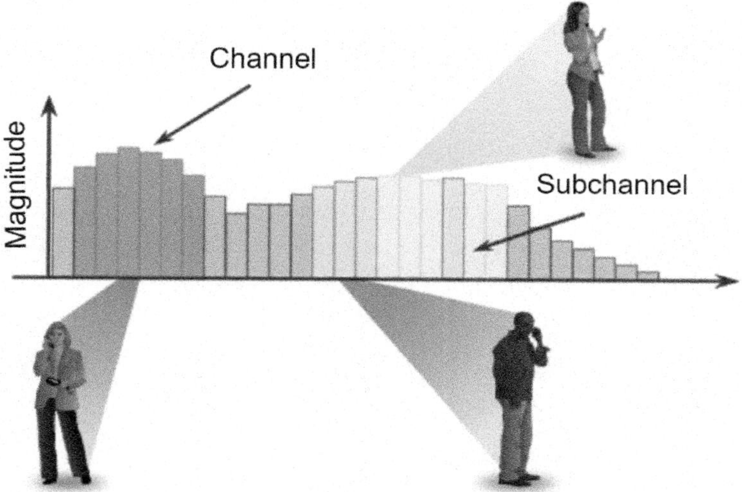

Figure 12-28 OFDM

Channel Selection (12.5.2)

A best practice for WLANs requiring multiple APs is to use non-overlapping channels. For example, the 802.11b/g/n standards operate in the 2.4 GHz to 2.5 GHz spectrum. The 2.4 GHz band is subdivided into multiple channels. Each channel is allotted 22 MHz bandwidth and is separated from the next channel by 5 MHz. The 802.11b standard identifies 11 channels for North America, as shown in Figure 12-29 (13 in Europe and 14 in Japan).

Note

Search the Internet for 2.4 GHz channels to learn more about the variations for different countries.

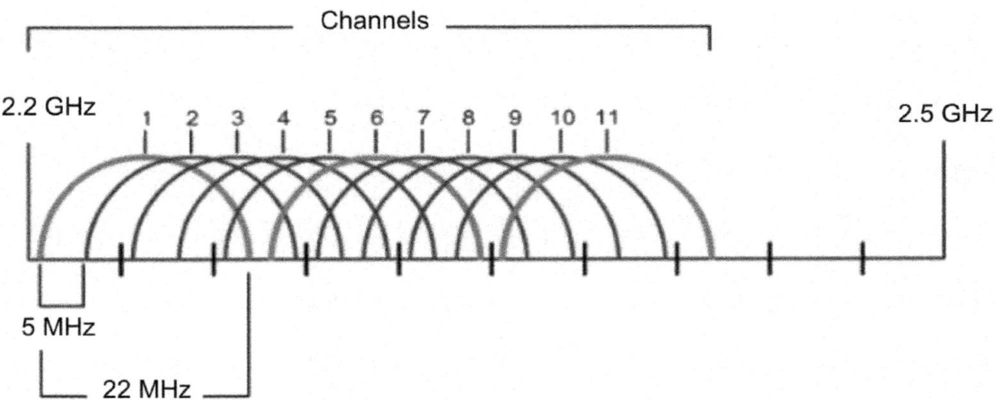

Figure 12-29 2.4 GHz Overlapping Channels in North America

Interference occurs when one signal overlaps a channel reserved for another signal, causing possible distortion. The best practice for 2.4 GHz WLANs that require multiple APs is to use non-overlapping channels, although most modern APs will do this automatically. If there are three adjacent APs, use channels 1, 6, and 11, as shown in Figure 12-30.

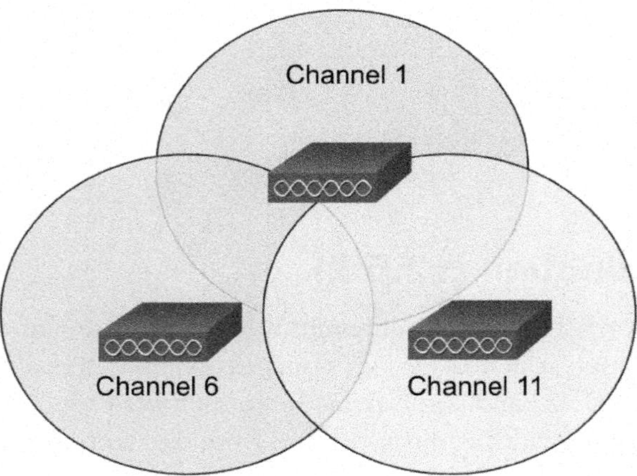

Figure 12-30 2.4 GHz Non-Overlapping Channels for 802.11b/g/n

For the 5 GHz standards 802.11a/n/ac, there are 24 channels. The 5 GHz band is divided into three sections. Each channel is separated from the next channel by 20 MHz. Figure 12-31 shows the first section of eight channels for the 5 GHz band. Although there is a slight overlap, the channels do not interfere with one another. 5 GHz wireless can provide faster data transmission for wireless clients in heavily populated wireless networks because of the large amount of non-overlapping wireless channels.

Note

Search the Internet for 5 GHz channels to learn more about the other 16 channels available and to learn more about the variations for different countries.

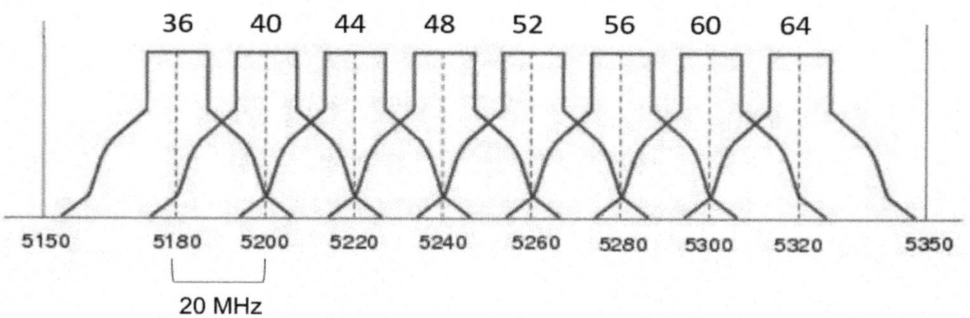

Figure 12-31 5 GHz First Eight Non-Interfering Channels

As with 2.4 GHz WLANs, choose non-interfering channels when configuring multiple 5 GHz APs that are adjacent to each other, as shown in Figure 12-32.

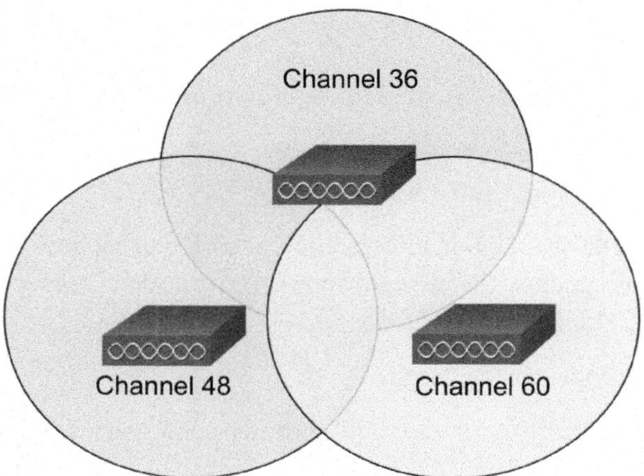

Figure 12-32 5 GHz Non-Interfering Channels for 802.11a/n/ac

Plan a WLAN Deployment (12.5.3)

The number of users supported by a WLAN depends on the geographical layout of the facility, including the number of bodies and devices that can fit in a space, the data rates users expect, the use of non-overlapping channels by multiple APs in an ESS, and transmit power settings.

When planning the location of APs, the approximate circular coverage area is important, as shown in Figure 12-33.

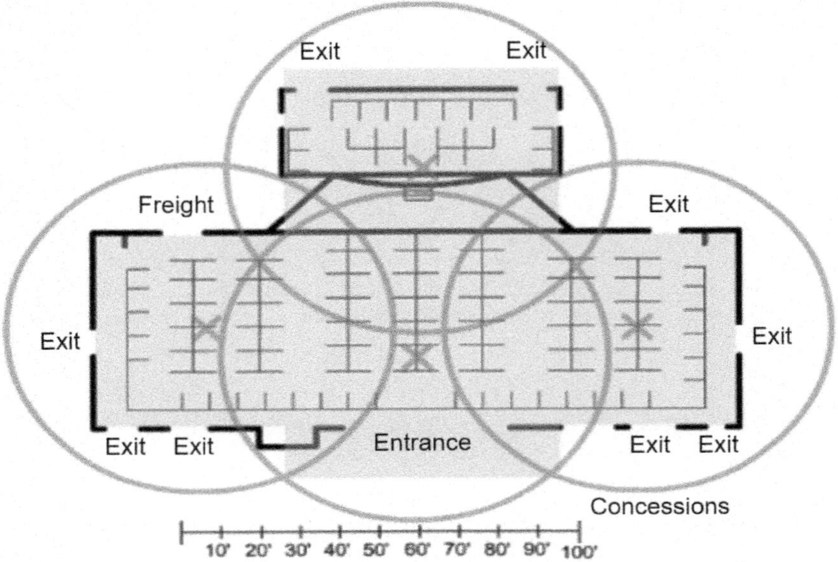

Figure 12-33 Example of a WLAN Deployment Plan

However, there are some additional recommendations:

- If APs are to use existing wiring, or if there are locations where APs cannot be placed, note these locations on the map.
- Note all potential sources of interference, which can include microwave ovens, wireless video cameras, fluorescent lights, motion detectors, or any other device that uses the 2.4 GHz range.
- Position APs above obstructions.
- Position APs vertically near the ceiling in the center of each coverage area, if possible.
- Position APs in locations where users are expected to be. For example, conference rooms are typically a better location for APs than a hallway.
- If an IEEE 802.11 network has been configured for mixed mode, the wireless clients may experience slower than normal speeds to support the older wireless standards.

When estimating the expected coverage area of an AP, realize that this value varies depending on the WLAN standard or mix of standards that are deployed, the nature of the facility, and the transmit power that the AP is configured for. Always consult the specifications for the AP when planning for coverage areas.

Interactive Graphic

Check Your Understanding—Channel Management (12.5.4)

Refer to the online course to complete this activity.

WLAN Threats (12.6)

This section discusses threats to WLANs.

Video

Video—WLAN Threats (12.6.1)

Refer to the online course to view this video.

Wireless Security Overview (12.6.2)

A WLAN is open to anyone within range of an AP and the appropriate credentials to associate to it. With a wireless NIC and knowledge of cracking techniques, an attacker may not have to physically enter the workplace to gain access to a WLAN.

Attacks can be generated by outsiders, disgruntled employees, and even unintentionally by employees. Wireless networks are specifically susceptible to several threats, including:

- **Interception of data:** Wireless data should be encrypted to prevent it from being read by eavesdroppers.
- **Wireless intruders:** Unauthorized users attempting to access network resources can be deterred through effective authentication techniques.
- **Denial of Service (DoS) Attacks:** Access to WLAN services can be compromised either accidentally or maliciously. Various solutions exist depending on the source of the DoS attack.
- **Rogue APs:** Unauthorized APs installed by a well-intentioned user or for malicious purposes can be detected using management software.

DoS Attacks (12.6.3)

Wireless DoS attacks can be the result of the following:

- **Improperly configured devices:** Configuration errors can disable the WLAN. For instance, an administrator could accidently alter a configuration and disable the network, or an intruder with administrator privileges could intentionally disable a WLAN.
- **A malicious user intentionally interfering with the wireless communication:** Their goal is to disable the wireless network completely or to the point where no legitimate device can access the medium.
- **Accidental interference:** WLANs are prone to interference from other wireless devices, including microwave ovens, cordless phones, baby monitors, and more, as shown in Figure 12-34. The 2.4 GHz band is more prone to interference than the 5 GHz band.

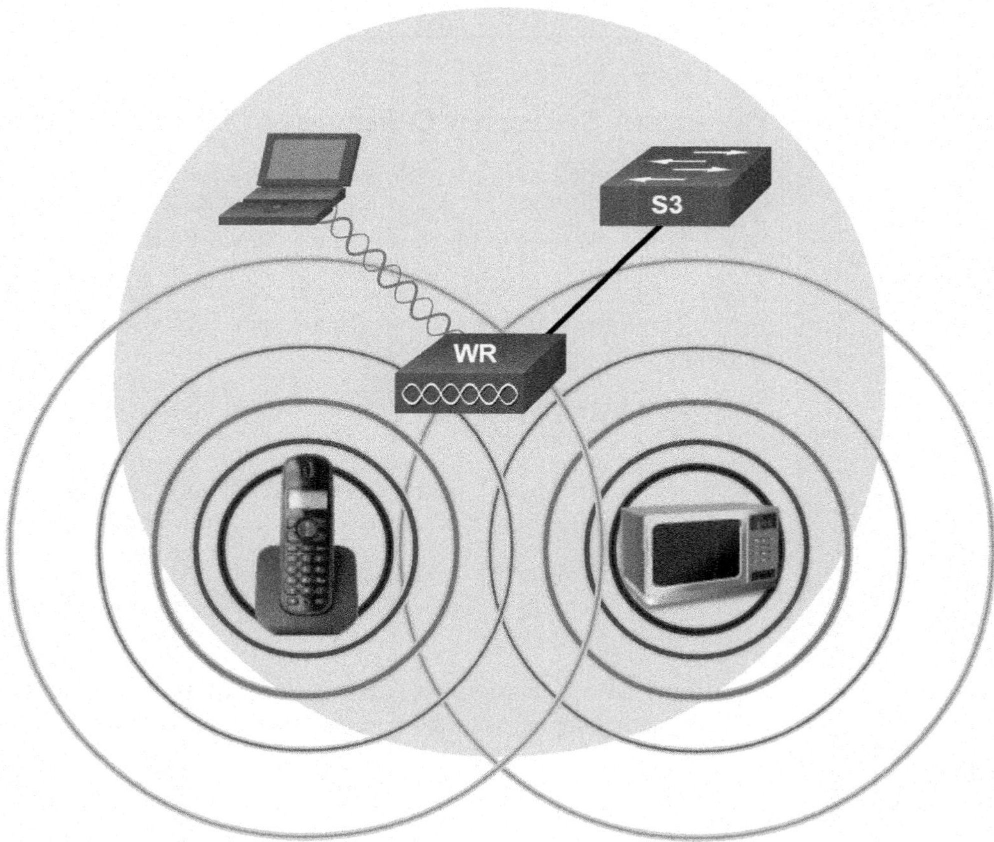

Figure 12-34 Example of Accidental Interference

Rogue Access Points (12.6.4)

A rogue AP is an AP or wireless router that has been connected to a corporate network without explicit authorization and against corporate policy. Anyone with access to the premises can install (maliciously or nonmaliciously) an inexpensive wireless router that can potentially allow access to a secure network resource.

Once connected, the rogue AP can be used by an attacker to capture MAC addresses, capture data packets, gain access to network resources, or launch a man-in-the-middle attack.

A personal network hotspot could also be used as a rogue AP. For example, a user with secure network access enables their authorized Windows host to become a Wi-Fi AP. Doing so circumvents the security measures, and other unauthorized devices can now access network resources as a shared device.

To prevent the installation of rogue APs, organizations must configure WLCs with rogue AP policies, as shown in Figure 12-35, and use monitoring software to actively monitor the radio spectrum for unauthorized APs.

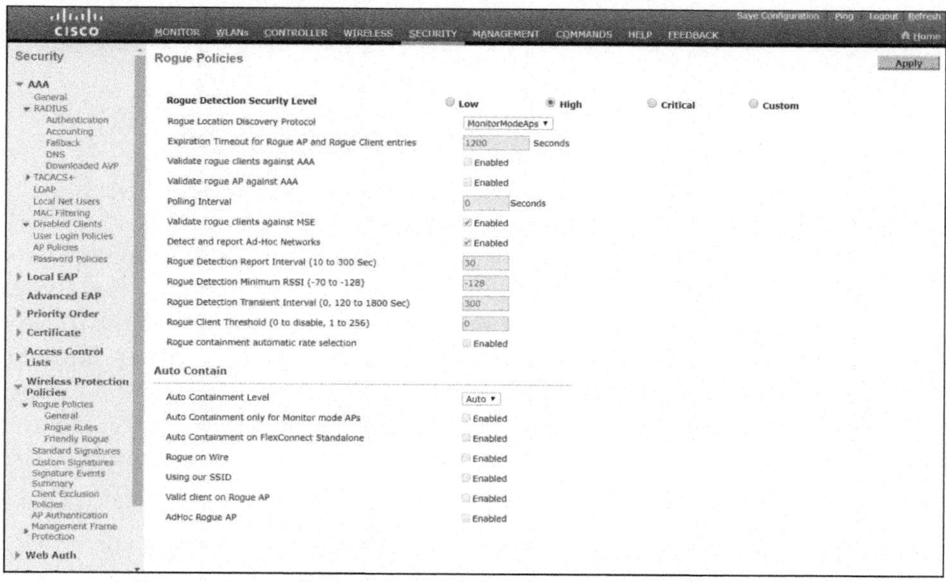

Figure 12-35 Rogue Policy Configuration Page on a Cisco WLC

Man-in-the-Middle Attack (12.6.5)

In a man-in-the-middle (MITM) attack, the hacker is positioned in between two legitimate entities in order to read or modify the data that passes between the two parties. There are many ways in which to create a MITM attack.

A popular wireless MITM attack is called the “evil twin AP” attack, where an attacker introduces a rogue AP and configures it with the same SSID as a legitimate AP, as shown in Figure 12-36. Locations offering free Wi-Fi, such as airports, cafes, and restaurants, are particularly popular spots for this type of attack due to the open authentication.

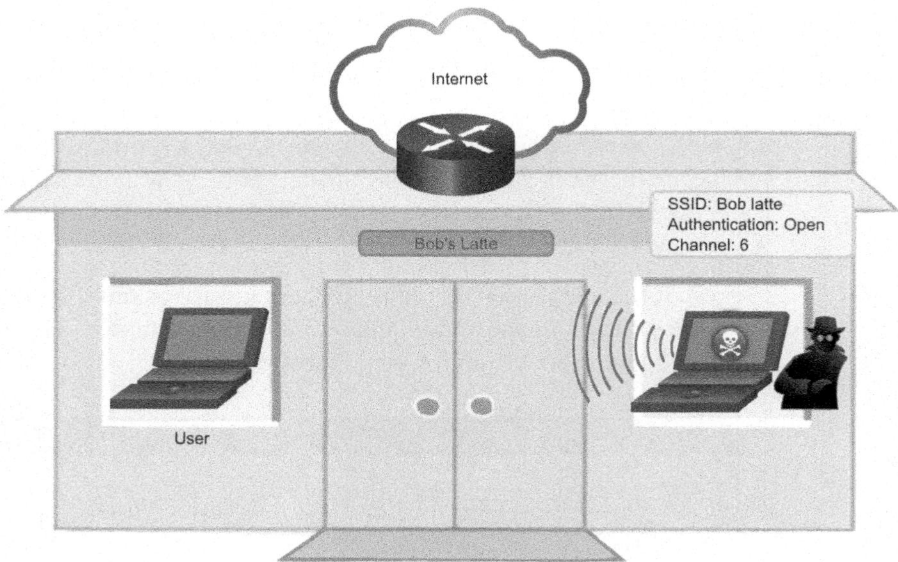

Figure 12-36 Threat Actor Broadcasting an Open Network

Wireless clients attempting to connect to a WLAN would see two APs with the same SSID offering wireless access. Those near the rogue AP find the stronger signal and most likely associate with it. User traffic is now sent to the rogue AP, which in turn captures the data and forwards it to the legitimate AP, as shown in Figure 12-37. Return traffic from the legitimate AP is sent to the rogue AP, captured, and then forwarded to the unsuspecting user. The attacker can steal the user’s passwords, personal information, gain access to their device, and compromise the system.

Defeating an attack like an MITM attack depends on the sophistication of the WLAN infrastructure and the vigilance in monitoring activity on the network. The process begins with identifying legitimate devices on the WLAN. To do this, users must be authenticated. After all the legitimate devices are known, the network can be monitored for abnormal devices or traffic.

Interactive Graphic

Check Your Understanding—WLAN Threats (12.6.6)

Refer to the online course to complete this activity.

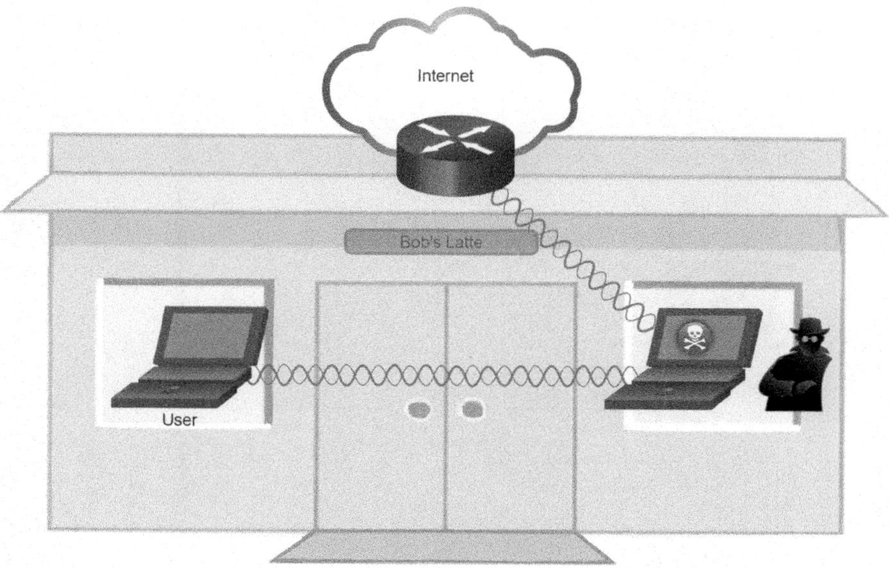

Figure 12-37 User Unknowingly Connects to Threat Actor's SSID

Secure WLANs (12.7)

This section explain WLAN security mechanisms.

Video

Video—Secure WLANs (12.7.1)

Refer to the online course to view this video.

SSID Cloaking and MAC Address Filtering (12.7.2)

Wireless signals can travel through solid matter, such as ceilings, floors, walls, outside of the home, or office space. Without stringent security measures in place, installing a WLAN can be the equivalent of putting Ethernet ports everywhere, even outside.

To address the threats of keeping wireless intruders out and protecting data, two early security features were used and are still available on most routers and APs: *SSID cloaking* and *MAC address filtering*.

SSID Cloaking

APs and some wireless routers allow the SSID beacon frame to be disabled, as shown Figure 12-38. Wireless clients must manually configure the SSID to connect to the network. This is referred to as active mode because the wireless clients must know the name of the SSID.

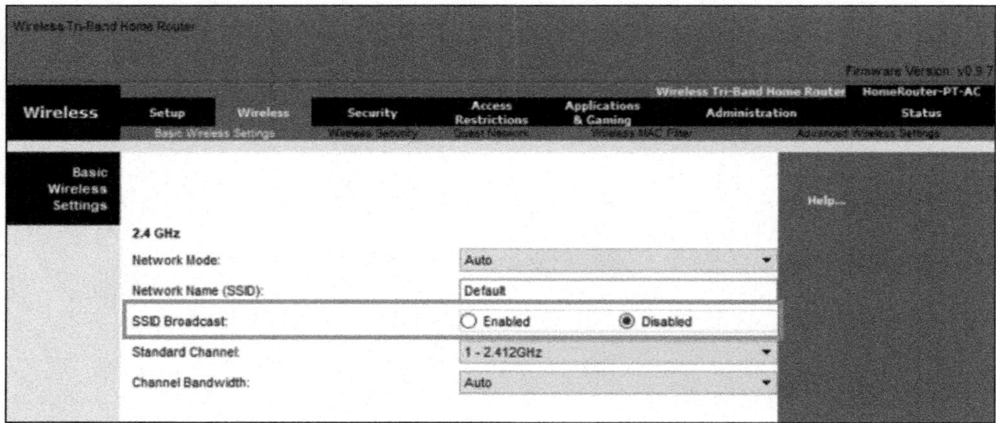

Figure 12-38 Disabling SSID Broadcast (SSID Cloaking) on a Wireless Router

MAC Addresses Filtering

An administrator can manually permit or deny clients wireless access based on their physical MAC hardware address. In Figure 12-39, the router is configured to permit two MAC addresses. Devices with different MAC addresses will not be able to join the 2.4GHz WLAN.

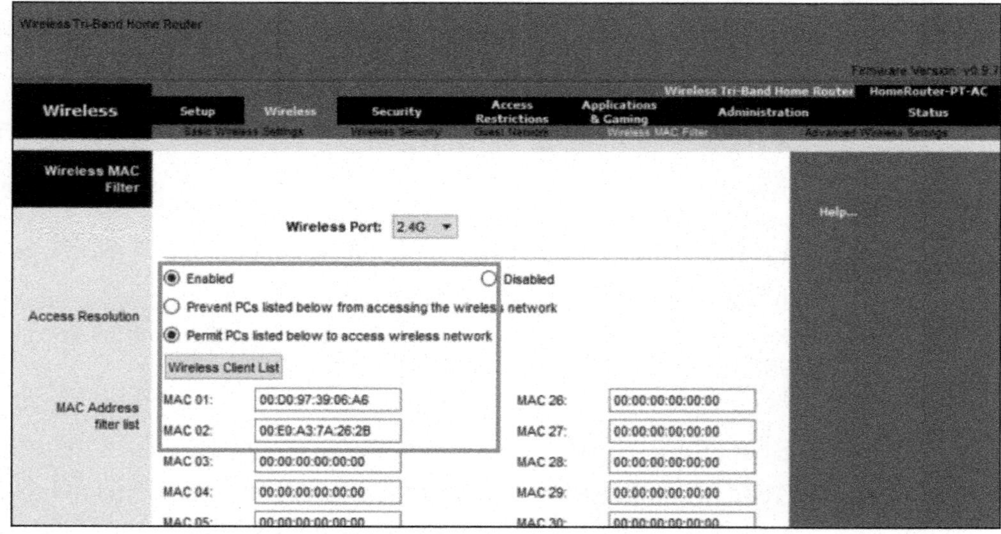

Figure 12-39 Configuring MAC Address Filtering on a Wireless Router

802.11 Original Authentication Methods (12.7.3)

Although these two features would deter most users, the reality is that neither SSID cloaking nor MAC address filtering would deter a crafty intruder. SSIDs are easily discovered even if APs do not broadcast them, and MAC addresses can be spoofed. The best way to secure a wireless network is to use authentication and encryption systems.

Two types of authentication were introduced with the original 802.11 standard:

- **Open system authentication:** Any wireless client should easily be able to connect and should only be used in situations where security is of no concern, such as those providing free Internet access, such as cafes, hotels, and in remote areas. The wireless client is responsible for providing security, such as using a virtual private network (VPN) to connect securely. VPNs provide authentication and encryption services. VPNs are beyond the scope of this section.
- **Shared key authentication:** Provides mechanisms, such as WEP, WPA, WPA2, and WPA3, to authenticate and encrypt data between a wireless client and AP. However, the password must be pre-shared between both parties to connect.

The chart in Figure 12-40 summarizes these authentication methods.

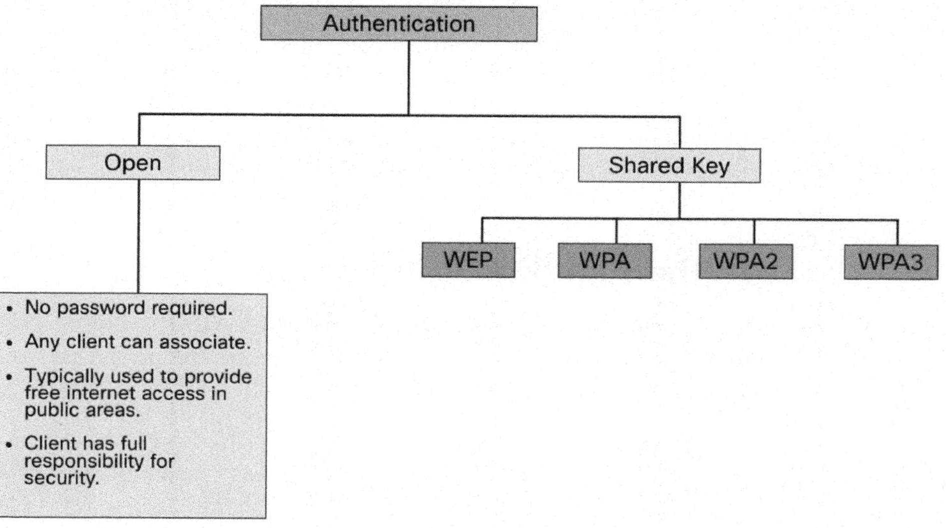

Figure 12-40 802.11 Authentication Methods

Shared Key Authentication Methods (12.7.4)

There are four shared key authentication techniques available, as described in Table 12-3. Until the availability of WPA3 devices becomes ubiquitous, wireless networks should use the WPA2 standard.

Table 12-3 Shared Key Authentication Methods

Authentication Method	Description
Wired Equivalent Privacy (WEP)	■ The original 802.11 specification designed to secure the data using the Rivest Cipher 4 (RC4) encryption method with a static key. ■ However, the key never changes when exchanging packets. ■ This makes it easy to hack. WEP is no longer recommended and should never be used.
Wi-Fi Protected Access (WPA)	■ A Wi-Fi Alliance standard that uses WEP, but secures the data with the much stronger *Temporal Key Integrity Protocol (TKIP)* encryption algorithm. ■ TKIP changes the key for each packet, making it much more difficult to hack.
WPA2	■ WPA2 is the current industry standard for securing wireless networks. ■ It uses the *Advanced Encryption Standard (AES)* for encryption. AES is currently considered the strongest encryption protocol.
WPA3	■ The next generation of Wi-Fi security. ■ All WPA3-enabled devices use the latest security methods, disallow outdated legacy protocols, and require the use of *Protected Management Frames (PMF)*. ■ However, devices with WPA3 are not yet readily available.

Authenticating a Home User (12.7.5)

Home routers typically have two choices for authentication: WPA and WPA2. WPA2 is the stronger of the two. Figure 12-41 shows the option to select one of two WPA2 authentication methods:

- **Personal:** Intended for home or small office networks, users authenticate using a *pre-shared key (PSK)*. Wireless clients authenticate with the wireless router using a pre-shared password. No special authentication server is required.
- **Enterprise:** Intended for enterprise networks but requires a Remote Authentication Dial-In User Service (RADIUS) authentication server. Although more complicated to set up, it provides additional security. The device must be authenticated by the RADIUS server and then users must authenticate using 802.1X standard, which uses the *Extensible Authentication Protocol (EAP)* for authentication.

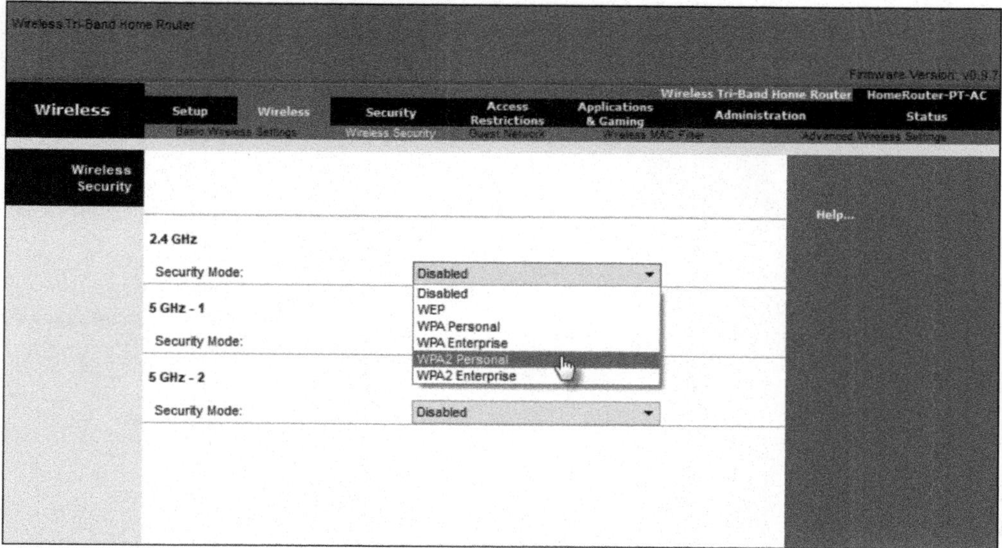

Figure 12-41 Selecting the Authentication Method on a Wireless Router

In the figure, the administrator is configuring the wireless router with WPA2 Personal authentication on the 2.4 GHz band.

Encryption Methods (12.7.6)

Encryption is used to protect data. If an intruder has captured encrypted data, the intruder would not be able to decipher it in any reasonable amount of time.

The WPA and WPA2 standards use the following encryption protocols:

- **Temporal Key Integrity Protocol (TKIP):** TKIP is the encryption method used by WPA. It provides support for legacy WLAN equipment by addressing the original flaws associated with the 802.11 WEP encryption method. It makes use of WEP, but encrypts the Layer 2 payload using TKIP, and carries out a *Message Integrity Check (MIC)* in the encrypted packet to ensure the message has not been altered.
- **Advanced Encryption Standard (AES):** AES is the encryption method used by WPA2. It is the preferred method because it is a far stronger method of encryption. It uses the Counter Cipher Mode with Block Chaining Message Authentication Code Protocol (CCMP) that allows destination hosts to recognize if the encrypted and non-encrypted bits have been altered.

In Figure 12-42, the administrator is configuring the wireless router to use WPA2 with AES encryption on the 2.4 GHz band.

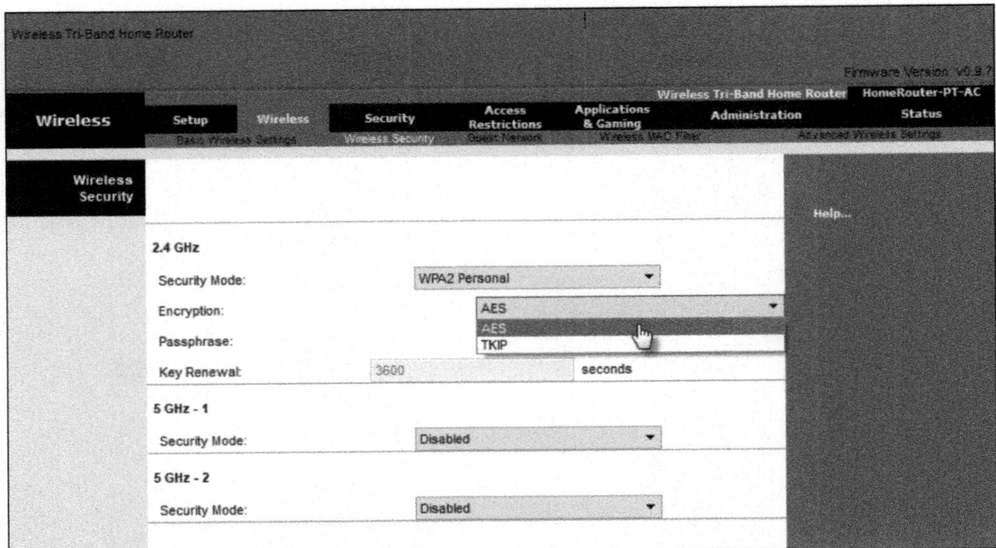

Figure 12-42 Setting the Encryption Method on a Wireless Router

Authentication in the Enterprise (12.7.7)

In networks that have stricter security requirements, an additional authentication or login is required to grant wireless clients such access. The Enterprise security mode choice requires an Authentication, Authorization, and Accounting (AAA) RADIUS server.

- **RADIUS Server IP address:** This is the reachable address of the RADIUS server.
- **UDP port numbers:** Officially assigned UDP ports 1812 for RADIUS Authentication, and 1813 for RADIUS Accounting, but can also operate using UDP ports 1645 and 1646.
- **Shared key:** Used to authenticate the AP with the RADIUS server.

In Figure 12-43, the administrator is configuring the wireless router with WPA2 Enterprise authentication using AES encryption. The RADIUS server IPv4 address is configured as well with a strong password to be used between the wireless router and the RADIUS server.

The shared key is not a parameter that must be configured on a wireless client. It is only required on the AP to authenticate with the RADIUS server. User authentication and authorization is handled by the 802.1X standard, which provides a centralized, server-based authentication of end users.

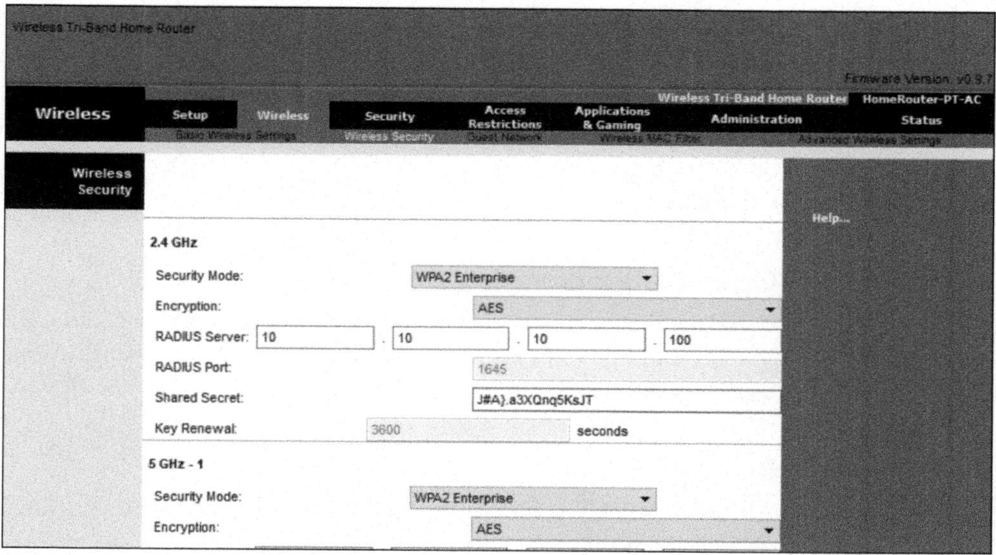

Figure 12-43 Configuring WPA2 Enterprise Authentication on a Wireless Router

The 802.1X login process uses EAP to communicate with the AP and RADIUS server. EAP is a framework for authenticating network access. It can provide a secure authentication mechanism and negotiate a secure private key, which can then be used for a wireless encryption session using TKIP or AES encryption.

WPA3 (12.7.8)

At the time of this writing, devices that support WPA3 authentication were not readily available. However, WPA2 is no longer considered secure. WPA3, if available, is the recommended 802.11 authentication method. WPA3 includes four features:

- WPA3-Personal
- WPA3-Enterprise
- Open Networks
- Internet of Things (IoT) Onboarding

WPA3-Personal

In WPA2-Personal, threat actors can listen in on the "handshake" between a wireless client and the AP and use a brute force attack to try to guess the PSK. WPA3-Personal thwarts this attack by using *Simultaneous Authentication of Equals (SAE)*, a feature specified in the IEEE 802.11-2016. The PSK is never exposed, making it impossible for the threat actor to guess.

WPA3-Enterprise

WPA3-Enterprise still uses 802.1X/EAP authentication. However, it requires the use of a 192-bit cryptographic suite and eliminates the mixing of security protocols for previous 802.11 standards. WPA3-Enterprise adheres to the Commercial National Security Algorithm (CNSA) Suite, which is commonly used in high security Wi-Fi networks.

Open Networks

Open networks in WPA2 send user traffic in unauthenticated, clear text. In WPA3, open or public Wi-Fi networks still do not use any authentication. However, they do use Opportunistic Wireless Encryption (OWE) to encrypt all wireless traffic.

IoT Onboarding

Although WPA2 included *Wi-Fi Protected Setup (WPS)* to quickly onboard devices without configuring them first, WPS is vulnerable to a variety of attacks and is not recommended. Furthermore, IoT devices are typically headless, meaning they have no built-in GUI for configuration and lack any easy way to get connected to the wireless network. The *Device Provisioning Protocol (DPP)* was designed to address this need. Each headless device has a hardcoded public key. The key is typically stamped on the outside of the device or its packaging as a Quick Response (QR) code. The network administrator can scan the QR code and quickly onboard the device. Although not strictly part of the WPA3 standard, DPP will replace WPS over time.

Interactive Graphic

Check Your Understanding—Secure WLANs (12.7.9)

Refer to the online course to complete this activity.

Summary (12.8)

A Wireless LAN (WLAN) is a type of wireless network that is commonly used in homes, offices, and campus environments. Wireless networks are based on IEEE standards and can be classified into four main types: WPAN, WLAN, WMAN, and WWAN. Wireless LAN technologies use the unlicensed radio spectrum to send and receive data. Examples of this technology are Bluetooth, WiMAX, Cellular Broadband, and Satellite Broadband. The IEEE 802.11 WLAN standards define how radio frequencies are used for wireless links. WLAN networks operate in the 2.4 GHz frequency band and the 5 GHz band. Standards ensure interoperability between devices that are made by different manufacturers. Internationally, the three organizations influencing WLAN standards are the ITU-R, the IEEE, and the Wi-Fi Alliance.

To communicate wirelessly, most devices include integrated wireless NICs that incorporate a radio transmitter/receiver. The wireless router serves as an access point, a switch, and a router. Wireless clients use their wireless NIC to discover nearby APs advertising their SSID. Clients then attempt to associate and authenticate with an AP. After being authenticated, wireless users have access to network resources. APs can be categorized as either autonomous APs or controller-based APs. There are three types of antennas for business class APs: omnidirectional, directional, and MIMO.

The 802.11 standard identifies two main wireless topology modes: Ad hoc mode and Infrastructure mode. Tethering is used to provide quick wireless access. Infrastructure mode defines two topology building blocks: a Basic Service Set (BSS) and an Extended Service Set (ESS). All 802.11 wireless frames contain the following fields: frame control, duration, address 1, address 2, address 3, sequence control, address 4, payload, and FCS. WLANs use CSMA/CA as the method to determine how and when to send data on the network. Part of the 802.11 process is discovering a WLAN and subsequently connecting to it. Wireless devices discover a wireless AP, authenticate with it, and then associate with it. Wireless clients connect to the AP using a scanning process, which may be passive or active.

CAPWAP is an IEEE standard protocol that enables a WLC to manage multiple APs and WLANs. The CAPWAP split MAC concept does all the functions normally performed by individual APs and distributes them between two functional components: AP MAC functions and WLC MAC functions. DTLS is a protocol that provides security between the AP and the WLC. FlexConnect is a wireless solution for branch office and remote office deployments. You configure and control access points in a branch office from the corporate office through a WAN link, without deploying a controller in each office. There are two modes of operation for the FlexConnect AP: connected and standalone.

Wireless LAN devices have transmitters and receivers tuned to specific frequencies of radio waves to communicate. Frequencies are allocated as ranges. Ranges are then split into smaller ranges called channels: DSSS, FHSS, and OFDM. The 802.11b/g/n standards operate in the 2.4 GHz to 2.5 GHz spectrum. The 2.4 GHz band is subdivided into multiple channels. Each channel is allotted 22 MHz bandwidth and is separated from the next channel by 5 MHz. When planning the location of APs, the approximate circular coverage area is important.

Wireless networks are susceptible to threats, including data interception, wireless intruders, DoS attacks, and rogue APs. Wireless DoS attacks can be the result of improperly configured devices, a malicious user intentionally interfering with the wireless communication, and accidental interference. A rogue AP is an AP or wireless router that has been connected to a corporate network without explicit authorization. When connected, a threat actor can use the rogue AP to capture MAC addresses, capture data packets, gain access to network resources, or launch a MITM attack. In a MITM attack, the threat actor is positioned in between two legitimate entities to read or modify the data that passes between the two parties. A popular wireless MITM attack is called the "evil twin AP" attack, where a threat actor introduces a rogue AP and configures it with the same SSID as a legitimate AP. To prevent the installation of rogue APs, organizations must configure WLCs with rogue AP policies.

To keep wireless intruders out and protect data, two early security features are still available on most routers and APs: SSID cloaking and MAC address filtering. There are four shared key authentication techniques available: WEP, WPA, WPA2, and WPA3 (devices with WPA3 are not yet readily available). Home routers typically have two choices for authentication: WPA and WPA2. WPA2 is the stronger of the two. Encryption is used to protect data. The WPA and WPA2 standards use the following encryption protocols: TKIP and AES. In networks that have stricter security requirements, an additional authentication or login is required to grant wireless clients access. The Enterprise security mode choice requires an Authentication, Authorization, and Accounting (AAA) RADIUS server.

Practice

There are no Labs or Packet Tracer activities in this module.

Check Your Understanding Questions

Complete all the review questions listed here to test your understanding of the sections and concepts in this chapter. The appendix "Answers to the 'Check Your Understanding' Questions" lists the answers.

1. An AP regularly broadcasts which type of management frame?

   A. Authentication
   B. Beacon
   C. Probe request
   D. Probe response

2. What type of wireless antenna is best suited for providing coverage in large open spaces, such as hallways or large conference rooms?

   A. Directional
   B. Omnidirectional
   C. Parabolic dish
   D. Yagi

3. Which wireless security method requires clients to manually identify the SSID to connect to the WLAN?

   A. MAC Address Filtering
   B. IP Address Filtering
   C. SSID cloaking
   D. SSID disclosing

4. What are the two methods that a wireless client can use to discover an AP? (Choose two.)

   A. Delivering a broadcast frame
   B. Initiating a three-way handshake
   C. Receiving a broadcast beacon frame probe response
   D. Sending an ARP request
   E. Transmitting a probe request

5. What type of wireless network topology would be used in a medium to large organization?

   A. Ad hoc
   B. Hotspot
   C. Infrastructure
   D. Mixed mode
   E. Tethering

6. What IEEE 802.11 wireless standards operate only in the 2.4 GHz range? (Choose two.)

   A. 802.11a
   B. 802.11b
   C. 802.11g
   D. 802.11n
   E. 802.11ac
   F. 802.11ad

7. Which IEEE wireless standard is backward compatible with older wireless protocols and supports data rates up to 1.3 Gb/s?

   A. 802.11
   B. 802.11a
   C. 802.11ac
   D. 802.11g
   E. 802.11n

8. Which feature of 802.11n wireless access points allows them to transmit data at faster speeds than previous versions of 802.11 Wi-Fi standards did?

   A. MIMO
   B. NEMO
   C. SPS
   D. WPS

9. In a 2.4 GHz wireless network, which three channels should be used to avoid interference from nearby wireless devices? (Choose three.)

   A. 0
   B. 1
   C. 3
   D. 6
   E. 9
   F. 11
   G. 14

10. Which authentication method is more secure but requires the services of a RADIUS server?

    A. WEP Enterprise

    B. WPA Personal

    C. WPA Enterprise

    D. WPA2 Personal

11. Which option correctly describes when an AP openly advertises its service by periodically sending broadcast beacon frames containing the SSID, supported standards, and security settings?

    A. Active mode

    B. Mixed mode

    C. Open authentication mode

    D. Passive mode

CHAPTER 13

WLAN Configuration

Objectives

Upon completion of this chapter, you will be able to answer the following questions:

- How do you configure a WLAN to support a remote site?
- How do you configure a WLC WLAN to use the management interface and WPA2 PSK authentication?
- How do you configure a WLC WLAN to use a VLAN interface, a DHCP server, and WPA2 Enterprise authentication?
- How do you troubleshoot common wireless configuration issues?

Key Terms

This chapter uses the following key terms. You can find the definitions in the Glossary.

wireless mesh network (WMN) Page 408

port forwarding Page 410

port triggering Page 411

rogue access points Page 414

Introduction (13.0)

Some of us remember getting on the Internet using dial up. Dial up involved using your landline phone. Your landline phone was unavailable to make or receive calls while you were on the Internet. Your dial-up connection to the Internet was very slow. It basically meant that, for most people, your computer was always in one place in your home or school.

Then we were able to connect to the Internet without using our landlines. But our computers were still hardwired to the devices that connected them to the Internet. Today we can connect to the Internet using wireless devices that let us take our phones, laptops, and tablets almost anywhere. It's nice to have this freedom of movement, but it requires special end and intermediary devices and a good understanding of wireless protocols. Want to know more? Then this is the module for you!

Remote Site WLAN Configuration (13.1)

In this section, you learn how to configure a wireless Local Area Network (WLAN) to support a remote site.

Video

Video—Configure a Wireless Network (13.1.1)

Refer to the online course to view this video.

The Wireless Router (13.1.2)

Remote workers, small branch offices, and home networks often use a small office and home router. These routers are sometimes called an integrated router because they typically include a switch for wired clients, a port for an Internet connection (sometimes labeled "WAN"), and wireless components for wireless client access, as shown for the Cisco Meraki MX64W in Figure 13-1. For the rest of this module, small office and home routers are referred to as wireless routers.

Figure 13-2 shows a topology depicting the physical connection of a wired laptop to the wireless router, which is then connected to a cable or DSL modem for Internet connectivity.

These wireless routers typically provide WLAN security, dynamic host configuration protocol (DHCP) services, integrated Name Address Translation (NAT), quality of service (QoS), and a variety of other features. The feature set will vary based on the router model.

Figure 13-1 Cisco Meraki MX64W Wireless Router

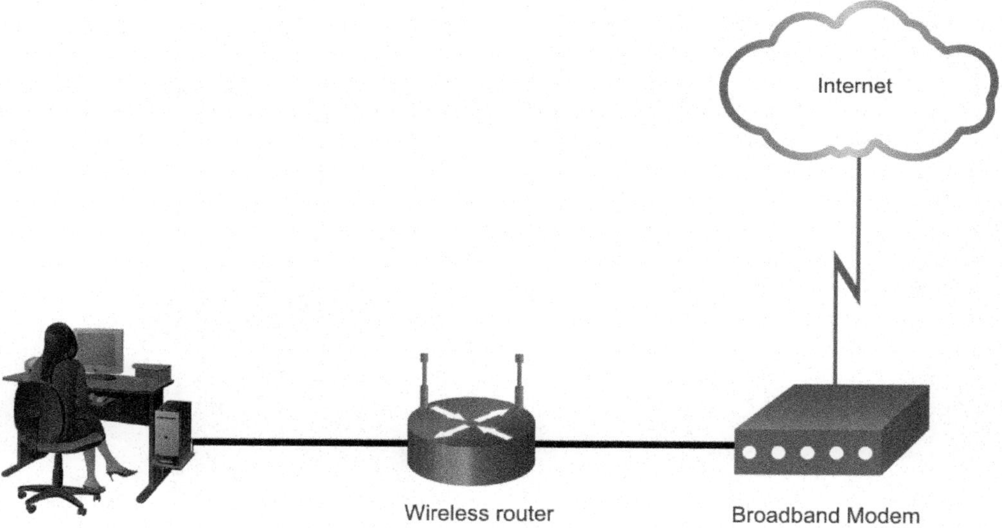

Figure 13-2 Wireless Router Connecting a WLAN to the Internet

Note

Cable or DSL modem configuration is usually done by the service provider's representative either onsite or remotely through a walkthrough with you on the phone. If you buy the modem, it comes with documentation for how to connect it to your service provider, which will most likely include contacting your service provider for more information.

Log in to the Wireless Router (13.1.3)

Most wireless routers are ready for service out of the box. They are preconfigured to be connected to the network and provide services. For example, the wireless router uses DHCP to automatically provide addressing information to connected devices. However, wireless router default IP addresses, usernames, and passwords can easily be found on the Internet. Just enter the search phrase "default wireless router IP address" or "default wireless router passwords" to see a listing of many websites that provide this information. For example, username and password for the wireless router in the figure is "admin". Therefore, your first priority should be to change these defaults for security reasons.

To gain access to the wireless router's configuration GUI, open a web browser. In the address field, enter the default IP address for your wireless router. The default IP address can be found in the documentation that came with the wireless router, or you can search the Internet. Figure 13-3 shows the IPv4 address 192.168.0.1, which is a common default for many manufacturers. A security window prompts for authorization to access the router GUI. The word admin is commonly used as the default username and password. Again, check your wireless router's documentation or search the Internet.

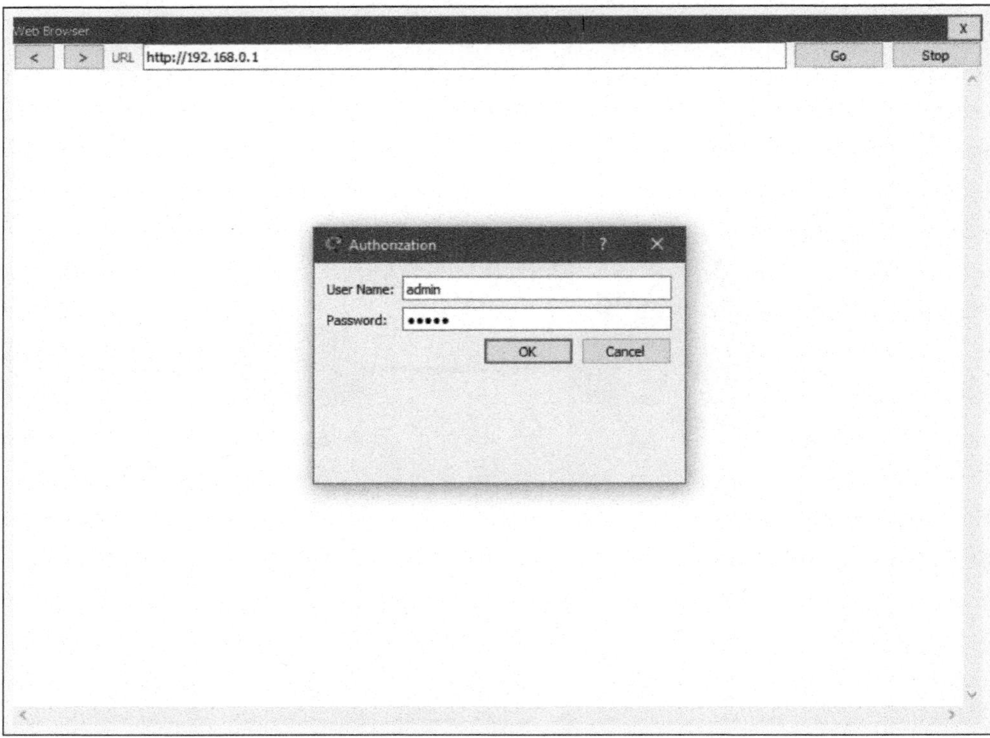

Figure 13-3 Connecting to a Wireless Router Using a Browser

Basic Network Setup (13.1.4)

Basic network setup includes the following steps:

Step 1. **Log in to the router from a web browser.** After logging in, a GUI opens, as shown in Figure 13-4. The GUI will have tabs or menus to help you navigate to various router configuration tasks. It is often necessary to save the settings changed in one window before proceeding to another window. At this point, it is a best practice to make changes to the default settings.

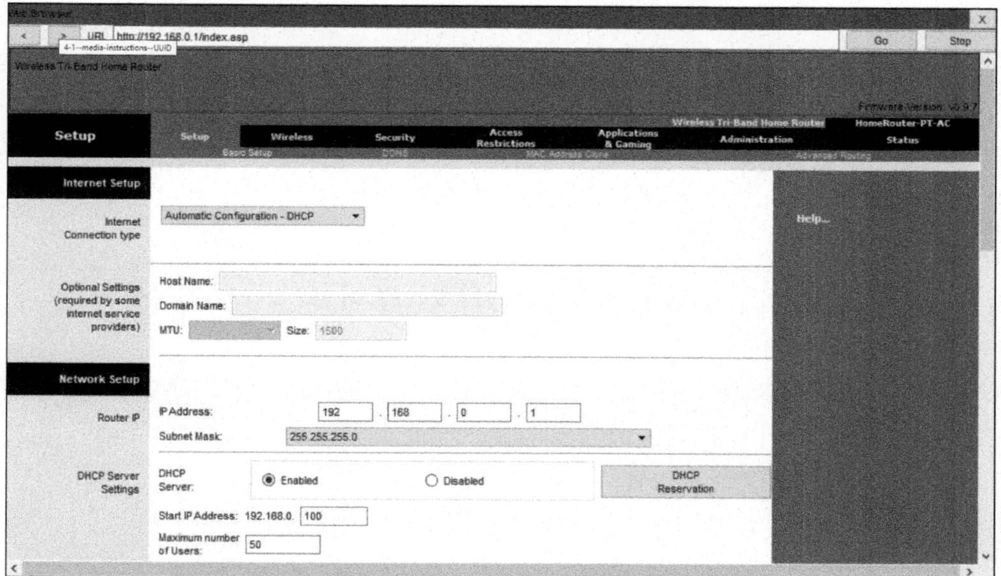

Figure 13-4 Basic Network Setup—Step 1

Step 2. **Change the default administrative password.** To change the default login password, find the administration portion of the router's GUI. In this example, the Administration tab was selected. This is where the router password can be changed. On some devices, such as the one in Figure 13-5, you can change only the password. The username remains admin or whatever the default username is for the router you are configuring.

Step 3. **Log in with the new administrative password.** After you save the new password, the wireless router will request authorization again. Enter the username and new password, as shown in Figure 13-6.

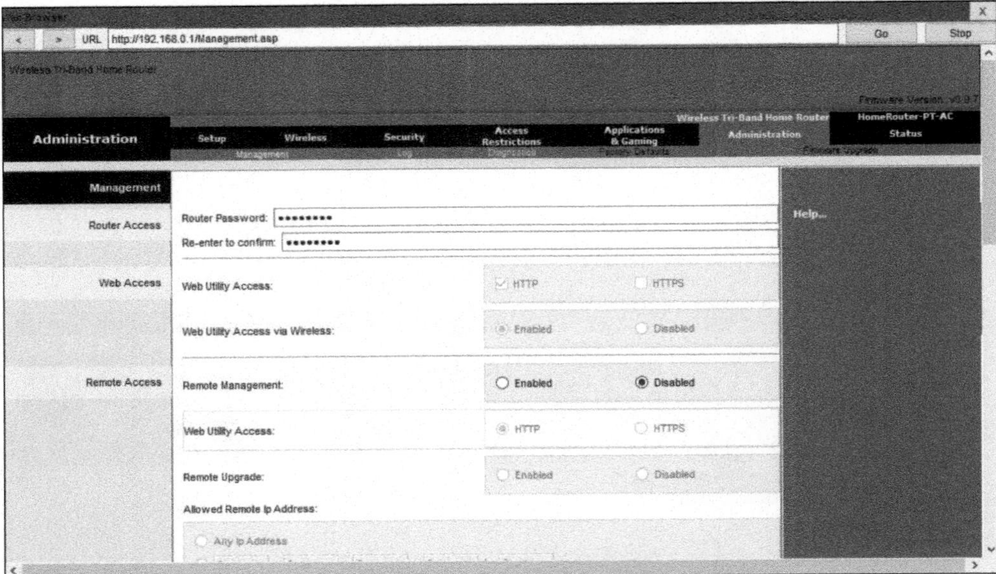

Figure 13-5 Basic Network Setup—Step 2

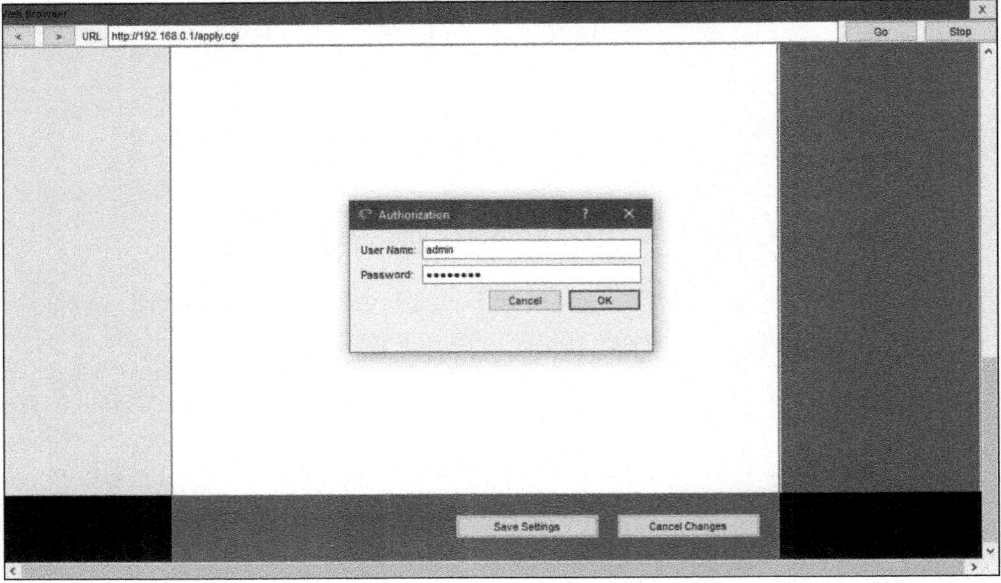

Figure 13-6 Basic Network Setup—Step 3

Step 4. **Change the default DHCP IPv4 addresses.** Change the default router IPv4 address. It is a best practice to use private IPv4 addressing inside your network. The IPv4 address 10.10.10.1 is used in Figure 13-7, but it could be any private IPv4 address you choose.

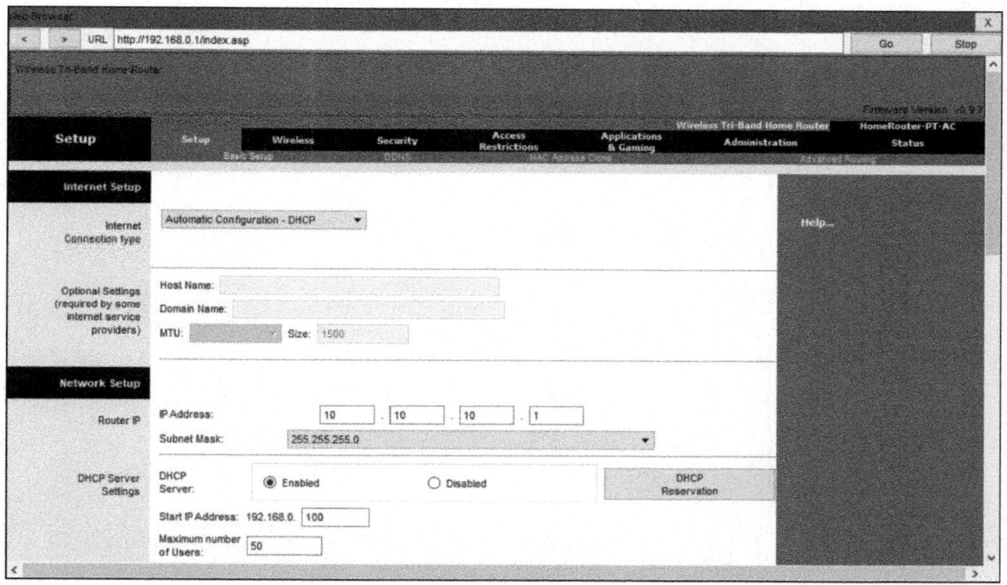

Figure 13-7 Basic Network Setup—Step 4

Step 5. **Renew the IP address.** When you click Save, you will temporarily lose access to the wireless router. Open a command window and renew your IP address with the **ipconfig /renew** command, as shown in Example 13-1.

Example 13-1 Basic Network Setup—Step 5

```
Packet Tracer PC Command Line 1.0
C:\> ipconfig /renew

   IP Address......................: 10.10.10.100
   Subnet Mask.....................: 255.255.255.0
   Default Gateway.................: 10.10.10.1
   DNS Server......................: 0.0.0.0

C:\>
```

Step 6. **Log in to the router with the new IP address.** Enter the router's new IP address to regain access to the router configuration GUI, as shown in

Figure 13-8. You are now ready to continue configuring the router for wireless access.

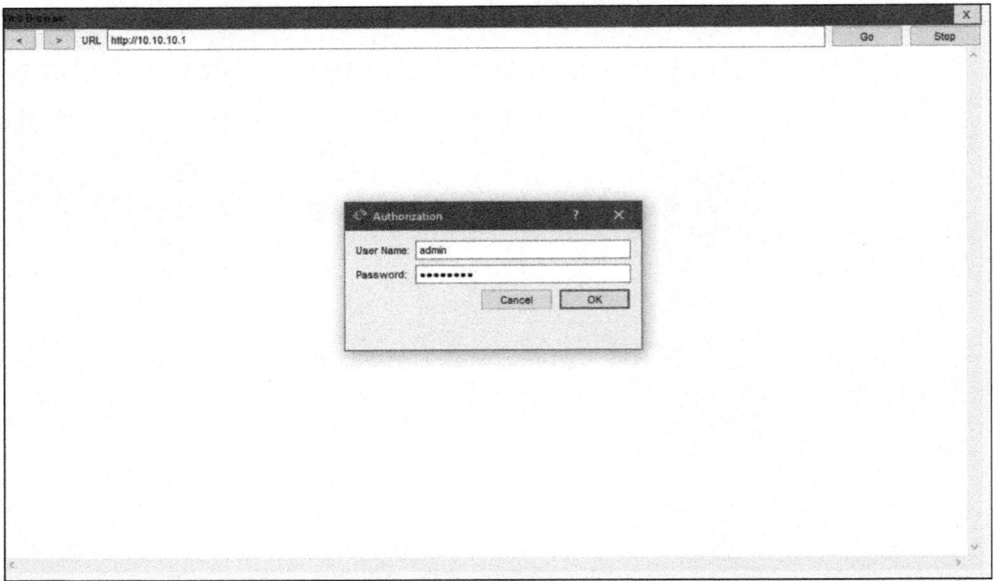

Figure 13-8 Basic Network Setup—Step 6

Basic Wireless Setup (13.1.5)

Basic wireless setup includes the following steps:

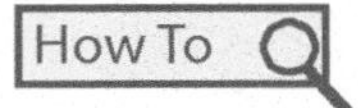

Step 1. **View the WLAN defaults.** Out of the box, a wireless router provides wireless access to devices using a default wireless network name and password. The network name is called the Service Set Identifier (SSID). Locate the basic wireless settings for your router to change these defaults, as shown in Figure 13-9.

Step 2. **Change the network mode.** Some wireless routers allow you to select which 802.11 standard to implement. Figure 13-10 shows that Legacy has been selected. This means wireless devices connecting to the wireless router can have a variety of wireless network interface cards (NICs) installed. Today's wireless routers configured for legacy or mixed mode most likely support 802.11a, 802.11n, and 802.11ac NICs.

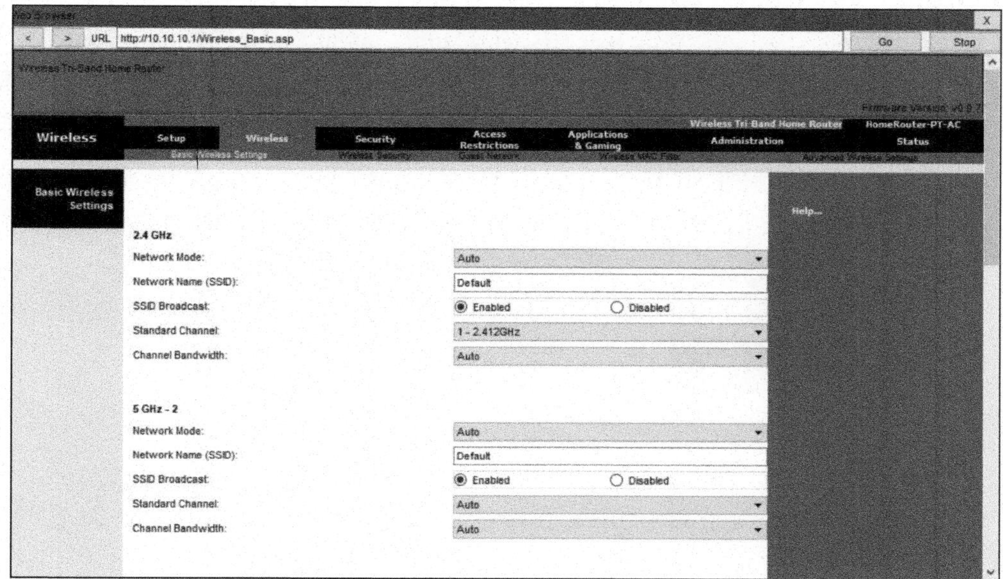

Figure 13-9 Basic Wireless Setup—Step 1

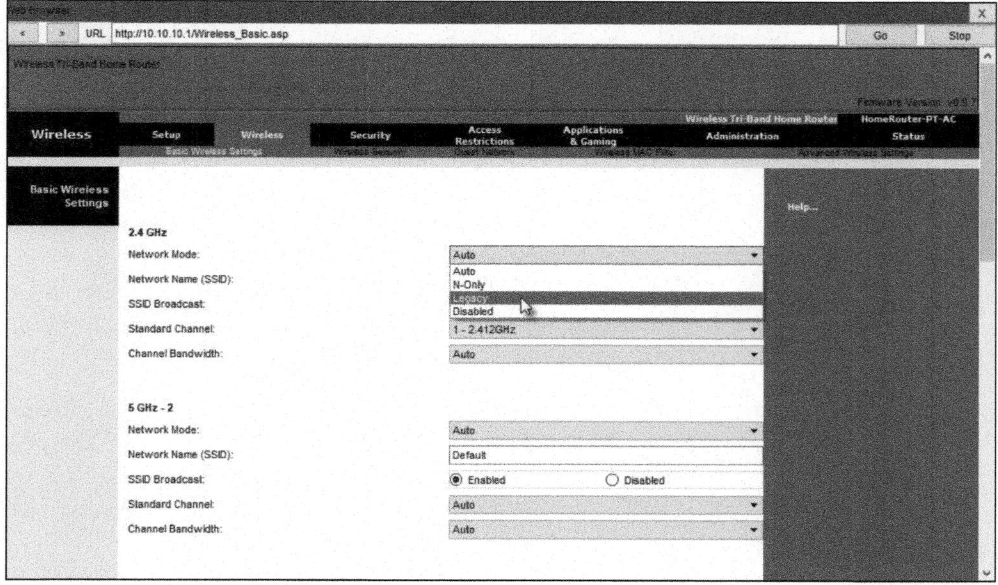

Figure 13-10 Basic Wireless Setup—Step 2

Step 3. **Configure the SSID.** Assign a SSID to the WLANs. OfficeNet is used in Figure 13-11 for all three WLANs (the third WLAN is not shown).

The wireless router announces its presence by sending broadcasts advertising its SSID. This allows wireless hosts to automatically discover the name of the wireless network. If the SSID broadcast is disabled, you must manually enter the SSID on each wireless device that connects to the WLAN.

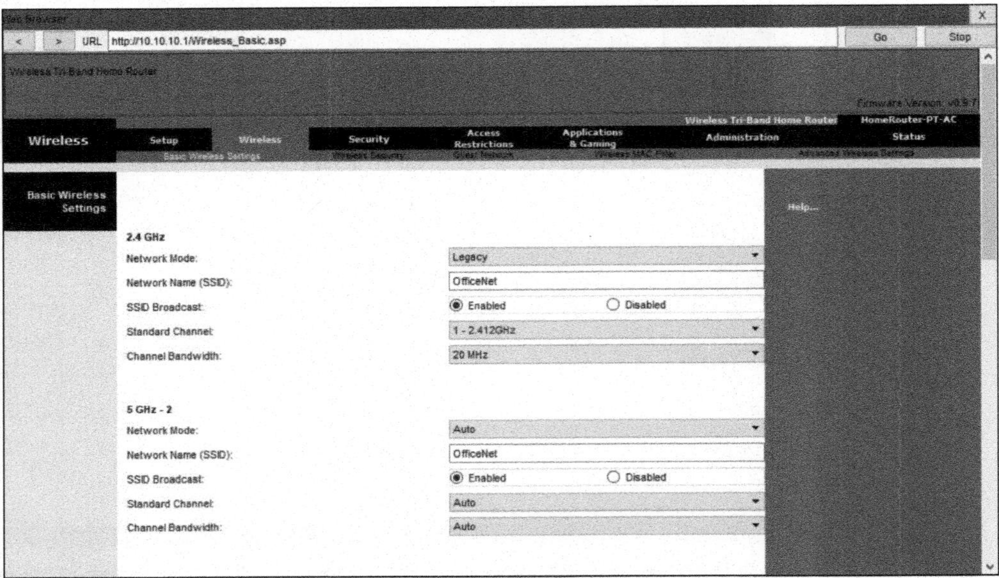

Figure 13-11 Basic Wireless Setup—Step 3

Step 4. **Configure the channel.** Devices configured with the same channel within the 2.4 GHz band may overlap and cause distortion, slowing down the wireless performance and potentially breaking network connections. The solution to avoid interference is to configure non-overlapping channels on the wireless routers and access points that are near to each other. Specifically, channels 1, 6, and 11 are non-overlapping. In Figure 13-12, the wireless router is configured to use channel 6.

Step 5. **Configure the security mode.** Out of the box, a wireless router may have no WLAN security configured. In Figure 13-13, the personal version of Wi-Fi Protected Access version 2 (WPA2 Personal) is selected for all three WLANs. WPA2 with Advanced Encryption Standard (AES) encryption is currently the strongest security mode.

Step 6. **Configure the passphrase.** WPA2 personal uses a passphrase to authenticate wireless clients, as shown in Figure 13-14. WPA2 Personal is easier to use in a small office or home environment because it does not require an authentication server. Larger organizations implement WPA2 enterprise and require wireless clients to authenticate with a username and password.

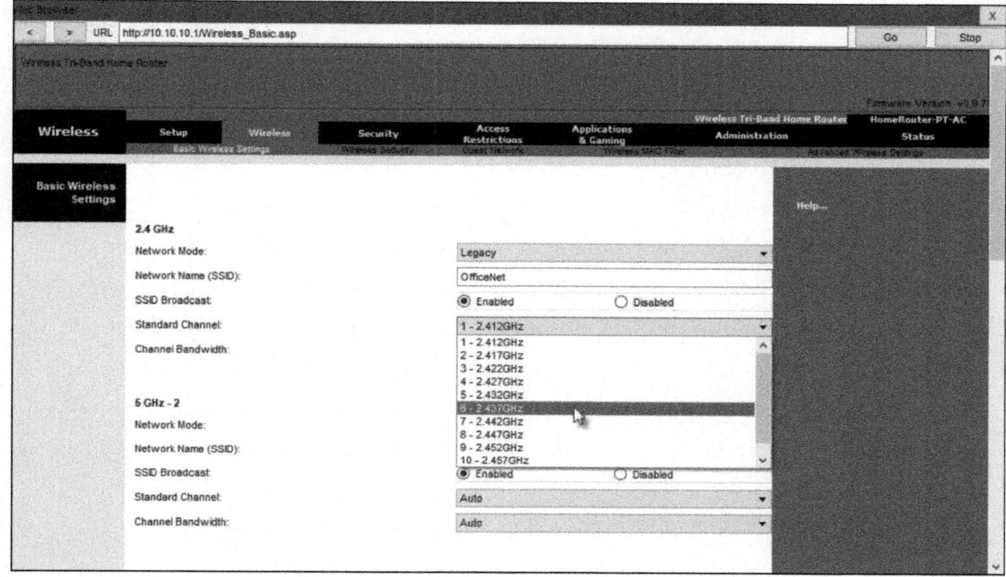

Figure 13-12 Basic Wireless Setup—Step 4

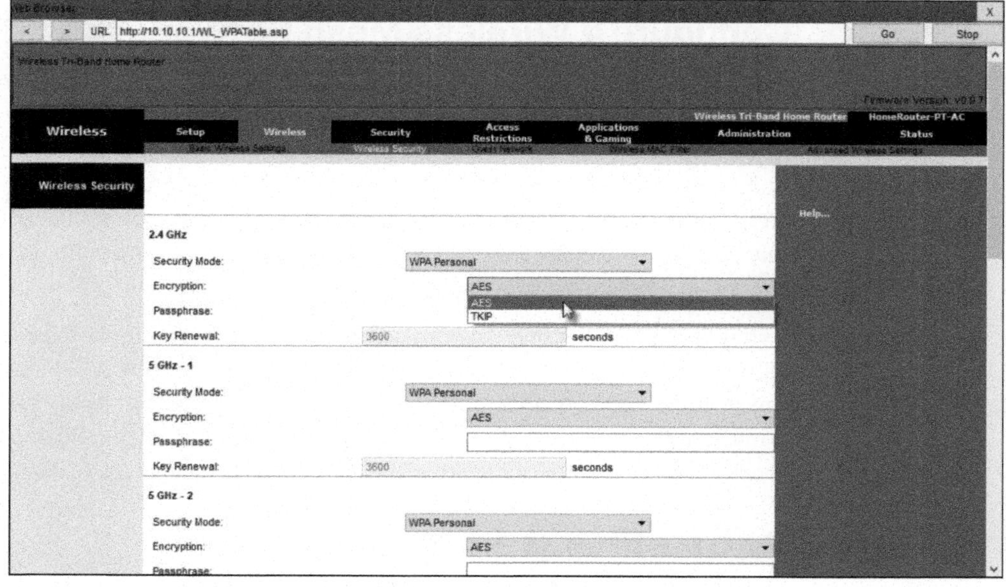

Figure 13-13 Basic Wireless Setup—Step 5

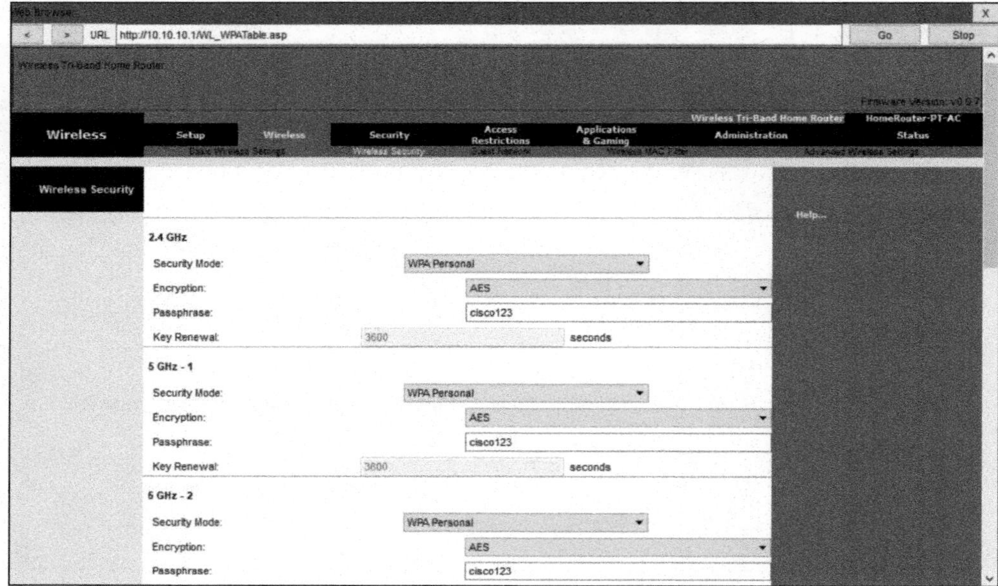

Figure 13-14 Basic Wireless Setup—Step 6

Configure a Wireless Mesh Network (13.1.6)

In a small office or home network, one wireless router may suffice to provide wireless access to all the clients. However, if you want to extend the range beyond approximately 45 meters indoors and 90 meters outdoors, you can add wireless access points. As shown in the wireless mesh network in Figure 13-15, two access points are configured with the same WLAN settings from our previous example. Notice that the channels selected are 1 and 11 so that the access points do not interfere with channel 6 configured previously on the wireless router.

Extending a WLAN in a small office or home has become increasingly easier. Manufacturers have made creating a *wireless mesh network (WMN)* simple through smartphone apps. You buy the system, disperse the access points, plug them in, download the app, and configure your WMN in a few steps. Search the Internet for "best wi-fi mesh network system" to find reviews of current offerings.

NAT for IPv4 (13.1.7)

On a wireless router, if you look for a page like the Status page shown in Figure 13-16, you will find the IPv4 addressing information that the router uses to send data to the Internet. Notice that the IPv4 address, 209.165.201.11, is a different network than the 10.10.10.1 address assigned to the router's LAN interface. All the devices on the router's LAN will get assigned addresses with the 10.10.10 prefix.

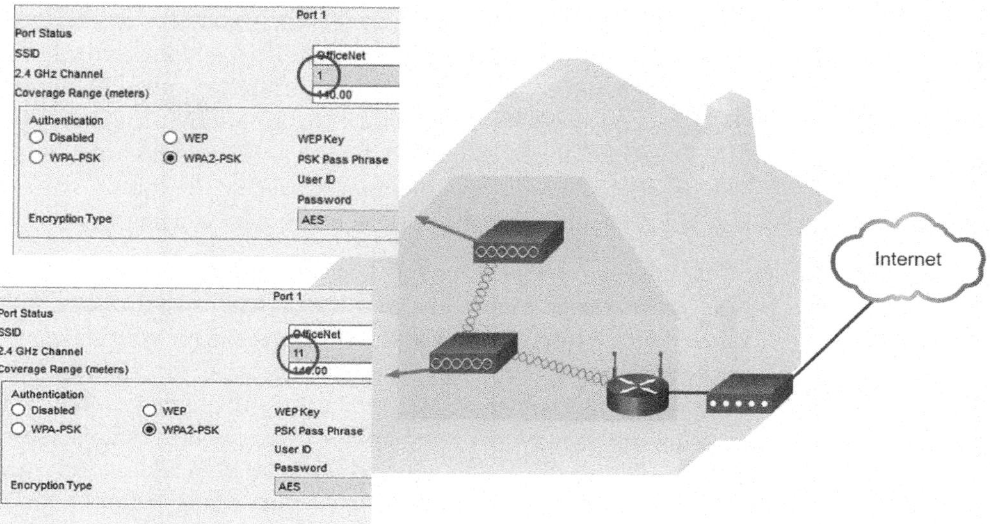

Figure 13-15 Wireless Mesh Network with a Wireless Router and Two APs

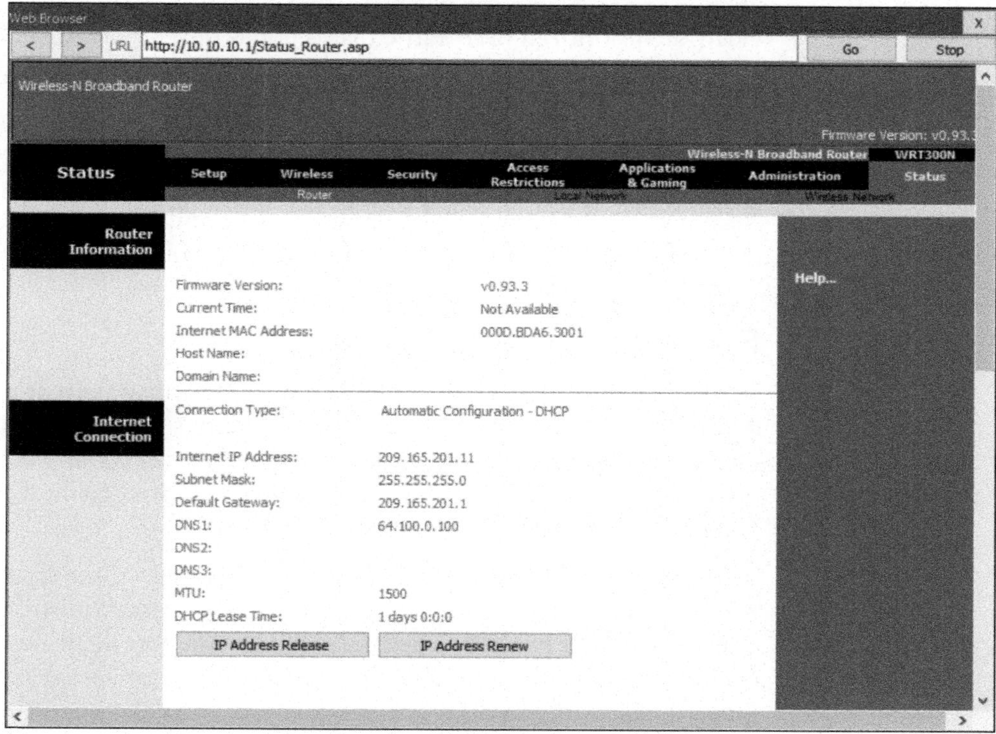

Figure 13-16 Verifying the Status of a Wireless Router

The 209.165.201.11 IPv4 address is publicly routable on the Internet. Any address with the 10 in the first octet is a private IPv4 address and cannot be routed on the Internet. Therefore, the router will use a process called Network Address Translation (NAT) to convert private IPv4 addresses to Internet-routable IPv4 addresses. With NAT, a private (local) source IPv4 address is translated to a public (global) address. The process is reversed for incoming packets. The router is able to translate many internal IPv4 addresses into public addresses by using NAT.

Some Internet service providers (ISPs) use private addressing to connect to customer devices. However, eventually your traffic will leave the provider's network and be routed on the Internet. To see the IP addresses for your devices, search the Internet for "what is my IP address." Do this for other devices on the same network and you will see that they all share the same public IPv4 address. NAT makes this possible by tracking the source port numbers for every session established by a device. If your ISP has IPv6 enabled, you will see a unique IPv6 address for each device.

Quality of Service (13.1.8)

Many wireless routers have an option for configuring Quality of Service (QoS). By configuring QoS, you can guarantee that certain traffic types, such as voice and video, are prioritized over traffic that is not as time-sensitive, such as email and web browsing. On some wireless routers, traffic can also be prioritized on specific ports.

Figure 13-17 is a simplified mockup of a QoS interface based on a Netgear GUI. You will usually find the QoS settings in the advanced menus. If you have a wireless router available, investigate the QoS settings. Sometimes, these might be listed under "bandwidth control" or something similar. Consult the wireless router's documentation or search the Internet for "qos settings" for your router's make and model.

Port Forwarding (13.1.9)

Wireless routers typically block TCP and UDP ports to prevent unauthorized access in and out of a LAN. However, there are situations when specific ports must be opened so that certain programs and applications can communicate with devices on different networks. *Port forwarding* is a rule-based method of directing traffic between devices on separate networks.

When traffic reaches the router, the router determines if the traffic should be forwarded to a certain device based on the port number found with the traffic. For example, a router might be configured to forward port 80, which is associated with HTTP. When the router receives a packet with the destination port of 80, the router forwards the traffic to the server inside the network that serves web pages. In Figure 13-18, port forwarding is enabled for port 80 and is associated with the web server at IPv4 address 10.10.10.50.

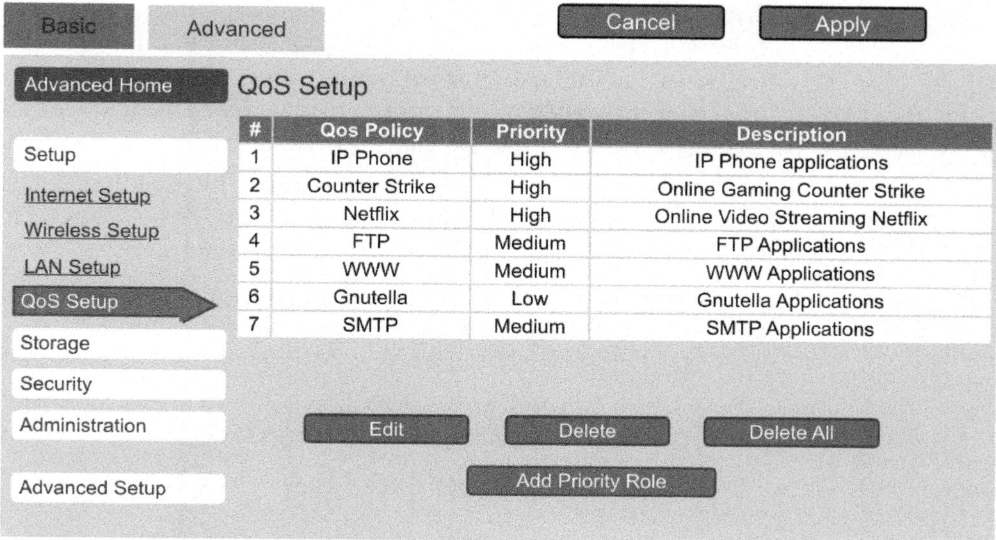

#	Qos Policy	Priority	Description
1	IP Phone	High	IP Phone applications
2	Counter Strike	High	Online Gaming Counter Strike
3	Netflix	High	Online Video Streaming Netflix
4	FTP	Medium	FTP Applications
5	WWW	Medium	WWW Applications
6	Gnutella	Low	Gnutella Applications
7	SMTP	Medium	SMTP Applications

Figure 13-17 QoS Settings on a Wireless Router

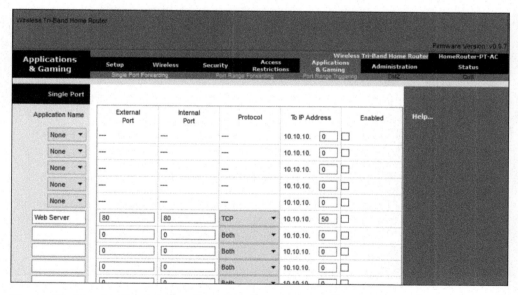

Applications & Gaming

Setup | Wireless | Security | Access Restrictions | Applications & Gaming | Administration | Status

Wireless Tri-Band Home Router HomeRouter-PT-AC

Single Port

Application Name	External Port	Internal Port	Protocol	To IP Address	Enabled
None	—	—	—	10.10.10. 0	☐
None	—	—	—	10.10.10. 0	☐
None	—	—	—	10.10.10. 0	☐
None	—	—	—	10.10.10. 0	☐
None	—	—	—	10.10.10. 0	☐
Web Server	80	80	TCP	10.10.10. 50	☐
	0	0	Both	10.10.10. 0	☐
	0	0	Both	10.10.10. 0	☐
	0	0	Both	10.10.10. 0	☐

Help...

Figure 13-18 Configuring Port Forwarding on a Wireless Router

Port triggering allows the router to temporarily forward data through inbound ports to a specific device. You can use port triggering to forward data to a computer only when a designated port range is used to make an outbound request. For example, a video game might use ports 27000 to 27100 for connecting with other players.

These are the trigger ports. A chat client might use port 56 for connecting the same players so that they can interact with each other. In this instance, if there is gaming traffic on an outbound port within the triggered port range, inbound chat traffic on port 56 is forwarded to the computer that is being used to play the video game and chat with friends. When the game is over and the triggered ports are no longer in use, port 56 is no longer allowed to send traffic of any type to this computer.

Packet Tracer—Configure a Wireless Network (13.1.10)

In this activity, you configure a wireless router and an access point to accept wireless clients and route IP packets.

Lab—Configure a Wireless Network (13.1.11)

In this lab, you configure basic settings on a wireless router and connect a PC to a router wirelessly.

Configure a Basic WLAN on the WLC (13.2)

In this section, you learn how to configure a Wireless LAN Controller (WLC) WLAN to use the management interface and WPA2 pre-shared key (PSK) authentication.

Video

Video—Configure a Basic WLAN on the WLC (13.2.1)

Refer to the online course to view this video.

WLC Topology (13.2.2)

The topology and addressing scheme used for the videos and this section are shown in Figure 13-19 and Table 13-1. The access point (AP) is a controller-based AP, as opposed to an autonomous AP. Recall that controller-based APs require no initial configuration and are often called lightweight APs (LAPs). LAPs use the Lightweight Access Point Protocol (LWAPP) to communicate with a WLAN controller (WLC). Controller-based APs are useful in situations where many APs are required in the network. As more APs are added, each AP is automatically configured and managed by the WLC.

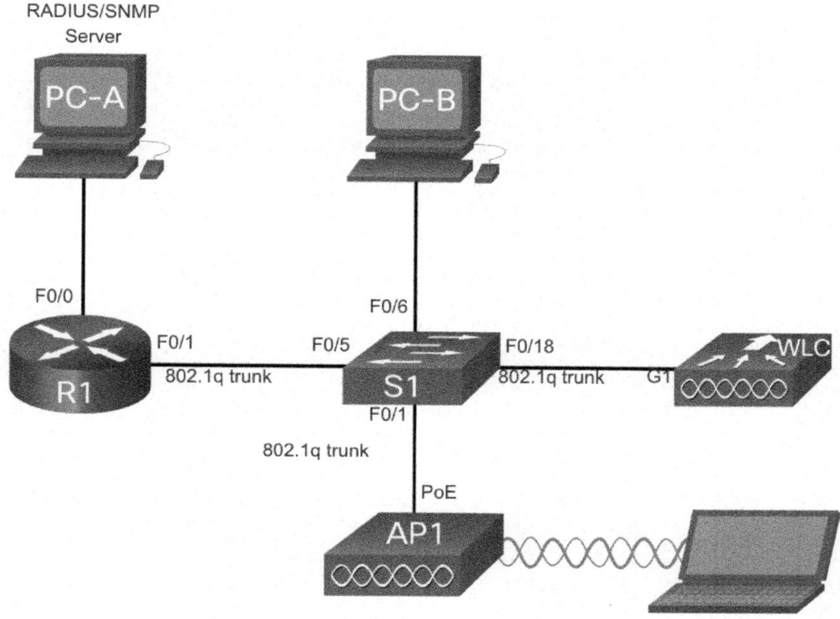

The AP is PoE, which means it is powered over the Ethernet cable that is attached to the switch.

Figure 13-19 WLC Reference Topology

Table 13-1 Addressing Table

Device	Interface	IP Address	Subnet Mask
R1	F0/0	172.16.1.1	255.255.255.0
R1	F0/1.1	192.168.200.1	255.255.255.0
S1	VLAN 1	DHCP	
WLC	Management	192.168.200.254	255.255.255.0
AP1	Wired 0	192.168.200.3	255.255.255.0
PC-A	NIC	172.16.1.254	255.255.255.0
PC-B	NIC	DHCP	
Wireless Laptop	NIC	DHCP	

Log in to the WLC (13.2.3)

Configuring a wireless LAN controller (WLC) is not that much different from configuring a wireless router. The big difference is that a WLC controls APs and provides more services and management capabilities, many of which are beyond the scope of this module.

Note

The figures in this section and the next that show the graphical user interface (GUI) and menus are from a Cisco 3504 Wireless Controller. However, other WLC models will have similar menus and features.

Figure 13-20 shows the user logging in to the WLC with credentials that were configured during initial setup.

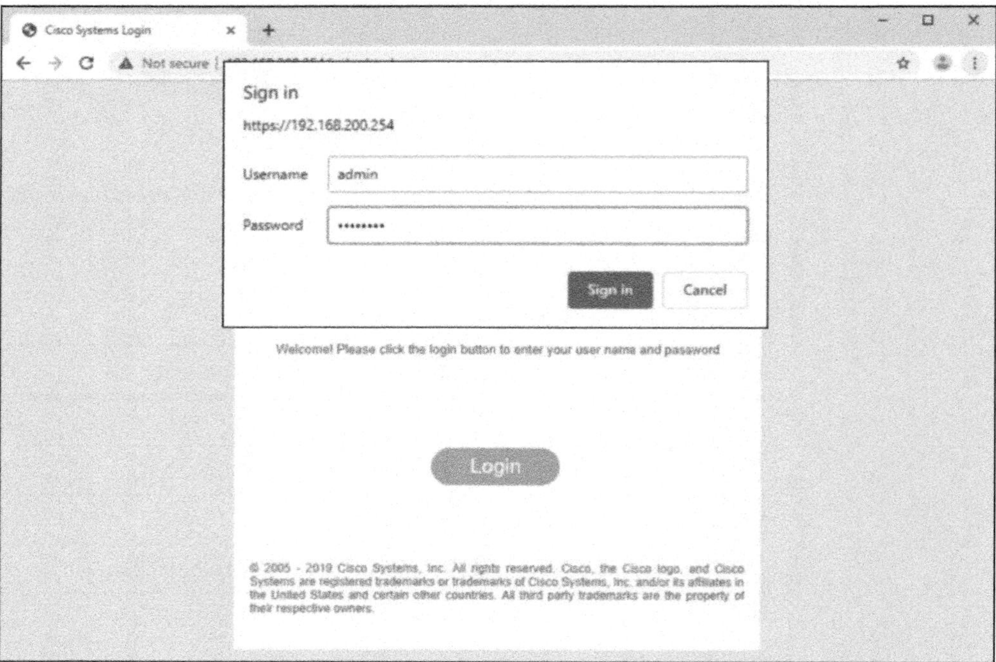

Figure 13-20 Logging In to the WLC

The Network Summary page is a dashboard that provides a quick overview of the number of configured wireless networks, associated access points (APs), and active clients. You can also see the number of *rogue access points* and clients, as shown in Figure 13-21.

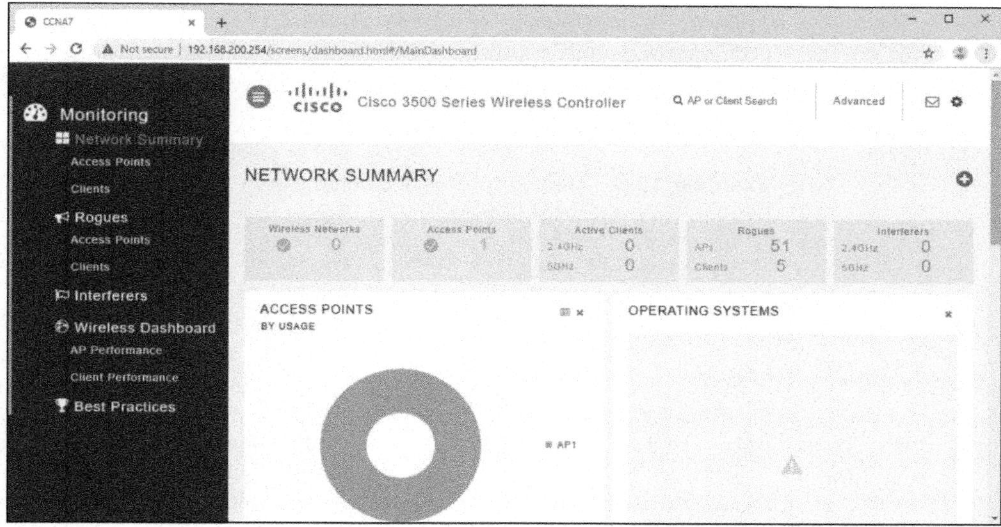

Figure 13-21 WLC Network Summary Dashboard

View AP Information (13.2.4)

Click **Access Points** from the left menu to view an overall picture of the AP's system information and performance, as shown in Figure 13-22. The AP is using IP address 192.168.200.3. Because Cisco Discovery Protocol (CDP) is active on this network, the WLC knows that the AP is connected to the FastEthernet 0/1 port on the switch.

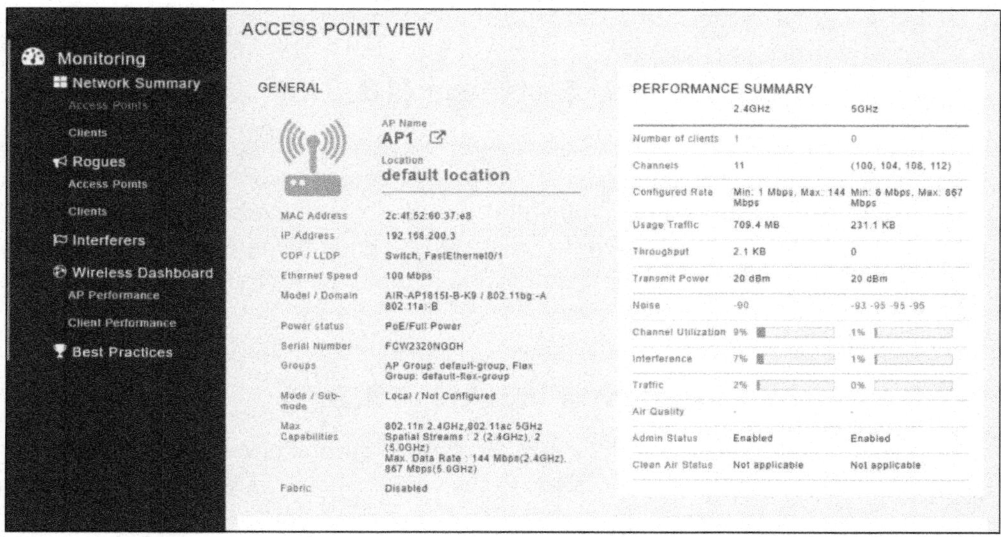

Figure 13-22 Access Point View Page

This AP in the topology is a Cisco Aironet 1815i, which means you can use the command line and a limited set of familiar IOS commands. In Example 13-2, the network administrator pinged the default gateway, pinged the WLC, and verified the wired interface.

Example 13-2 Verifying AP Has Connectivity

```
AP1# ping 192.168.200.1
Sending 5, 100-byte ICMP Echos to 192.168.200.1, timeout is 2 seconds
!!!!!
Success rate is 100 percent (5/5), round-trip min/avg/max = 1069812.242/
  1071814.785/1073817.215 ms
AP1#
AP1# ping 192.168.200.254
Sending 5, 100-byte ICMP Echos to 192.168.200.254, timeout is 2 seconds
!!!!!
Success rate is 100 percent (5/5), round-trip min/avg/max = 1055820.953/
  1057820.738/1059819.928 ms
AP1#
AP1# show interface wired 0
wired0     Link encap:Ethernet  HWaddr 2C:4F:52:60:37:E8
           inet addr:192.168.200.3  Bcast:192.168.200.255  Mask:255.255.255.255
           UP BROADCAST RUNNING PROMISC MULTICAST  MTU:1500  Metric:1
           RX packets:2478 errors:0 dropped:3 overruns:0 frame:0
           TX packets:1494 errors:0 dropped:0 overruns:0 carrier:0
           collisions:0 txqueuelen:80
           RX bytes:207632 (202.7 KiB)  TX bytes:300872 (293.8 KiB)
AP1#
```

Advanced Settings (13.2.5)

Most WLC will come with some basic settings and menus that users can quickly access to implement a variety of common configurations. However, as a network administrator, you will typically access the advanced settings. For the Cisco 3504 Wireless Controller, click **Advanced** in the upper-right corner to access the advanced Summary page, as shown in Figure 13-23. From here, you can access all the features of the WLC.

Configure a WLAN (13.2.6)

Wireless LAN Controllers have ports and interfaces. Ports are the sockets for the physical connections to the wired network. They resemble switch ports. Interfaces are virtual. They are created in software and are very similar to VLAN interfaces. In fact, each interface that will carry traffic from a WLAN is configured on the WLC as a different VLAN.

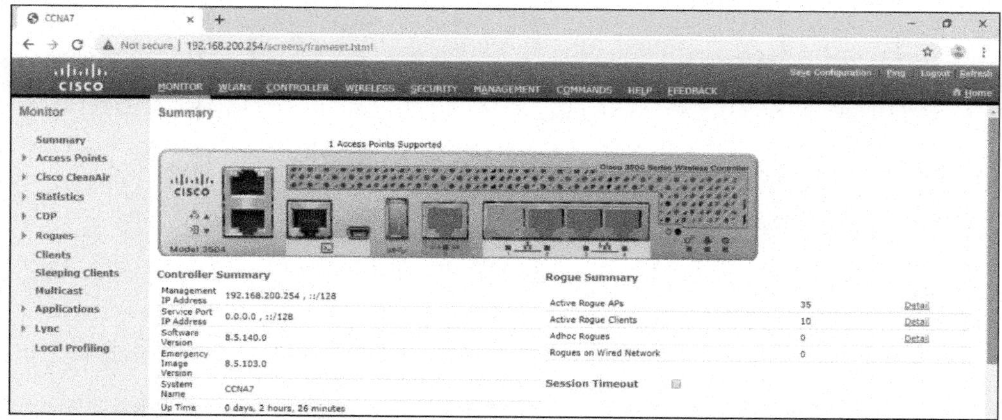

Figure 13-23 Viewing the Summary Page of Advanced Settings

The Cisco 3504 WLC can support 150 access points and 4096 VLANs; however, it has only five physical ports, as shown in Figure 13-24. This means that each physical port can support many APs and WLANs. The ports on the WLC are essentially trunk ports that can carry traffic from multiple VLANs to a switch for distribution to multiple APs. Each AP can support multiple WLANs.

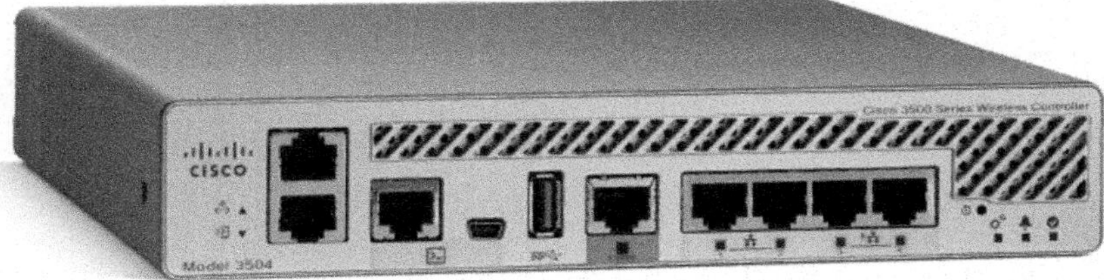

Figure 13-24 Backplane of a Cisco 3504 WLC

Basic WLAN configuration on the WLC includes the following steps:

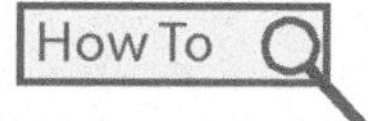

Step 1. **Create the WLAN.** In Figure 13-25, the administrator is creating a new WLAN that will use Wireless_LAN as the name and service set identifier (SSID). The ID is an arbitrary value that is used to identify the WLAN in display output on the WLC.

Step 2. **Apply and Enable the WLAN.** After clicking **Apply**, the network administrator must enable the WLAN before it can be accessed by users, as shown in Figure 13-26. The Enable check box allows the network administrator to configure a variety of features for the WLAN, as well as additional WLANs, before enabling them for wireless client access. From here, the network administrator can configure a variety of settings for the WLAN, including security, QoS, policies, and other advanced settings.

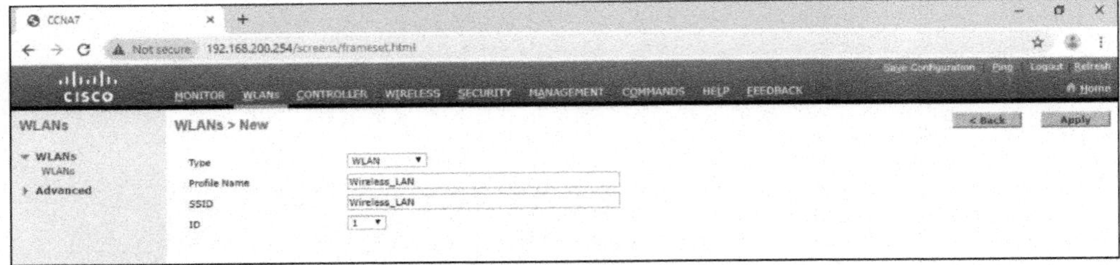

Figure 13-25 Configure a WLAN—Step 1

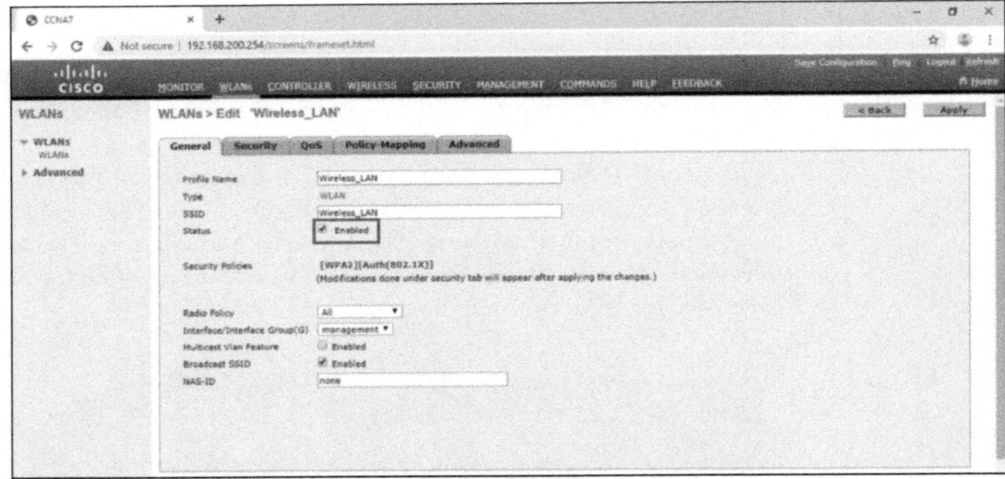

Figure 13-26 Configure a WLAN—Step 2

Step 3. **Select the Interface.** When you create a WLAN, you must select the interface that will carry the WLAN traffic. Figure 13-27 shows the selection of an interface that has already been created on the WLC. You learn how to create interfaces later in this module.

Step 4. **Secure the WLAN.** Click the **Security** tab to access all the available options for securing the LAN. The network administrator wants to secure Layer 2 with WPA2-PSK. WPA2 and 802.1X are set by default. In the Layer 2 Security drop-down box, verify that **WPA+WPA2** is selected (not shown). Click PSK and enter the pre-shared key, as shown Figure 13-28. Then click **Apply.** This will enable the WLAN with WPA2-PSK authentication. Wireless clients that know the pre-shared key can now associate and authenticate with the AP.

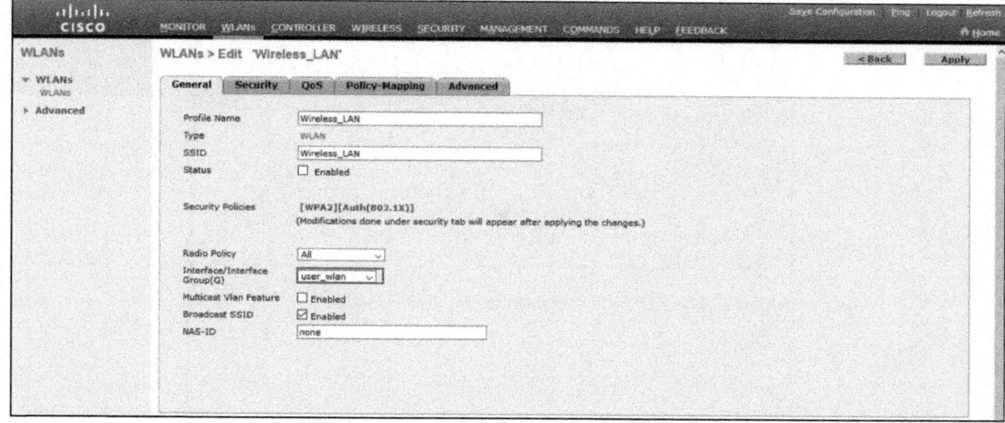

Figure 13-27 Configure a WLAN—Step 3

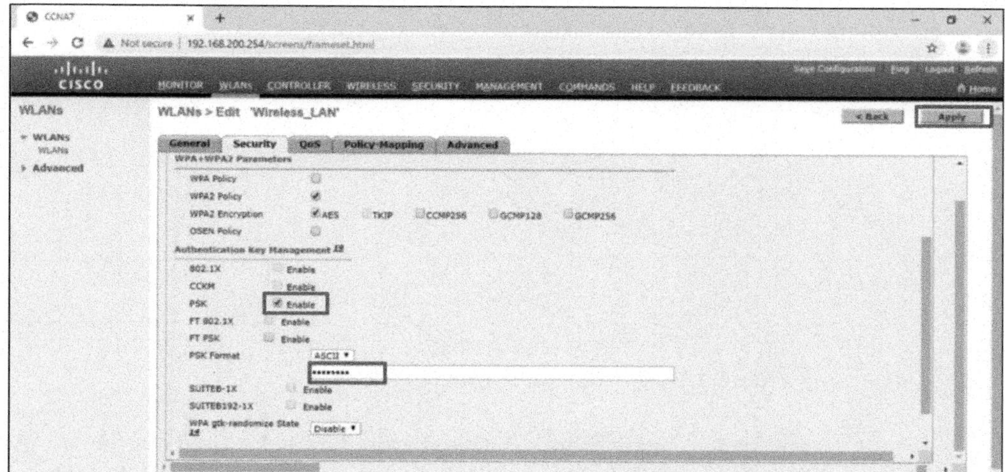

Figure 13-28 Configure a WLAN—Step 4

Step 5. **Verify the WLAN is operational.** Click **WLANs** in the menu on the left to view the newly configured WLAN. In Figure 13-29, you can verify that WLAN ID 1 is configured with Wireless_LAN as the name and SSID; it is enabled and is using WPA2 PSK security.

Step 6. **Monitor the WLAN.** Click the **MONITOR** tab at the top to access the advanced Summary page again. Here you can see that the Wireless_LAN now has one client using its services, as shown in Figure 13-30.

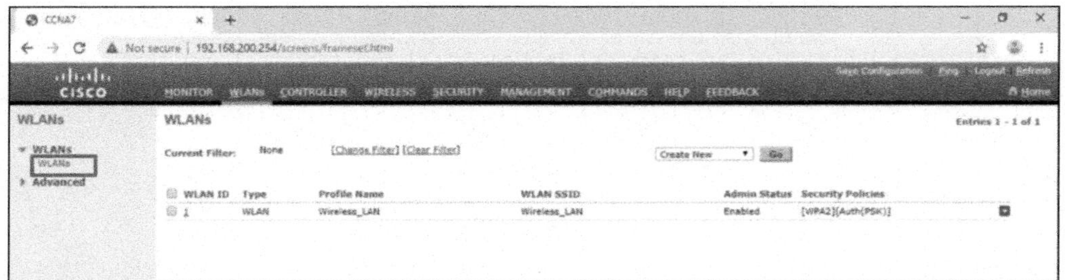

Figure 13-29 Configure a WLAN—Step 5

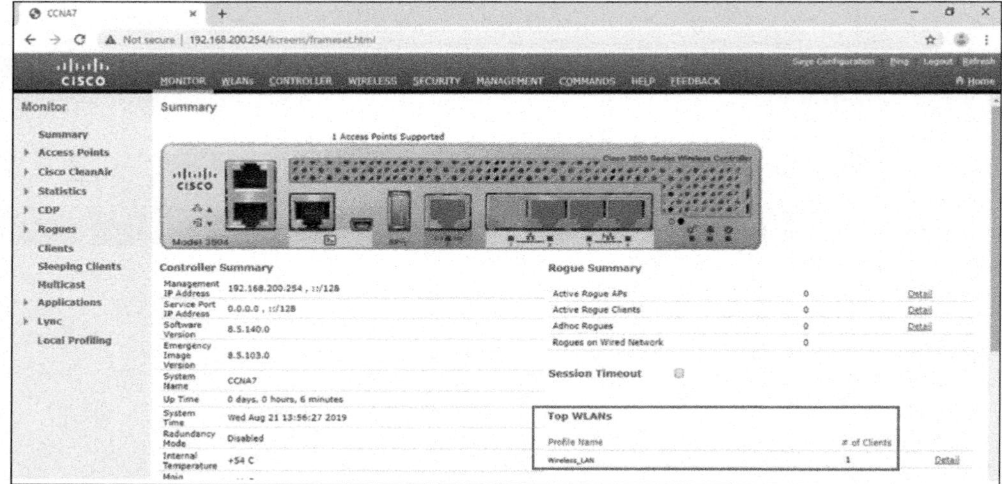

Figure 13-30 Configure a WLAN—Step 6

Step 7. **View Wireless Client Details.** Click **Clients** in the left menu to view more information about the clients connected to the WLAN, as shown in Figure 13-31. One client is attached to Wireless_LAN through AP1 and was given the IP address 192.168.5.2. DHCP services in this topology are provided by the router.

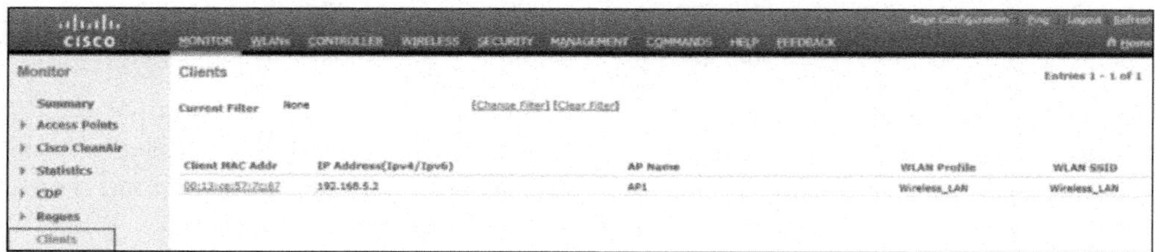

Figure 13-31 View Wireless Client Details

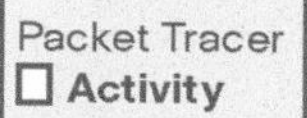

Packet Tracer—Configure a Basic WLAN on the WLC (13.2.7)

In this lab, you explore some of the features of a wireless LAN controller. You create a new WLAN on the controller and implement security on that LAN. Then you configure a wireless host to connect to the new WLAN through an AP that is under the control of the WLC. Finally, you verify connectivity.

Configure a WPA2 Enterprise WLAN on the WLC (13.3)

In this section, you learn how to configure a WLC WLAN to use a VLAN interface, a DHCP server, and WPA2 Enterprise authentication.

Video

Video—Define an SNMP and RADIUS Server on the WLC (13.3.1)

Refer to the online course to view this video.

SNMP and RADIUS (13.3.2)

In Figure 13-32, PC-A is running Simple Network Management Protocol (SNMP) and Remote Authentication Dial-In User Service (RADIUS) server software. SNMP is used to monitor the network. The network administrator wants the WLC to forward all SNMP log messages, called traps, to the SNMP server.

In addition, for WLAN user authentication, the network administrator wants to use a RADIUS server for authentication, authorization, and accounting (AAA) services. Instead of entering a publicly known pre-shared key to authenticate, as they do with WPA2-PSK, users will enter their own username and password credentials. The credentials will be verified by the RADIUS server. This way, individual user access can be tracked and audited if necessary, and user accounts can be added or modified from a central location. The RADIUS server is required for WLANs that are using WPA2 Enterprise authentication.

Note

SNMP server and RADIUS server configuration is beyond the scope of this module.

Configure SNMP Server Information (13.3.3)

Click the **MANAGEMENT** tab to access a variety of management features. SNMP is listed at the top of the menu on the left. Click **SNMP** to expand the submenus, and then click **Trap Receivers**. Click **New** to configure a new SNMP trap receiver, as shown in Figure 13-33.

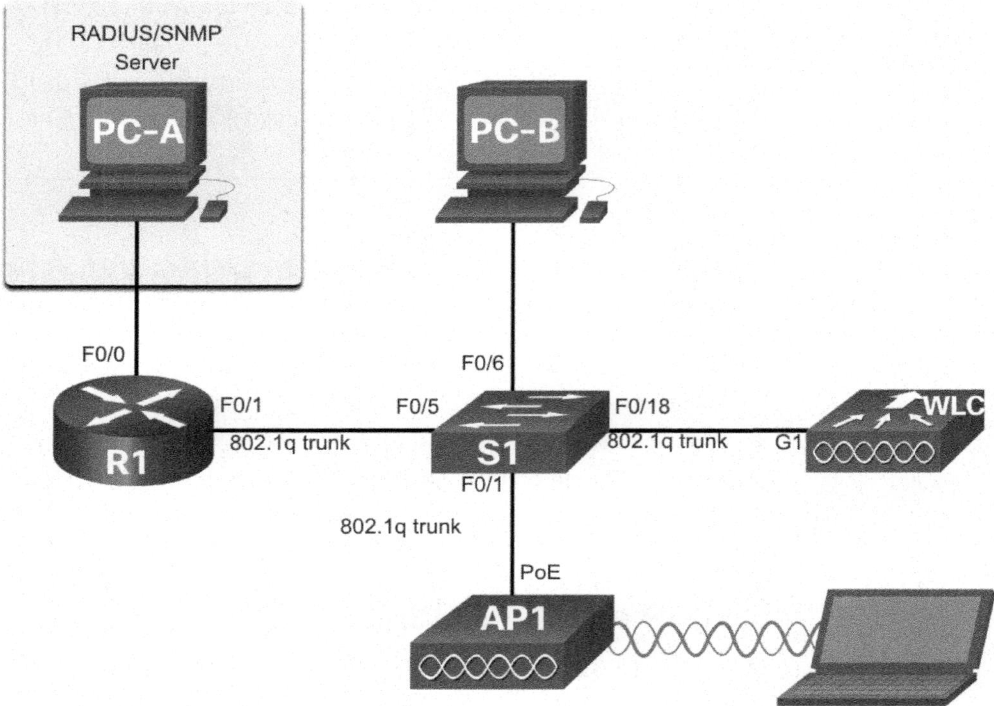

Figure 13-32 WLC Reference Topology

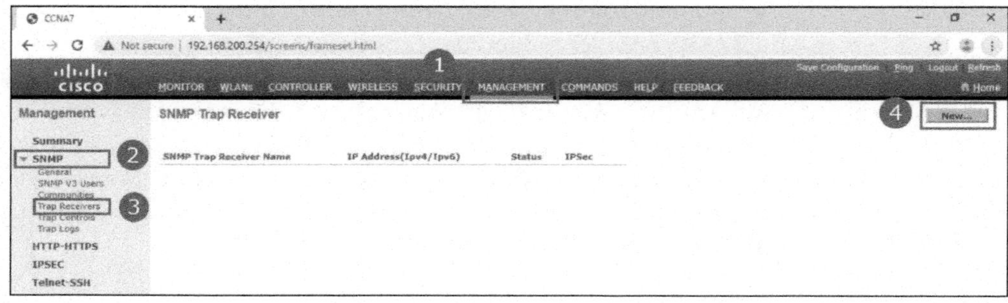

Figure 13-33 Creating a New SNMP Trap Receiver

Enter the SNMP Community name and the IP address (IPv4 or IPv6) for the SNMP server. Click **Apply**. The WLC will now forward SNMP log messages to the SNMP server, as shown in Figure 13-34.

Configure RADIUS Server Information (13.3.4)

In our example configuration, the network administrator wants to configure a WLAN using WPA2 Enterprise, as opposed to WPA2 Personal or WPA2 PSK. Authentication will be handled by the RADIUS server running on PC-A.

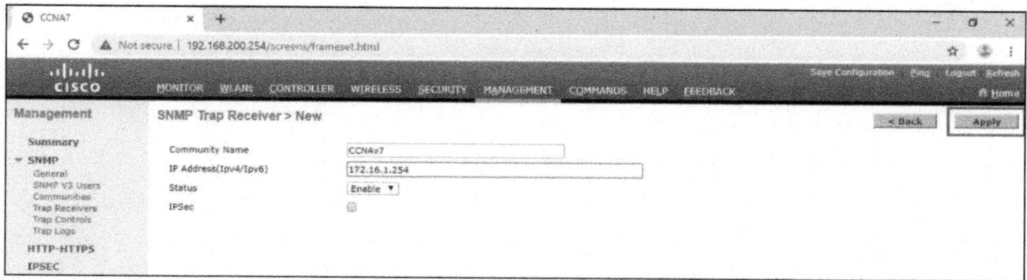

Figure 13-34 Configuring the SNMP Community Name and IPv4 Address

To configure the WLC with the RADIUS server information, click the **SECURITY** tab, **RADIUS**, **Authentication.** No RADIUS servers are currently configured. Click **New** to add PC-A as the RADIUS server, as shown in Figure 13-35.

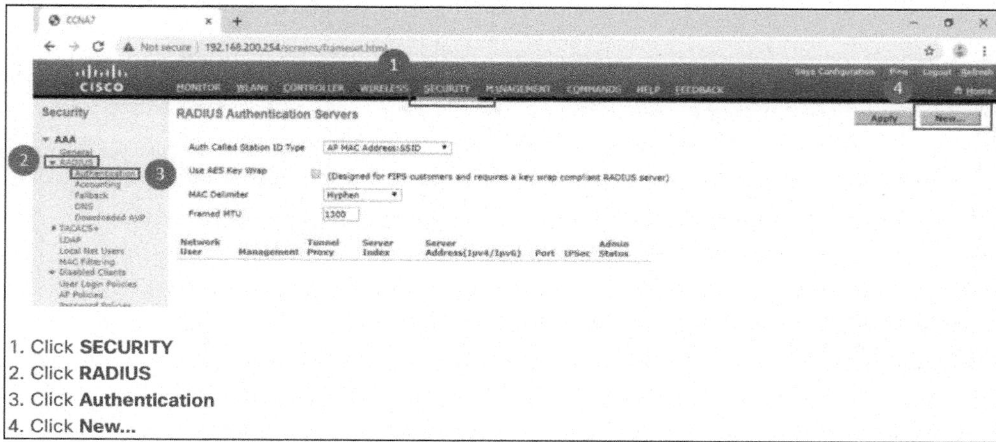

Figure 13-35 Creating a New RADIUS Server

Enter the IPv4 address for PC-A and the shared secret. This is the password used between the WLC and the RADIUS server. It is not for users. Click **Apply**, as shown in Figure 13-36.

After clicking Apply, the list of configured RADIUS Authentication Servers refreshes with the new server listed, as shown in Figure 13-37.

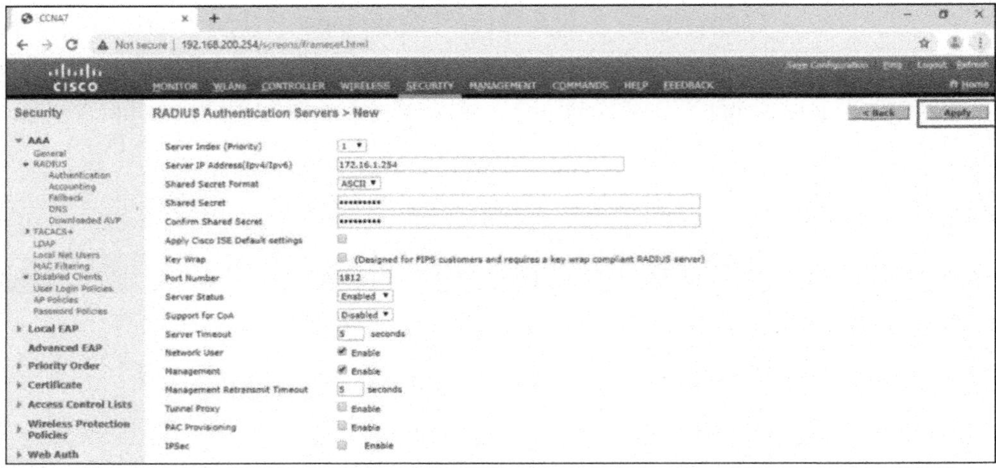

Figure 13-36 Configuring RADIUS Server Information

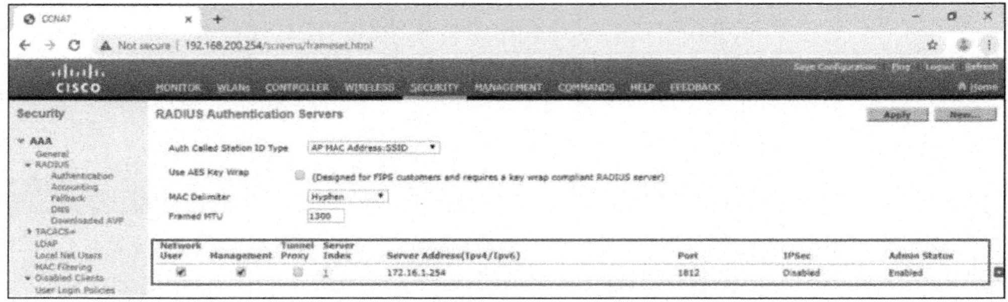

Figure 13-37 Verifying the RADIUS Server Configuration

Video

Video—Configure a VLAN for a New WLAN (13.3.5)

Refer to the online course to view this video.

Topology with VLAN 5 Addressing (13.3.6)

Each WLAN configured on the WLC needs its own virtual interface. The WLC has five physical ports for data traffic. Each physical port can be configured to support multiple WLANs, each on its own virtual interface. Physical ports can also be aggregated to create high-bandwidth links.

The network administrator has decided that the new WLAN will use interface VLAN 5 and network 192.168.5.0/24. R1 already has a subinterface configured and active for VLAN 5, as shown in Figure 13-38 and the **show ip interface brief** output in Example 13-3.

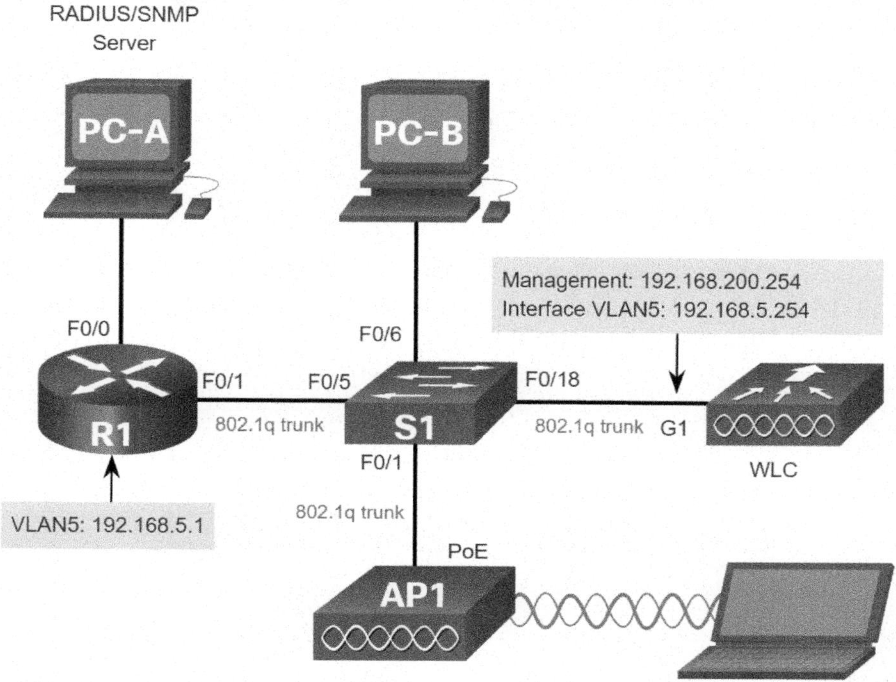

Figure 13-38 WLC Reference Topology

Example 13-3 Verifying VLAN 5 Interface on R1

```
R1# show ip interface brief
Interface              IP-Address      OK? Method Status                Protocol
FastEthernet0/0        172.16.1.1      YES manual up                    up
FastEthernet0/1        unassigned      YES unset  up                    up
FastEthernet0/1.1      192.168.200.1   YES manual up                    up
FastEthernet0/1.5      192.168.5.254   YES manual up                    up
(output omitted)
R1#
```

Configure a New Interface (13.3.7)

VLAN interface configuration on the WLC includes the following steps:

Step 1. **Create a new interface.** To add a new interface, click **CONTROLLER, Interfaces, New**, as shown in Figure 13-39.

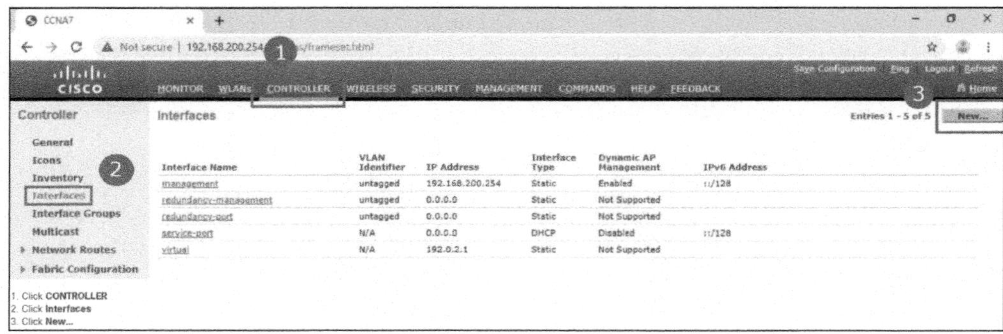

Figure 13-39 Configure a New Interface—Step 1

Step 2. **Configure the VLAN name and ID.** In Figure 13-40, the network administrator configures the interface name as vlan5 and the VLAN ID as 5. Clicking **Apply** will create the new interface.

Figure 13-40 Configure a New Interface—Step 2

Step 3. **Configure the port and interface address.** On the Edit page for the interface, configure the physical port number. G1 in the topology is Port Number 1 on the WLC. Then configure the VLAN 5 interface addressing. In Figure 13-41, VLAN 5 is assigned IPv4 address 192.168.5.254/24. R1 is the default gateway at IPv4 address 192.168.5.1.

Step 4. **Configure the DHCP server address.** In larger enterprises, WLCs will be configured to forward DHCP messages to a dedicated DHCP server. Scroll down the page to configure the primary DHCP server as IPv4 address 192.168.5.1, as shown in Figure 13-42. This is the default gateway router address. The router is configured with a DHCP pool for the WLAN network. As hosts join the WLAN that is associated with the VLAN 5 interface, they will receive addressing information from this pool.

Step 5. **Apply and Confirm.** Scroll to the top and click **Apply**, as shown in Figure 13-43. Click **OK** for the warning message.

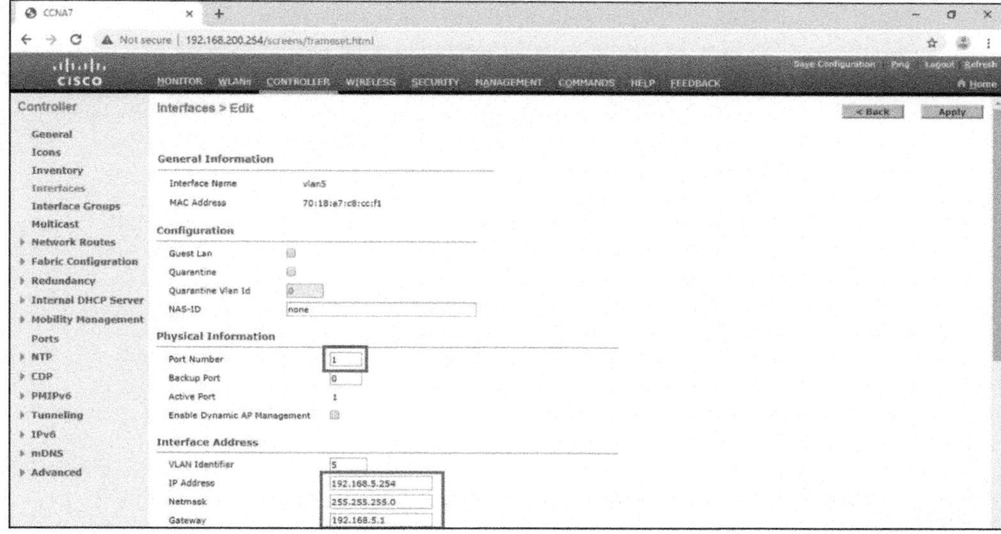

Figure 13-41 Configure a New Interface—Step 3

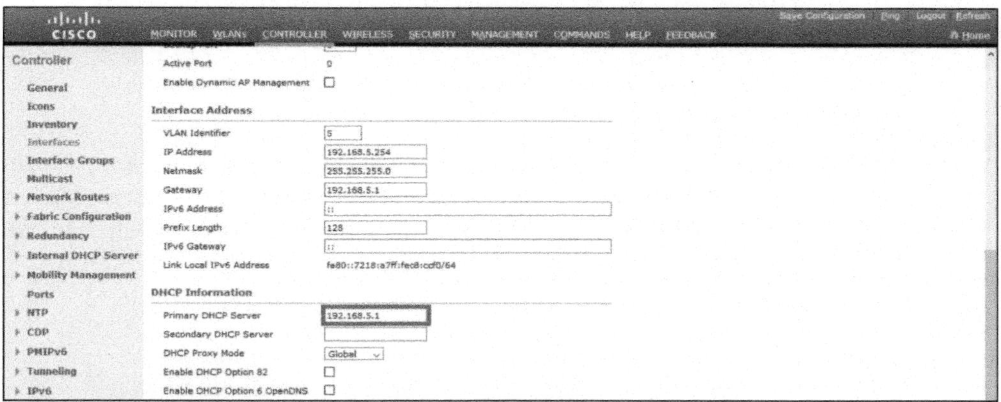

Figure 13-42 Configure a New Interface—Step 4

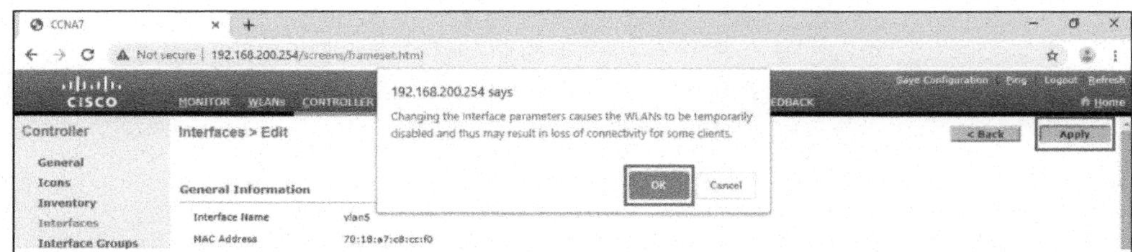

Figure 13-43 Configure a New Interface—Step 5

Step 6. **Verify Interfaces.** Click **Interfaces.** The new vlan5 interface is now shown in the list of interfaces with its IPv4 address, as shown in Figure 13-44.

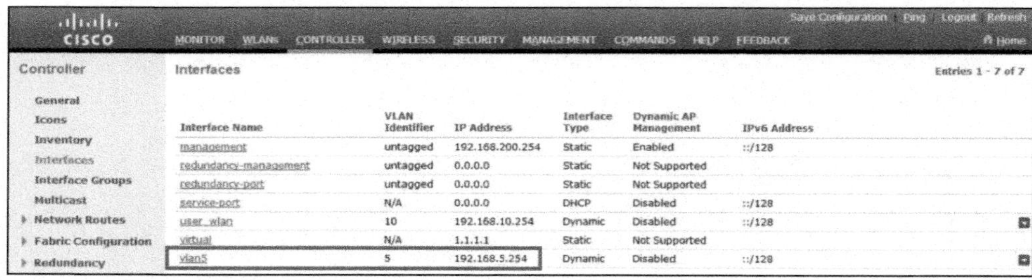

Figure 13-44 Configure a New Interface—Step 6

Video—Configure a DHCP Scope (13.3.8)

Refer to the online course to view this video.

Configure a DHCP Scope (13.3.9)

DHCP scope configuration includes the following steps:

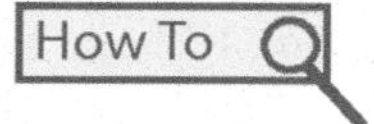

Step 1. **Create a new DHCP scope.** A DHCP scope is very similar to a DHCP pool on a router. It can include a variety of information, including a pool of addresses to assign to DHCP clients, DNS server information, lease times, and more. To configure a new DHCP scope, click **Internal DHCP Server, DHCP Scope, New,** as shown in Figure 13-45.

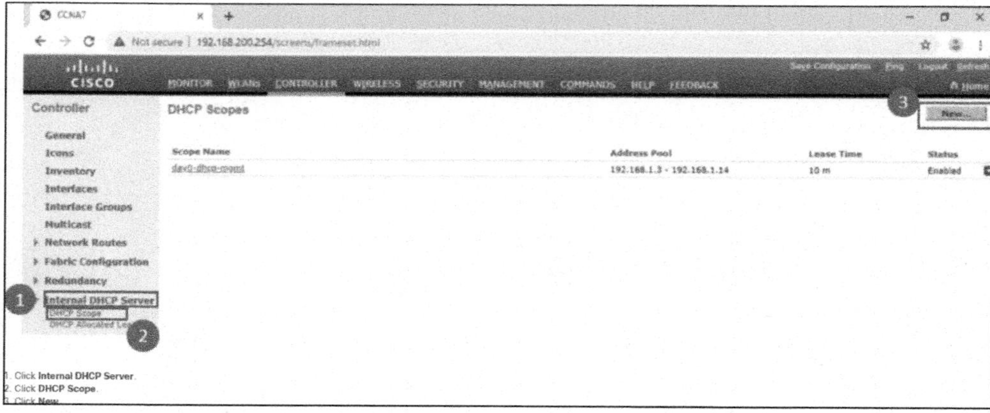

Figure 13-45 Configure a DHCP Scope—Step 1

Step 2. **Name the DHCP scope.** On the next screen, name the scope, as shown in Figure 13-46. Because this scope will apply to the wireless management network, the network administrator uses Wireless_Management as the Scope Name and clicks **Apply**.

Figure 13-46 Configure a DHCP Scope—Step 2

Step 3. **Verify the new DHCP scope.** You are returned to the DHCP Scopes page and can verify the scope is ready to be configured, as shown in Figure 13-47. Click the new scope name to configure the DHCP scope.

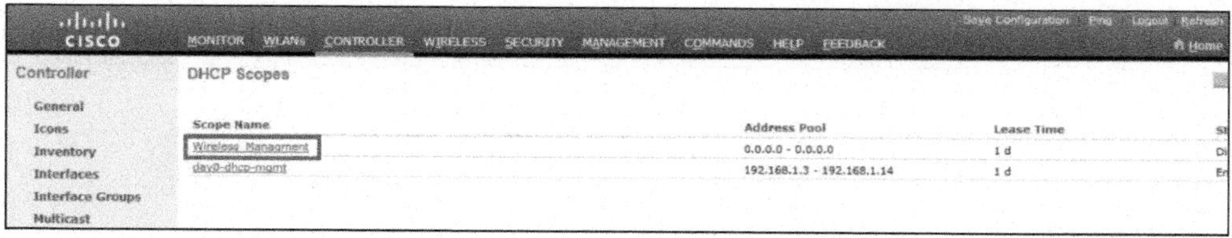

Figure 13-47 Configure a DHCP Scope—Step 3

Step 4. **Configure and enable the new DHCP scope.** On the Edit screen for the Wireless_Management scope, configure a pool of addresses for the 192.168.200.0/24 network starting at .240 and ending at .249. The network address and subnet mask are configured. The default router IPv4 address is configured, which is the subinterface for R1 at 192.168.200.1. For this example, in Figure 13-48, the rest of the scope is left unchanged. The network administrator selects **Enabled** from the Status drop down and clicks **Apply**.

Step 5. **Verify the enable DHCP scope.** The network administrator is returned to the DHCP Scopes page and can verify the scope is ready to be allocated to a new WLAN, as shown in Figure 13-49.

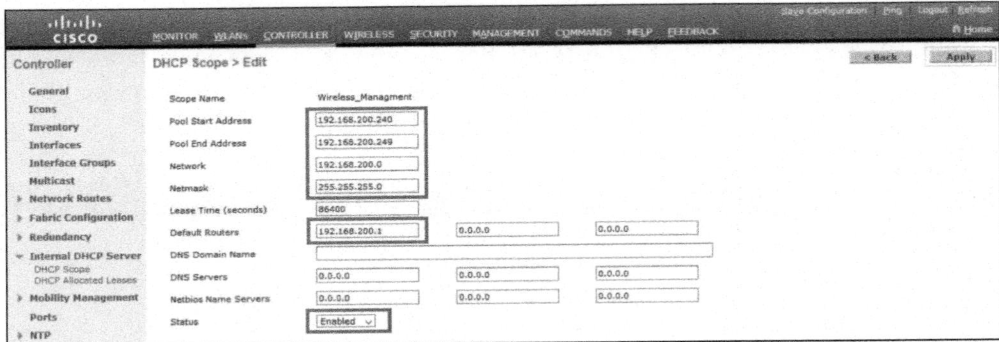

Figure 13-48 Configure a DHCP Scope—Step 4

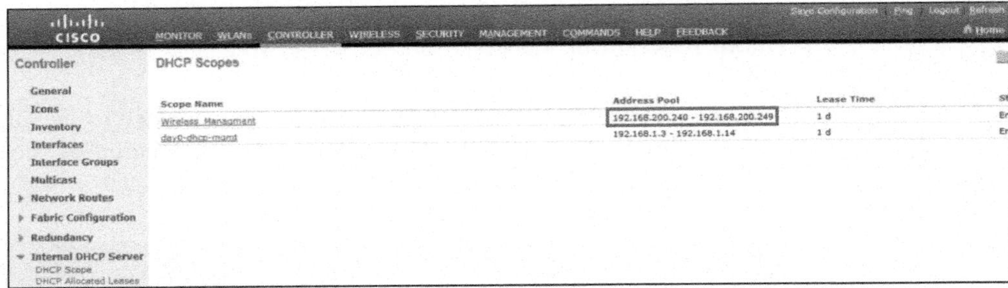

Figure 13-49 Configure a DHCP Scope—Step 5

Video—Configure a WPA2 Enterprise WLAN (13.3.10)

Refer to the online course to view this video.

Configure a WPA2 Enterprise WLAN (13.3.11)

By default, all newly created WLANs on the WLC will use WPA2 with Advanced Encryption System (AES). 802.1X is the default key management protocol used to communicate with the RADIUS server. Because the network administrator already configured the WLC with the IPv4 address of the RADIUS server running on PC-A, the only configuration left to do is to create a new WLAN to use interface vlan5.

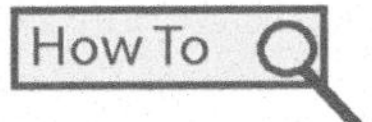

Step 1. **Create a new WLAN.** Click the **WLANs** tab and then **Go** to create a new WLAN, as shown in Figure 13-50.

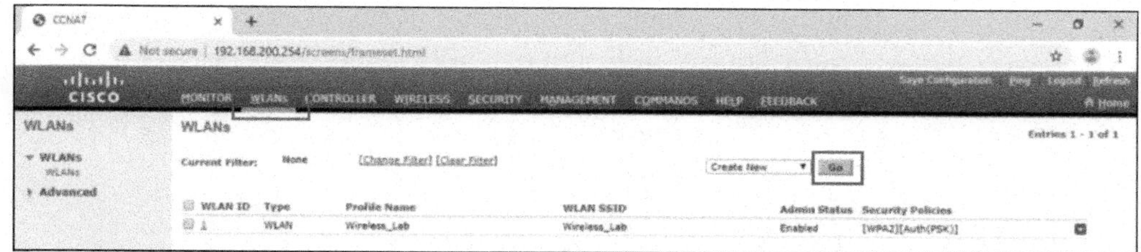

Figure 13-50 Configure a WPA2 Enterprise WLAN—Step 1

Step 2. **Configure the WLAN name and SSID.** Fill in the profile name and SSID. To be consistent with the VLAN that was previously configured, choose an ID of 5. However, any available value can be used. Click **Apply** to create the new WLAN, as shown in Figure 13-51.

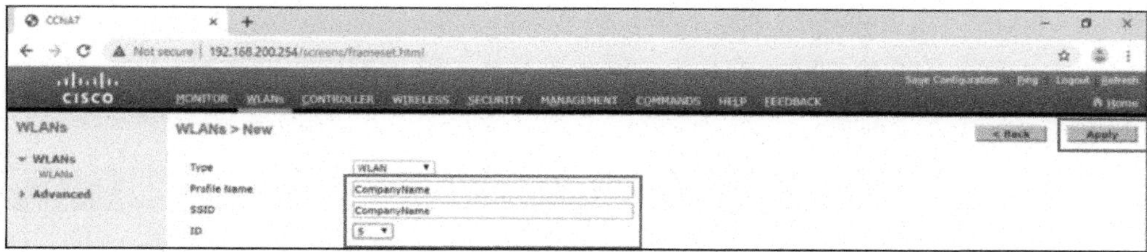

Figure 13-51 Configure a WPA2 Enterprise WLAN—Step 2

Step 3. **Enable the WLAN for VLAN 5.** The WLAN is created but it still needs to be enabled and associated with the correct VLAN interface. Change the status to **Enabled** and choose **vlan5** from the Interface/Interface Group(G) drop-down list. Click **Apply**, and click **OK** to accept the popup message, as shown in Figure 13-52.

Step 4. **Verify AES and 802.1X defaults.** Click the **Security** tab to view the default security configuration for the new WLAN, as shown in Figure 13-53. The WLAN will use WPA2 security with AES encryption. Authentication traffic is handled by 802.1X between the WLC and the RADIUS server.

Step 5. **Configure the RADIUS server.** You now need to select the RADIUS server that will be used to authenticate users for this WLAN. Click the **AAA Servers** tab. In the drop-down box select the RADIUS server that was configured on the WLC previously. Apply your changes, as shown in Figure 13-54.

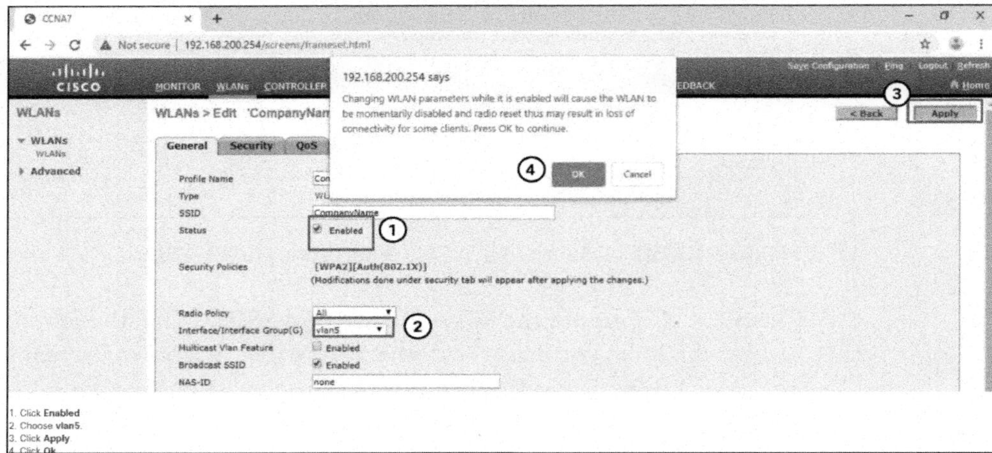

Figure 13-52 Configure a WPA2 Enterprise WLAN—Step 3

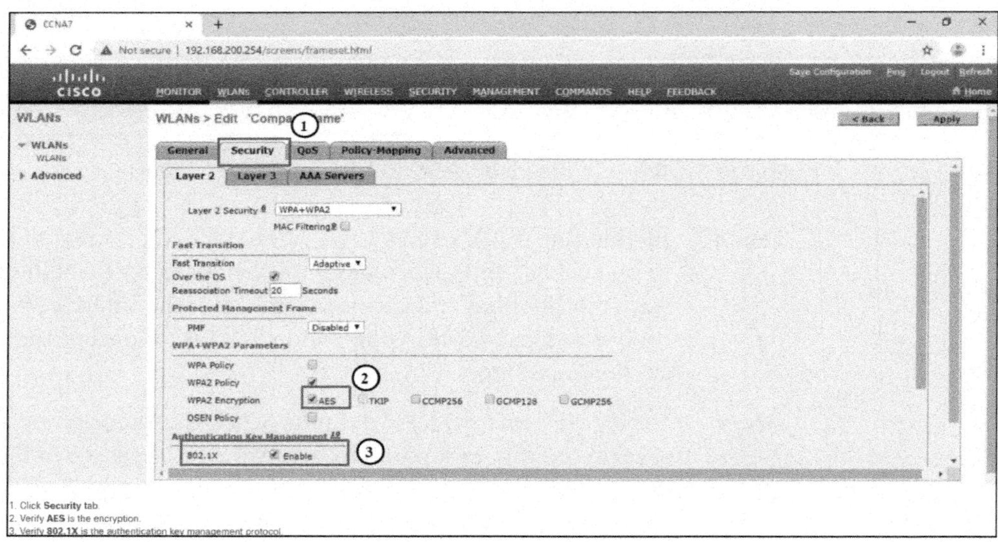

Figure 13-53 Configure a WPA2 Enterprise WLAN—Step 4

Step 6. **Verify that the new WLAN is available.** To verify the new WLAN is listed and enabled, click **Back** or the WLANs submenu on the left. Both the Wireless_LAN WLAN and the CompanyName WLAN are listed. In Figure 13-55, notice that both are enabled. Wireless_LAN is using WPA2 with PSK authentication. CompanyName is using WPA2 security with 802.1X authentication.

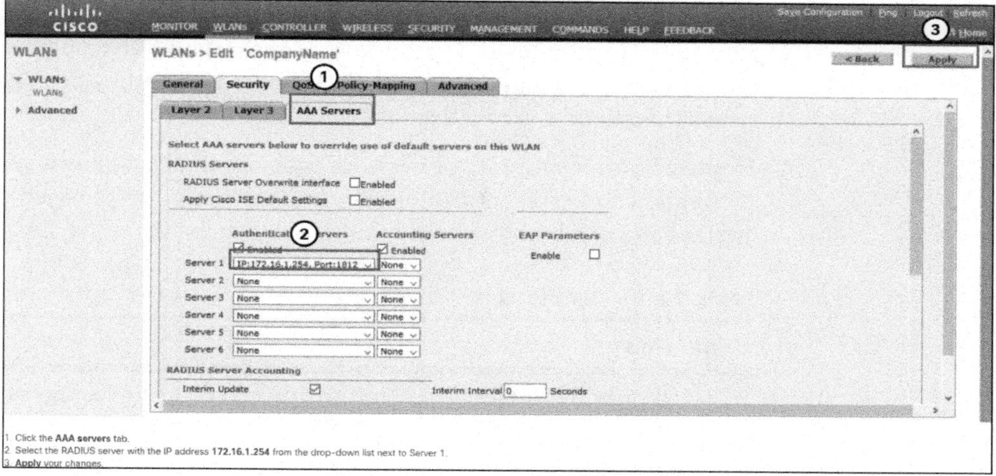

Figure 13-54 Configure a WPA2 Enterprise WLAN—Step 5

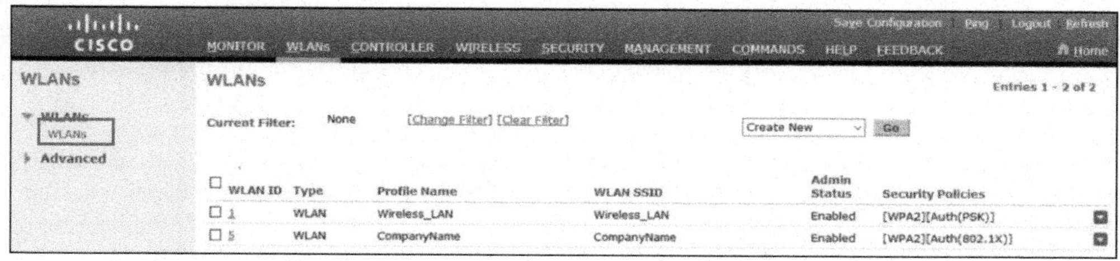

Figure 13-55 Configure a WPA2 Enterprise WLAN—Step 6

Packet Tracer—Configure a WPA2 Enterprise WLAN on the WLC (13.3.12)

In this activity, you configure a new WLAN on a wireless LAN controller (WLC), including the VLAN interface that it will use. You configure the WLAN to use a RADIUS server and WPA2-Enterprise to authenticate users. You also configure the WLC to use an SNMP server.

Troubleshoot WLAN Issues (13.4)

In this section, you learn how to troubleshoot common wireless configuration issues.

Troubleshooting Approaches (13.4.1)

In the previous sections, you learned about WLAN configuration. Here we discuss troubleshooting WLAN issues.

Network problems can be simple or complex and can result from a combination of hardware, software, and connectivity issues. Technicians must be able to analyze the problem and determine the cause of the error before they can resolve the network issue. This process is called troubleshooting.

Troubleshooting any sort of network problem should follow a systematic approach. A common and efficient troubleshooting methodology is based on the scientific method and can be broken into the six main steps shown in Table 13-2.

Table 13-2 Six Steps for Troubleshooting

Step	Title	Description
1	Identify the Problem	The first step in the troubleshooting process is to identify the problem. Although tools can be used in this step, a conversation with the user is often very helpful.
2	Establish a Theory of Probable Causes	After you have talked to the user and identified the problem, you can try to establish a theory of probable causes. This step often yields more than a few probable causes to the problem.
3	Test the Theory to Determine Cause	Based on the probable causes, test your theories to determine which one is the cause of the problem. A technician will often apply a quick procedure to test to see if it solves the problem. If a quick procedure does not correct the problem, you might need to research the problem further to establish the exact cause.
4	Establish a Plan of Action to Resolve the Problem and Implement the Solution	After you have determined the exact cause of the problem, establish a plan of action to resolve the problem and implement the solution.
5	Verify Full System Functionality and Implement Preventive Measures	After you have corrected the problem, verify full functionality and, if applicable, implement preventive measures.
6	Document Findings, Actions, and Outcomes	In the final step of the troubleshooting process, document your findings, actions, and outcomes. This is very important for future reference.

To assess the problem, determine how many devices on the network are experiencing the problem. If there is a problem with one device on the network, start the troubleshooting process at that device. If there is a problem with all devices on the network, start the troubleshooting process at the device where all other devices are connected. You should develop a logical and consistent method for diagnosing network problems by eliminating one problem at a time.

Wireless Client Not Connecting (13.4.2)

When troubleshooting a WLAN, a process of elimination is recommended.

In Figure 13-56, a wireless client is not connecting to the WLAN.

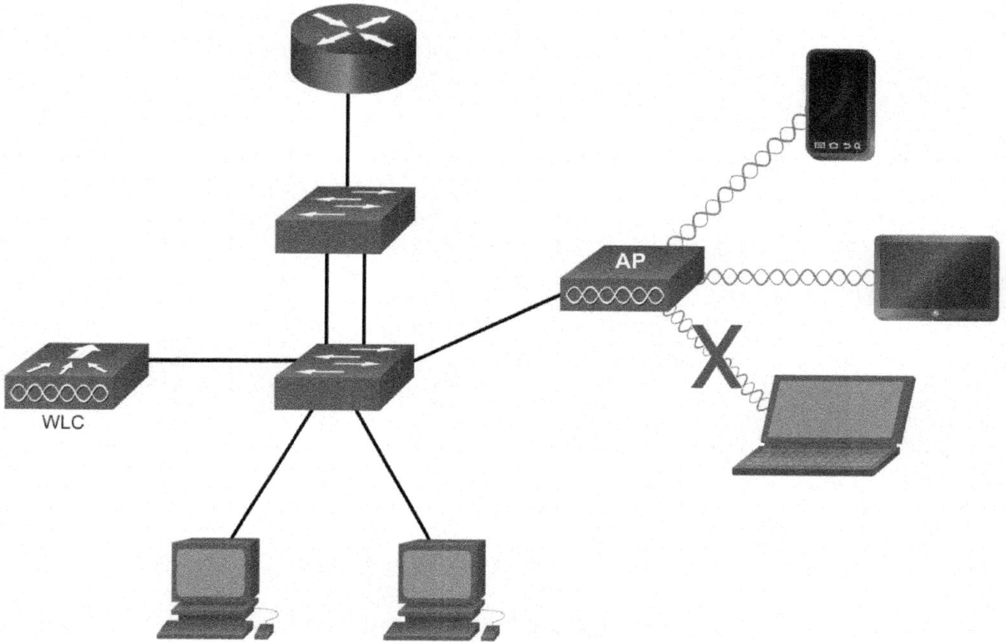

Figure 13-56 WLC Reference Topology with a Client Connectivity Issue

If there is no connectivity, check the following:

- Confirm the network configuration on the PC using the **ipconfig** command. Verify that the PC has received an IP address via DHCP or is configured with a static IP address.
- Confirm that the device can connect to the wired network. Connect the device to the wired LAN and ping a known IP address.
- If necessary, reload drivers as appropriate for the client. It may be necessary to try a different wireless NIC.
- If the wireless NIC of the client is working, check the security mode and encryption settings on the client. If the security settings do not match, the client cannot gain access to the WLAN.

If the PC is operational but the wireless connection is performing poorly, check the following:

- How far is the PC from an AP? Is the PC out of the planned coverage area (BSA)?
- Check the channel settings on the wireless client. The client software should detect the appropriate channel as long as the SSID is correct.
- Check for the presence of other devices in the area that may be interfering with the 2.4 GHz band. Examples of other devices are cordless phones, baby monitors, microwave ovens, wireless security systems, and potentially rogue APs. Data from these devices can cause interference in the WLAN and intermittent connection problems between a wireless client and AP.

Next, ensure that all the devices are actually in place. Consider a possible physical security issue. Is there power to all devices and are they powered on?

Finally, inspect links between cabled devices looking for bad connectors or damaged or missing cables. If the physical plant is in place, verify the wired LAN by pinging devices, including the AP. If connectivity still fails at this point, perhaps something is wrong with the AP or its configuration.

When the user PC is eliminated as the source of the problem, and the physical status of devices is confirmed, begin investigating the performance of the AP. Check the power status of the AP.

Troubleshooting When the Network Is Slow (13.4.3)

To optimize and increase the bandwidth of 802.11 dual-band routers and APs, either

- **Upgrade your wireless clients:** Older 802.11b, 802.11g, and even 802.11n devices can slow the entire WLAN. For the best performance, all wireless devices should support the same highest acceptable standard. Although 802.11ax was released in 2019, 802.11ac is most likely the highest standard that enterprises can currently enforce.
- **Split the traffic:** The easiest way to improve wireless performance is to split the wireless traffic between the 802.11n 2.4 GHz band and the 5 GHz band. Therefore, 802.11n (or better) can use the two bands as two separate wireless networks to help manage the traffic. For example, use the 2.4 GHz network for basic Internet tasks, such as web browsing, email, and downloads, and use the 5 GHz band for streaming multimedia, as shown in Figure 13-57.

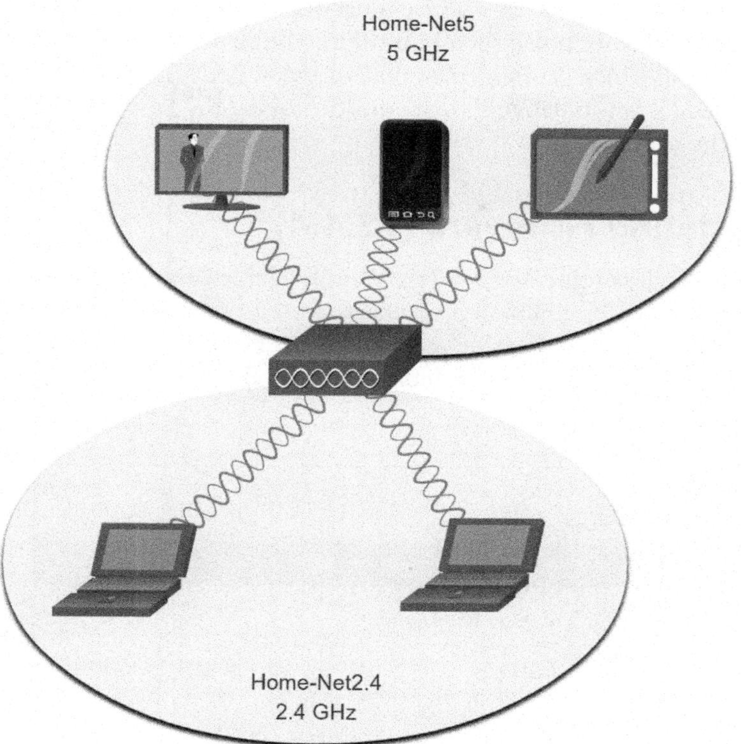

Figure 13-57 AP with Traffic Split Between 2.4 GHz and 5 GHz Band

There are several reasons for using a split-the-traffic approach:

- The 2.4 GHz band may be suitable for basic Internet traffic that is not time-sensitive.
- The bandwidth may still be shared with other nearby WLANs.
- The 5 GHz band is much less crowded than the 2.4 GHz band; ideal for streaming multimedia.
- The 5 GHz band has more channels; therefore, the channel chosen is likely interference-free.

By default, dual-band routers and APs use the same network name on both the 2.4 GHz band and the 5 GHz band. The simplest way to segment traffic is to rename one of the wireless networks. With a separate, descriptive name, it is easier to connect to the right network.

To improve the range of a wireless network, ensure the wireless router or AP location is free of obstructions, such as furniture, fixtures, and tall appliances. These block the signal, which shortens the range of the WLAN. If this still does not solve the problem, a Wi-Fi Range Extender or deploying the Powerline wireless technology may be used.

Updating Firmware (13.4.4)

Most wireless routers and APs offer upgradable firmware. Firmware releases may contain fixes for common problems reported by customers as well as security vulnerabilities. You should periodically check the router or AP for updated firmware. In Figure 13-58, the network administrator is verifying that the firmware is up to date on a Cisco Meraki AP.

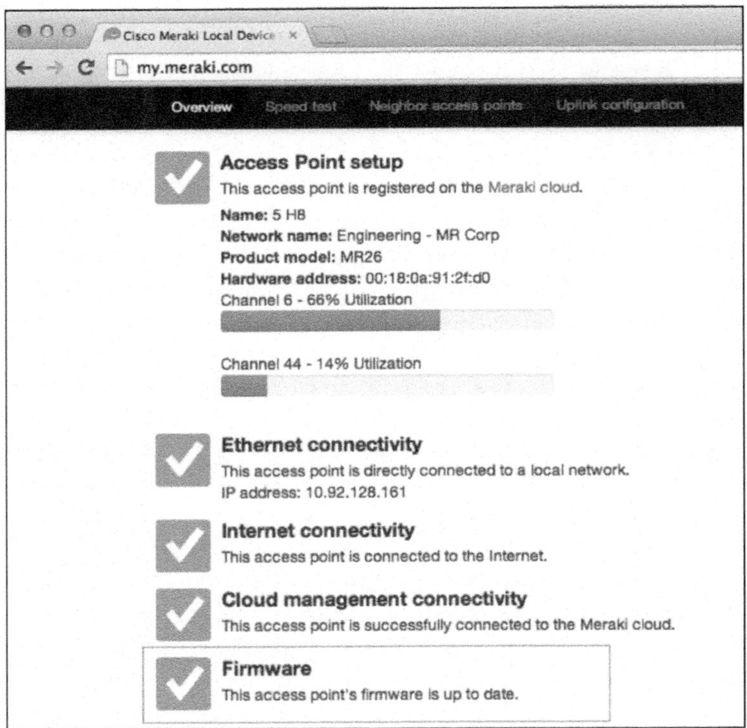

Figure 13-58 Verifying Firmware on a Cisco Meraki AP

On a WLC, there will most likely be the ability to upgrade the firmware on all APs that the WLC controls. In Figure 13-59, the network administrator is downloading the firmware image that will be used to upgrade all the APs.

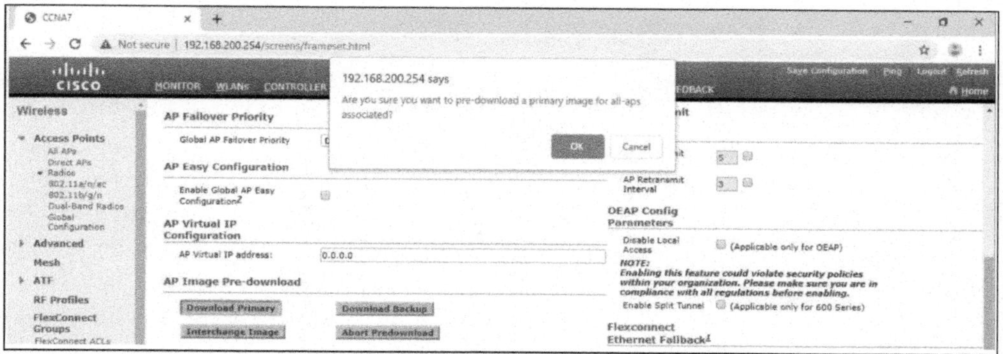

Figure 13-59 Firmware Download on the Cisco 3504 WLC

Users will be disconnected from the WLAN and the Internet until the upgrade finishes. The wireless router may need to reboot several times before normal network operations are restored.

Packet Tracer—Troubleshoot WLAN Issues (13.4.5)

Now that you have learned how to configure wireless in home and enterprise networks, you need to learn how to troubleshoot in both wireless environments. Your goal is to enable connectivity between hosts on the networks to the Web Server by both IP address and URL. Connectivity between the home and enterprise networks is not required.

Summary (13.5)

Remote workers, small branch offices, and home networks often use a wireless router, which typically includes a switch for wired clients, a port for an Internet connection (sometimes labeled "WAN"), and wireless components for wireless client access. Most wireless routers are preconfigured to be connected to the network and provide services. The wireless router uses DHCP to automatically provide addressing information to connected devices. Your first priority should be to change the username and password of your wireless router. Use your router's interface to complete basic network and wireless setup. If you want to extend the range beyond approximately 45 meters indoors and 90 meters outdoors, you can add wireless access points. The router will use a process called Network Address Translation (NAT) to convert private IPv4 addresses to Internet-routable IPv4 addresses. By configuring QoS, you can guarantee that certain traffic types, such as voice and video, are prioritized over traffic that is not as time-sensitive, such as email and web browsing.

Lightweight APs (LAPs) use the Lightweight Access Point Protocol (LWAPP) to communicate with a WLAN controller (WLC). Configuring a wireless LAN controller (WLC) is similar to configuring a wireless router except that a WLC controls APs and provides more services and management capabilities. Use the WLC interface to view an overall picture of the AP's system information and performance, to access advanced settings, and to configure a WLAN.

SNMP is used to monitor the network. The WLC is set to forward all SNMP log messages, called traps, to the SNMP server. For WLAN user authentication, a RADIUS server is used for authentication, accounting, and auditing (AAA) services. Individual user access can be tracked and audited. Use the WLC interface to configure SNMP server and RADIUS server information, VLAN interfaces, DHCP scope, and a WPA2 Enterprise WLAN.

There are six steps to the troubleshooting process. When troubleshooting a WLAN, a process of elimination is recommended. Common problems are the following: no connectivity and poorly performing wireless connection when the PC is operational. To optimize and increase the bandwidth of 802.11 dual-band routers and APs, either upgrade your wireless clients or split the traffic. Most wireless routers and APs offer upgradable firmware. Firmware releases may contain fixes for common problems reported by customers as well as security vulnerabilities. You should periodically check the router or AP for updated firmware.

Packet Tracer
☐ **Activity**

Packet Tracer—WLAN Configuration (13.5.1)

In this activity, you configure both a wireless home router and a WLC-based network. You will implement both WPA2-PSK and WPA2-Enterprise security.

Practice

The following activities provide practice with the topics introduced in this chapter. The Labs are available in the companion *Switching, Routing, and Wireless Essentials Labs and Study Guide (CCNAv7)* (ISBN 9780136634386). The Packet Tracer Activity instructions are also in the Labs & Study Guide. The PKA files are found in the online course.

Lab

Lab 13.1.11: Configure a Wireless Network

Packet Tracer Activities

Packet Tracer 13.1.10: Configure a Wireless Network

Packet Tracer 13.2.7: Configure a Basic WLAN on the WLC

Packet Tracer 13.3.12: Configure a WPA2 Enterprise WLAN on the WLC

Packet Tracer 13.4.5: Troubleshoot WLAN Issues

Packet Tracer 13.5.1: WLAN Configuration

Check Your Understanding Questions

Complete all the review questions listed here to test your understanding of the sections and concepts in this chapter. The appendix "Answers to the 'Check Your Understanding' Questions" lists the answers.

1. What is the first security setting that should be applied when connecting a wireless router in a small network?

   A. Change the default administrative username and password.

   B. Enable encryption on the wireless router.

   C. Disable the wireless network SSID broadcast beacon.

   D. Enable MAC address filtering on the wireless router.

2. Which option is an easy way to improve wireless performance on an 802.11n wireless router?

    A. Connect a Wi-Fi range extender on the 2.4 GHz band to a wireless router on the 5 GHz band.

    B. Require all wireless devices to use the 802.11g standard.

    C. Use different SSID names for the 2.4 GHz and 5 GHz bands.

    D. Use the same SSID name for all wireless bands.

3. Which Cisco 3504 WLC dashboard menu option provides an overview of the number of configured wireless networks, associated access points (APs), and active clients?

    A. Access Points

    B. Advanced

    C. Network Summary

    D. Rogues

4. Which protocol is used to monitor a network?

    A. LWAPP

    B. RADIUS

    C. SNMP

    D. WLC

5. Which service on a wireless router enables a host with an internal private IPv4 address to access an outside network using a public IPv4 address?

    A. DHCP

    B. DNS

    C. LWAPP

    D. NAT

6. Which service available on some wireless routers can be used to prioritize email over web data traffic?

    A. DHCP

    B. DNS

    C. NAT

    D. QoS

7. What must be done before creating a new WLAN on a Cisco 3500 series WLC?

   A. Build or have a RADIUS server available.

   B. Build or have an SNMP server available.

   C. Create a new SSID.

   D. Create a new VLAN interface.

8. Which frequency band SSID name should users with time-sensitive applications connect to?

   A. The 2.4 GHz band, because it is less crowded than the 5 GHz band.

   B. The 2.4 GHz band, because it has more channels than the 5 GHz band.

   C. The 2.4 GHz band, because the channel is likely interference-free.

   D. The 5 GHz band, because it has more channels than the 2.4 GHz band.

9. A Cisco 3500 series WLC is configured to access a RADIUS server. The configuration requires a shared secret password. What is the purpose for the shared secret password?

   A. It allows users to authenticate and access the WLAN.

   B. It is used by the RADIUS server to authenticate WLAN users.

   C. It is used to authenticate and encrypt user data on the WLAN.

   D. It is used to encrypt messages between the WLC and the server.

10. Which type of WLAN extends wireless coverage using a few APs controlled using a smartphone app?

   A. Lightweight access point (LWAP)

   B. Wi-Fi Extender

   C. Wireless LAN Controller (WLC)

   D. Wireless Mesh Network (WMN)

CHAPTER 14

Routing Concepts

Objectives

Upon completion of this chapter, you will be able to answer the following questions:

- How do routers determine the best path?
- How do routers forward packets to the destination?
- How do you configure basic settings on a router?
- What is the structure of a routing table?
- What are the differences between static and dynamic routing concepts?

Key Terms

This chapter uses the following key terms. You can find the definitions in the Glossary.

routing table Page 447

directly connected networks Page 450

directly connected interface Page 450

remote networks Page 450

static routes Page 450

dynamic routing protocols Page 450

Enhanced Interior Gateway Routing Protocol (EIGRP) Page 450

Open Shortest Path First (OSPF) Page 450

Point-to-Point protocol (PPP) Page 453

High-Level Data Link Control (HDLC) Page 453

exit interface Page 455

process switching Page 455

fast switching Page 455

Cisco Express Forwarding (CEF) Page 455

fast-switching cache Page 456

Forwarding Information Base (FIB) Page 458

adjacency table Page 458

converged Page 458

default static route Page 467

administrative distance (AD) Page 469

metric Page 470

next-hop IP address Page 470

local route interface Page 471

stub network Page 472

neighbor Page 472

stub router Page 472

next-hop router Page 475

route lookup process Page 477

child route Page 477

parent route Page 477

Routing Information Protocol (RIP) Page 482

RIPv1 Page 482

RIPv2 Page 482

Intermediate System-to-Intermediate System (IS-IS) Page 482

Interior Gateway Routing Protocol (IGRP) Page 482

Border Gateway Protocol (BGP) Page 482

Exterior Gateway Protocol (EGP) Page 482

Interior Gateway Protocols (IGPs) Page 482

autonomous system (AS) Page 482

distance vector routing protocols Page 482

link-state routing protocols Page 482

path vector routing protocols Page 482

routing algorithm Page 482

RIPng Page 483

data structures Page 483

routing protocol messages Page 483

distance Page 485

cost Page 485

equal cost load balancing Page 486

load balancing Page 486

unequal cost load balancing Page 486

Introduction (14.0)

No matter how effectively you set up your network, something will always stop working correctly, or even stop working completely. This is a simple truth about networking. So, even though you already know quite a bit about routing, you still need to know *how* your routers actually work. This knowledge is critical if you want to be able to troubleshoot your network. This module goes into detail about the workings of a router. Jump in!

Path Determination (14.1)

A router refers to its *routing table* when making best path decisions. In this section, we examine the path determination function of a router.

Two Functions of Router (14.1.1)

Before a router forwards a packet anywhere, it has to determine the best path for the packet to take. This section explains how routers make this determination.

Ethernet switches are used to connect end devices and other intermediary devices, such as other Ethernet switches, to the same network. A router connects multiple networks, which means that it has multiple interfaces that each belong to a different IP network.

When a router receives an IP packet on one interface, it determines which interface to use to forward the packet to the destination. This is known as routing. The interface that the router uses to forward the packet may be the final destination, or it may be a network connected to another router that is used to reach the destination network. Each network that a router connects to typically requires a separate interface, but this may not always be the case.

The primary functions of a router are to determine the best path to forward packets based on the information in its routing table, and to forward packets toward their destination.

Router Functions Example (14.1.2)

The router uses its IP routing table to determine which path (route) to use to forward a packet. In Figure 14-1, R1 is routing a packet from the source PC to the destination PC. Both R1 and R2 use their respective IP routing tables to first determine the best path, and then forward the packet.

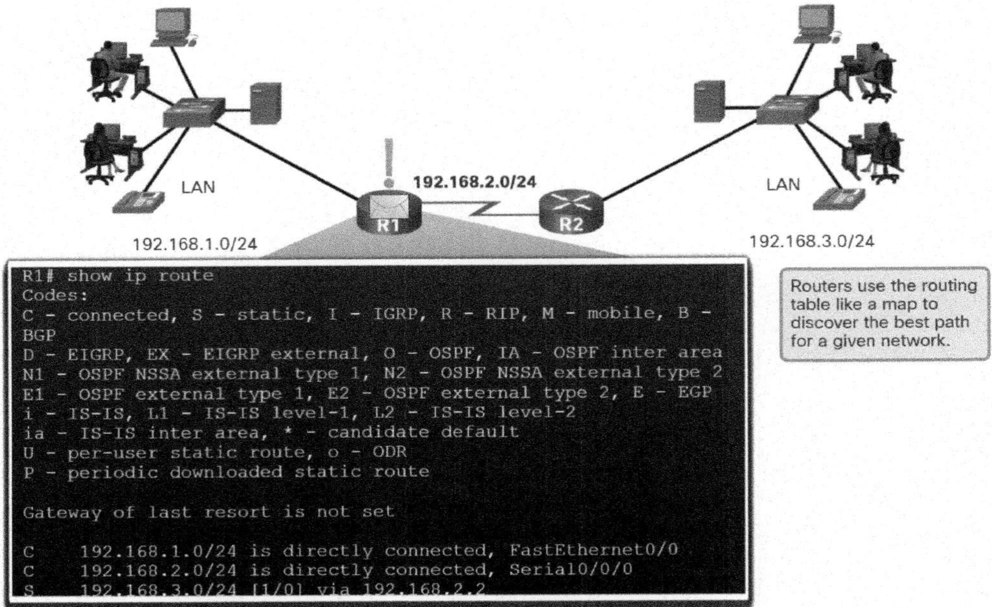

LAN
192.168.2.0/24
LAN
R1
R2
192.168.1.0/24
192.168.3.0/24
R1# show ip route
Codes:
C - connected, S - static, I - IGRP, R - RIP, M - mobile, B -
BGP
D - EIGRP, EX - EIGRP external, O - OSPF, IA - OSPF inter area
N1 - OSPF NSSA external type 1, N2 - OSPF NSSA external type 2
E1 - OSPF external type 1, E2 - OSPF external type 2, E - EGP
i - IS-IS, L1 - IS-IS level-1, L2 - IS-IS level-2
ia - IS-IS inter area, * - candidate default
U - per-user static route, o - ODR
P - periodic downloaded static route
Gateway of last resort is not set
C 192.168.1.0/24 is directly connected, FastEthernet0/0
C 192.168.2.0/24 is directly connected, Serial0/0/0
S 192.168.3.0/24 [1/0] via 192.168.2.2
Routers use the routing table like a map to discover the best path for a given network.

Figure 14-1 Routing a Packet from Source to Destination

Best Path Equals Longest Match (14.1.3)

What is meant by saying that the router must determine the best path in the routing table? The best path in the routing table is also known as the longest match. The longest match is a process the router uses to find a match between the destination IP address of the packet and a routing entry in the routing table.

The routing table contains route entries consisting of a prefix (network address) and prefix length. For there to be a match between the destination IP address of a packet and a route in the routing table, a minimum number of far-left bits must match between the IP address of the packet and the route in the routing table. The prefix length of the route in the routing table is used to determine the minimum number of far-left bits that must match. Remember that an IP packet contains only the destination IP address and not the prefix length.

The longest match is the route in the routing table that has the greatest number of far-left matching bits with the destination IP address of the packet. The route with the greatest number of equivalent far-left bits, or the longest match, is always the preferred route.

Note

The term *prefix length* will be used to refer to the network portion of both IPv4 and IPv6 addresses.

IPv4 Address Longest Match Example (14.1.4)

In Table 14-1, an IPv4 packet has the destination IPv4 address 172.16.0.10. The router has three route entries in its IPv4 routing table that match this packet: 172.16.0.0/12, 172.16.0.0/18, and 172.16.0.0/26. Of the three routes, 172.16.0.0/26 has the longest match and would be chosen to forward the packet. Remember, for any of these routes to be considered a match, there must be at least the number of matching bits indicated by the subnet mask of the route.

Table 14-1 IPv4 Address Longest Match

Destination IPv4 Address		Address in Binary
172.16.0.10		**10101100.00010000.00000000.00**001010
Route Entry	**Prefix/Prefix Length**	**Address in Binary**
1	172.16.0.0/12	**10101100.0001**0000.00000000.00001010
2	172.16.0.0/18	**10101100.00010000.00**000000.00001010
3	172.16.0.0/26	**10101100.00010000.00000000.00**001010

IPv6 Address Longest Match Example (14.1.5)

In Table 14-2, an IPv6 packet has the destination IPv6 address 2001:db8:c000::99. This example shows three route entries, but only two of them are a valid match, with one of those being the longest match. The first two route entries have prefix lengths that have the required number of matching bits as indicated by the prefix length. The first route entry with a prefix length of /40 matches the 40 far-left bits in the IPv6 address. The second route entry has a prefix length of /48, with all 48 bits matching the destination IPv6 address, and is the longest match. The third route entry is not a match because its /64 prefix requires 64 matching bits. For the prefix 2001:db8:c000:5555::/64 to be a match, the first 64 bits must the destination IPv6 address of the packet. Only the first 48 bits match, so this route entry is not considered a match.

For the destination IPv6 packet with the address **2001:db8:c000:**:99, consider the following three route entries:

Table 14-2 IPv6 Address Longest Match

Route Entry	Prefix/Prefix Length	Does it match?
1	**2001:db8:c000::/40**	Match of 40 bits
2	**2001:db8:c000::/48**	Match of 48 bits (longest match)
3	**2001:db8:c000**:5555::/**64**	Does not match 64 bits

Build the Routing Table (14.1.6)

A routing table consists of prefixes and their prefix lengths. But how does the router learn about these networks? How does R1 in Figure 14-2 populate its routing table?

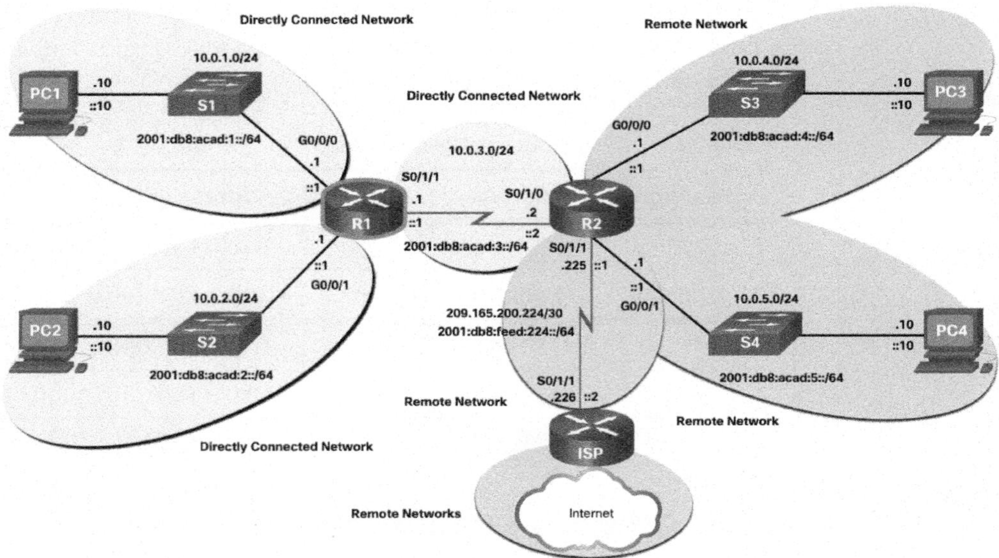

Figure 14-2 Routing Reference Topology

Directly Connected Networks

Directly connected networks are networks that are configured on the active interfaces of a router. A directly connected network is added to the routing table when a *directly connected interface* is configured with an IP address and subnet mask (prefix length) and is active (up and up).

Remote Networks

Remote networks are networks that are not directly connected to the router. Routers learn about remote networks in two ways:

- *Static routes*: Added to the routing table when a route is manually configured.
- *Dynamic routing protocols*: Added to the routing table when routing protocols dynamically learn about the remote network. Dynamic routing protocols include *Enhanced Interior Gateway Routing Protocol (EIGRP)*, *Open Shortest Path First (OSPF)*, as well as several others.

Default Route

A default route specifies a next-hop router to use when the routing table does not contain a specific route that matches the destination IP address. The default route can be entered manually as a static route or can be learned automatically from a dynamic routing protocol.

A default route over IPv4 has a route entry of 0.0.0.0/0 and a default route over IPv6 has a route entry of ::/0. The /0 prefix length indicates that zero bits or no bits need to match the destination IP address for this route entry to be used. If there are no routes with a longer match, more than 0 bits, the default route is used to forward the packet. The default route is sometimes referred to as a gateway of last resort.

Interactive Graphic

Check Your Understanding—Path Determination (14.1.7)

Refer to the online course to complete this activity.

Packet Forwarding (14.2)

In this section, you learn how a router makes packet forwarding decisions.

Packet Forwarding Decision Process (14.2.1)

Now that the router has determined the best path for a packet based on the longest match, it must determine how to encapsulate the packet and forward it out the correct egress interface.

Figure 14-3 demonstrates how a router first determines the best path, and then forwards the packet.

The following steps describe the packet forwarding process shown in the figure:

Step 1. The data link frame with an encapsulated IP packet arrives on the ingress interface.

Step 2. The router examines the destination IP address in the packet header and consults its IP routing table.

Step 3. The router finds the longest matching prefix in the routing table.

Step 4. The router encapsulates the packet in a data link frame and forwards it out the egress interface. The destination could be a device connected to the network or a next-hop router.

Step 5. However, if there is no matching route entry, the packet is dropped.

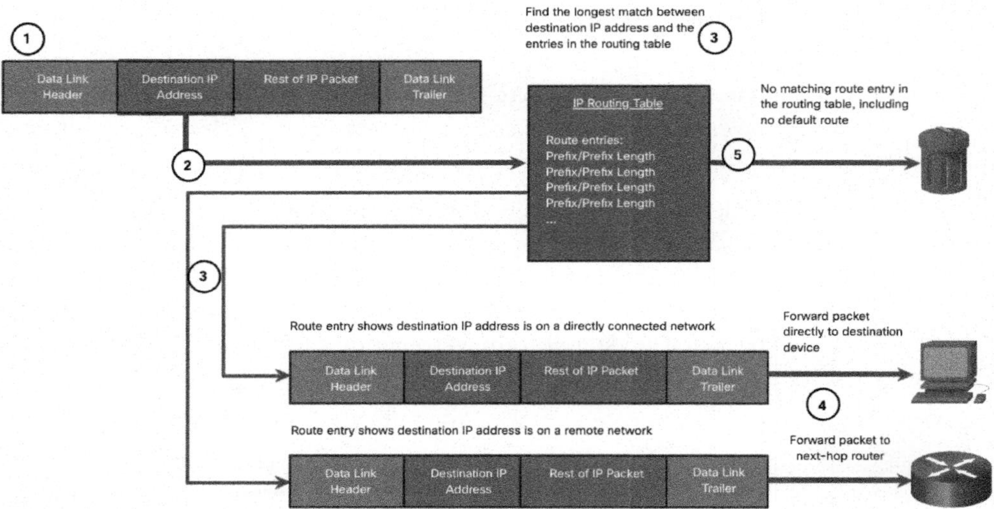

Figure 14-3 The Packet Forwarding Decision Process

Forwards the Packet to a Device on a Directly Connected Network

If the route entry indicates that the egress interface is a directly connected network, this means that the destination IP address of the packet belongs to a device on the directly connected network. Therefore, the packet can be forwarded directly to the destination device. The destination device is typically an end device on an Ethernet LAN, which means the packet must be encapsulated in an Ethernet frame.

To encapsulate the packet in the Ethernet frame, the router needs to determine the destination MAC address associated with the destination IP address of the packet. The process varies based on whether the packet is an IPv4 or IPv6 packet:

- **IPv4 packet:** The router checks its address resolution protocol (ARP) table for the destination IPv4 address and an associated Ethernet MAC address. If there is no match, the router sends an ARP Request. The destination device will return an ARP Reply with its MAC address. The router can now forward the IPv4 packet in an Ethernet frame with the proper destination MAC address.
- **IPv6 packet:** The router checks its neighbor cache for the destination IPv6 address and an associated Ethernet MAC address. If there is no match, the router sends an Internet Control Message Protocol version 6 (ICMPv6) Neighbor Solicitation (NS) message. The destination device will return an ICMPv6 Neighbor Advertisement (NA) message with its MAC address. The router can now forward the IPv6 packet in an Ethernet frame with the proper destination MAC address.

Forwards the Packet to a Next-Hop Router

If the route entry indicates that the destination IP address is on a remote network, this means the destination IP address of the packet belongs to a device on network that is not directly connected. Therefore, the packet must be forwarded to another router, specifically a next-hop router. The next-hop address is indicated in the route entry.

If the forwarding router and the next-hop router are on an Ethernet network, a similar process (ARP and ICMPv6 Neighbor Discovery) will occur for determining the destination MAC address of the packet as described previously. The difference is that the router will search for the IP address of the next-hop router in its ARP table or neighbor cache, instead of the destination IP address of the packet.

Note

This process will vary for other types of Layer 2 networks.

Drops the Packet—No Match in Routing Table

If there is no match between the destination IP address and a prefix in the routing table, and if there is no default route, the packet will be dropped.

End-to-End Packet Forwarding (14.2.2)

The primary responsibility of the packet forwarding function is to encapsulate packets in the appropriate data link frame type for the outgoing interface. For example, the data link frame format for a serial link could be *Point-to-Point protocol (PPP)*, *High-Level Data Link Control (HDLC)* protocol, or some other Layer 2 protocol.

Figures 14-4 through 14-7 and the accompanying text describe the forwarding processes used by the routers to forward a packet from PC1 to PC2.

PC1 Sends Packet to PC2

In Figure 14-4, PC1 sends a packet to PC2. Because PC2 is on a different network, PC1 will forward the packet to its default gateway. PC1 will look in its ARP cache for the MAC address of the default gateway and add the indicated frame information.

Note

If an ARP entry does not exist in the ARP table for the default gateway of 192.168.1.1, PC1 sends an ARP request. Router R1 would then return an ARP reply with its MAC address.

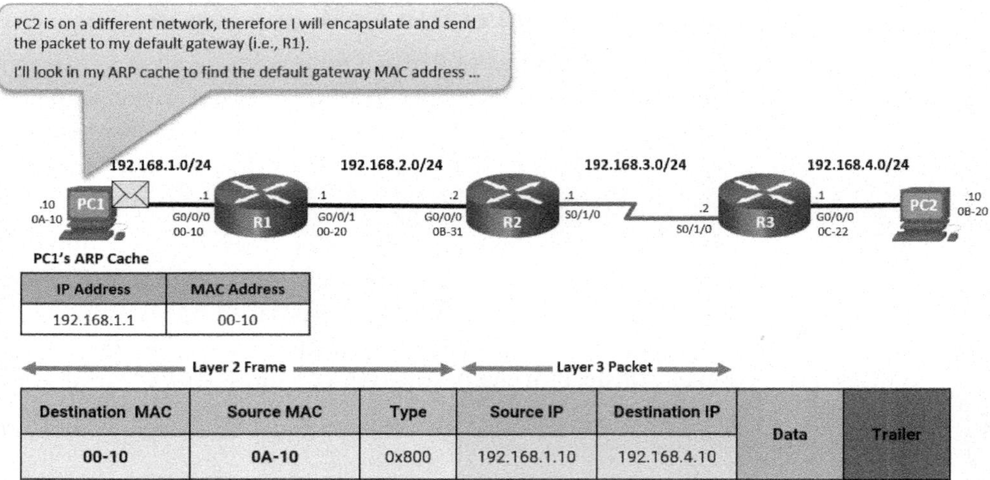

IP Address	MAC Address
192.168.1.1	00-10

Destination MAC	Source MAC	Type	Source IP	Destination IP	Data	Trailer
00-10	0A-10	0x800	192.168.1.10	192.168.4.10		

Figure 14-4 PC1 Sends Packet to PC2

R1 Forwards the Packet to PC2

R1 now forwards the packet to PC2. Because the exit interface is on an Ethernet network, R1 must resolve the next-hop IPv4 address with a destination MAC address using its ARP table, as shown in Figure 14-5. If an ARP entry does not exist in the ARP table for the next-hop interface of 192.168.2.2, R1 sends an ARP request. R2 would then return an ARP Reply.

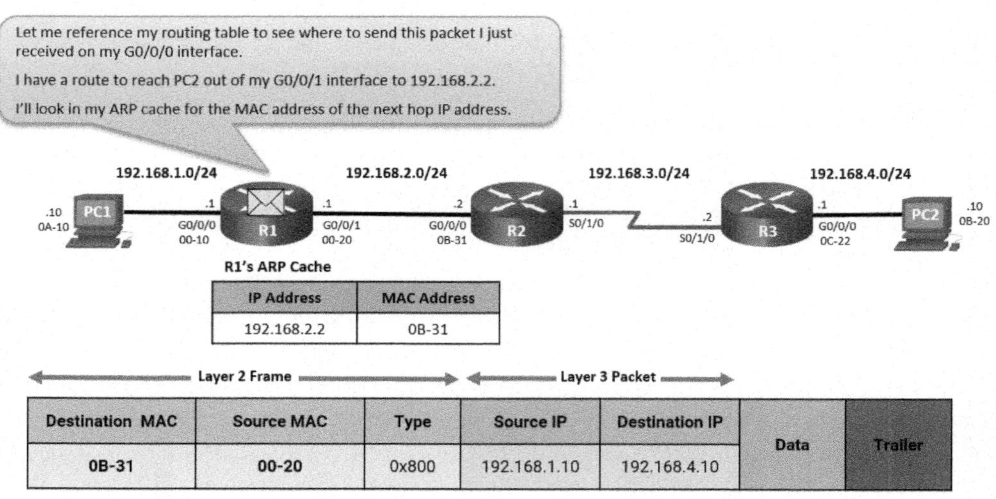

IP Address	MAC Address
192.168.2.2	0B-31

Destination MAC	Source MAC	Type	Source IP	Destination IP	Data	Trailer
0B-31	00-20	0x800	192.168.1.10	192.168.4.10		

Figure 14-5 R1 Forwards the Packet to PC2

R2 Forwards the Packet to R3

R2 now forwards the packet to R3. Because the *exit interface* is not an Ethernet network, R2 does not have to resolve the next-hop IPv4 address with a destination MAC address. When the interface is a point-to-point (P2P) serial connection, the router encapsulates the IPv4 packet into the proper data link frame format used by the exit interface (HDLC, PPP, and so on). Because there are no MAC addresses on serial interfaces, R2 sets the data link destination address to an equivalent of a broadcast, as shown in Figure 14-6.

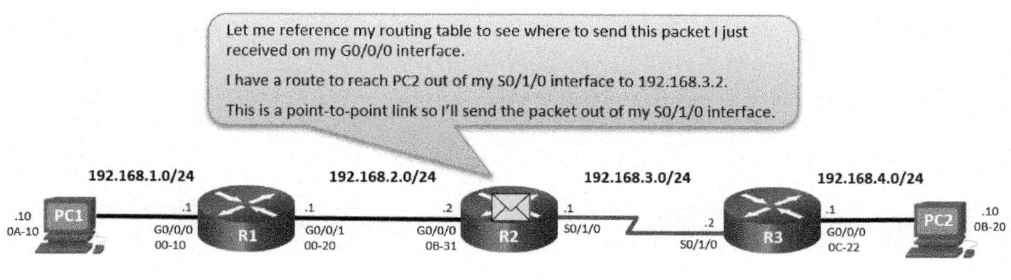

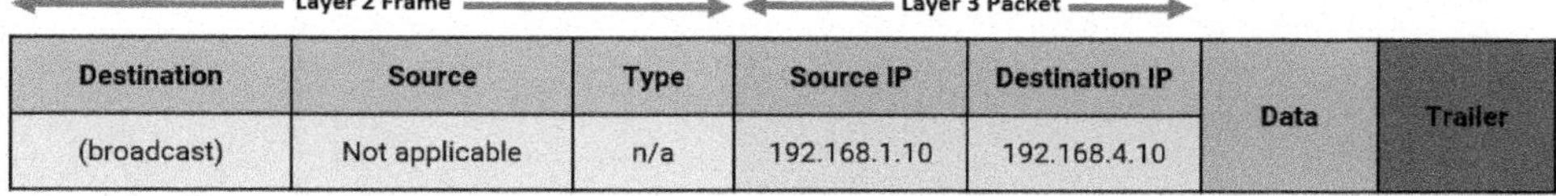
Layer 2 Frame / Layer 3 Packet

Destination	Source	Type	Source IP	Destination IP	Data	Trailer
(broadcast)	Not applicable	n/a	192.168.1.10	192.168.4.10		

Figure 14-6 R2 Forwards the Packet to R3

R3 Forwards the Packet to PC2

R3 now forwards the packet to PC2. Because the destination IPv4 address is on a directly connected Ethernet network, R3 must resolve the destination IPv4 address of the packet with its associated MAC address, as shown in Figure 14-7. If the entry is not in the ARP table, R3 sends an ARP request out of its FastEthernet 0/0 interface. PC2 would then return an ARP reply with its MAC address.

Packet Forwarding Mechanisms (14.2.3)

As mentioned previously, the primary responsibility of the packet forwarding function is to encapsulate packets in the appropriate data link frame type for the outgoing interface. The more efficiently a router can perform this task, the faster packets can be forwarded by the router. Routers support the following three packet forwarding mechanisms:

- *Process switching*
- *Fast switching*
- *Cisco Express Forwarding (CEF)*

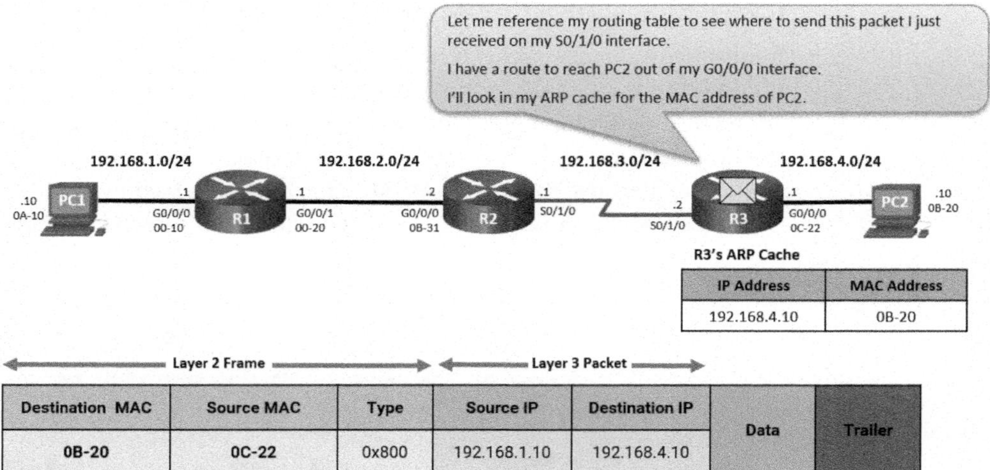

Figure 14-7 R3 Forwards the Packet to PC2

Assume that there is a traffic flow that consists of five packets. They are all going to the same destination. The following provides more information about these packet forwarding mechanisms.

Process Switching

Process switching is an older packet forwarding mechanism still available for Cisco routers. When a packet arrives on an interface, it is forwarded to the control plane where the CPU matches the destination address with an entry in its routing table, and then determines the exit interface and forwards the packet, as shown in Figure 14-8. It is important to understand that the router does this for every packet, even if the destination is the same for a stream of packets. This process-switching mechanism is very slow and is rarely implemented in modern networks. Contrast this with fast switching.

Fast Switching

Fast switching is another, older packet forwarding mechanism that was the successor to process switching. Fast switching uses a *fast-switching cache* to store next-hop information. When a packet arrives on an interface, it is forwarded to the control plane where the CPU searches for a match in the fast-switching cache. If it is not there, it is process-switched and forwarded to the exit interface. The flow information for the packet is also stored in the fast-switching cache. If another packet going to the same destination arrives on an interface, the next-hop information in the cache is reused without CPU intervention.

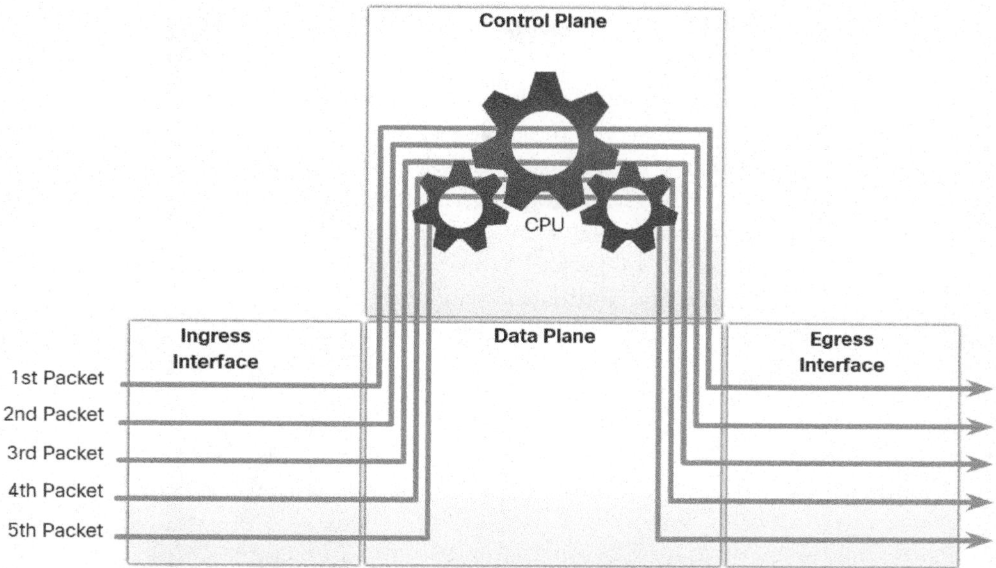

Figure 14-8 Process Switching

With fast switching in Figure 14-9, notice how only the first packet of a flow is process-switched and added to the fast-switching cache. The next four packets are quickly processed based on the information in the fast-switching cache.

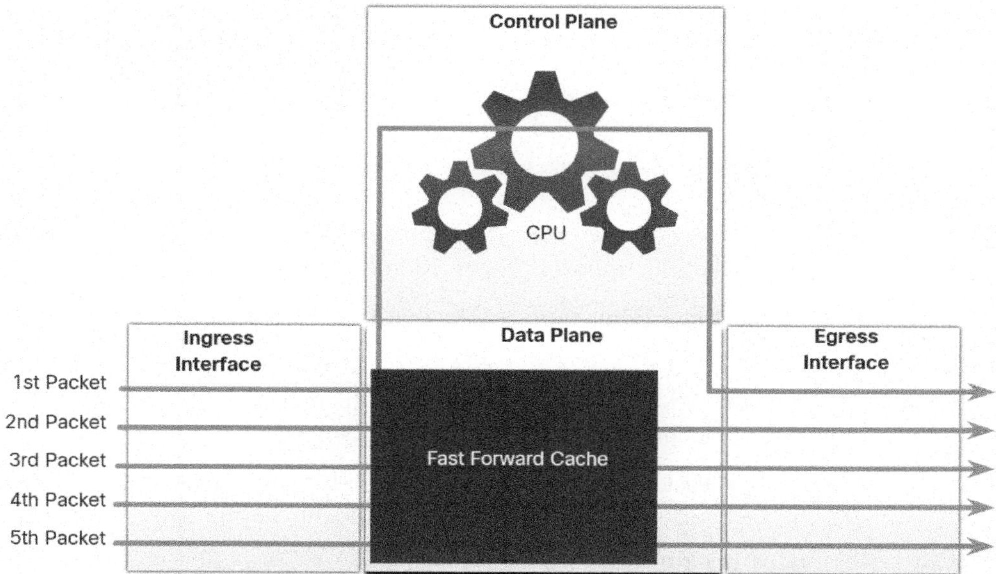

Figure 14-9 Fast Switching

Cisco Express Forwarding (CEF)

CEF is the most recent and default Cisco IOS packet-forwarding mechanism. Like fast switching, CEF builds a *Forwarding Information Base (FIB)* and an *adjacency table*. However, the table entries are not packet-triggered like fast switching, but change-triggered, such as when something changes in the network topology. Therefore, when a network has *converged*, the FIB and adjacency tables contain all the information that a router would have to consider when forwarding a packet, as shown in Figure 14-10. CEF is the fastest forwarding mechanism and the default on Cisco routers and multilayer switches.

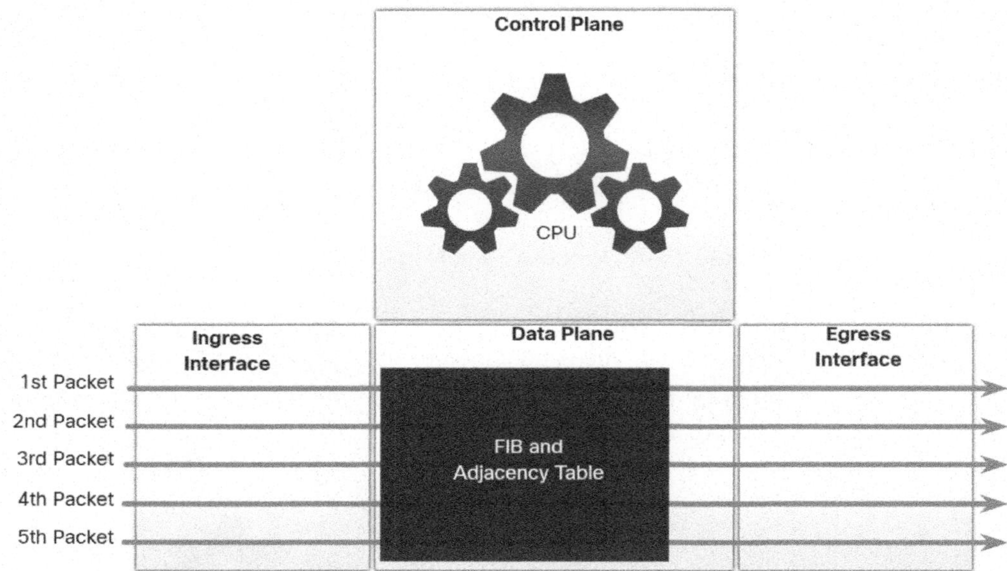

Figure 14-10 Cisco Express Forwarding (CEF)

CEF builds the FIB and adjacency tables after the network has converged. All five packets are quickly processed in the data plane.

A common analogy used to describe these three packet-forwarding mechanisms is as follows:

- Process switching solves a problem by doing math longhand, even if it is the identical problem that was just solved.
- Fast switching solves a problem by doing math longhand one time and remembering the answer for subsequent identical problems.
- CEF solves every possible problem ahead of time in a spreadsheet.

Interactive Graphic

Check Your Understanding—Packet Forwarding (14.2.4)

Refer to the online course to complete this activity.

Basic Router Configuration Review (14.3)

A router must be configured with specific settings before it can be deployed. New routers are not configured. They must be initially configured using the console port.

In this section, you learn how to configure basic settings on a router.

Topology (14.3.1)

A router creates a routing table to help it determine where to forward packets. But before diving into the details of the IP routing table, this section reviews basic router configuration and verification tasks. You will also complete a Packet Tracer activity to refresh your skills.

The topology in Figure 14-11 will be used for configuration and verification examples. It will also be used in the next section to discuss the IP routing table.

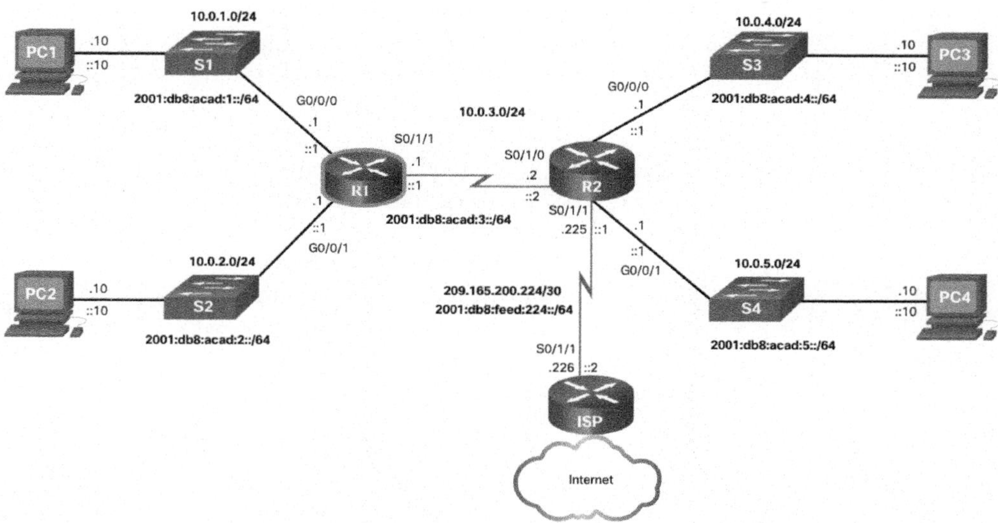

Figure 14-11 Reference Topology

Configuration Commands (14.3.2)

Example 14-1 shows the full configuration for R1.

Example 14-1 R1 Configuration

```
Router> enable
Router# configure terminal
Enter configuration commands, one per line. End with CNTL/Z.
Router(config)# hostname R1
R1(config)# enable secret class
R1(config)# line console 0
R1(config-line)# logging synchronous
R1(config-line)# password cisco
R1(config-line)# login
R1(config-line)# exit
R1(config)#
R1(config)# line vty 0 4
R1(config-line)# password cisco
R1(config-line)# login
R1(config-line)# transport input ssh telnet
R1(config-line)# exit
R1(config)#
R1(config)# service password-encryption
R1(config)# banner motd #
Enter TEXT message. End with a new line and the #
***********************************************
WARNING: Unauthorized access is prohibited!
***********************************************
#
R1(config)# ipv6 unicast-routing
R1(config)#
R1(config)# interface gigabitethernet 0/0/0
R1(config-if)# description Link to LAN 1
R1(config-if)# ip address 10.0.1.1 255.255.255.0
R1(config-if)# ipv6 address 2001:db8:acad:1::1/64
R1(config-if)# ipv6 address fe80::1:a link-local
R1(config-if)# no shutdown
R1(config-if)# exit
R1(config)#
R1(config)# interface gigabitethernet 0/0/1
R1(config-if)# description Link to LAN 2
R1(config-if)# ip address 10.0.2.1 255.255.255.0
R1(config-if)# ipv6 address 2001:db8:acad:2::1/64
R1(config-if)# ipv6 address fe80::1:b link-local
R1(config-if)# no shutdown
R1(config-if)# exit
R1(config)#
R1(config)# interface serial 0/1/1
R1(config-if)# description Link to R2
```

```
R1(config-if)# ip address 10.0.3.1 255.255.255.0
R1(config-if)# ipv6 address 2001:db8:acad:3::1/64
R1(config-if)# ipv6 address fe80::1:c link-local
R1(config-if)# no shutdown
R1(config-if)# exit
R1#
R1# copy running-config startup-config
Destination filename [startup-config]?
Building configuration...
[OK]
R1#
```

Verification Commands (14.3.3)

Common verification commands include the following:

- **show ip interface brief**
- **show running-config interface** *interface-type number*
- **show interfaces**
- **show ip interface**
- **show ip route**
- **ping**

In each case, replace **ip** with **ipv6** for the IPv6 version of the command. Refer back to Figure 14-11 for the topology used for the command output in Examples 14-2 through 14-10.

Example 14-2 The **show ip interface brief** Command

```
R1# show ip interface brief
Interface              IP-Address      OK? Method Status                Protocol
GigabitEthernet0/0/0   10.0.1.1        YES manual up                    up
GigabitEthernet0/0/1   10.0.2.1        YES manual up                    up
Serial0/1/0            unassigned      YES unset  administratively down down
Serial0/1/1            10.0.3.1        YES manual up                    up
GigabitEthernet0       unassigned      YES unset  down                  down
R1#
```

Example 14-3 The **show ipv6 interface brief** Command

```
R1# show ipv6 interface brief
GigabitEthernet0/0/0   [up/up]
    FE80::1:A
    2001:DB8:ACAD:1::1
```

```
GigabitEthernet0/0/1    [up/up]
    FE80::1:B
    2001:DB8:ACAD:2::1
Serial0/1/0             [administratively down/down]
    unassigned
Serial0/1/1             [up/up]
    FE80::1:C
    2001:DB8:ACAD:3::1
GigabitEthernet0        [down/down]
    unassigned
R1#
```

Example 14-4 The **show running-config interface** Command

```
R1# show running-config interface gigabitethernet 0/0/0
Building configuration...
Current configuration : 189 bytes
!
interface GigabitEthernet0/0/0
 description Link to LAN 1
 ip address 10.0.1.1 255.255.255.0
 negotiation auto
 ipv6 address FE80::1:A link-local
 ipv6 address 2001:DB8:ACAD:1::1/64
end
R1#
```

Example 14-5 The **show interfaces** Command

```
R1# show interfaces gigabitEthernet 0/0/0
GigabitEthernet0/0/0 is up, line protocol is up
  Hardware is ISR4321-2x1GE, address is a0e0.af0d.e140 (bia a0e0.af0d.e140)
  Internet address is 10.0.1.1/24
  MTU 1500 bytes, BW 100000 Kbit/sec, DLY 100 usec,
     reliability 255/255, txload 1/255, rxload 1/255
  Encapsulation ARPA, loopback not set
  Keepalive not supported
  Full Duplex, 100Mbps, link type is auto, media type is RJ45
  output flow-control is off, input flow-control is off
  ARP type: ARPA, ARP Timeout 04:00:00
  Last input 00:00:00, output 00:00:06, output hang never
  Last clearing of "show interface" counters never
  Input queue: 0/375/0/0 (size/max/drops/flushes); Total output drops: 0
  Queueing strategy: fifo
```

```
  Output queue: 0/40 (size/max)
  5 minute input rate 2000 bits/sec, 1 packets/sec
  5 minute output rate 0 bits/sec, 0 packets/sec
     57793 packets input, 10528767 bytes, 0 no buffer
     Received 19711 broadcasts (0 IP multicasts)
     0 runts, 0 giants, 0 throttles
     0 input errors, 0 CRC, 0 frame, 0 overrun, 0 ignored
     0 watchdog, 36766 multicast, 0 pause input
     10350 packets output, 1280030 bytes, 0 underruns
     0 output errors, 0 collisions, 1 interface resets
     0 unknown protocol drops
     0 babbles, 0 late collision, 0 deferred
     0 lost carrier, 0 no carrier, 0 pause output
     0 output buffer failures, 0 output buffers swapped out
R1#
```

Example 14-6 The **show ip interface** Command

```
R1# show ip interface gigabitethernet 0/0/0
GigabitEthernet0/0/0 is up, line protocol is up
  Internet address is 10.0.1.1/24
  Broadcast address is 255.255.255.255
  Address determined by setup command
  MTU is 1500 bytes
  Helper address is not set
  Directed broadcast forwarding is disabled
  Multicast reserved groups joined: 224.0.0.5 224.0.0.6
  Outgoing Common access list is not set
  Outgoing access list is not set
  Inbound Common access list is not set
  Inbound  access list is not set
  Proxy ARP is enabled
  Local Proxy ARP is disabled
  Security level is default
  Split horizon is enabled
  ICMP redirects are always sent
  ICMP unreachables are always sent
  ICMP mask replies are never sent
  IP fast switching is enabled
  IP Flow switching is disabled
  IP CEF switching is enabled
  IP CEF switching turbo vector
  IP Null turbo vector
  Associated unicast routing topologies:
        Topology "base", operation state is UP
```

```
  IP multicast fast switching is enabled
  IP multicast distributed fast switching is disabled
  IP route-cache flags are Fast, CEF
  Router Discovery is disabled
  IP output packet accounting is disabled
  IP access violation accounting is disabled
  TCP/IP header compression is disabled
  RTP/IP header compression is disabled
  Probe proxy name replies are disabled
  Policy routing is disabled
  Network address translation is disabled
  BGP Policy Mapping is disabled
  Input features: MCI Check
  IPv4 WCCP Redirect outbound is disabled
  IPv4 WCCP Redirect inbound is disabled
  IPv4 WCCP Redirect exclude is disabled
R1#
```

Example 14-7 The **show ipv6 interface** Command

```
R1# show ipv6 interface gigabitethernet 0/0/0
GigabitEthernet0/0/0 is up, line protocol is up
  IPv6 is enabled, link-local address is FE80::1:A
  No Virtual link-local address(es):
  Global unicast address(es):
    2001:DB8:ACAD:1::1, subnet is 2001:DB8:ACAD:1::/64
  Joined group address(es):
    FF02::1
    FF02::2
    FF02::5
    FF02::6
    FF02::1:FF00:1
    FF02::1:FF01:A
  MTU is 1500 bytes
  ICMP error messages limited to one every 100 milliseconds
  ICMP redirects are enabled
  ICMP unreachables are sent
  ND DAD is enabled, number of DAD attempts: 1
  ND reachable time is 30000 milliseconds (using 30000)
  ND advertised reachable time is 0 (unspecified)
  ND advertised retransmit interval is 0 (unspecified)
  ND router advertisements are sent every 200 seconds
  ND router advertisements live for 1800 seconds
  ND advertised default router preference is Medium
  Hosts use stateless autoconfig for addresses.
R1#
```

Example 14-8 The **show ip route** Command

```
R1# show ip route | begin Gateway
Gateway of last resort is not set
      10.0.0.0/8 is variably subnetted, 6 subnets, 2 masks
C        10.0.1.0/24 is directly connected, GigabitEthernet0/0/0
L        10.0.1.1/32 is directly connected, GigabitEthernet0/0/0
C        10.0.2.0/24 is directly connected, GigabitEthernet0/0/1
L        10.0.2.1/32 is directly connected, GigabitEthernet0/0/1
C        10.0.3.0/24 is directly connected, Serial0/1/1
L        10.0.3.1/32 is directly connected, Serial0/1/1
R1#
```

Example 14-9 The **show ipv6 route** Command

```
R1# show ipv6 route
 (Output omitted)
C   2001:DB8:ACAD:1::/64 [0/0]
     via GigabitEthernet0/0/0, directly connected
L   2001:DB8:ACAD:1::1/128 [0/0]
     via GigabitEthernet0/0/0, receive
C   2001:DB8:ACAD:2::/64 [0/0]
     via GigabitEthernet0/0/1, directly connected
L   2001:DB8:ACAD:2::1/128 [0/0]
     via GigabitEthernet0/0/1, receive
C   2001:DB8:ACAD:3::/64 [0/0]
     via Serial0/1/1, directly connected
L   2001:DB8:ACAD:3::1/128 [0/0]
     via Serial0/1/1, receive
L   FF00::/8 [0/0]
     via Null0, receive
R1#
```

Example 14-10 The **ping** Command

```
R1# ping 10.0.3.2
Type escape sequence to abort.
Sending 5, 100-byte ICMP Echos to 10.0.3.2, timeout is 2 seconds:
!!!!!
Success rate is 100 percent (5/5), round-trip min/avg/max = 2/2/2 ms
R1# ping 2001:db8:acad:3::2
Type escape sequence to abort.
Sending 5, 100-byte ICMP Echos to 2001:DB8:ACAD:3::2, timeout is 2 seconds:
!!!!!
Success rate is 100 percent (5/5), round-trip min/avg/max = 2/2/2 ms
R1#
```

Filter Command Output (14.3.4)

Another useful feature that improves user experience in the command-line interface (CLI) is filtering **show** output. Filtering commands can be used to display specific sections of output. To enable the filtering command, enter a pipe (|) character after the **show** command and then enter a filtering parameter and a filtering expression.

The filtering parameters that can be configured after the pipe include the following:

- **section:** This displays the entire section that starts with the filtering expression.
- **include:** This includes all output lines that match the filtering expression.
- **exclude:** This excludes all output lines that match the filtering expression.
- **begin:** This displays all the output lines from a certain point, starting with the line that matches the filtering expression.

Note

Output filters can be used in combination with any **show** command.

Example 14-11 demonstrate some of the more common uses of filtering parameters.

Example 14-11 Filtering Output Examples

```
R1# show running-config | section line vty
line vty 0 4
 password 7 121A0C0411044C
 login
 transport input telnet ssh
R1#
R1# show ipv6 interface brief | include up
GigabitEthernet0/0/0    [up/up]
GigabitEthernet0/0/1    [up/up]
Serial0/1/1             [up/up]
R1#
R1# show ip interface brief | exclude unassigned
Interface               IP-Address      OK? Method Status                Protocol
GigabitEthernet0/0/0    192.168.10.1    YES manual up                    up
GigabitEthernet0/0/1    192.168.11.1    YES manual up                    up
Serial0/1/1             209.165.200.225 YES manual up                    up
R1#
R1# show ip route | begin Gateway
Gateway of last resort is not set
      192.168.10.0/24 is variably subnetted, 2 subnets, 2 masks
C        192.168.10.0/24 is directly connected, GigabitEthernet0/0/0
```

```
L        192.168.10.1/32 is directly connected, GigabitEthernet0/0/0
      192.168.11.0/24 is variably subnetted, 2 subnets, 2 masks
C        192.168.11.0/24 is directly connected, GigabitEthernet0/0/1
L        192.168.11.1/32 is directly connected, GigabitEthernet0/0/1
      209.165.200.0/24 is variably subnetted, 2 subnets, 2 masks
C        209.165.200.224/30 is directly connected, Serial0/1/1
L        209.165.200.225/32 is directly connected, Serial0/1/1
R1#
```

Packet Tracer
☐ Activity

Packet Tracer—Basic Router Configuration Review (14.3.5)

Routers R1 and R2 each have two LANs. R1 is already configured. Your task is to configure the appropriate addressing for R2 and verify connectivity between the LANs.

IP Routing Table (14.4)

The routing table keeps track of all the available networks that a router can forward packets to. In this section, you learn how the routing table is used to make forwarding decisions.

Route Sources (14.4.1)

How does a router know where it can send packets? It creates a routing table that is based on the network in which it is located.

A routing table contains a list of routes to known networks (prefixes and prefix lengths). The source of this information is derived from the following:

- Directly connected networks
- Static routes
- Dynamic routing protocols

In the previous Figure 14-11, R1 and R2 are using the dynamic routing protocol OSPF to share routing information. In addition, R2 is configured with a *default static route* to the ISP.

The full routing table for each router after directly connected networks, static routing, and dynamic routing is configured is shown in Examples 14-12 and 14-13. The rest of this section will demonstrate how these tables are populated.

Example 14-12 R1 Routing Table

```
R1# show ip route | begin Gateway
Gateway of last resort is 10.0.3.2 to network 0.0.0.0
O*E2  0.0.0.0/0 [110/1] via 10.0.3.2, 00:51:34, Serial0/1/1
      10.0.0.0/8 is variably subnetted, 8 subnets, 2 masks
C        10.0.1.0/24 is directly connected, GigabitEthernet0/0/0
L        10.0.1.1/32 is directly connected, GigabitEthernet0/0/0
C        10.0.2.0/24 is directly connected, GigabitEthernet0/0/1
L        10.0.2.1/32 is directly connected, GigabitEthernet0/0/1
C        10.0.3.0/24 is directly connected, Serial0/1/1
L        10.0.3.1/32 is directly connected, Serial0/1/1
O        10.0.4.0/24 [110/50] via 10.0.3.2, 00:24:22, Serial0/1/1
O        10.0.5.0/24 [110/50] via 10.0.3.2, 00:24:15, Serial0/1/1
R1#
```

Example 14-13 R2 Routing Table

```
R2# show ip route | begin Gateway
Gateway of last resort is 209.165.200.226 to network 0.0.0.0
S*    0.0.0.0/0 [1/0] via 209.165.200.226
      10.0.0.0/8 is variably subnetted, 8 subnets, 2 masks
O        10.0.1.0/24 [110/65] via 10.0.3.1, 00:31:38, Serial0/1/0
O        10.0.2.0/24 [110/65] via 10.0.3.1, 00:31:38, Serial0/1/0
C        10.0.3.0/24 is directly connected, Serial0/1/0
L        10.0.3.2/32 is directly connected, Serial0/1/0
C        10.0.4.0/24 is directly connected, GigabitEthernet0/0/0
L        10.0.4.1/32 is directly connected, GigabitEthernet0/0/0
C        10.0.5.0/24 is directly connected, GigabitEthernet0/0/1
L        10.0.5.1/32 is directly connected, GigabitEthernet0/0/1
      209.165.200.0/24 is variably subnetted, 2 subnets, 2 masks
C        209.165.200.224/30 is directly connected, Serial0/1/1
L        209.165.200.225/32 is directly connected, Serial0/1/1
R2#
```

In the routing tables for R1 and R2, notice that the sources for each route are identified by a code. The code identifies how the route was learned. For instance, common codes include the following:

- L: Identifies the address assigned to a router interface. This allows the router to efficiently determine when it receives a packet for the interface instead of being forwarded.
- C: Identifies a directly connected network.
- S: Identifies a static route created to reach a specific network.

- O: Identifies a dynamically learned network from another router using the OSPF routing protocol.
- * : This route is a candidate for a default route.

Routing Table Principles (14.4.2)

There are three routing table principles as described in Table 14-3. These are issues that are addressed by the proper configuration of dynamic routing protocols or static routes on all the routers between the source and destination devices.

Table 14-3 Routing Table Principles

Routing Table Principle	Example
Every router makes its decision alone, based on the information it has in its own routing table.	■ R1 can only forward packets using its own routing table. ■ R1 does not know what routes are in the routing tables of other routers (for example, R2).
The information in a routing table of one router does not necessarily match the routing table of another router.	■ Just because R1 has route in its routing table to a network in the Internet via R2, that does not mean that R2 knows about that same network.
Routing information about a path does not provide return routing information.	■ R1 receives a packet with the destination IP address of PC1 and the source IP address of PC3. ■ Just because R1 knows to forward the packet out its G0/0/0 interface doesn't necessarily mean that it knows how to forward packets originating from PC1 back to the remote network of PC3.

Routing Table Entries (14.4.3)

As a network administrator, it is imperative to know how to interpret the content of IPv4 and IPv6 routing tables. Figure 14-12 displays IPv4 and IPv6 routing table entries on R1 for the route to remote network 10.0.4.0/24 and 2001:db8:acad:4::/64. Both these routes were learned dynamically from the OSPF routing protocol.

In the figure, the numbers identify the following information:

1. **Route source:** This identifies how the route was learned.
2. **Destination network (prefix and prefix length):** This identifies the address of the remote network.
3. *Administrative distance (AD)*: This identifies the trustworthiness of the route source. Lower values indicate preferred route source.

IPv4 Routing Table

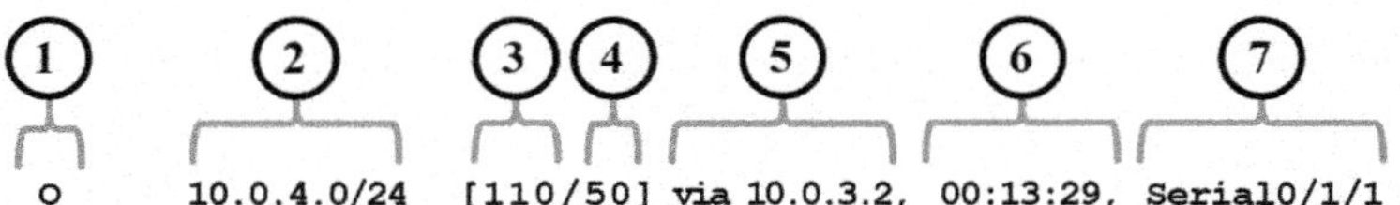

IPv6 Routing Table

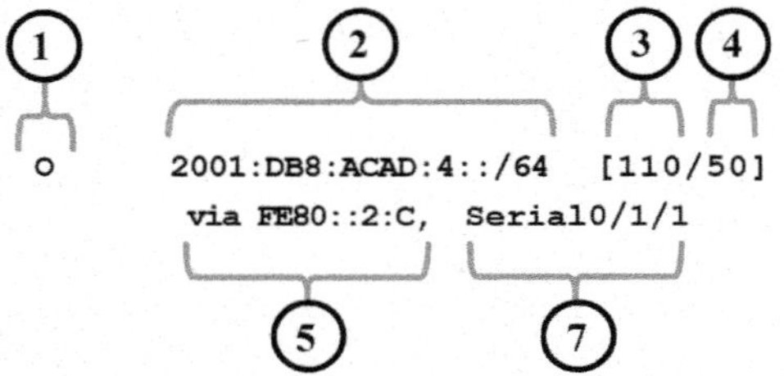

Figure 14-12 IPv4 and IPv6 Routing Table Entries

4. *Metric*: This identifies the value assigned to reach the remote network. Lower values indicate preferred routes.
5. *Next-hop IP address*: This identifies the IP address of the next router to which the packet would be forwarded.
6. **Route timestamp:** This identifies how much time has passed since the route was learned.
7. **Exit interface:** This identifies the egress interface to use for outgoing packets to reach their final destination.

Note

The prefix length of the destination network specifies the minimum number of far-left bits that must match between the IP address of the packet and the destination network (prefix) for this route to be used.

Directly Connected Networks (14.4.4)

Before a router can learn about any remote networks, it must have at least one active interface configured with an IP address and subnet mask (prefix length). This is known as a directly connected network or a directly connected route. Routers add a directly connected route to its routing table when an interface is configured with an IP address and is activated.

A directly connected network is denoted by a status code of C in the routing table. The route contains a network prefix and prefix length.

The routing table also contains a *local route interface* for each of its directly connected networks, indicated by the status code of L. This is the IP address that is assigned to the interface on that directly connected network. For IPv4 local routes, the prefix length is /32, and for IPv6 local routes, the prefix length is /128. This means the destination IP address of the packet must match all the bits in the local route for this route to be a match. The purpose of the local route is to efficiently determine when it receives a packet for the interface instead of a packet that needs to be forwarded.

Directly connected networks and local routes are shown in Example 14-14.

Example 14-14 Directly Connected IPv4 and IPv6 Networks on R1

```
R1# show ip route
Codes: L - local, C - connected, S - static, R - RIP, M - mobile, B - BGP
(Output omitted)
C         10.0.1.0/24 is directly connected, GigabitEthernet0/0/0
L         10.0.1.1/32 is directly connected, GigabitEthernet0/0/0
R1#
R1# show ipv6 route
IPv6 Routing Table - default - 10 entries
Codes: C - Connected, L - Local, S - Static, U - Per-user Static route
(Output omitted)

C    2001:DB8:ACAD:1::/64 [0/0]
      via GigabitEthernet0/0/0, directly connected
L    2001:DB8:ACAD:1::1/128 [0/0]
      via GigabitEthernet0/0/0, receive
R1#
```

Static Routes (14.4.5)

After directly connected interfaces are configured and added to the routing table, static or dynamic routing can be implemented for accessing remote networks.

Static routes are manually configured. They define an explicit path between two networking devices. Unlike a dynamic routing protocol, static routes are not automatically updated and must be manually reconfigured if the network topology changes. The benefits of using static routes include improved security and resource efficiency. Static routes use less bandwidth than dynamic routing protocols, and no CPU cycles are used to calculate and communicate routes. The main disadvantage to using static routes is the lack of automatic reconfiguration if the network topology changes.

Static routing has three primary uses:

- It provides ease of routing table maintenance in smaller networks that are not expected to grow significantly.
- It uses a single default route to represent a path to any network that does not have a more specific match with another route in the routing table. Default routes are used to send traffic to any destination beyond the next upstream router.
- It routes to and from *stub networks*. A stub network is a network accessed by a single route, and the router has only one *neighbor*.

Figure 14-13 shows an example of stub networks. Notice that any network attached to R1 would have only one way to reach other destinations, whether to networks attached to R2, or to destinations beyond R2. This means that networks 10.0.1.0/24 and 10.0.2.0/24 are stub networks, and R1 is a *stub router*.

In this example, a static route can be configured on R2 to reach the R1 networks. Additionally, because R1 has only one way to send out nonlocal traffic, a default static route can be configured on R1 to point to R2 as the next hop for all other networks.

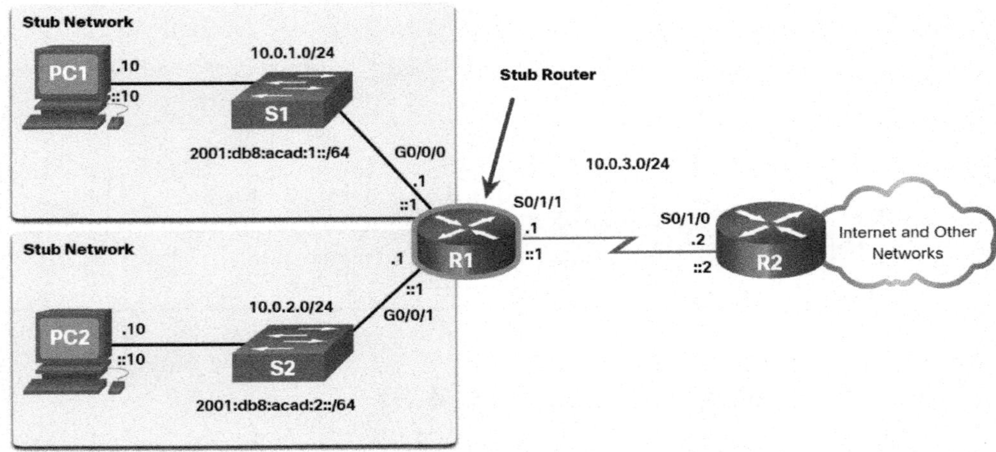

Figure 14-13 Stub Networks Attached to R1

Static Routes in the IP Routing Table (14.4.6)

For demonstrating static routing, the topology in Figure 14-14 is simplified to show only one LAN attached to each router. The figure shows IPv4 and IPv6 static routes configured on R1 to reach the 10.0.4.0/24 and 2001:db8:acad:4::/64 networks on

R2. The configuration commands are for demonstration only and are discussed in another module.

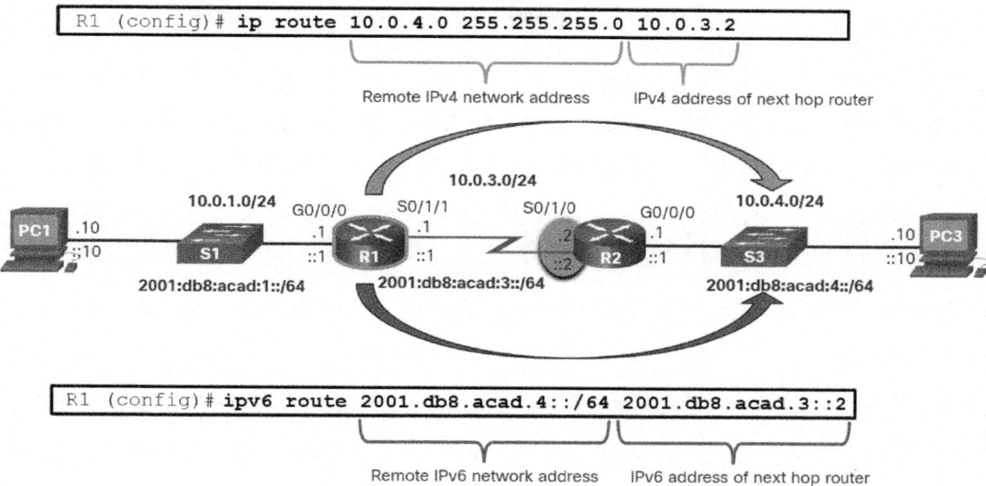

Figure 14-14 IPv4 and IPv6 Static Routes Configured on R1

The output in Example 14-15 shows the IPv4 and IPv6 static routing entries on R1 that can reach the 10.0.4.0/24 and 2001:db8:acad:4::/64 networks on R2. Notice that both routing entries use the status code of S, indicating that the route was learned by a static route. Both entries also include the IP address of the next hop router, via *ip-address*. The **static** parameter at the end of the command displays only static routes.

Example 14-15 Static IPv4 and IPv6 Routes on R1

```
R1# show ip route static
Codes: L - local, C - connected, S - static, R - RIP, M - mobile, B - BGP
(output omitted)

      10.0.0.0/8 is variably subnetted, 8 subnets, 2 masks
S        10.0.4.0/24 [1/0] via 10.0.3.2
R1# show ipv6 route static
IPv6 Routing Table - default - 8 entries
Codes: C - Connected, L - Local, S - Static, U - Per-user Static route
(output omitted)

S   2001:DB8:ACAD:4::/64 [1/0]
     via 2001:DB8:ACAD:3::2
```

Dynamic Routing Protocols (14.4.7)

Dynamic routing protocols are used by routers to automatically share information about the reachability and status of remote networks. Dynamic routing protocols perform several activities, including network discovery and maintaining routing tables.

Important advantages of dynamic routing protocols are the ability to select a best path and the ability to automatically discover a new best path when there is a change in the topology.

Network discovery is the ability of a routing protocol to share information about the networks that it knows about with other routers that are also using the same routing protocol. Instead of depending on manually configured static routes to remote networks on every router, a dynamic routing protocol allows the routers to automatically learn about these networks from other routers. These networks, and the best path to each, are added to the routing table of the router and identified as a network learned by a specific dynamic routing protocol.

Figure 14-15 shows routers R1 and R2 using a common routing protocol to share network information.

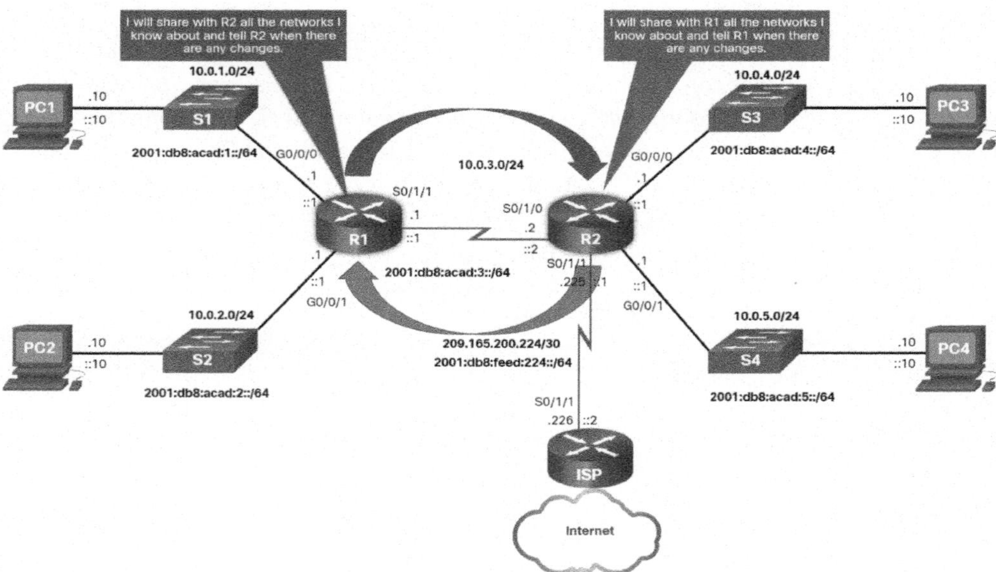

Figure 14-15 Dynamic Routing Between R1 and R2

A previous example used static routes to the 10.0.4.0/24 and 2001:db8:acad:4::/64 networks. These static routes are no longer configured, and OSPF is now being used to dynamically learn all the networks connected to R1 and R2. Example 14-16 shows the IPv4 and IPv6 OSPF routing entries on R1 that can reach these networks on R2.

Example 14-16 Dynamic IPv4 and IPv6 Routes on R1

```
R1# show ip route
Codes: L - local, C - connected, S - static, R - RIP, M - mobile, B - BGP
       D - EIGRP, EX - EIGRP external, O - OSPF, IA - OSPF inter area
(output omitted for brevity)
O        10.0.4.0/24 [110/50] via 10.0.3.2, 00:24:22, Serial0/1/1
O        10.0.5.0/24 [110/50] via 10.0.3.2, 00:24:15, Serial0/1/1
R1#
R1# show ipv6 route
IPv6 Routing Table - default - 10 entries
(Output omitted)
       NDr - Redirect, RL - RPL, O - OSPF Intra, OI - OSPF Inter
O   2001:DB8:ACAD:4::/64 [110/50]
     via FE80::2:C, Serial0/1/1
O   2001:DB8:ACAD:5::/64 [110/50]
     via FE80::2:C, Serial0/1/1
```

Notice that both routing entries use the status code of **O** to indicate the route was learned by the OSPF routing protocol. Both entries also include the IP address of the *next-hop router*, via *ip-address*.

Note

IPv6 routing protocols use the link-local address of the next-hop router.

Note

OSPF routing configuration for IPv4 and IPv6 is beyond the scope of this course.

Default Route (14.4.9)

A default route is similar to a default gateway on a host. The default route specifies a next-hop router to use when the routing table does not contain a specific route that matches the destination IP address.

A default route can be either a static route or learned automatically from a dynamic routing protocol. A default route has an IPv4 route entry of 0.0.0.0/0 or an IPv6 route entry of ::/0. This means that zero or no bits need to match between the destination IP address and the default route.

Most enterprise routers have a default route in their routing table. This is to reduce the number of routes in a routing table.

A router, such as a home or small office router that only has one LAN, may reach all its remote networks through a default route. This is useful when the router has only directly connected networks and one exit point to a service provider router.

In Figure 14-16, routers R1 and R2 are using OSPF to share routing information about their own networks (10.0.x.x/24 and 2001:db8:acad:x::/64 networks). R2 has a static default route to the ISP router. R2 will forward any packets with a destination IP address that does not specifically match one of the networks in its routing table to the ISP router. This would include all packets destined for the Internet.

R2 has shared its default route with R1 using OSPF. R1 will now have a default route in its routing table that it learned dynamically from OSPF. R1 will also forward any packets with a destination IP address that does not specifically match one of the networks in its routing table to R2.

Example 14-17 shows the IPv4 and IPv6 routing table entries for the static default routes configured on R2.

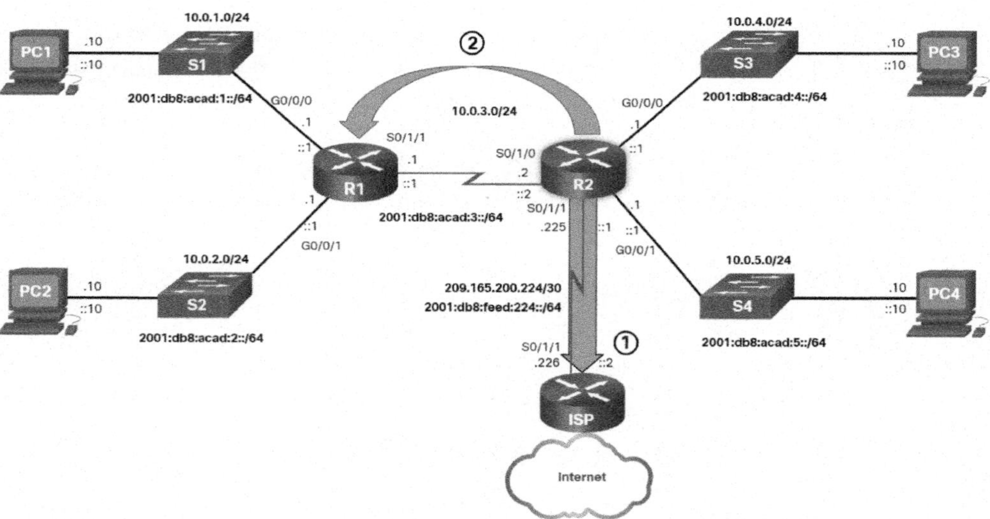

Figure 14-16 Default Routing to ISP

Example 14-17 Default IPv4 and IPv6 Routes on R1

```
R2# show ip route static
S*     0.0.0.0/0 [1/0] via 209.165.200.226
R2#
R2# show ipv6 route static
S    ::/0 [1/0]
       via 2001:DB8:FEED:224::2
R2#
```

Structure of an IPv4 Routing Table (14.4.10)

IPv4 was standardized in the early 1980s using the now obsolete classful addressing architecture. The IPv4 routing table is organized using this same classful structure. In the **show ip route** output in Example 14-18, notice that some route entries are left justified while others are indented. This is based on how the routing process searches the IPv4 routing table for the longest match. This was all because of classful addressing. Although the *route lookup process* no longer uses classes, the structure of the IPv4 routing table still retains this format.

Example 14-18 Obsolete Routing Table with Classful Addressing Architecture

```
Router# show ip route
(Output omitted)
      192.168.1.0/24 is variably subnetted, 2 subnets, 2 masks
C        192.168.1.0/24 is directly connected, GigabitEthernet0/0
L        192.168.1.1/32 is directly connected, GigabitEthernet0/0
O     192.168.2.0/24 [110/65] via 192.168.12.2, 00:32:33, Serial0/0/0
O     192.168.3.0/24 [110/65] via 192.168.13.2, 00:31:48, Serial0/0/1
      192.168.12.0/24 is variably subnetted, 2 subnets, 2 masks
C        192.168.12.0/30 is directly connected, Serial0/0/0
L        192.168.12.1/32 is directly connected, Serial0/0/0
      192.168.13.0/24 is variably subnetted, 2 subnets, 2 masks
C        192.168.13.0/30 is directly connected, Serial0/0/1
L        192.168.13.1/32 is directly connected, Serial0/0/1
      192.168.23.0/30 is subnetted, 1 subnets
O        192.168.23.0/30 [110/128] via 192.168.12.2, 00:31:38, Serial0/0/0
Router#
```

Note

The IPv4 routing table in the example is not from any router in the topology used in this module.

Although the details of the structure are beyond the scope of this module, it is helpful to recognize the structure of the table. An indented entry is known as a *child route*. A route entry is indented if it is the subnet of a classful address (class A, B or C network). Directly connected networks will always be indented (child routes) because the local address of the interface is always entered in the routing table as a /32. The child route will include the route source and all the forwarding information such as the next-hop address. The classful network address of this subnet will be shown above the route entry, less indented, and without a source code. That route is known as a *parent route*.

Note

This is just a brief introduction to the structure of an IPv4 routing table and does not cover details or specifics of this architecture.

Example 14-19 shows the IPv4 routing table of R1 in the topology. Notice that all the networks in the topology are subnets, which are child routes, of the class A network and parent route 10.0.0.0/8.

Example 14-19 Child Routes in the R1 Routing Table

```
R1# show ip route
(output omitted for brevity)
O*E2   0.0.0.0/0 [110/1] via 10.0.3.2, 00:51:34, Serial0/1/1
       10.0.0.0/8 is variably subnetted, 8 subnets, 2 masks
C         10.0.1.0/24 is directly connected, GigabitEthernet0/0/0
L         10.0.1.1/32 is directly connected, GigabitEthernet0/0/0
C         10.0.2.0/24 is directly connected, GigabitEthernet0/0/1
L         10.0.2.1/32 is directly connected, GigabitEthernet0/0/1
C         10.0.3.0/24 is directly connected, Serial0/1/1
L         10.0.3.1/32 is directly connected, Serial0/1/1
O         10.0.4.0/24 [110/50] via 10.0.3.2, 00:24:22, Serial0/1/1
O         10.0.5.0/24 [110/50] via 10.0.3.2, 00:24:15, Serial0/1/1
R1#
```

Structure of an IPv6 Routing Table (14.4.11)

The concept of classful addressing was never part of IPv6, so the structure of an IPv6 routing table is very straightforward, as shown in Example 14-20. Every IPv6 route entry is formatted and aligned the same way.

Example 14-20 The IPv6 Routing Table Structure

```
R1# show ipv6 route
(output omitted for brevity)
OE2 ::/0 [110/1], tag 2
     via FE80::2:C, Serial0/0/1
C    2001:DB8:ACAD:1::/64 [0/0]
     via GigabitEthernet0/0/0, directly connected
L    2001:DB8:ACAD:1::1/128 [0/0]
     via GigabitEthernet0/0/0, receive
C    2001:DB8:ACAD:2::/64 [0/0]
     via GigabitEthernet0/0/1, directly connected
L    2001:DB8:ACAD:2::1/128 [0/0]
     via GigabitEthernet0/0/1, receive
```

```
C     2001:DB8:ACAD:3::/64 [0/0]
       via Serial0/1/1, directly connected
L     2001:DB8:ACAD:3::1/128 [0/0]
       via Serial0/1/1, receive
O     2001:DB8:ACAD:4::/64 [110/50]
       via FE80::2:C, Serial0/1/1
O     2001:DB8:ACAD:5::/64 [110/50]
       via FE80::2:C, Serial0/1/1
L     FF00::/8 [0/0]
       via Null0, receive
R1#
```

Administrative Distance (14.4.12)

A route entry for a specific network address (prefix and prefix length) can appear only once in the routing table. However, it is possible that the routing table learns about the same network address from more than one routing source.

Except for very specific circumstances, only one dynamic routing protocol should be implemented on a router. However, it is possible to configure both OSPF and EIGRP on a router, and both routing protocols may learn of the same destination network. Each routing protocol may decide on a different path to reach the destination based on the metric of that routing protocol.

This raises a few questions, such as the following:

- How does the router know which source to use?
- Which route should it install in the routing table? The route learned from OSPF or the route learned from EIGRP?

Cisco IOS uses what is known as the administrative distance (AD) to determine the route to install into the IP routing table. The AD represents the "trustworthiness" of the route. The lower the AD, the more trustworthy the route source. Because EIGRP has an AD of 90 and OSPF has an AD of 110, the EIGRP route entry would be installed in the routing table.

Note

The AD does not necessarily represent which dynamic routing protocol is best.

A more common example is a router learning the same network address from a static route and a dynamic routing protocol, such as OSPF. A static route has an AD of 1, whereas an OSPF-discovered route has an AD of 110. Given two separate route sources to the same destination, the router chooses to install the route with the lowest AD. When a router has the choice of a static route and an OSPF route, the static route takes precedence.

Note

Directly connected networks have the lowest AD of 0. Only a directly connected network can have an AD of 0.

Table 14-4 lists various routing protocols and their associated ADs.

Table 14-4 Administrative Distance Values

Route Source	Administrative Distance
Directly connected	0
Static route	1
EIGRP summary route	5
External BGP	20
Internal EIGRP	90
OSPF	110
IS-IS	115
RIP	120
External EIGRP	170
Internal BGP	200

Interactive Graphic

Check Your Understanding—IP Routing Table (14.4.13)

Refer to the online course to complete this activity.

Static and Dynamic Routing (14.5)

In this section, you learn the differences between static routes and dynamic routing protocols.

Static or Dynamic? (14.5.1)

The previous section discussed the ways that a router creates its routing table. So, you now know that routing, like IP addressing, can be either static or dynamic. Should you use static or dynamic routing? The answer is both! Static and dynamic routing are not mutually exclusive. Rather, most networks use a combination of dynamic routing protocols and static routes.

Static Routes

Static routes are commonly used in the following scenarios:

- As a default route forwarding packets to a service provider
- For routes outside the routing domain and not learned by the dynamic routing protocol
- When the network administrator wants to explicitly define the path for a specific network
- For routing between stub networks

Static routes are useful for smaller networks with only one path to an outside network. They also provide security in a larger network for certain types of traffic, or links to other networks that need more control.

Dynamic Routing Protocols

Dynamic routing protocols help the network administrator manage the time-consuming and exacting process of configuring and maintaining static routes. Dynamic routing protocols are implemented in any type of network consisting of more than just a few routers. Dynamic routing protocols are scalable and automatically determine better routes if there is a change in the topology.

Dynamic routing protocols are commonly used in the following scenarios:

- In networks consisting of more than just a few routers.
- When a change in the network topology requires the network to automatically determine another path.
- For scalability. As the network grows, the dynamic routing protocol automatically learns about any new networks.

Table 14-5 shows a comparison of some the differences between dynamic and static routing.

Table 14-5 Comparison of Dynamic and Static Routing

Feature	Dynamic Routing	Static Routing
Configuration complexity	Independent of network size	Increases with network size
Topology changes	Automatically adapts to topology changes	Administrator intervention required

Feature	Dynamic Routing	Static Routing
Scalability	Suitable for simple to complex network topologies	Suitable for simple topologies
Security	Security must be configured	Security is inherent
Resource Usage	Uses CPU, memory, and link bandwidth	No additional resources needed
Path Predictability	Route depends on topology and routing protocol used	Explicitly defined by the administrator

Dynamic Routing Evolution (14.5.2)

Dynamic routing protocols have been used in networks since the late 1980s. One of the first routing protocols was *Routing Information Protocol (RIP)*. *RIPv1* was released in 1988, but some of the basic algorithms within the protocol were used on the Advanced Research Projects Agency Network (ARPANET) as early as 1969.

As networks evolved and became more complex, new routing protocols emerged. The RIP protocol was updated to *RIPv2* to accommodate growth in the network environment. However, RIPv2 still does not scale to the larger network implementations of today. To address the needs of larger networks, two advanced routing protocols were developed: OSPF and *Intermediate System-to-Intermediate System (IS-IS)*. Cisco developed the *Interior Gateway Routing Protocol (IGRP)*, which was later replaced by Enhanced IGRP (EIGRP), which also scales well in larger network implementations.

Additionally, there was the need to connect the different routing domains of different organizations and provide routing between them. The *Border Gateway Protocol (BGP)*, the successor of *Exterior Gateway Protocol (EGP)* is used between Internet Service Providers (ISPs). BGP is also used between ISPs and some private organizations to exchange routing information.

Figure 14-17 displays the timeline of when the various protocols were introduced.

To support IPv6 communication, newer versions of the IP routing protocols have been developed, as shown in the IPv6 row in the table.

Table 14-6 classifies the current routing protocols. *Interior Gateway Protocols (IGPs)* are routing protocols used to exchange routing information within a routing domain administered by a single organization. There is only one EGP and it is BGP. BGP is used to exchange routing information between different organizations, known as *autonomous systems (AS)*. BGP is used by ISPs to route packets over the Internet. *Distance vector routing protocols*, *link-state routing protocols*, and *path vector routing protocols* refer to the type of *routing algorithm* used to determine best path.

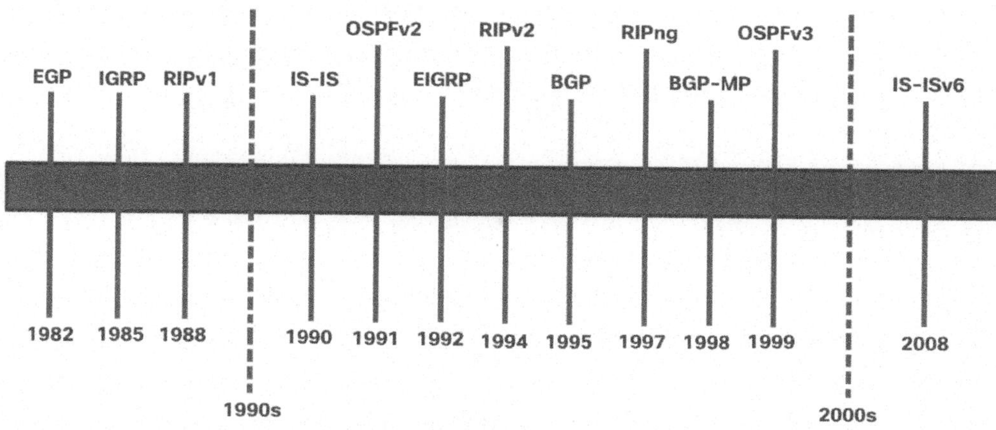

Figure 14-17 History Timeline of Routing Protocols

Table 14-6 Classification of Dynamic Routing Protocols

	Interior Gateway Protocols				**Exterior Gateway Protocols**
	Distance Vector		**Link-State**		**Path Vector**
IPv4	RIPv2	EIGRP	OSPFv2	IS-IS	BGP-4
IPv6	*RIPng*	EIGRP for IPv6	OSPFv3	IS-IS for IPv6	BGP-MP

Dynamic Routing Protocol Concepts (14.5.3)

A routing protocol is a set of processes, algorithms, and messages that are used to exchange routing information and populate the routing table with the choice of best paths. The purpose of dynamic routing protocols includes the following:

- Discovery of remote networks
- Maintaining up-to-date routing information
- Choosing the best path to destination networks
- Ability to find a new best path if the current path is no longer available

The main components of dynamic routing protocols include the following:

- *Data structures*: Routing protocols typically use tables or databases for their operations. This information is kept in RAM.
- *Routing protocol messages*: Routing protocols use various types of messages to discover neighboring routers, exchange routing information, and other tasks to learn and maintain accurate information about the network.

- **Algorithm:** An algorithm is a finite list of steps used to accomplish a task. Routing protocols use algorithms for facilitating routing information and for the best path determination.

Routing protocols allow routers to dynamically share information about remote networks and automatically offer this information to their own routing tables. In Figure 14-18, R1 is sending routing updates to R2 and R3.

Routing protocols determine the best path, or route, to each network. That route is then offered to the routing table. The route will be installed in the routing table if there is not another routing source with a lower AD. A primary benefit of dynamic routing protocols is that routers exchange routing information when there is a topology change. This exchange allows routers to automatically learn about new networks and to find alternate paths when there is a link failure to a current network.

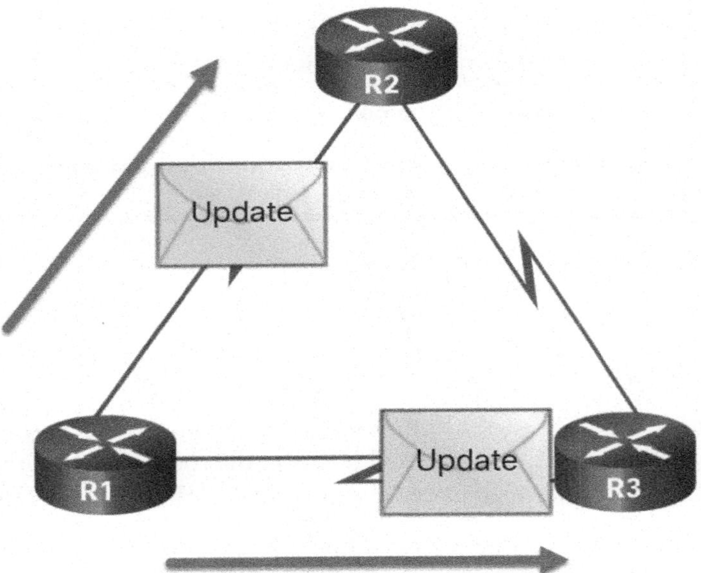

Figure 14-18 Dynamic Routing

Best Path (14.5.4)

Before a path to a remote network is offered to the routing table, the dynamic routing protocol must determine the best path to that network. Determining the best path may involve the evaluation of multiple paths to the same destination network and selecting the optimum or shortest path to reach that network. Whenever multiple paths to the same network exist, each path uses a different exit interface on the router to reach that network.

The best path is selected by a routing protocol based on the value or metric it uses to determine the *distance* to reach a network. A metric is the quantitative value used to measure the distance to a given network. The best path to a network is the path with the lowest metric.

Dynamic routing protocols typically use their own rules and metrics to build and update routing tables. The routing algorithm generates a value, or a metric, for each path through the network. Metrics can be based on either a single characteristic or several characteristics of a path. Some routing protocols can base route selection on multiple metrics, combining them into a single metric.

Table 14-7 lists common dynamic protocols and their metrics.

Table 14-7 Common Dynamic Routing Protocol Metrics

Routing Protocol	Metric
RIP	■ The metric is "hop count." ■ Each router along a path adds a hop to the hop count. ■ A maximum of 15 hops allowed.
OSPF	■ The metric is "*cost*," which is based on the cumulative bandwidth from source to destination. ■ Faster links are assigned lower costs compared to slower (higher cost) links.
EIGRP	■ It calculates a metric based on the slowest bandwidth and delay values. ■ It could also include load and reliability into the metric calculation.

Figure 14-19 highlights how the path may be different depending on the metric being used. If the best path fails, the dynamic routing protocol will automatically select a new best path if one exists.

Load Balancing (14.5.5)

What happens if a routing table has two or more paths with identical metrics to the same destination network?

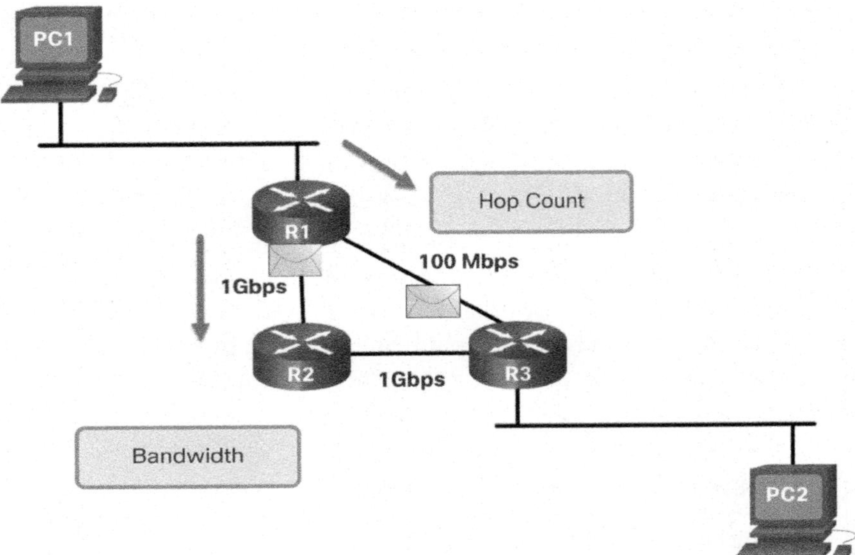

Figure 14-19 Routing Protocols with Different Metric

When a router has two or more paths to a destination with equal cost metrics, the router forwards the packets using both paths equally. This is called *equal cost load balancing*. The routing table contains the single destination network, but has multiple exit interfaces, one for each equal cost path. The router forwards packets using the multiple exit interfaces listed in the routing table.

If configured correctly, *load balancing* can increase the effectiveness and performance of the network.

Equal cost load balancing is implemented automatically by dynamic routing protocols. It is enabled with static routes when there are multiple static routes to the same destination network using different next-hop routers.

Note

Only EIGRP supports *unequal cost load balancing*.

Figure 14-20 provides an example of equal cost load balancing.

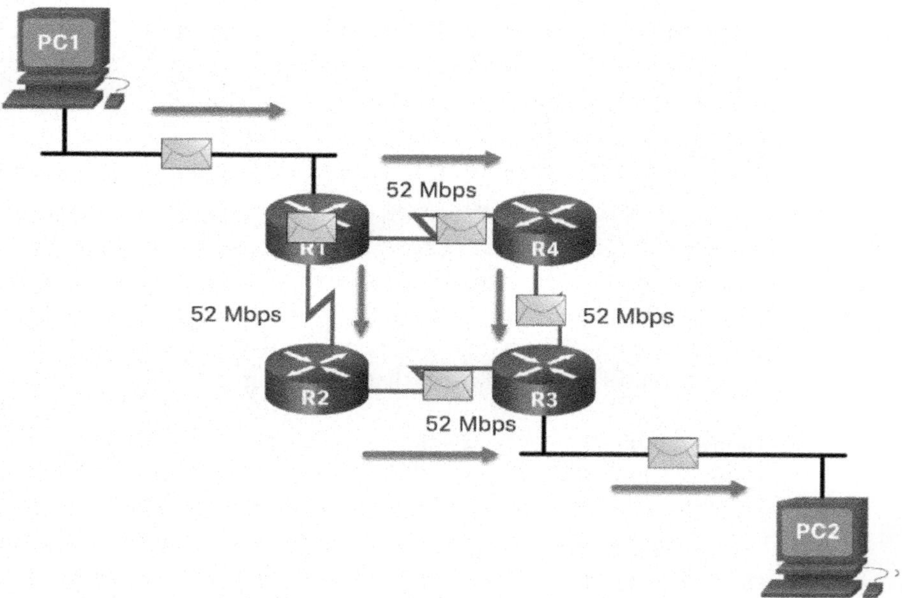

Figure 14-20 Equal-Cost Load Balancing

Interactive Graphic

Check Your Understanding—Dynamic and Static Routing (14.5.6)

Refer to the online course to complete this activity.

Summary (14.6)

The following is a summary of each section in the module:

Path Determination

The primary functions of a router are to determine the best path to forward packets based on the information in its routing table, and to forward packets toward their destination. The best path in the routing table is also known as the longest match. The longest match is the route in the routing table that has the greatest number of far-left matching bits with the destination IP address of the packet. Directly connected networks are networks that are configured on the active interfaces of a router. A directly connected network is added to the routing table when an interface is configured with an IP address and subnet mask (prefix length) and is active (up and up). Routers learn about remote networks in two ways: static routes are added to the routing table when a route is manually configured, and with dynamic routing protocols. Using dynamic routing protocols such as EIGRP and OSPF, routes are added to the routing table when routing protocols dynamically learn about the remote network.

Packet Forwarding

After a router determines the correct path, it can forward the packet on a directly connected network, it can forward the packet to a next-hop router, or it can drop the packet. The primary responsibility of the packet forwarding function is to encapsulate packets in the appropriate data link frame type for the outgoing interface. Routers support three packet forwarding mechanisms: process switching, fast switching, and CEF. The following steps describe the packet forwarding process:

Step 1. The data link frame with an encapsulated IP packet arrives on the ingress interface.

Step 2. The router examines the destination IP address in the packet header and consults its IP routing table.

Step 3. The router finds the longest matching prefix in the routing table.

Step 4. The router encapsulates the packet in a data link frame and forwards it out the egress interface. The destination could be a device connected to the network or a next-hop router.

Step 5. However, if there is no matching route entry, the packet is dropped.

Basic Router Configuration Review

There are several configuration and verification commands for routers, including **show ip route**, **show ip interface**, **show ip interface brief**, and **show running-config**.

To reduce the amount of command output, use a filter. Filtering commands can be used to display specific sections of output. To enable the filtering command, enter a pipe (|) character after the **show** command and then enter a filtering parameter and a filtering expression. The filtering parameters that can be configured after the pipe include the following:

- **section:** Shows entire section that starts with the filtering expression
- **include:** Includes all output lines that match the filtering expression
- **exclude:** Excludes all output lines that match the filtering expression
- **begin:** Shows all the output lines from a certain point, starting with the line that matches the filtering expression

IP Routing Table

A routing table contains a list of routes of known networks (prefixes and prefix lengths). The source of this information is derived from directly connected networks, static routes, and dynamic routing protocols. Common routing table codes include the following:

- **L:** Identifies the address assigned to a router interface. This allows the router to efficiently determine when it receives a packet for the interface instead of being forwarded.
- **C:** Identifies a directly connected network.
- **S:** Identifies a static route created to reach a specific network.
- **O:** Identifies a dynamically learned network from another router using the OSPF routing protocol.
- ***:** This route is a candidate for a default route.

Every router makes its decision alone, based on the information it has in its own routing table. The information in a routing table of one router does not necessarily match the routing table of another router. Routing information about a path does not provide return routing information. Routing table entries include the route source, destination network, AD, metric, next-hop, route timestamp, and exit interface. To learn about remote networks, a router must have at least one active interface configured with an IP address and subnet mask (prefix length), called a directly connected network. Static routes are manually configured and define an explicit path between two networking devices. Dynamic routing protocols can discover a network, maintain routing tables, select a best path, and automatically discover a new best path if the topology changes. The default route specifies a next-hop router to use when the routing table does not contain a specific route that matches the destination IP address.

A default route can be either a static route or learned automatically from a dynamic routing protocol. A default route has an IPv4 route entry of 0.0.0.0/0 or an IPv6 route entry of ::/0. IPv4 routing tables still have a structure based on classful addressing represented by levels of indentation. IPv6 routing tables do not use the IPv4 routing table structure. Cisco IOS uses what is known as the administrative distance (AD) to determine the route to install into the IP routing table. The AD represents the "trustworthiness" of the route. The lower the AD, the more trustworthy the route source.

Static and Dynamic Routing

Static routes are commonly used as follows:

- As a default route forwarding packets to a service provider.
- For routes outside the routing domain and not learned by the dynamic routing protocol.
- When the network administrator wants to explicitly define the path for a specific network.
- For routing between stub networks.

Dynamic routing protocol are commonly used as follows:

- In networks consisting of more than just a few routers.
- When a change in the network topology requires the network to automatically determine another path.
- For scalability. As the network grows, the dynamic routing protocol automatically learns about any new networks.

Current routing protocols include IGPs and EGPs. IGPs exchange routing information within a routing domain administered by a single organization. The only EGP is BGP. BGP exchanges routing information between different organizations. BGP is used by ISPs to route packets over the Internet. Distance vector, link-state, and path vector routing protocols refer to the type of routing algorithm used to determine best path. The main components of dynamic routing protocols are data structures, routing protocol messages, and algorithms. The best path is selected by a routing protocol based on the value or metric it uses to determine the distance to reach a network. A metric is the quantitative value used to measure the distance to a given network. The best path to a network is the path with the lowest metric. When a router has two or more paths to a destination with equal cost metrics, the router forwards the packets using both paths equally. This is called equal cost load balancing.

Practice

The following activities provide practice with the topics introduced in this chapter. The Labs are available in the companion *Switching, Routing, and Wireless Essentials Labs and Study Guide (CCNAv7)* (ISBN 9780136634386). The Packet Tracer Activity instructions are also in the Labs & Study Guide. The PKA files are found in the online course.

Packet Tracer Activity

Packet Tracer Activity

Packet Tracer 14.3.5: Basic Router Configuration Review

Check Your Understanding Questions

Complete all the review questions listed here to test your understanding of the sections and concepts in this chapter. The appendix "Answers to the 'Check Your Understanding' Questions" lists the answers.

1. What are two advantages of static routing over dynamic routing? (Choose two.)

   A. Static routing is more secure because it does not advertise routes to other routers.

   B. Static routing is relatively easy to configure for large networks.

   C. Static routing requires very little knowledge of the network for correct implementation.

   D. Static routing scales well with expanding networks.

   E. Static routing uses fewer router resources than dynamic routing.

2. What type of route allows a router to forward packets even though its routing table contains no specific route to the destination network?

   A. default route

   B. destination route

   C. dynamic route

   D. generic route

3. The network administrator configures the router with the **ip route 172.16.1.0 255.255.255.0 172.16.2.2** command. How will this route appear in the routing table?

   A. C 172.16.1.0 [1/0] via 172.16.2.2

   B. C 172.16.1.0 is directly connected, Serial0/0

   C. S 172.16.1.0 [1/0] via 172.16.2.2

   D. S 172.16.1.0 is directly connected, Serial0/0

4. R1 is configured with the **ip route 10.1.0.0 255.255.0.0 g0/0/0** command. What happens if the G0/0/0 interface goes down?

   A. The manually configured route remains in the routing table.

   B. The router polls neighbors for a replacement route.

   C. The router redirects the static route to a backup interface.

   D. The static route is removed from the routing table.

5. What static route identifies the gateway IP address to which the router sends all IP packets for which it does not have a learned route?

   A. Default static route

   B. Floating static route

   C. Generic static route

   D. Summary static route

6. On which type of network would a dynamic routing protocol be implemented in?

   A. On a home network router with wireless clients.

   B. On a network that commonly experiences topology changes.

   C. On a small two router network.

   D. On a stub network connecting to a provider.

7. Which value in a routing table is used to compare routes received from a routing protocol?

   A. administrative distance

   B. metric

   C. outgoing interface

   D. routing protocol

8. Which two route source codes are automatically created within a routing table whenever a router interface is configured with an IP address and activated? (Choose two.)

   A. C

   B. D

   C. L

   D. O

   E. S

9. A network administrator configures the G0/0/0 interface with the **ip address 10.1.1.0 255.255.255.0** command. However, when the administrator issues the **show ip route** command, the routing table does not show the directly connected network. What is the possible cause of the problem?

   A. No packets with a destination network of 172.16.1.0 have been sent to R1.

   B. The configuration needs to be saved first.

   C. The G0/0/0 interface has not been activated.

   D. The subnet mask is incorrect for the IPv4 address.

CHAPTER 15

IP Static Routing

Objectives

Upon completion of this chapter, you will be able to answer the following questions:

- What is the command syntax for static routes?
- How do you configure IPv4 and IPv6 static routes?
- How do you configure IPv4 and IPv6 default static routes?
- How do you configure a floating static route to provide a backup connection?
- How do you configure IPv4 and IPv6 static host routes that direct traffic to a specific host?

Key Terms

This chapter uses the following key terms. You can find the definitions in the Glossary.

standard static route *Page 496*

floating static route *Page 496*

summary static route *Page 496*

next-hop static route *Page 497*

directly connected static route *Page 497*

fully specified static route *Page 497*

recursive lookup *Page 497*

dual-stack *Page 499*

IPv6 link-local address *Page 509*

link-local address *Page 509*

edge router *Page 513*

quad-zero *Page 514*

host route *Page 521*

local host route *Page 522*

Introduction (15.0)

There are so many ways to dynamically route a packet that you might wonder why anybody would take the time to manually configure a static route. It is kind of like handwashing all your clothes when you have a perfectly good washing machine. But you know that some clothing items cannot go into the washing machine. Some items benefit from being washed by hand. There is a similarity in networking. As it turns out, there are many situations where a manually configured a static route is your best option.

There are different kinds of static routes, and each is perfect for solving (or avoiding) a specific type of network problem. Many networks use both dynamic and static routing, so network administrators need to know how to configure, verify, and troubleshoot static routes. You are taking this course because you want to become a network administrator, or you want to improve your existing network administrator skills. You will be glad you took this module, because you will use these skills frequently! And because this module is about configuring static routes, there are several Syntax Checker activities, followed by a Packet Tracer and a Lab where you can hone your skills!

Static Routes (15.1)

In this section, you will learn about the different types of static routes and the command syntax for IPv4 and IPv6 static routes.

Types of Static Routes (15.1.1)

Static routes are commonly implemented on a network. This is true even when there is a dynamic routing protocol configured. For instance, an organization could configure a default static route to the service provider and advertise this route to other corporate routers using the dynamic routing protocol.

Static routes can be configured for IPv4 and IPv6. Both protocols support the following types of static routes:

- *Standard static route*
- Default static route
- *Floating static route*
- *Summary static route*

Static routes are configured using the **ip route** and **ipv6 route** global configuration commands.

Next-Hop Options (15.1.2)

When configuring a static route, the next hop can be identified by an IP address, exit interface, or both. How the destination is specified creates one of the three following types of static route:

- *Next-hop static route*: Only the next-hop IP address is specified.
- *Directly connected static route*: Only the router exit interface is specified.
- *Fully specified static route*: The next-hop IP address and exit interface are specified.

IPv4 Static Route Command (15.1.3)

IPv4 static routes are configured using the following global configuration command:

```
Router(config)# ip route network-address subnet-mask {ip-address | exit-intf [ip-
   address]} [distance]
```

Note

Either the *ip-address*, *exit-intf*, or the *ip-address* and *exit-intf* parameters must be configured.

Table 15-1 describes the **ip route** command parameters.

Table 15-1 IPv4 Static Route Parameters

Parameter	Description
network-address	■ Identifies the destination IPv4 network address of the remote network to add to the routing table.
subnet-mask	■ Identifies the subnet mask of the remote network. ■ The subnet mask can be modified to summarize a group of networks and create a summary static route.
ip-address	■ Identifies the next-hop router IPv4 address. ■ Typically used with broadcast networks (that is, Ethernet). ■ Could create a *recursive lookup* static route where the router performs an additional lookup to find the exit interface.

Parameter	Description
exit-intf	■ Identifies the exit interface to forward packets. ■ Creates a directly connected static route. ■ Typically used in a point-to-point configuration.
exit-intf *ip-address*	■ Creates a fully specified static route because it specifies the exit interface and next-hop IPv4 address.
distance	■ Optional command that can be used to assign an administrative distance value between 1 and 255. ■ Typically used to configure a floating static route by setting an administrative distance that is higher than a dynamically learned route.

IPv6 Static Route Command (15.1.4)

IPv6 static routes are configured using the following global configuration command:

```
Router(config)# ipv6 route ipv6-prefix/prefix-length {ipv6-address | exit-intf
   [ipv6-address]} [distance]
```

Most of the parameters are identical to the IPv4 version of the command.

Table 15-2 shows the various **ipv6 route** command parameters and their descriptions.

Table 15-2 IPv6 Static Route Parameters

Parameter	Description
ipv6-prefix	■ Identifies the destination IPv6 network address of the remote network to add to the routing table.
/prefix-length	■ Identifies the prefix length of the remote network.
ipv6-address	■ Identifies the next-hop router IPv6 address. ■ Typically used with broadcast networks (that is, Ethernet). ■ Could create a recursive lookup static route where the router performs an additional lookup to find the exit interface.
exit-intf	■ Identifies the exit interface to forward packets. ■ Creates a directly connected static route. ■ Typically used in a point-to-point configuration.
exit-intf *ipv6-address*	■ Creates a fully specified static route because it specifies the exit interface and next-hop IPv6 address.

Parameter	Description
distance	■ Optional command that can be used to assign an administrative distance value between 1 and 255. ■ Typically used to configure a floating static route by setting an administrative distance that is higher than a dynamically learned route.

Note

The **ipv6 unicast-routing** global configuration command must be configured to enable the router to forward IPv6 packets.

Dual-Stack Topology (15.1.5)

Figure 15-1 shows a *dual-stack* network topology used throughout this module. Currently, no static routes are configured for either IPv4 or IPv6.

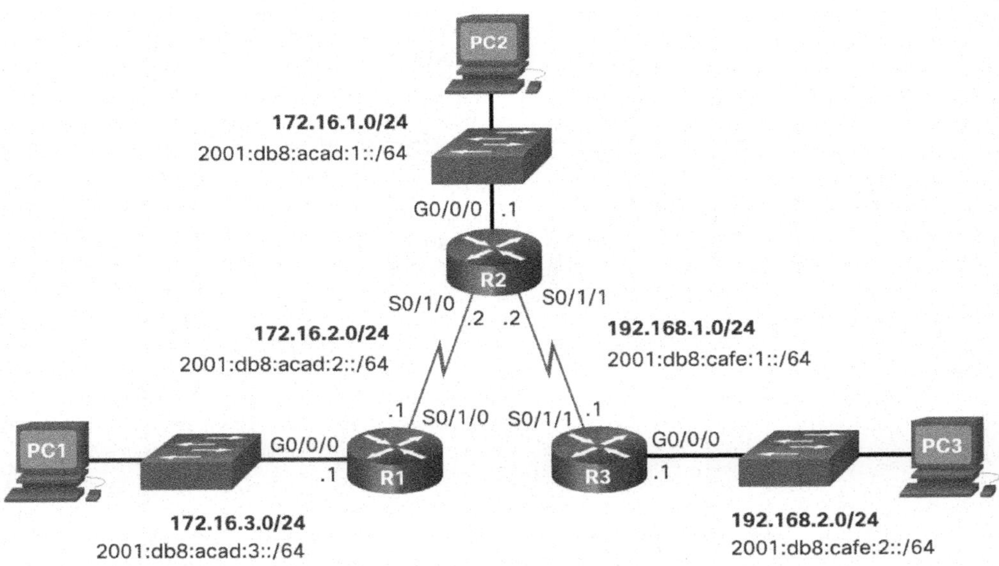

Figure 15-1 Dual-Stack Reference Topology

IPv4 Starting Routing Tables (15.1.6)

Examples 15-1 through 15-3 show the IPv4 routing table of each router. Notice that each router has entries only for directly connected networks and associated local addresses.

Example 15-1 R1 IPv4 Routing Table

```
R1# show ip route | begin Gateway
Gateway of last resort is not set
      172.16.0.0/16 is variably subnetted, 4 subnets, 2 masks
C        172.16.2.0/24 is directly connected, Serial0/1/0
L        172.16.2.1/32 is directly connected, Serial0/1/0
C        172.16.3.0/24 is directly connected, GigabitEthernet0/0/0
L        172.16.3.1/32 is directly connected, GigabitEthernet0/0/0
R1#
```

Example 15-2 R2 IPv4 Routing Table

```
R2# show ip route | begin Gateway
Gateway of last resort is not set
      172.16.0.0/16 is variably subnetted, 4 subnets, 2 masks
C        172.16.1.0/24 is directly connected, GigabitEthernet0/0/0
L        172.16.1.1/32 is directly connected, GigabitEthernet0/0/0
C        172.16.2.0/24 is directly connected, Serial0/1/0
L        172.16.2.2/32 is directly connected, Serial0/1/0
      192.168.1.0/24 is variably subnetted, 2 subnets, 2 masks
C        192.168.1.0/24 is directly connected, Serial0/1/1
L        192.168.1.2/32 is directly connected, Serial0/1/1
R2#
```

Example 15-3 R3 IPv4 Routing Table

```
R3# show ip route | begin Gateway
Gateway of last resort is not set
      192.168.1.0/24 is variably subnetted, 2 subnets, 2 masks
C        192.168.1.0/24 is directly connected, Serial0/1/1
L        192.168.1.1/32 is directly connected, Serial0/1/1
      192.168.2.0/24 is variably subnetted, 2 subnets, 2 masks
C        192.168.2.0/24 is directly connected, GigabitEthernet0/0/0
L        192.168.2.1/32 is directly connected, GigabitEthernet0/0/0
R3#
```

None of the routers have knowledge of any networks beyond the directly connected interfaces. This means each router can only reach directly connected networks, as demonstrated in the ping tests.

In Example 15-4, a ping from R1 to the Serial 0/1/0 interface of R2 should be successful because it is a directly connected network.

Example 15-4 R1 Can Ping R2

```
R1# ping 172.16.2.2
Type escape sequence to abort.
Sending 5, 100-byte ICMP Echos to 172.16.2.2, timeout is 2 seconds:
!!!!!
```

However, as shown in Example 15-5, a ping from R1 to the R3 LAN should fail because R1 does not have an entry in its routing table for the R3 LAN network.

Example 15-5 R1 Cannot Ping R3 LAN

```
R1# ping 192.168.2.1
Type escape sequence to abort.
Sending 5, 100-byte ICMP Echos to 192.168.2.1, timeout is 2 seconds:
.....
Success rate is 0 percent (0/5)
```

IPv6 Starting Routing Tables (15.1.7)

Examples 15-6 through 15-8 show the IPv6 routing table of each router. Notice that each router has entries only for directly connected networks and associated local addresses.

Example 15-6 R1 IPv6 Routing Table

```
R1# show ipv6 route | begin C
C    2001:DB8:ACAD:2::/64 [0/0]
      via Serial0/1/0, directly connected
L    2001:DB8:ACAD:2::1/128 [0/0]
      via Serial0/1/0, receive
C    2001:DB8:ACAD:3::/64 [0/0]
      via GigabitEthernet0/0/0, directly connected
L    2001:DB8:ACAD:3::1/128 [0/0]
      via GigabitEthernet0/0/0, receive
L    FF00::/8 [0/0]
      via Null0, receive
R1#
```

Example 15-7 R2 IPv6 Routing Table

```
R2# show ipv6 route | begin C
C    2001:DB8:ACAD:1::/64 [0/0]
      via GigabitEthernet0/0/0, directly connected
L    2001:DB8:ACAD:1::1/128 [0/0]
      via GigabitEthernet0/0/0, receive
C    2001:DB8:ACAD:2::/64 [0/0]
      via Serial0/1/0, directly connected
L    2001:DB8:ACAD:2::2/128 [0/0]
      via Serial0/1/0, receive
C    2001:DB8:CAFE:1::/64 [0/0]
      via Serial0/1/1, directly connected
L    2001:DB8:CAFE:1::2/128 [0/0]
      via Serial0/1/1, receive
L    FF00::/8 [0/0]
      via Null0, receive
R2#
```

Example 15-8 R3 IPv6 Routing Table

```
R3# show ipv6 route | begin C
C    2001:DB8:CAFE:1::/64 [0/0]
      via Serial0/1/1, directly connected
L    2001:DB8:CAFE:1::1/128 [0/0]
      via Serial0/1/1, receive
C    2001:DB8:CAFE:2::/64 [0/0]
      via GigabitEthernet0/0/0, directly connected
L    2001:DB8:CAFE:2::1/128 [0/0]
      via GigabitEthernet0/0/0, receive
L    FF00::/8 [0/0]
      via Null0, receive
R3#
```

None of the routers have knowledge of any networks beyond the directly connected interfaces.

As shown Example 15-9, a ping from R1 to the Serial 0/1/0 interface on R2 should be successful.

Example 15-9 R1 Can Ping R2

```
R1# ping 2001:db8:acad:2::2
Type escape sequence to abort.
Sending 5, 100-byte ICMP Echos to 2001:DB8:ACAD:2::2, timeout is 2 seconds:
!!!!!
Success rate is 100 percent (5/5), round-trip min/avg/max = 2/2/3 ms
R1#
```

However, a ping to the R3 LAN is unsuccessful, as shown in Example 15-10. This is because R1 does not have an entry in its routing table for that network.

Example 15-10 R1 Cannot Ping R3 LAN

```
R1# ping 2001:DB8:cafe:2::1
Type escape sequence to abort.
Sending 5, 100-byte ICMP Echos to 2001:DB8:CAFE:2::1, timeout is 2 seconds:
% No valid route for destination
Success rate is 0 percent (0/1)
R1#
```

Interactive Graphic

Check Your Understanding—Static Routes (15.1.8)

Refer to the online course to complete this activity.

Configure IP Static Routes (15.2)

Static routes are manually entered; therefore, careful consideration and attention must be given when configuring them.

In this section, you learn how to configure IPv4 and IPv6 static routes to enable remote network connectivity in a small- to medium-sized business network.

IPv4 Next-Hop Static Route (15.2.1)

The commands to configure standard static routes vary slightly between IPv4 and IPv6. This section shows you how to configure standard next-hop, directly connected, and fully specified static routes for both IPv4 and IPv6.

In a next-hop static route, only the next-hop IP address is specified. The exit interface is derived from the next hop. For example, three next-hop IPv4 static routes are configured on R1 using the IP address of the next hop, R2.

The commands to configure R1 with the IPv4 static routes to the three remote networks are shown in Example 15-11.

Example 15-11 IPv4 Next-Hop Static Route Configuration on R1

```
R1(config)# ip route 172.16.1.0 255.255.255.0 172.16.2.2
R1(config)# ip route 192.168.1.0 255.255.255.0 172.16.2.2
R1(config)# ip route 192.168.2.0 255.255.255.0 172.16.2.2
R1(config)#
```

The routing table for R1 now has routes to the three remote IPv4 networks, as shown in Example 15-12.

Example 15-12 R1 IPv4 Routing Table

```
R1# show ip route | begin Gateway
Gateway of last resort is not set
      172.16.0.0/16 is variably subnetted, 5 subnets, 2 masks
S        172.16.1.0/24 [1/0] via 172.16.2.2
C        172.16.2.0/24 is directly connected, Serial0/1/0
L        172.16.2.1/32 is directly connected, Serial0/1/0
C        172.16.3.0/24 is directly connected, GigabitEthernet0/0/0
L        172.16.3.1/32 is directly connected, GigabitEthernet0/0/0
S     192.168.1.0/24 [1/0] via 172.16.2.2
S     192.168.2.0/24 [1/0] via 172.16.2.2
R1#
```

IPv6 Next-Hop Static Route (15.2.2)

The commands to configure R1 with the IPv6 static routes to the three remote networks are shown in Example 15-13.

Example 15-13 IPv6 Next-Hop Static Route Configuration on R1

```
R1(config)# ipv6 unicast-routing
R1(config)# ipv6 route 2001:db8:acad:1::/64 2001:db8:acad:2::2
R1(config)# ipv6 route 2001:db8:cafe:1::/64 2001:db8:acad:2::2
R1(config)# ipv6 route 2001:db8:cafe:2::/64 2001:db8:acad:2::2
R1(config)#
```

The routing table for R1 now has routes to the three remote IPv6 networks, as shown in Example 15-14.

Example 15-14 R1 IPv6 Routing Table

```
R1# show ipv6 route
IPv6 Routing Table - default - 8 entries
Codes: C - Connected, L - Local, S - Static, U - Per-user Static route
       B - BGP, R - RIP, H - NHRP, I1 - ISIS L1
       I2 - ISIS L2, IA - ISIS interarea, IS - ISIS summary, D - EIGRP
       EX - EIGRP external, ND - ND Default, NDp - ND Prefix, DCE - Destination
       NDr - Redirect, RL - RPL, O - OSPF Intra, OI - OSPF Inter
       OE1 - OSPF ext 1, OE2 - OSPF ext 2, ON1 - OSPF NSSA ext 1
       ON2 - OSPF NSSA ext 2, la - LISP alt, lr - LISP site-registrations
       ld - LISP dyn-eid, lA - LISP away, le - LISP extranet-policy
       a - Application
S    2001:DB8:ACAD:1::/64 [1/0]
      via 2001:DB8:ACAD:2::2
C    2001:DB8:ACAD:2::/64 [0/0]
      via Serial0/1/0, directly connected
L    2001:DB8:ACAD:2::1/128 [0/0]
      via Serial0/1/0, receive
C    2001:DB8:ACAD:3::/64 [0/0]
      via GigabitEthernet0/0/0, directly connected
L    2001:DB8:ACAD:3::1/128 [0/0]
      via GigabitEthernet0/0/0, receive
S    2001:DB8:CAFE:1::/64 [1/0]
      via 2001:DB8:ACAD:2::2
S    2001:DB8:CAFE:2::/64 [1/0]
      via 2001:DB8:ACAD:2::2
L    FF00::/8 [0/0]
      via Null0, receive
R1#
```

IPv4 Directly Connected Static Route (15.2.3)

When configuring a static route, another option is to use the exit interface to specify the next-hop address. Three directly connected IPv4 static routes are configured on R1 using the exit interface, as shown in Example 15-15.

Example 15-15 IPv4 Directly Connected Static Route Configuration on R1

```
R1(config)# ip route 172.16.1.0 255.255.255.0 s0/1/0
R1(config)# ip route 192.168.1.0 255.255.255.0 s0/1/0
R1(config)# ip route 192.168.2.0 255.255.255.0 s0/1/0
```

In Example 15-16, the IPv4 routing table for R1 shows that when a packet is destined for the 192.168.2.0/24 network, R1 looks for a match in the routing table and finds that it can forward the packet out of its Serial 0/1/0 interface.

Note

Using a next-hop address is generally recommended. Directly connected static routes should be used only with point-to-point (P2P) serial interfaces, as in this example.

Example 15-16 R1 IPv4 Routing Table

```
R1# show ip route | begin Gateway
Gateway of last resort is not set
      172.16.0.0/16 is variably subnetted, 5 subnets, 2 masks
S        172.16.1.0/24 is directly connected, Serial0/1/0
C        172.16.2.0/24 is directly connected, Serial0/1/0
L        172.16.2.1/32 is directly connected, Serial0/1/0
C        172.16.3.0/24 is directly connected, GigabitEthernet0/0/0
L        172.16.3.1/32 is directly connected, GigabitEthernet0/0/0
S     192.168.1.0/24 is directly connected, Serial0/1/0
S     192.168.2.0/24 is directly connected, Serial0/1/0
R1#
```

IPv6 Directly Connected Static Route (15.2.4)

In Example 15-17, three directly connected IPv6 static routes are configured on R1 using the exit interface.

Example 15-17 IPv6 Directly Connected Static Route Configuration on R1

```
R1(config)# ipv6 route 2001:db8:acad:1::/64 s0/1/0
R1(config)# ipv6 route 2001:db8:cafe:1::/64 s0/1/0
R1(config)# ipv6 route 2001:db8:cafe:2::/64 s0/1/0
R1(config)#
```

The IPv6 routing table for R1 in Example 15-18 shows that when a packet is destined for the 2001:db8:cafe:2::/64 network, R1 looks for a match in the routing table and finds that it can forward the packet out of its Serial 0/1/0 interface.

Note

Using a next-hop address is generally recommended. Directly connected static routes should be used only with point-to-point serial interfaces, as in this example.

Example 15-18 R1 IPv6 Routing Table

```
R1# show ipv6 route
IPv6 Routing Table - default - 8 entries
Codes: C - Connected, L - Local, S - Static, U - Per-user Static route
       B - BGP, R - RIP, H - NHRP, I1 - ISIS L1
       I2 - ISIS L2, IA - ISIS interarea, IS - ISIS summary, D - EIGRP
       EX - EIGRP external, ND - ND Default, NDp - ND Prefix, DCE - Destination
       NDr - Redirect, RL - RPL, O - OSPF Intra, OI - OSPF Inter
       OE1 - OSPF ext 1, OE2 - OSPF ext 2, ON1 - OSPF NSSA ext 1
       ON2 - OSPF NSSA ext 2, la - LISP alt, lr - LISP site-registrations
       ld - LISP dyn-eid, lA - LISP away, le - LISP extranet-policy
       a - Application
S    2001:DB8:ACAD:1::/64 [1/0]
      via Serial0/1/0, directly connected
C    2001:DB8:ACAD:2::/64 [0/0]
      via Serial0/1/0, directly connected
L    2001:DB8:ACAD:2::1/128 [0/0]
      via Serial0/1/0, receive
C    2001:DB8:ACAD:3::/64 [0/0]
      via GigabitEthernet0/0/0, directly connected
L    2001:DB8:ACAD:3::1/128 [0/0]
      via GigabitEthernet0/0/0, receive
S    2001:DB8:CAFE:1::/64 [1/0]
      via Serial0/1/0, directly connected
S    2001:DB8:CAFE:2::/64 [1/0]
      via Serial0/1/0, directly connected
L    FF00::/8 [0/0]
      via Null0, receiveIPv6 Routing Table - default - 8 entries
R1#
```

IPv4 Fully Specified Static Route (15.2.5)

In a fully specified static route, both the exit interface and the next-hop IP address are specified. This form of static route is used when the exit interface is a multiaccess interface and it is necessary to explicitly identify the next hop. The next hop must be directly connected to the specified exit interface. Using an exit interface is optional; however, it is necessary to use a next-hop address.

Suppose that the network link between R1 and R2 is an Ethernet link and that the GigabitEthernet 0/0/1 interface of R1 is connected to that network, as shown in Figure 15-2.

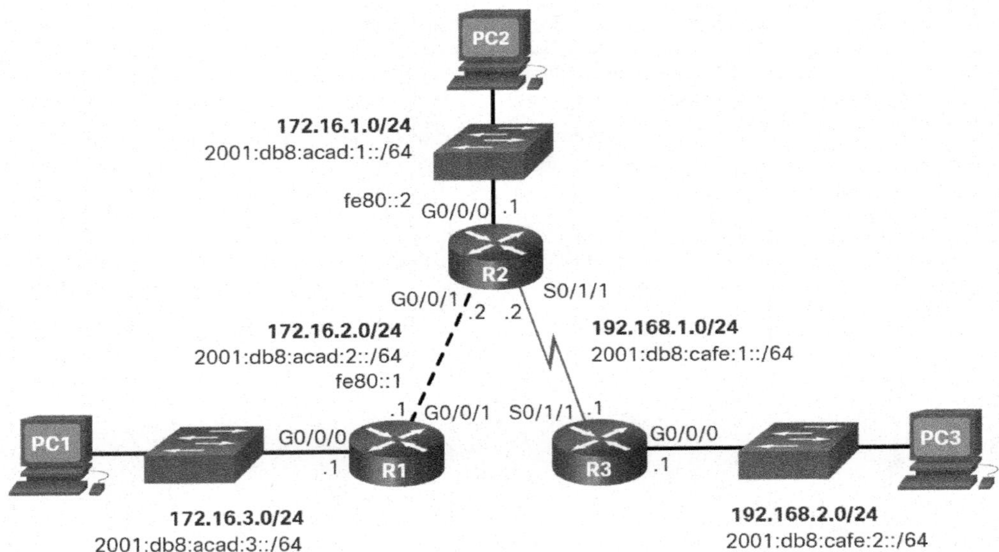

Figure 15-2 Dual-Stack Reference Topology with Ethernet Link Between R1 and R2

The difference between an Ethernet multiaccess network and a point-to-point serial network is that a point-to-point serial network has only one other device on that network, the router at the other end of the link. With Ethernet networks, there may be many different devices sharing the same multiaccess network, including hosts and even multiple routers.

It is recommended that when the exit interface is an Ethernet network, the static route includes a next-hop address. You can also use a fully specified static route that includes both the exit interface and the next-hop address, as shown for R1 fully specified static routes in Example 15-19.

Example 15-19 IPv4 Fully Specified Static Route Configuration on R1

```
R1(config)# ip route 172.16.1.0 255.255.255.0 GigabitEthernet 0/0/1 172.16.2.2
R1(config)# ip route 192.168.1.0 255.255.255.0 GigabitEthernet 0/0/1 172.16.2.2
R1(config)# ip route 192.168.2.0 255.255.255.0 GigabitEthernet 0/0/1 172.16.2.2
R1(config)#
```

When forwarding packets to R2, the exit interface is GigabitEthernet 0/0/1 and the next-hop IPv4 address is 172.16.2.2, as shown in the **show ip route** output from R1 in Example 15-20.

Example 15-20 R1 IPv4 Routing Table

```
R1# show ip route | begin Gateway
Gateway of last resort is not set
      172.16.0.0/16 is variably subnetted, 5 subnets, 2 masks
S        172.16.1.0/24 [1/0] via 172.16.2.2, GigabitEthernet0/0/1
C        172.16.2.0/24 is directly connected, GigabitEthernet0/0/1
L        172.16.2.1/32 is directly connected, GigabitEthernet0/0/1
C        172.16.3.0/24 is directly connected, GigabitEthernet0/0/0
L        172.16.3.1/32 is directly connected, GigabitEthernet0/0/0
S     192.168.1.0/24 [1/0] via 172.16.2.2, GigabitEthernet0/0/1
S     192.168.2.0/24 [1/0] via 172.16.2.2, GigabitEthernet0/0/1
R1#
```

IPv6 Fully Specified Static Route (15.2.6)

In a fully specified IPv6 static route, both the exit interface and the next-hop IPv6 address are specified. There is a situation in IPv6 when a fully specified static route must be used. If the IPv6 static route uses an *IPv6 link-local address* as the next-hop address, use a fully specified static route. Figure 15-3 shows a topology of just R1 and R2 to demonstrate a fully specified IPv6 static route configuration using an IPv6 link-local address as the next-hop address.

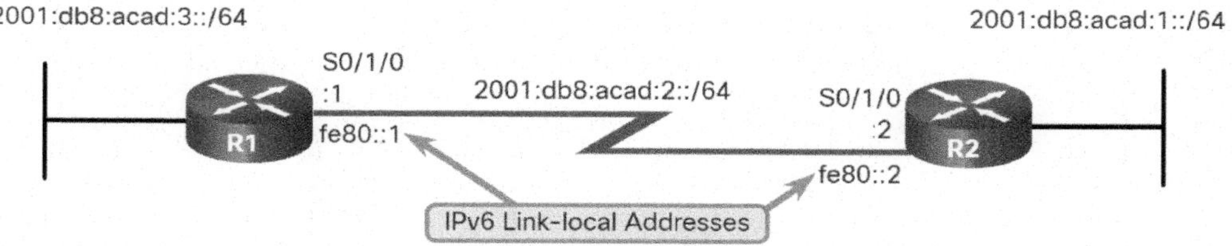

Figure 15-3 IPv6 Topology for Fully Specified Static Route Configuration

In Example 15-21, a fully specified static route is configured using the *link-local address* of R2 as the next-hop address. Notice that Internetwork Operating System (IOS) requires that an exit interface be specified.

Example 15-21 IPv6 Fully Specified Static Route on R1

```
R1(config)# ipv6 route 2001:db8:acad:1::/64 fe80::2
%Interface has to be specified for a link-local nexthop
R1(config)# ipv6 route 2001:db8:acad:1::/64 s0/1/0 fe80::2
R1(config)#
```

The reason a fully specified static route must be used is because IPv6 link-local addresses are not contained in the IPv6 routing table. Link-local addresses are unique only on a given link or network. The next-hop link-local address may be a valid address on multiple networks connected to the router. Therefore, it is necessary that the exit interface be included.

Example 15-22 shows the IPv6 routing table entry for this route. Notice that both the next-hop link-local address and the exit interface are included.

Example 15-22 R1 IPv6 Routing Table

```
R1# show ipv6 route static | begin 2001:db8:acad:1::/64
S    2001:DB8:ACAD:1::/64 [1/0]
     via FE80::2, Seria0/1/0
R1#
```

Verify a Static Route (15.2.7)

Along with **show ip route**, **show ipv6 route**, **ping**, and **traceroute**, other useful commands to verify static routes include the following:

- **show ip route static**
- **show ip route** *network*
- **show running-config | section ip route**

Replace **ip** with **ipv6** for the IPv6 versions of the command. Figure 15-4 repeats the dual-stack reference topology that is used for the following command examples.

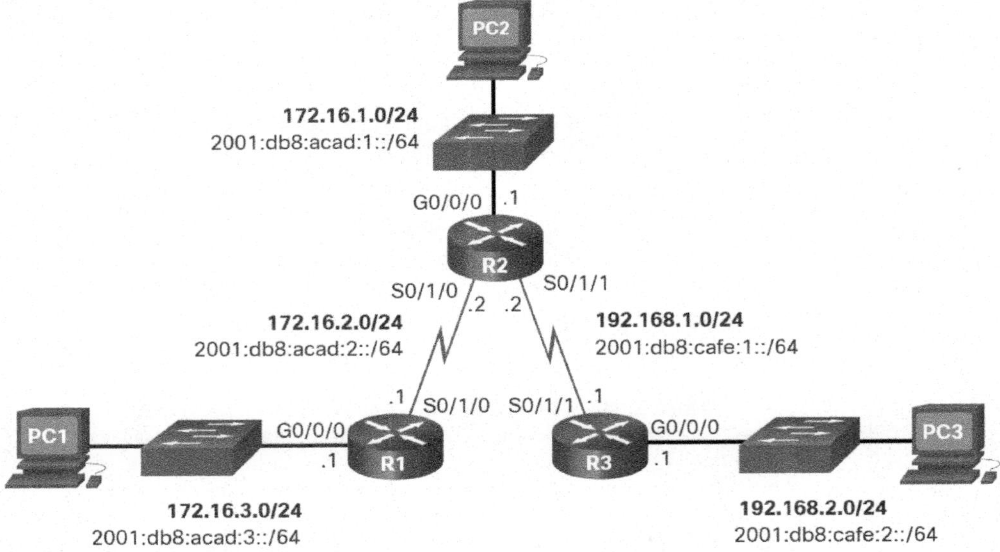

Figure 15-4 Dual-Stack Reference Topology

Display Only IPv4 Static Routes

The output in Example 15-23 shows only the IPv4 static routes in the routing table. Also note where the filter begins the output, excluding all the codes.

Example 15-23 Display Only IPv4 Static Routes

```
R1# show ip route static | begin Gateway
Gateway of last resort is not set
      172.16.0.0/16 is variably subnetted, 5 subnets, 2 masks
S        172.16.1.0/24 [1/0] via 172.16.2.2
S     192.168.1.0/24 [1/0] via 172.16.2.2
S     192.168.2.0/24 [1/0] via 172.16.2.2
R1#
```

Display a Specific IPv4 Network

Example 15-24 shows only the output for only the specified network in the routing table.

Example 15-24 Display a Specific IPv4 Network

```
R1# show ip route 192.168.2.1
Routing entry for 192.168.2.0/24
  Known via "static", distance 1, metric 0
  Routing Descriptor Blocks:
  * 172.16.2.2
      Route metric is 0, traffic share count is 1
R1#
```

Display the IPv4 Static Route Configuration

The command in Example 15-25 filters the running configuration for only IPv4 static routes.

Example 15-25 Display the IPv4 Static Route Configuration

```
R1# show running-config | section ip route
ip route 172.16.1.0 255.255.255.0 172.16.2.2
ip route 192.168.1.0 255.255.255.0 172.16.2.2
ip route 192.168.2.0 255.255.255.0 172.16.2.2
R1#
```

Display Only IPv6 Static Routes

The output in Example 15-26 shows only the IPv6 static routes in the routing table.

Example 15-26 Display Only IPv6 Static Routes

```
R1# show ipv6 route static
IPv6 Routing Table - default - 8 entries
Codes: C - Connected, L - Local, S - Static, U - Per-user Static route
       B - BGP, R - RIP, H - NHRP, I1 - ISIS L1
       I2 - ISIS L2, IA - ISIS interarea, IS - ISIS summary, D - EIGRP
       EX - EIGRP external, ND - ND Default, NDp - ND Prefix, DCE - Destination
       NDr - Redirect, RL - RPL, O - OSPF Intra, OI - OSPF Inter
       OE1 - OSPF ext 1, OE2 - OSPF ext 2, ON1 - OSPF NSSA ext 1
       ON2 - OSPF NSSA ext 2, la - LISP alt, lr - LISP site-registrations
       ld - LISP dyn-eid, lA - LISP away, le - LISP extranet-policy
       a - Application
S   2001:DB8:ACAD:1::/64 [1/0]
     via 2001:DB8:ACAD:2::2
S   2001:DB8:CAFE:1::/64 [1/0]
     via 2001:DB8:ACAD:2::2
S   2001:DB8:CAFE:2::/64 [1/0]
     via 2001:DB8:ACAD:2::2
R1#
```

Display a Specific IPv6 Network

The command in Example 15-27 shows output for only the specified network in the routing table.

Example 15-27 Display a Specific IPv6 Network

```
R1# show ipv6 route 2001:db8:cafe:2::
Routing entry for 2001:DB8:CAFE:2::/64
  Known via "static", distance 1, metric 0
  Route count is 1/1, share count 0
  Routing paths:
    2001:DB8:ACAD:2::2
      Last updated 00:23:55 ago

R1#
```

Display the IPv6 Static Route Configuration

The command in Example 15-28 filters the running configuration for only IPv6 static routes.

Example 15-28 Display the IPv6 Static Route Configuration

```
R1# show running-config | section ipv6 route
ipv6 route 2001:DB8:ACAD:1::/64 2001:DB8:ACAD:2::2
ipv6 route 2001:DB8:CAFE:1::/64 2001:DB8:ACAD:2::2
ipv6 route 2001:DB8:CAFE:2::/64 2001:DB8:ACAD:2::2
R1#
```

Interactive Graphic

Syntax Checker—Configure Static Routes (15.2.8)

Refer to the online course to complete this activity.

Configure IP Default Static Routes (15.3)

In this section, you learn how to configure an IPv4 and IPv6 default static route.

Default Static Route (15.3.1)

This section shows you how to configure a default route for IPv4 and IPv6. It also explains the situations in which a default route is a good choice. A default route is a static route that matches all packets. Instead of routers storing routes for all the networks in the Internet, they can store a single default route to represent any network that is not in the routing table.

Routers commonly use default routes that are either configured locally or learned from another router, using a dynamic routing protocol. A default route does not require any far-left bits to match between the default route and the destination IP address. A default route is used when no other routes in the routing table match the destination IP address of the packet. In other words, if a more specific match does not exist, the default route is used as the Gateway of Last Resort.

Default static routes are commonly used when connecting an *edge router* to a service provider network, or a stub router (a router with only one upstream neighbor router).

Figure 15-5 shows a typical default static route scenario.

IPv4 Default Static Route

The command syntax for an IPv4 default static route is similar to any other IPv4 static route, except that the network address is 0.0.0.0 and the subnet mask is 0.0.0.0. The 0.0.0.0 0.0.0.0 in the route will match any network address.

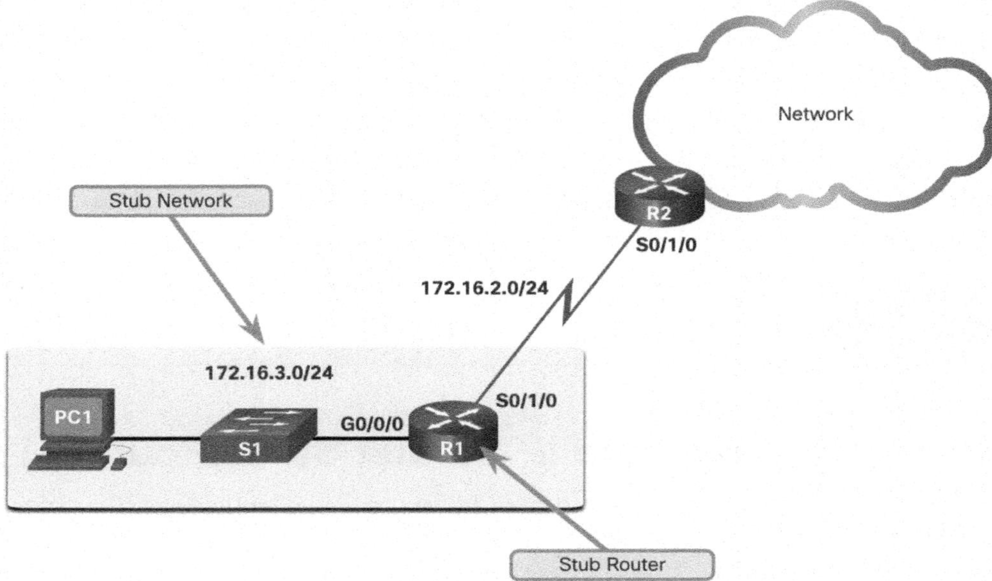

Figure 15-5 Default Static Route Stub Network Topology

Note

An IPv4 default static route is commonly referred to as a *quad-zero* route.

The basic command syntax for an IPv4 default static route is as follows:

```
Router(config)# ip route 0.0.0.0 0.0.0.0 {ip-address | exit-intf}
```

IPv6 Default Static Route

The command syntax for an IPv6 default static route is similar to any other IPv6 static route, except that the ipv6-prefix/prefix-length is ::/0, which matches all routes.

The basic command syntax for an IPv6 default static route is as follows:

```
Router(config)# ipv6 route ::/0 {ipv6-address | exit-intf}
```

Configure a Default Static Route (15.3.2)

In Figure 15-6, R1 could be configured with three static routes, one to reach each of the remote networks in the example topology. However, R1 is a stub router because it is only connected to R2. Therefore, it would be more efficient to configure a single default static route.

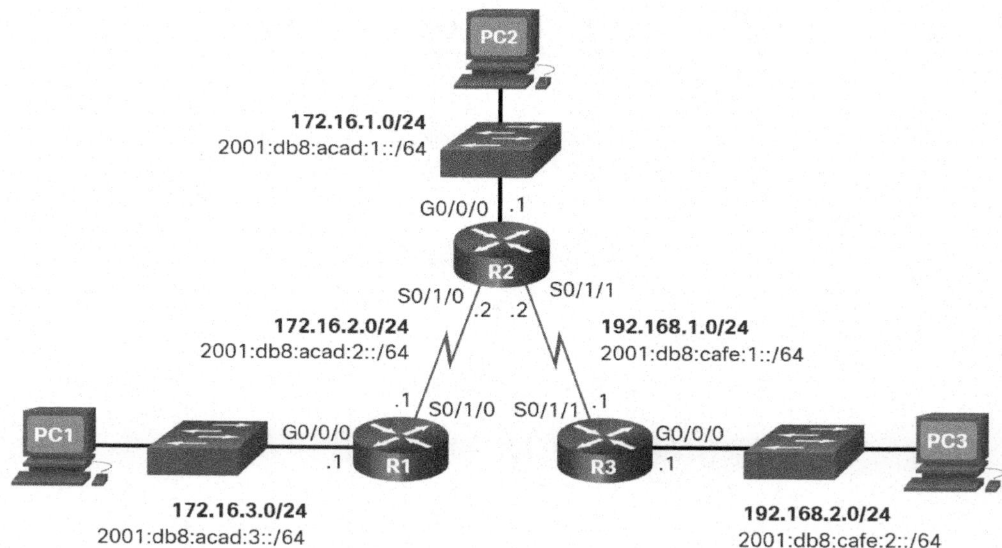

Figure 15-6 Dual Stack Reference Topology

Example 15-29 shows an IPv4 default static route configured on R1. With the configuration shown in the example, any packets not matching more specific route entries are forwarded to R2 at 172.16.2.2.

Example 15-29 IPv4 Default Static Route on R1

```
R1(config)# ip route 0.0.0.0 0.0.0.0 172.16.2.2
```

An IPv6 default static route is configured in similar fashion, as shown in Example 15-30. With this configuration any packets not matching more specific IPv6 route entries are forwarded to R2 at 2001:db8:acad:2::2.

Example 15-30 IPv6 Default Static Route on R1

```
R1(config)# ipv6 route ::/0 2001:db8:acad:2::2
```

Verify a Default Static Route (15.3.3)

In Example 15-31, the **show ip route static** command output from R1 displays the contents of the static routes in the routing table. Note the asterisk (*) next to the route with code S. As displayed in the codes table in the **show ip route** output, the asterisk indicates that this static route is a candidate default route, which is why it is selected as the Gateway of Last Resort.

Example 15-31 R1 IPv4 Default Route in the Routing Table

```
R1# show ip route static
Codes: L - local, C - connected, S - static, R - RIP, M - mobile, B - BGP
D - EIGRP, EX - EIGRP external, O - OSPF, IA - OSPF inter area
N1 - OSPF NSSA external type 1, N2 - OSPF NSSA external type 2
E1 - OSPF external type 1, E2 - OSPF external type 2
i - IS-IS, su - IS-IS summary, L1 - IS-IS level-1, L2 - IS-IS level-2
ia - IS-IS inter area, * - candidate default, U - per-user static route
o - ODR, P - periodic downloaded static route, H - NHRP, l - LISP
+ - replicated route, % - next hop override

Gateway of last resort is 172.16.2.2 to network 0.0.0.0

S* 0.0.0.0/0 [1/0] via 172.16.2.2

R1#
```

Example 15-32 shows the **show ipv6 route static** command output to display the contents of the routing table.

Example 15-32 R1 IPv6 Default Route in the Routing Table

```
R1# show ipv6 route static
IPv6 Routing Table - default - 8 entries
Codes: C - Connected, L - Local, S - Static, U - Per-user Static route
       B - BGP, R - RIP, H - NHRP, I1 - ISIS L1
       I2 - ISIS L2, IA - ISIS interarea, IS - ISIS summary, D - EIGRP
       EX - EIGRP external, ND - ND Default, NDp - ND Prefix, DCE - Destination
       NDr - Redirect, RL - RPL, O - OSPF Intra, OI - OSPF Inter
       OE1 - OSPF ext 1, OE2 - OSPF ext 2, ON1 - OSPF NSSA ext 1
       ON2 - OSPF NSSA ext 2, la - LISP alt, lr - LISP site-registrations
       ld - LISP dyn-eid, lA - LISP away, le - LISP extranet-policy
       a - Application
S   ::/0 [1/0]
     via 2001:DB8:ACAD:2::2
R1#
```

Notice that the static default route configuration uses the /0 mask for IPv4 default routes and the ::/0 prefix for IPv6 default routes. Remember that the IPv4 subnet mask and IPv6 prefix-length in a routing table determine how many bits must match between the destination IP address of the packet and the route in the routing table. A /0 mask or ::/0 prefix indicates that none of the bits are required to match. As long as a more specific match does not exist, the default static route matches all packets.

Interactive Graphic

Syntax Checker—Configure Default Static Routes (15.3.4)

Refer to the online course to complete this activity.

Configure Floating Static Routes (15.4)

In this section, you learn how to configure IPv4 and IPv6 floating static routes to provide a backup connection.

Floating Static Routes (15.4.1)

As with the other sections in this module, you learn how to configure IPv4 and IPv6 floating static routes and when to use them.

Another type of static route is a floating static route. Floating static routes are static routes that are used to provide a backup path to a primary static or dynamic route, in the event of a link failure. The floating static route is used only when the primary route is not available.

To accomplish this, the floating static route is configured with a higher administrative distance than the primary route. The administrative distance represents the trustworthiness of a route. If multiple paths to the destination exist, the router will choose the path with the lowest administrative distance.

For example, assume that an administrator wants to create a floating static route as a backup to an EIGRP-learned route. The floating static route must be configured with a higher administrative distance than Enhanced Interior Gateway Routing Protocol (EIGRP). EIGRP has an administrative distance of 90. If the floating static route is configured with an administrative distance of 95, the dynamic route learned through EIGRP is preferred to the floating static route. If the EIGRP-learned route is lost, the floating static route is used in its place.

In Figure 15-7, the branch router typically forwards all traffic to the HQ router over the private WAN link. In this example, the routers exchange route information using EIGRP. A floating static route, with an administrative distance of 91 or higher, could be configured to serve as a backup route. If the private WAN link fails and the EIGRP route disappears from the routing table, the router selects the floating static route as the best path to reach the HQ LAN.

- A route learned through dynamic routing is preferred.
- If a dynamic route is lost, the floating static route will be used.

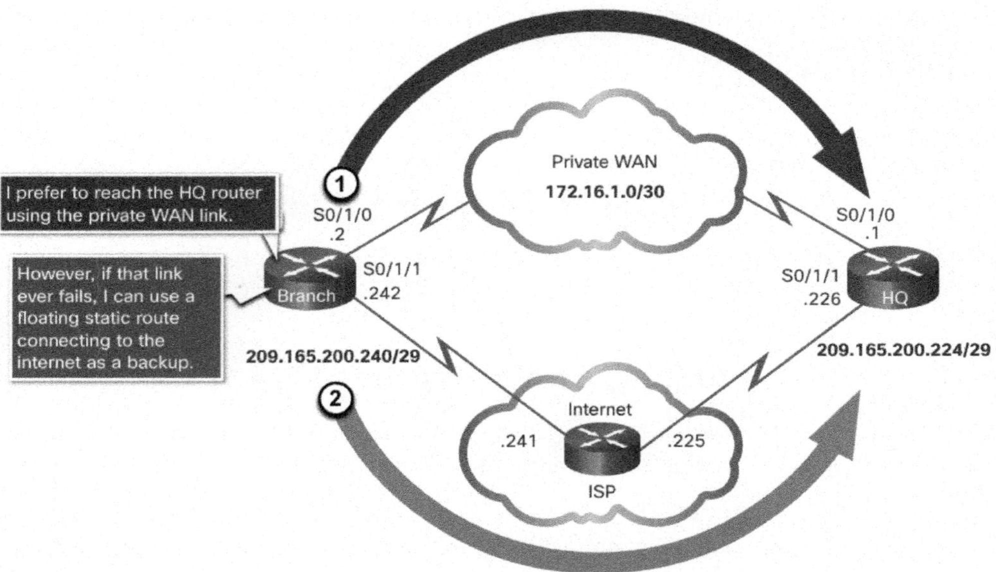

Figure 15-7 Floating Static Route Topology

By default, static routes have an administrative distance of 1, making them preferable to routes learned from dynamic routing protocols. For example, the administrative distances of some common interior gateway dynamic routing protocols are as follows:

- EIGRP = 90
- OSPF = 110
- IS-IS = 115

The administrative distance of a static route can be increased to make the route less desirable than that of another static route or a route learned through a dynamic routing protocol. In this way, the static route "floats" and is not used when the route with the better administrative distance is active. However, if the preferred route is lost, the floating static route can take over, and traffic can be sent through this alternate route.

Configure IPv4 and IPv6 Floating Static Routes (15.4.2)

IP floating static routes are configured by using the **distance** argument to specify an administrative distance. If no administrative distance is configured, the default value (1) is used.

Refer to the topology in Figure 15-8. In this scenario, the preferred default route from R1 is to R2. The connection to R3 should be used for backup only.

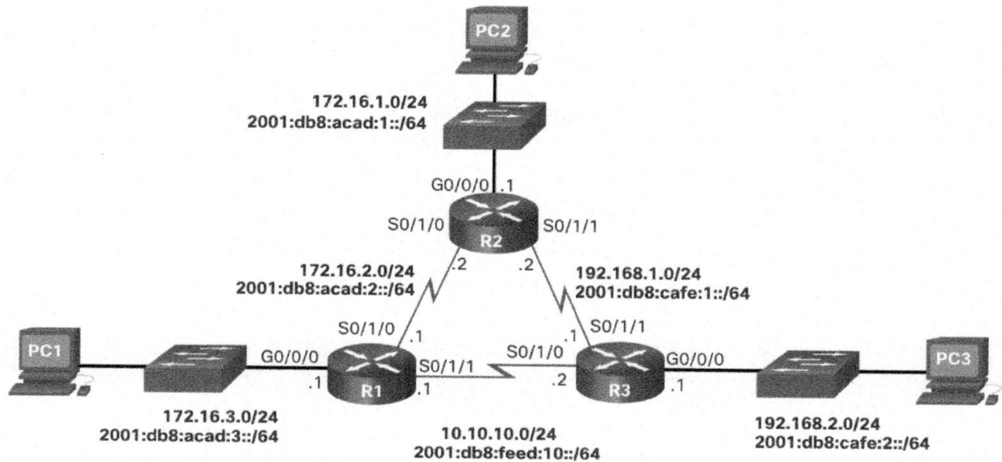

Figure 15-8 Full Mesh Dual-Stack Reference Topology

In Example 15-33, R1 is configured with IPv4 and IPv6 default static routes pointing to R2. Because no administrative distance is configured, the default value (1) is used for these static routes. R1 is also configured with IPv4 and IPv6 floating static default routes pointing to R3 with an administrative distance of 5. This value is greater than the default value of 1; therefore, this route floats and is not present in the routing table unless the preferred route fails.

Example 15-33 IPv4 and IPv6 Floating Static Route Configuration on R1

```
R1(config)# ip route 0.0.0.0 0.0.0.0 172.16.2.2
R1(config)# ip route 0.0.0.0 0.0.0.0 10.10.10.2 5
R1(config)# ipv6 route ::/0 2001:db8:acad:2::2
R1(config)# ipv6 route ::/0 2001:db8:feed:10::2 5
R1(config)#
```

The **show ip route static** and **show ipv6 route** static output in Example 15-34 verifies that the default routes to R2 are installed in the routing table. Note that the IPv4 floating static route to R3 is not present in the routing table.

Example 15-34 R1 IPv4 and IPv6 Routing Tables

```
R1# show ip route static | begin Gateway
Gateway of last resort is 172.16.2.2 to network 0.0.0.0

S*    0.0.0.0/0 [1/0] via 172.16.2.2
R1#
```

```
R1# show ipv6 route static | begin S :
S    ::/0 [1/0]
      via 2001:DB8:ACAD:2::2
R1#
```

Use the **show run** command to verify that floating static routes are in the configuration. The command output in Example 15-35 verifies that both IPv6 static default routes are in the running configuration.

Example 15-35 Verifying an IPv6 Floating Static Route Is in the Configuration

```
R1# show run | include ipv6 route
ipv6 route ::/0 2001:db8:feed:10::2 5
ipv6 route ::/0 2001:db8:acad:2::2
R1#
```

Test the Floating Static Route (15.4.3)

What would happen if R2 failed? To simulate this failure, both serial interfaces of R2 are shut down, as shown in Example 15-36.

Example 15-36 Testing a Floating Static Route

```
R2(config)# interface s0/1/0
R2(config-if)# shut
*Sep 18 23:36:27.000: %LINK-5-CHANGED: Interface Serial0/1/0, changed state to
  administratively down
*Sep 18 23:36:28.000: %LINEPROTO-5-UPDOWN: Line protocol on Interface Serial0/1/0,
  changed state to down
R2(config-if)# interface s0/1/1
R2(config-if)# shut
*Sep 18 23:36:41.598: %LINK-5-CHANGED: Interface Serial0/1/1, changed state to
  administratively down
*Sep 18 23:36:42.598: %LINEPROTO-5-UPDOWN: Line protocol on Interface Serial0/1/1,
  changed state to down
R1(config-if)# end
R1#
```

Notice that R1 automatically generates messages indicating that the serial interface to R2 is down, as shown in Example 15-37.

Example 15-37 Log Messages on R1

```
R1#
*Sep 18 23:35:48.810: %LINK-3-UPDOWN: Interface Serial0/1/0, changed state to down
R1#
*Sep 18 23:35:49.811: %LINEPROTO-5-UPDOWN: Line protocol on Interface Serial0/1/0,
  changed state to down
R1#
```

A look at the IP routing tables of R1 verifies that the floating static default routes are now installed as the default routes and are pointing to R3 as the next-hop router, as shown in Example 15-38.

Example 15-38 Verifying Floating Static Routes are Now Installed on R1

```
R1# show ip route static | begin Gateway
Gateway of last resort is 10.10.10.2 to network 0.0.0.0
S*     0.0.0.0/0 [5/0] via 10.10.10.2
R1#
R1# show ipv6 route static | begin ::
S    ::/0 [5/0]
      via 2001:DB8:FEED:10::2
R1#
```

Interactive Graphic

Syntax Checker—Configure Floating Static Route (15.4.4)

Refer to the online course to complete this activity.

Configure Static Host Routes (15.5)

In this section, you will learn how to configure IPv4 and IPv6 static host routes that direct traffic to a specific host.

Host Routes (15.5.1)

This section shows you how to configure an IPv4 and IPv6 static *host route* and when to use them.

A host route is an IPv4 address with a 32-bit mask, or an IPv6 address with a 128-bit mask. The following shows the three ways a host route can be added to the routing table:

- Automatically installed when an IP address is configured on the router (as shown in the figures)
- Configured as a static host route
- Host route automatically obtained through other methods (discussed in later courses)

Automatically Installed Host Routes (15.5.2)

Cisco IOS automatically installs a host route, also known as a *local host route*, when an interface address is configured on the router. A host route allows for a more efficient process for packets that are directed to the router itself, rather than for packet forwarding. This is in addition to the connected route, designated with a C in the routing table for the network address of the interface.

When an active interface on a router is configured with an IP address, a local host route is automatically added to the routing table. The local routes are marked with L in the output of the routing table.

For example, refer to the topology in Figure 15-9.

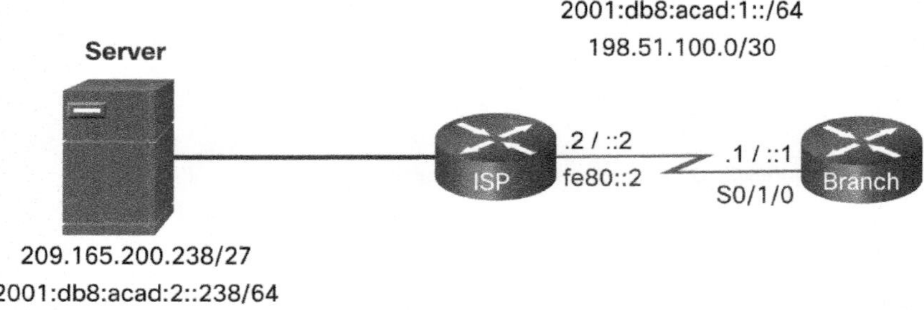

Figure 15-9 Host Route Reference Topology

The IP addresses assigned to the Branch Serial 0/1/0 interface are 198.51.100.1/30 and 2001:db8:acad:1::1/64. The local routes for the interface are installed by the IOS in the IPv4 and IPv6 routing tables, as shown in Example 15-39.

Example 15-39 Local IPv4 and IPv6 Routes

```
Branch# show ip route | begin Gateway
Gateway of last resort is not set
      198.51.100.0/24 is variably subnetted, 2 subnets, 2 masks
C        198.51.100.0/30 is directly connected, Serial0/1/0
L        198.51.100.1/32 is directly connected, Serial0/1/0
Branch# show ipv6 route | begin ::
C   2001:DB8:ACAD:1::/64 [0/0]
     via Serial0/1/0, directly connected
L   2001:DB8:ACAD:1::1/128 [0/0]
     via Serial0/1/0, receive
L   FF00::/8 [0/0]
     via Null0, receive
Branch#
```

Static Host Routes (15.5.3)

A host route can be a manually configured static route to direct traffic to a specific destination device, such as the server shown in Figure 15-9. The static route uses a destination IP address and a 255.255.255.255 (/32) mask for IPv4 host routes and a /128 prefix length for IPv6 host routes.

Configure Static Host Routes (15.5.4)

Example 15-40 shows the IPv4 and IPv6 static host route configuration on the Branch router to access the server.

Example 15-40 IPv4 and IPv6 Static Host Route Configuration

```
Branch(config)# ip route 209.165.200.238 255.255.255.255 198.51.100.2
Branch(config)# ipv6 route 2001:db8:acad:2::238/128 2001:db8:acad:1::2
Branch(config)# exit
Branch#
```

Verify Static Host Routes (15.5.5)

A review of both the IPv4 and IPv6 route tables verifies that the routes are active, as shown in Example 15-41.

Example 15-41 Verifying the IPv4 and IPv6 Static Host Routes

```
Branch# show ip route | begin Gateway
Gateway of last resort is not set
      198.51.100.0/24 is variably subnetted, 2 subnets, 2 masks
C        198.51.100.0/30 is directly connected, Serial0/1/0
L        198.51.100.1/32 is directly connected, Serial0/1/0
      209.165.200.0/32 is subnetted, 1 subnets
S        209.165.200.238 [1/0] via 198.51.100.2
Branch#
Branch# show ipv6 route
(Output omitted)
C    2001:DB8:ACAD:1::/64 [0/0]
      via Serial0/1/0, directly connected
L    2001:DB8:ACAD:1::1/128 [0/0]
      via Serial0/1/0, receive
S    2001:DB8:ACAD:2::238/128 [1/0]
      via 2001:DB8:ACAD:1::2
Branch#
```

Configure IPv6 Static Host Route with Link-Local Next-Hop (15.5.6)

For IPv6 static routes, the next-hop address can be the link-local address of the adjacent router. However, you must specify an interface type and an interface number when using a link-local address as the next hop, as shown in Example 15-42. First, the original IPv6 static host route is removed, then a fully specified route is configured with the IPv6 address of the server and the IPv6 link-local address of the ISP router.

Example 15-42 Configuring and Verifying an IPv6 Static Host Route with Link-Local as the Next-Hop

```
Branch(config)# no ipv6 route 2001:db8:acad:2::238/128 2001:db8:acad:1::2
Branch(config)# ipv6 route 2001:db8:acad:2::238/128 serial 0/1/0 fe80::2
Branch#
Branch# show ipv6 route | begin ::
C    2001:DB8:ACAD:1::/64 [0/0]
      via Serial0/1/0, directly connected
L    2001:DB8:ACAD:1::1/128 [0/0]
      via Serial0/1/0, receive
S    2001:DB8:ACAD:2::238/128 [1/0]
      via FE80::2, Serial0/1/0
Branch#
```

Interactive Graphic

Syntax Checker—Configure Static Host Routes (15.5.7)

Refer to the online course to complete this activity.

Summary (15.6)

The following is a summary of each section in the module:

Static Routes

Static routes can be configured for IPv4 and IPv6. Both protocols support the following types of static routes: standard static route, default static route, floating static route, and summary static route. Static routes are configured using the **ip route** and **ipv6 route** global configuration commands. When configuring a static route, the next hop can be identified by an IP address, exit interface, or both. How the destination is specified creates one of the three following types of static route: next-hop, directly connected, and fully specified. IPv4 static routes are configured using the following global configuration command: **ip route** *network-address subnet-mask* { *ip-address* | *exit-intf* [*ip-address*] } [**distance**]. IPv6 static routes are configured using the following global configuration command: **ipv6 route** *ipv6-prefix/prefix-length* { *ipv6-address* | *exit-intf* [*ipv6-address*] } [**distance**]. The command to display an IPv4 routing table is **show ip route** and the command to display an IPv6 routing table is **show ipv6 route**.

Configure IP Static Routes

In a next-hop static route, only the next-hop IP address is specified. The exit interface is derived from the next hop. When configuring a static route, another option is to use the exit interface to specify the next-hop address. Directly connected static routes should be used only with point-to-point serial interfaces. In a fully specified static route, both the exit interface and the next-hop IP address are specified. This form of static route is used when the exit interface is a multiaccess interface and it is necessary to explicitly identify the next hop. The next hop must be directly connected to the specified exit interface. In a fully specified IPv6 static route, both the exit interface and the next-hop IPv6 address are specified. Along with **show ip route**, **show ipv6 route**, **ping**, and **traceroute**, other useful commands to verify static routes include **show ip route static**, **show ip route** *network*, and **show running-config | section ip route**. Replace **ip** with **ipv6** for the IPv6 versions of the command.

Configure IP Default Static Routes

A default route is a static route that matches all packets. A default route does not require any far-left bits to match between the default route and the destination IP address. Default static routes are commonly used when connecting an edge router to a service provider network, and a stub router. The command syntax for an IPv4 default static route is similar to any other IPv4 static route, except that the network address is 0.0.0.0 and the subnet mask is 0.0.0.0. The 0.0.0.0 0.0.0.0 in the route will match any network address. The command syntax for an IPv6 default static route is similar to any other IPv6 static route, except that the *ipv6-prefix/prefix-length* is ::/0, which matches all routes. To verify an IPv4 default static route, use the **show ip route static** command. For IPV6 use the **show ipv6 route static** command.

Configure Floating Static Routes

Floating static routes are static routes that are used to provide a backup path to a primary static or dynamic route in the event of a link failure. The floating static route is configured with a higher administrative distance than the primary route. By default, static routes have an administrative distance of 1, making them preferable to routes learned from dynamic routing protocols. The administrative distances of some common interior gateway dynamic routing protocols are EIGRP = 90, OSPF = 110, and IS-IS = 115. IP floating static routes are configured by using the **distance** argument to specify an administrative distance. If no administrative distance is configured, the default value (1) is used. The **show ip route** and **show ipv6 route** output verifies that the default routes to a router are installed in the routing table.

Configure Static Host Routes

A host route is an IPv4 address with a 32-bit mask or an IPv6 address with a 128-bit mask. There are three ways a host route can be added to the routing table: automatically installed when an IP address is configured on the router, configured as a static host route, or automatically obtained through other methods not covered in this module. Cisco IOS automatically installs a host route, also known as a local host route, when an interface address is configured on the router. A host route can be a manually configured static route to direct traffic to a specific destination device. For IPv6 static routes, the next-hop address can be the link-local address of the adjacent router; however, you must specify an interface type and an interface number when using a link-local address as the next hop. To do this, the original IPv6 static host route is removed, then a fully specified route is configured with the IPv6 address of the server and the IPv6 link-local address of the ISP router.

Packet Tracer
☐ **Activity**

Packet Tracer—Configure IPv4 and IPv6 Static and Default Routes (15.6.1)

In this Packet Tracer summary activity, you will configure static, default, and floating static routes for both the IPv4 and IPv6 protocols.

Lab—Configure IPv4 and IPv6 Static and Default Routes (15.6.2)

In this lab, you complete the following objectives:

Part 1: Build the Network and Configure Basic Device Settings

Part 2: Configure and Verify IPv4 and IPv6 Addressing on R1 and R2

Part 3: Configure and Verify Static and Default Routing for IPv4 on R1 and R2

Part 4: Configure and Verify Static and Default Routing for IPv6 on R1 and R2

Practice

The following activities provide practice with the topics introduced in this chapter. The Labs are available in the companion *Switching, Routing, and Wireless Essentials Labs and Study Guide (CCNAv7)* (ISBN 9780136634386). The Packet Tracer Activity instructions are also in the Labs & Study Guide. The PKA files are found in the online course.

Lab

Lab 15.6.2: Configure IPv4 and IPv6 Static and Default Routes

Packet Tracer Activity

Packet Tracer 15.6.1: Configure IPv4 and IPv6 Static and Default Routes

Check Your Understanding Questions

Complete all the review questions listed here to test your understanding of the sections and concepts in this chapter. The appendix "Answers to the 'Check Your Understanding' Questions" lists the answers.

1. Assume the administrator has entered the **ip route 192.168.10.0 255.255.255.0 10.10.10.2 5** command. How would an administrator test this configuration?

   A. Delete the default gateway route on the router.

   B. Manually shut down the router interface used as a primary route.

   C. Ping any valid address on the 192.168.10.0/24 network.

   D. Ping from the 192.168.10.0 network to the 10.10.10.2 address.

2. What route has the highest administrative distance?

   A. A directly connected network

   B. A route received through the EIGRP routing protocol

   C. A route received through the OSPF routing protocol

   D. A static route

3. Which route would be used to forward a packet with a source IP address of 10.10.10.1 and a destination IP address of 172.16.1.1?

   A. C 10.10.10.0/30 is directly connected, GigabitEthernet 0/1

   B. O 172.16.1.0/24 [110/65] via 10.10.200.2, 00:01:20, Serial 0/1/0

   C. S* 0.0.0.0/0 [1/0] via 172.16.1.1

   D. S 172.16.0.0/16 is directly connected, GigabitEthernet 0/0

4. Which type of static route that is configured on a router uses only the exit interface?

   A. Default static route

   B. Directly connected static route

   C. Fully specified static route

   D. Recursive static route

5. Which static route is a fully specified static route?

   A. **ip route 10.1.1.0 255.255.0.0 G0/0/1 172.16.2.2**

   B. **ip route 10.1.1.0 255.255.0.0 172.16.2.2**

   C. **ip route 10.1.1.0 255.255.0.0 172.16.2.2 5**

   D. **ip route 10.1.1.0 255.255.0.0 G0/0/1**

6. Which type of route could be configured to be a backup route for a dynamic routing protocol?

   A. Backup static route

   B. Floating static route

   C. Generic static route

   D. Summary static route

7. What is the correct syntax of a floating static route?

   A. **ip route 0.0.0.0 0.0.0.0 Serial 0/0/0**
   B. **ip route 172.16.0.0 255.248.0.0 10.0.0.1**
   C. **ip route 209.165.200.228 255.255.255.248 Serial 0/0/0**
   D. **ip route 209.165.200.228 255.255.255.248 10.0.0.1 120**

8. On which router would a default static route be configured?

   A. A router providing DHCP services to clients.
   B. A router that is connected to multiple providers.
   C. A stub router connecting to a service provider.
   D. On all routers in the network.

9. What network prefix and prefix-length combination is used to create a default static route that will match any IPv6 destination?

   A. **ipv6 route ::/0 2001:db8:acad:2::2**
   B. **ipv6 route ::/128 2001:db8:acad:2::2**
   C. **ipv6 route ::1/64 2001:db8:acad:2::2**
   D. **ipv6 route FFFF::/128 2001:db8:acad:2::2**

10. How would you test a floating static route?

   A. Delete the default gateway route on the router.
   B. Manually shut down the router interface used as a primary route.
   C. Ping any valid address on the 192.168.10.0/24 network.
   D. Ping from the 192.168.10.0 network to the 10.10.10.2 address.

CHAPTER 16

Troubleshoot Static and Default Routes

Objectives

Upon completion of this chapter, you will be able to answer the following questions:

- How does a router processes packets when a static route is configured?
- How do you troubleshoot common static and default route configuration issues?

Introduction (16.0)

Well done! You have come to the final module in the Switching, Routing, and Wireless Essentials v7.0 (SRWE) course. This course gave you the in-depth knowledge and skills you need to set up switches and routers (including wireless devices) on your growing network. You really are good at network administration!

But what makes a good network administrator into a great one? The ability to effectively troubleshoot. The best way to gain network troubleshooting skills is simple: always be troubleshooting. In this module, you will troubleshoot both static and default routes. There is a Syntax Checker, a Packet Tracer, and a hands-on Lab where you can hone your troubleshooting skills. Let's get to it!

Packet Processing with Static Routes (16.1)

In this section, you learn how a router processes packets when a static route is configured.

Static Routes and Packet Forwarding (16.1.1)

Before diving into the troubleshooting portion of this module, this section provides a brief review of how packets are forwarded in static routes. In Figure 16-1, PC1 is sending a packet to PC3.

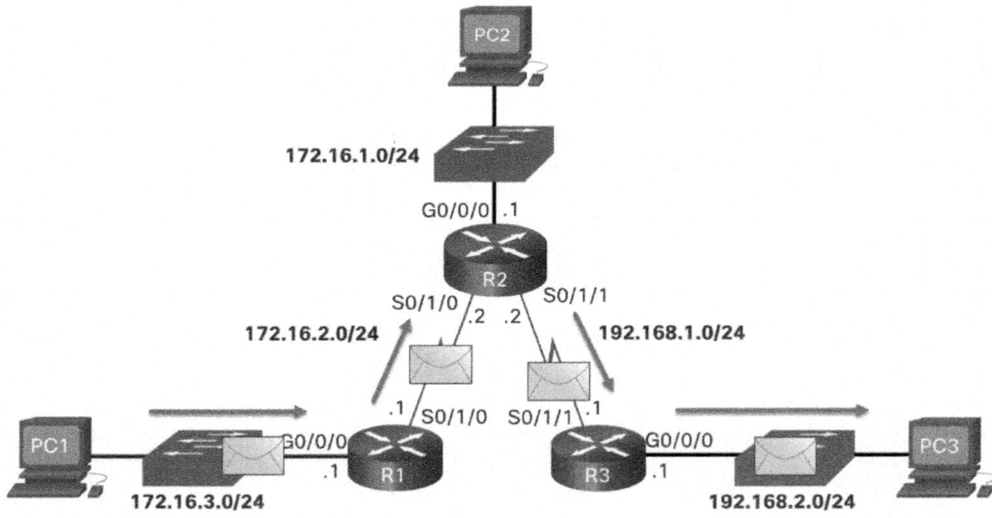

Figure 16-1 Packet Forwarding from PC1 to PC3

The following describes the packet forwarding process with static routes in the figure:

1. The packet arrives on the GigabitEthernet 0/0/0 interface of R1.
2. R1 does not have a specific route to the destination network, 192.168.2.0/24. Therefore, R1 uses the default static route.
3. R1 encapsulates the packet in a new frame. Because the link to R2 is a point-to-point link, R1 adds an "all 1s" address for the Layer 2 destination address.
4. The frame is forwarded out of the Serial 0/1/0 interface. The packet arrives on the Serial 0/1/0 interface on R2.
5. R2 de-encapsulates the frame and looks for a route to the destination. R2 has a static route to 192.168.2.0/24 out of the Serial 0/1/1 interface.
6. R2 encapsulates the packet in a new frame. Because the link to R3 is a point-to-point link, R2 adds an "all 1s" address for the Layer 2 destination address.
7. The frame is forwarded out of the Serial 0/1/1 interface. The packet arrives on the Serial 0/1/1 interface on R3.
8. R3 de-encapsulates the frame and looks for a route to the destination. R3 has a connected route to 192.168.2.0/24 out of the GigabitEthernet 0/0/0 interface.
9. R3 looks up the ARP table entry for 192.168.2.10 to find the Layer 2 Media Access Control (MAC) address for PC3. If no entry exists, R3 sends an Address Resolution Protocol (ARP) request out of the GigabitEthernet 0/0/0 interface, and PC3 responds with an ARP reply, which includes the PC3 MAC address.
10. R3 encapsulates the packet in a new frame with the MAC address of the GigabitEthernet 0/0/0 interface as the source Layer 2 address, and the MAC address of PC3 as the destination MAC address.
11. The frame is forwarded out of GigabitEthernet 0/0/0 interface. The packet arrives on the network interface card (NIC) interface of PC3.

Interactive Graphic

Check Your Understanding—Packet Processing with Static Routes (16.1.2)

Refer to the online course to complete this activity.

Troubleshoot IPv4 Static and Default Route Configuration (16.2)

In this section, you gain troubleshooting skills by solving common static and default route configuration issues.

Network Changes (16.2.1)

No matter how well you set up your network, you will have to be ready to troubleshoot some problem. Networks are frequently subject to events that can cause their status to change. For example, an interface can fail, or a service provider drops a connection. Links can become oversaturated, or an administrator may enter a wrong configuration.

When there is a change in the network, connectivity may be lost. Network administrators are responsible for pinpointing and solving the problem. To find and solve these issues, a network administrator must be familiar with tools to help isolate routing problems quickly.

Common Troubleshooting Commands (16.2.2)

Common Internetwork Operating System (IOS) troubleshooting commands include the following:

- **ping**
- **traceroute**
- **show ip route**
- **show ip interface brief**
- **show cdp neighbors detail**

Figure 16-2 shows the topology used to demonstrate these commands.

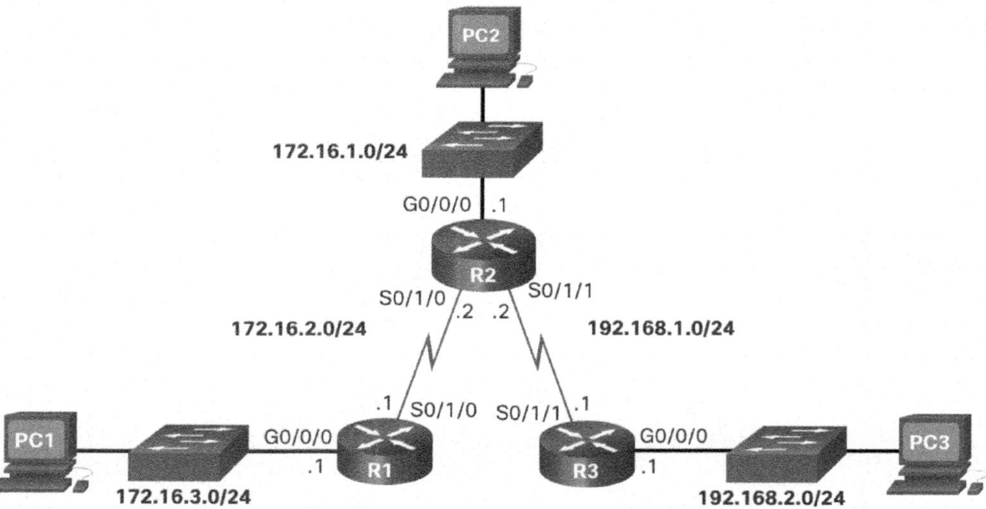

Figure 16-2 Reference Topology for Troubleshooting Static and Default Routes

Example 16-1 displays the result of an extended ping from the source interface of R1 to the LAN interface of R3. An extended ping is an enhanced version of the ping utility. Extended ping enables you to specify the source IP address for the ping packets.

Example 16-1 The Extended Ping Command

```
R1# ping 192.168.2.1 source 172.16.3.1
Type escape sequence to abort.
Sending 5, 100-byte ICMP Echos to 192.168.2.1, timeout is 2 seconds:
Packet sent with a source address of 172.16.3.1
!!!!!
Success rate is 100 percent (5/5), round-trip min/avg/max = 3/3/5 ms
R1#
```

Example 16-2 displays the result of a **traceroute** from R1 to the R3 LAN. Note that each hop route returns an Internet Control Message Protocol (ICMP) reply.

Example 16-2 The **traceroute** Command

```
R1# traceroute 192.168.2.1
Type escape sequence to abort.
Tracing the route to 192.168.2.1
VRF info: (vrf in name/id, vrf out name/id)
  1 172.16.2.2 1 msec 2 msec 1 msec
  2 192.168.1.1 2 msec 3 msec *
R1#
```

The **show ip route** command in Example 16-3 displays the routing table of R1.

Example 16-3 The **show ip route** Command

```
R1# show ip route | begin Gateway
Gateway of last resort is not set
      172.16.0.0/16 is variably subnetted, 5 subnets, 2 masks
S        172.16.1.0/24 [1/0] via 172.16.2.2
C        172.16.2.0/24 is directly connected, Serial0/1/0
L        172.16.2.1/32 is directly connected, Serial0/1/0
C        172.16.3.0/24 is directly connected, GigabitEthernet0/0/0
L        172.16.3.1/32 is directly connected, GigabitEthernet0/0/0
S     192.168.1.0/24 [1/0] via 172.16.2.2
S     192.168.2.0/24 [1/0] via 172.16.2.2
R1#
```

A quick status of all interfaces on the router is shown using the **show ip interface brief** command in Example 16-4.

Example 16-4 The **show ip interface brief** Command

```
R1# show ip interface brief
Interface              IP-Address      OK? Method Status                Protocol
GigabitEthernet0/0/0   172.16.3.1      YES manual up                    up
GigabitEthernet0/0/1   unassigned      YES unset  up                    up
Serial0/1/0            172.16.2.1      YES manual up                    up
Serial0/1/1            unassigned      YES unset  up                    up
R1#
```

The **show cdp neighbors** command shown in Example 16-5 provides a list of directly connected Cisco devices. This command validates Layer 2 (and therefore Layer 1) connectivity. For example, if a neighbor device is listed in the command output, but it cannot be pinged, Layer 3 addressing should be investigated.

Example 16-5 The **show cdp neighbors** Command

```
R1# show cdp neighbors
Capability Codes: R - Router, T - Trans Bridge, B - Source Route Bridge
                  S - Switch, H - Host, I - IGMP, r - Repeater, P - Phone,
                  D - Remote, C - CVTA, M - Two-port Mac Relay
Device ID         Local Intrfce     Holdtme    Capability  Platform  Port ID
Switch            Gig 0/0/1         129              S I   WS-C3560- Fas 0/5
R2                Ser 0/1/0         156            R S I   ISR4221/K Ser 0/1/0
R3                Ser 0/1/1         124            R S I   ISR4221/K Ser 0/1/0
Total cdp entries displayed : 3
R1#
```

Solve a Connectivity Problem (16.2.3)

Finding a missing (or misconfigured) route is a relatively straightforward process if the right tools are used in a methodical manner.

For instance, the user at PC1 in Figure 16-2 reports that he cannot access resources on the R3 LAN. This can be confirmed by pinging the LAN interface of R3 using the LAN interface of R1 as the source.

The following troubleshooting commands can be used to solve the connectivity problem.

Ping the Remote LAN

The network administrator can test connectivity between the two LANs from R1 instead of PC1. This can be done by sourcing the ping from the G0/0/0 interface on R1 to the G0/0/0 interface on R3, as shown in Example 16-6.

Example 16-6 Ping the Remote LAN

```
R1# ping 192.168.2.1 source g0/0/0
Type escape sequence to abort.
Sending 5, 100-byte ICMP Echos to 192.168.2.1, timeout is 2 seconds:
Packet sent with a source address of 172.16.3.1
.....
Success rate is 0 percent (0/5)
R1#
```

The ping results show that there is no connectivity between these LANs.

Ping the Next-Hop Router

Next, a ping to the S0/1/0 interface on R2 is successful, as shown in Example 16-7.

Example 16-7 Ping the Next-Hop Router

```
R1# ping 172.16.2.2
Type escape sequence to abort.
Sending 5, 100-byte ICMP Echos to 172.16.2.1, timeout is 2 seconds:
!!!!!
Success rate is 100 percent (5/5), round-trip min/avg/max = 3/3/4 ms
R1#
```

This ping is sourced from the S0/1/0 interface of R1. Therefore, the issue is not loss of connectivity between R1 and R2.

Ping R3 LAN from S0/1/0

A ping from R1 to the R3 interface 192.168.2.1 is successful as well, as shown in Example 16-8.

Example 16-8 Ping R3 LAN from S0/1/0

```
R1# ping 192.168.2.1
Type escape sequence to abort.
Sending 5, 100-byte ICMP Echos to 192.168.2.1, timeout is 2 seconds:
!!!!!
Success rate is 100 percent (5/5), round-trip min/avg/max = 3/3/4 ms
R1#
```

This ping is sourced from the S0/1/0 interface on R1. R3 has a route back to the network between R1 and R2, 172.16.2.0/24. This confirms that R1 can reach the remote LAN on R3. However, packets sourced from the LAN on R1 cannot. This indicates that either R2 or R3 may have an incorrect or missing route to the LAN on R1.

Verify the R2 Routing Table

The next step is to investigate the routing tables of R2 and R3. The routing table for R2 is shown in Example 16-9. Notice that the 172.16.3.0/24 network is configured incorrectly. The static route to the 172.16.3.0/24 network has been configured using the next-hop address 192.168.1.1. Therefore, packets destined for the 172.16.3.0/24 network are sent back to R3 instead of to R1.

Example 16-9 Verify the R2 Routing Table

```
R2# show ip route | begin Gateway
Gateway of last resort is not set
      172.16.0.0/16 is variably subnetted, 5 subnets, 2 masks
C        172.16.1.0/24 is directly connected, GigabitEthernet0/0/0
L        172.16.1.1/32 is directly connected, GigabitEthernet0/0/0
C        172.16.2.0/24 is directly connected, Serial0/1/0
L        172.16.2.2/32 is directly connected, Serial0/1/0
S        172.16.3.0/24 [1/0] via 192.168.1.1
      192.168.1.0/24 is variably subnetted, 2 subnets, 2 masks
C        192.168.1.0/24 is directly connected, Serial0/1/1
L        192.168.1.2/32 is directly connected, Serial0/1/1
S     192.168.2.0/24 [1/0] via 192.168.1.1
R2#
```

Correct the R2 Static Route Configuration

Next, the running configuration does, in fact, reveal the incorrect **ip route** statement. In Example 16-10, the incorrect route is removed, and the correct route is then entered.

Example 16-10 Correct the R2 Static Route Configuration

```
R2# show running-config | include ip route
ip route 172.16.3.0 255.255.255.0 192.168.1.1
ip route 192.168.2.0 255.255.255.0 192.168.1.1
R2#
R2# configure terminal
Enter configuration commands, one per line.  End with CNTL/Z.
R2(config)# no ip route 172.16.3.0 255.255.255.0 192.168.1.1
R2(config)# ip route 172.16.3.0 255.255.255.0 172.16.2.1
R2(config)#
```

Verify New Static Route Is Installed

In Example 16-11, the routing table on R2 is checked again to confirm that the route entry to the LAN on R1, 172.16.3.0, is correct and pointing toward R1.

Example 16-11 Verify New Static Route Is Installed

```
R2(config)# exit
R2#
*Sep 20 02:21:51.812: %SYS-5-CONFIG_I: Configured from console by console
R2# show ip route | begin Gateway
Gateway of last resort is not set
      172.16.0.0/16 is variably subnetted, 5 subnets, 2 masks
C        172.16.1.0/24 is directly connected, GigabitEthernet0/0/0
L        172.16.1.1/32 is directly connected, GigabitEthernet0/0/0
C        172.16.2.0/24 is directly connected, Serial0/1/0
L        172.16.2.2/32 is directly connected, Serial0/1/0
S        172.16.3.0/24 [1/0] via 172.16.2.1
      192.168.1.0/24 is variably subnetted, 2 subnets, 2 masks
C        192.168.1.0/24 is directly connected, Serial0/1/1
L        192.168.1.2/32 is directly connected, Serial0/1/1
S     192.168.2.0/24 [1/0] via 192.168.1.1
R2#
```

Ping the Remote LAN Again

Next, in Example 16-12, a ping from R1 sourced from G0/0/0 is used to verify that R1 can now reach the LAN interface of R3. As a last step in confirmation, the user on PC1 should also test connectivity to the 192.168.2.0/24 LAN.

Example 16-12 Ping the Remote LAN Again

```
R1# ping 192.168.2.1 source g0/0/0
Type escape sequence to abort.
Sending 5, 100-byte ICMP Echos to 192.168.2.1, timeout is 2 seconds:
Packet sent with a source address of 172.16.3.1
!!!!!
Success rate is 100 percent (5/5), round-trip min/avg/max = 4/4/4 ms
R1#
```

Interactive Graphic

Syntax Checker—Troubleshoot IPv4 Static and Default Routes (16.2.4)

Refer to the online course to complete this activity.

Summary (16.3)

The following is a summary of each section in the module:

Packet Processing with Static Routes

1. The packet arrives on the interface of R1.
2. R1 does not have a specific route to the destination network; therefore, R1 uses the default static route.
3. R1 encapsulates the packet in a new frame. Because the link to R2 is a point-to-point link, R1 adds an "all 1s" address for the Layer 2 destination address.
4. The frame is forwarded out of the appropriate interface. The packet arrives on the interface on R2.
5. R2 de-encapsulates the frame and looks for a route to the destination. R2 has a static route to the destination network out of one of its interfaces.
6. R2 encapsulates the packet in a new frame. Because the link to R3 is a point-to-point link, R2 adds an "all 1s" address for the Layer 2 destination address.
7. The frame is forwarded out of the appropriate interface. The packet arrives on the interface on R3.
8. R3 de-encapsulates the frame and looks for a route to the destination. R3 has a connected route to the destination network out of one of its interfaces.
9. R3 looks up the ARP table entry for the destination network to find the Layer 2 MAC address for PC3. If no entry exists, R3 sends an ARP request out of one of its interfaces, and PC3 responds with an ARP reply, which includes the PC3 MAC address.
10. R3 encapsulates the packet in a new frame with the MAC address of the appropriate interface as the source Layer 2 address and the MAC address of PC3 as the destination MAC address.
11. The frame is forwarded out of the appropriate interface. The packet arrives on the network interface card (NIC) interface of PC3.

Troubleshoot IPv4 Static and Default Route Configuration

Networks are frequently subject to events that can cause their status to change. An interface can fail, or a service provider drops a connection. Links can become oversaturated, or an administrator may enter a wrong configuration. Common IOS troubleshooting commands include the following:

- **ping**
- **traceroute**
- **show ip route**
- **show ip interface brief**
- **show cdp neighbors detail**

Packet Tracer—Troubleshoot Static and Default Routes (16.3.1)

In this activity you troubleshoot static and default routes and repair any errors that you find.

Troubleshoot IPv4 static routes.

Troubleshoot IPv6 static routes.

Configure IPv4 static routes.

Configure IPv4 default routes.

Configure IPv6 static routes.

Lab—Troubleshoot Static and Default Routes (16.3.2)

In this lab, you complete the following objectives:

- Evaluate Network Operation.
- Gather information, create an action plan, and implement corrections.

Practice

The following activities provide practice with the topics introduced in this chapter. The Labs are available in the companion *Switching, Routing, and Wireless Essentials Labs and Study Guide (CCNAv7)* (ISBN 9780136634386). The Packet Tracer Activity instructions are also in the Labs & Study Guide. The PKA files are found in the online course.

Lab

Lab 16.3.2: Troubleshoot IPv4 and IPv6 Static and Default Routes

Packet Tracer Activity

Packet Tracer 16.3.1: Troubleshoot Static and Default Routes

Check Your Understanding Questions

Complete all the review questions listed here to test your understanding of the sections and concepts in this chapter. The appendix "Answers to the 'Check Your Understanding' Questions" lists the answers.

1. Which three IOS troubleshooting commands can help to isolate problems with a static route? (Choose three.)
   A. **ping**
   B. **show arp**
   C. **show ip interface brief**
   D. **show ip route**
   E. **show version**
   f. **tracert**

2. What happens to a static route entry in a routing table when the outgoing interface associated with that route goes into the down state?
   A. The router automatically redirects the static route to use another interface.
   B. The router polls neighbors for a replacement route.
   C. The static route is removed from the routing table.
   D. The static route remains in the table because it was defined as static.

3. What action will a router take to forward a frame if it does not have an entry in the ARP table to resolve a destination MAC address?
   A. Sends a DNS request
   B. Drops the frame
   C. Sends an ARP request
   D. Sends frame to the default gateway

4. You cannot ping a directly connected host. Which IOS command can be used to validate Layer 1 and Layer 2 connectivity?

   A. **ping**
   B. **show cdp neighbors detail**
   C. **show ip interface brief**
   D. **show ip route**
   E. **traceroute**

5. A network administrator has entered a static route to an Ethernet LAN that is connected to an adjacent router. However, the route is not shown in the routing table. Which command would the administrator use to verify that the exit interface is up?

   A. **ping**
   B. **show cdp neighbors detail**
   C. **show ip interface brief**
   D. **show ip route**
   E. **traceroute**

Answers to the "Check Your Understanding" Questions

Chapter 1

1. C. Interface VLAN 1 is the default management SVI.
2. A and B. The Switch prompt typically occurs after a switch boots normally but does not have or has failed to load a startup configuration file.
3. A and E. In full-duplex operation, the NIC does not process frames any faster, the data flow is bidirectional, and there are no collisions.
4. C. The port speed LED indicates that the port speed mode is selected. When selected, the port LEDs will display colors with different meanings. If the LED is off, the port is operating at 10 Mbps. If the LED is green, the port is operating at 100 Mbps. If the LED is blinking green, the port is operating at 1000 Mbps.
5. B. The switch boot loader environment is presented when the switch cannot locate a valid operating system. The boot loader environment provides a few basic commands that allow a network administrator to reload the operating system or provide an alternate location of the operating system.
6. C. The **show interfaces** command is useful to detect media errors, to see if packets are being sent and received, and to determine if any runts, giants, CRCs, interface resets, or other errors have occurred. Problems with reachability to a remote network would likely be caused by a misconfigured default gateway or other routing issue, not a switch issue. The **show mac address-table** command shows the MAC address of a directly attached device.
7. B. SSH provides security for remote management connections to a network device. SSH does so through encryption for session authentication (username and password) as well as for data transmission. Telnet sends a username and password in plain text, which can be targeted to obtain the username and password through data capture. Both Telnet and SSH use TCP, support authentication, and connect to hosts in CLI.
8. A. The loopback interface is a logical interface internal to the router and is automatically placed in an UP state, as long as the router is functioning. It is not assigned to a physical port and can therefore never be connected to any other device. Multiple loopback interfaces can be enabled on a router.
9. A and B. The **show ip interface brief** command displays the IPv4 address of each interface, as well as the operational status of the interfaces at both Layer 1 and Layer 2. In order to see interface descriptions and speed and duplex settings, use the **show running-config interface** command. Next-hop addresses are displayed in the routing table with the **show ip route** command, and the MAC address of an interface can be seen with the **show interfaces** command.
10. D. When connecting to switches without the auto-MDIX feature, straight-through cables must be used to connect to devices such as

servers, workstations, or routers. Crossover cables must be used to connect to other switches or repeaters.

11. A. The loopback interface is useful in testing and managing a Cisco IOS device because it ensures that at least one interface will always be available. For example, it can be used for testing purposes, such as testing internal routing processes, by emulating networks behind the router.

12. D. When authenticating SSH users with the **login local** command, a username and password pair must be created and added to the local database. Otherwise, authentication would never be successful.

Chapter 2

1. A. A switch builds a MAC address table of MAC addresses and associated Ethernet switch port numbers by examining the source MAC address found in inbound frames. To forward a frame onward, the switch makes its forwarding decision on Layer 2 information; therefore, the switch examines the destination MAC address, looks in the MAC address for a port number associated with that destination MAC address, and sends it to the specific port. If the destination MAC address is not in the table, the switch forwards the frame out all ports except the inbound port that originated the frame.

2. B. Cisco LAN switches use the MAC address table to make traffic forwarding decisions. The decisions are based on the ingress port and the destination MAC address of the frame. The ingress port information is important because it carries the VLAN to which the port belongs.

3. D. When a switch receives a frame with a source MAC address that is not in the MAC address table, the switch will add that MAC address to the table and map that address to a specific port. Switches do not use IP addressing in the MAC address table.

4. D and F. A switch has the ability to create temporary point-to-point connections between the directly-attached transmitting and receiving network devices. The two devices have full-bandwidth, full-duplex connectivity during the transmission. Segmenting adds collision domains to reduce collisions.

5. D. If the destination MAC address is in the table, it will forward the frame out of the specified port.

6. C. If the destination MAC address is not in the table, the switch will forward the frame out all ports except the incoming port. This is called an unknown unicast.

7. C. If the destination MAC address is a broadcast or a multicast, the frame is also flooded out all ports except the incoming port.

8. D. In store-and-forward switching, the switch compares the frame check sequence (FCS) value in the last field of the datagram against its own FCS calculations. If the frame is error free, the switch forwards the frame. Otherwise, the frame is dropped.

9. B. Cut-through switching has the ability to perform rapid frame switching, which means the switch can make a forwarding decision as soon as it has looked up the destination MAC address of the frame in its MAC address table.

10. B. Full-duplex communication allows both ends to transmit and receive simultaneously,

offering 100 percent efficiency in both directions for a 200 percent potential use of stated bandwidth. Half-duplex communication is unidirectional, or one direction at a time. Gigabit Ethernet and 10 Gbps NICs require full-duplex to operate and do not support half-duplex operation.

Chapter 3

1. B, C, and D. A management VLAN is a VLAN that is configured to manage features of the switch. By default, all ports are members of the default VLAN. An 802.1Q trunk port supports both tagged and untagged traffic.
2. C. A native VLAN is the VLAN that does not receive a VLAN tag in the IEEE 802.1Q frame header. Cisco best practices recommend the use of an unused VLAN (not a data VLAN, the default VLAN of VLAN 1, or the management VLAN) as the native VLAN whenever possible.
3. B and C. Cost reduction and improved IT staff efficiency are all benefits of using VLANs, along with higher performance, broadcast storm mitigation, and simpler project and application management. End users are not usually aware of VLANs, and VLANs do require configuration. Because VLANs are assigned to access ports, they do not reduce the number of trunk links. VLANs increase security by segmenting traffic.
4. A. The **show interfaces switchport** command displays the following information for a given port: Switchport, Administrative Mode, Operational Mode, Administrative Trunking Encapsulation, Operational Trunking Encapsulation, Negotiation of Trunking, Access Mode VLAN, Trunking Native Mode VLAN, Administrative Native VLAN tagging, Voice VLAN.
5. C. Entering the **switchport access vlan 3** interface config command on Fa0/1 replaces the current port VLAN assignment from VLAN 2 to VLAN 3.
6. B. To restore a Catalyst switch to its factory default condition, unplug all cables except the console and power cable from the switch. Then enter the **erase startup-config** privileged EXEC mode command followed by the **delete vlan.dat** command and reboot the switch.
7. C and D. Extended range VLANs are stored in the running-configuration file by default and must be saved after being configured. Extended VLANs use the VLAN IDs from 1006 to 4094.
8. D. Any ports that are not moved to an active VLAN cannot communicate with other hosts after the VLAN is deleted. They must be assigned to an active VLAN or their VLAN must be created.
9. D and E. To enable trunking from a Cisco switch to a device that does not support DTP, use the **switchport mode trunk** and **switchport nonegotiate** interface configuration mode commands. This causes the interface to become a trunk, but it will not generate DTP frames.

Chapter 4

1. B. Using legacy inter-VLAN routing to interconnect four VLANs would require four separate physical interfaces. Therefore, the best router-based solution is to configure a router-on-a-stick.

2. C. Router-on-a-stick requires one interface configured as subinterfaces for each VLAN.

3. A. The subinterface must be assigned to VLAN 10 using the **encapsulation dot1q 10** command. The **encapsulation vlan 10** option is not a valid command and the **switchport mode** options are switch configuration commands.

4. A. host must have a default gateway configured. Hosts on VLANs must have their default gateway configured on a router subinterface to provide inter-VLAN routing services.

5. D. The switch port must be configured as a trunk, and the VLANs on the switch must have users connected to them.

6. A and B. Legacy (traditional) inter-VLAN routing would require more ports, and the configuration can be more complex than a router-on-a-stick solution.

7. D. The **encapsulation dot1q** *vlan_id* [**native**] command configures the subinterface to respond to 802.1Q encapsulated traffic from the specified *vlan-id*. The **native** keyword option is only appended to set the native VLAN to something other than VLAN 1.

8. A and B. The router-on-a-stick method requires one physical Ethernet router interface to route traffic between multiple VLANs on a network. The router interface is configured using software-based virtual subinterfaces to identify routable VLANs. Modern, enterprise networks rarely use router-on-a-stick because it does not scale easily to meet requirements, and multiple subinterfaces may impact the traffic flow speed. In these very large networks, network administrators use Layer 3 switches to configure inter-VLAN routing.

9. A and B. A routed port is created on a Layer 3 switch by disabling the switchport feature on a Layer 2 port using the **no switchport interface** configuration command. Then the interface can be configured with an IPv4 configuration to connect to a router or another Layer 3 switch. Only Layer 2 ports can be assigned to a VLAN or support trunking.

10. A and B. Modern, enterprise networks rarely implement inter-VLAN routing using the router-on-a-stick method. Instead, they use faster Layer 3 switches because they use hardware-based switching to achieve higher-packet processing rates than routers. Layer 3 switches provide a much more scalable method to provide inter-VLAN routing.

Chapter 5

1. A, C, and E. The three components that are combined to form a bridge ID are bridge priority, extended system ID, and MAC address.

2. D. The root port is the port with the lowest cost to reach the root bridge. Every non-root switch must have a root port.

3. C. Cisco switches running IOS 15.0 or later run PVST+ by default. Cisco Catalyst switches support PVST+, Rapid PVST+, and MSTP. However, only one version can be active at any time.

4. B. PVST+ results in optimum load balancing. However, this is accomplished by manually configuring switches to be elected as root bridges for different VLANs on the network. The root bridges are not automatically selected. Furthermore, having spanning tree instances for each VLAN actually consumes

more bandwidth, and it increases the CPU cycles for all the switches in the network.

5. C and D. Switches learn MAC addresses at the learning and forwarding port states. They receive and process BPDUs at the blocking, listening, learning, and forwarding port states.

6. C and D. Spanning Tree Protocol (STP) is required to ensure correct network operation when designing a network with multiple interconnected Layer 2 switches or using redundant links to eliminate single points of failure between Layer 2 switches. Routing is a Layer 3 function and does not relate to STP. VLANs do reduce the number of broadcast domains but relate to Layer 3 subnets, not STP.

7. E. When all switches are configured with the same default bridge priority (that is, 32,768), the lowest MAC address becomes the deciding factor for the election of the root bridge.

8. A. If switch access ports are configured as edge ports using PortFast, BPDUs should never be received on those ports. Cisco switches support a feature called BPDU guard. When it is enabled, BPDU guard will put an edge port in an error-disabled state if a BPDU is received by the port. This will prevent a Layer 2 loop occurring.

9. D. STP allows redundant physical connections between Layer 2 devices without creating Layer 2 loops by disabling ports that could create a loop.

10. A and E. PortFast-enabled ports immediately transition from blocking to forwarding state. PortFast should be enabled only on access ports connecting end devices. No BPDUs should ever be received through a port that is configured with PortFast.

Chapter 6

1. B. Increasing the link speed does not scale very well. Adding more VLANs will not reduce the amount of traffic that is flowing across the link. Inserting a router between the switches will not improve congestion.

2. E and F. Source MAC and destination MAC load balancing and source IP and destination IP load balancing are two implementation methods used in EtherChannel technology.

3. B. PAgP is used to automatically aggregate multiple ports into an EtherChannel bundle, but it works only between Cisco devices. LACP can be used for the same purpose between Cisco and non-Cisco devices. PAgP must have the same duplex mode at both ends and can use two ports or more. The number of ports depends on the switch platform or module. An EtherChannel aggregated link is seen as one port by the spanning tree algorithm.

4. A and C. The two protocols that can be used to form an EtherChannel are PAgP (Cisco proprietary) and LACP, also known as IEEE 802.3ad. STP (Spanning Tree Protocol) or RSTP (Rapid Spanning Tree Protocol) is used to avoid loops in a Layer 2 network. EtherChannel is the term that describes the bundling of two or more links that are treated as a single link for spanning tree and configuration.

5. C. Switch 1 and switch 2 will establish an EtherChannel if both sides are set to desirable, because both sides will negotiate the link. A channel can also be established if both sides are set to on, or if one side is set to auto and the other to desirable. Setting one switch to on will prevent that switch from negotiating the formation of an EtherChannel bundle.

6. A. The **channel-group mode active** command enables LACP unconditionally, and the **channel-group mode passive** command enables LACP only if the port receives an LACP packet from another device. The **channel-group mode desirable** command enables PAgP unconditionally, and the **channel-group mode auto** command enables PAgP only if the port receives a PAgP packet from another device.

7. B. The **channel-group mode active** command enables LACP unconditionally, and the **channel-group mode passive** command enables LACP only if the port receives an LACP packet from another device. The **channel-group mode desirable** command enables PAgP unconditionally, and the **channel-group mode auto** command enables PAgP only if the port receives a PAgP packet from another device.

8. D. An EtherChannel is formed by combining multiple (same type) Ethernet physical links so they are seen and configured as one logical link. It provides an aggregated link between two switches. Currently each EtherChannel can consist of up to eight compatibly configured Ethernet ports.

9. A and B. LACP is part of an IEEE specification (802.3ad) that enables several physical ports to automatically be bundled to form a single EtherChannel logical channel. LACP allows a switch to negotiate an automatic bundle by sending LACP packets to the peer. It performs a function similar to PAgP with Cisco EtherChannel, but it can be used to facilitate EtherChannels in multivendor environments. Cisco devices support both PAgP and LACP configurations.

10. A, C, and F. Speed and duplex settings must match for all interfaces in an EtherChannel. All interfaces in the EtherChannel must be in the same VLAN if the ports are not configured as trunks. Any ports may be used to establish an EtherChannel. SNMP community strings and port security settings are not relevant to EtherChannel.

Chapter 7

1. B. When a DHCP client receives DHCPOFFER messages, it will send a broadcast DHCPREQUEST message for two purposes. First, it indicates to the offering DHCP server that it would like to accept the offer and bind the IPv4 address. Second, it notifies any other responding DHCP servers that their offers are declined.

2. C. The DHCPREQUEST message is broadcast to inform other DHCP servers that an IPv4 address has been leased.

3. B. When a DHCPv4 client does not have an IPv4 address, a DHCPv4 server will reply with a broadcast DHCPOFFER or a unicast DHCPOFFER message back to the DHCPv4 client MAC address.

4. D. When a DHCP client lease is about to expire, the client sends a DHCPREQUEST message to the DHCPv4 server that originally provided the IPv4 address. This allows the client to request that the lease be extended.

5. B. By default, the **ip helper-address** command forwards the following eight UDP services:

   Port 37: Time

   Port 49: TACACS

   Port 53: DNS

Port 67: DHCP/BOOTP client

Port 68: DHCP/BOOTP server

Port 69: TFTP

Port 137: NetBIOS name service

Port 138: NetBIOS datagram service

6. C. The **ip address dhcp** command activates the DHCPv4 client on a given interface. By doing this, the router will obtain the IPv4 parameters from a DHCPv4 server.

7. B and D. SOHO routers are frequently required by the ISP to be configured as DHCPv4 clients in order to be connected to the provider.

8. C. The DHCP server is not on the same network as the hosts, so DHCP relay agent is required. This is achieved by issuing the **ip helper-address** command on the interface of the router that contains the DHCPv4 clients, in order to direct DHCP messages to the DHCPv4 server IPv4 address.

9. B. The router functioning as the DHCPv4 server assigns all IPv4 addresses in a DHCPv4 address pool except addresses specified by the **ip dhcp excluded-address** *low-address* [*high-address*] global config command.

10. D. The **ipconfig /release** Windows command releases the current host IPv4 configuration and the **ipconfig /renew** Windows command attempts to renew the IPv4 addressing with the DHCPv4 server.

11. A. The **show ip dhcp binding** command will show the leases, including IPv4 addresses, MAC addresses, lease expiration, type of lease, client ID, and username.

12. D. The client broadcasts a DHCPDISCOVER message to identify any available DHCP servers on the network. A DHCP server replies with a DHCPOFFER message. This message offers to the client a lease that contains such information as the IPv4 address and subnet mask to be assigned, the IPv4 address of the DNS server, and the IPv4 address of the default gateway. After the client receives the lease, the received information must be renewed through another DHCPREQUEST message prior to the lease expiration.

Chapter 8

1. C. When a PC is configured to use the SLAAC method for configuring IPv6 addresses, it will use the prefix and prefix-length information that is contained in the RA message, combined with a 64-bit interface ID (obtained by using the EUI-64 process or by using a random number that is generated by the client operating system), to form an IPv6 address. It uses the link-local address of the router interface that is attached to the LAN segment as its IPv6 default gateway address.

2. C. ICMPv6 RA messages contain flags to indicate whether a workstation should use SLAAC, a DHCPv6 server, or a combination to configure its IPv6 address. The A flag determines whether to use SLAAC. The O flag indicates whether to use a stateless DHCPv6 server. The M flag indicates whether to use stateful DHCPv6. The M and O flags are independent of SLAAC.

3. B. In stateless DHCPv6 configuration, a client configures its IPv6 address by using the prefix and prefix length in the RA message, combined with a self-generated interface ID. It then contacts a DHCPv6 server for additional configuration information via an INFORMATION-REQUEST message.

The DHCPv6 SOLICIT message is used by a client to locate a DHCPv6 server. The DHCPv6 ADVERTISE message is used by DHCPv6 servers to indicate their availability for DHCPv6 service. The DHCPv6 REQUEST message is used by a client, in the stateful DHCPv6 configuration, to request ALL configuration information from a DHCPv6 server.

4. D. SLAAC and stateless DHCPv6 enable clients to use ICMPv6 Router Advertisement (RA) messages to automatically assign IPv6 addresses to themselves, and also allow these clients to contact a stateless DHCPv6 server to obtain additional information, such as the domain name and address of DNS servers. Because the M flag is 0 by default, stateful DHCPv6 will not be used. RA messages are used to automatically create an interface IPv6 address.
5. D. SLAAC is a stateless allocation method and does not use a DHCP server to manage the IPv6 addresses. When a host generates an IPv6 address, it must verify that it is unique. The host will send an ICMPv6 Neighbor Solicitation message with its own IPv6 address as the target. As long as no other device responds with a Neighbor Advertisement message, the address is unique.
6. D. The EUI-64 process uses the MAC address of an Ethernet interface to construct an interface ID (IID). Because the MAC address is only 48 bits in length, 16 additional bits (FF:FE) must be added to the MAC address to create the full 64-bit interface ID. The 7th bit is flipped, which modifies the second hex digit of the interface id.
7. A. Under stateful DHCPv6 configuration, which is indicated by setting M flag as 1 (through the **ipv6 nd managed-config-flag** interface command), the dynamic IPv6 address assignments are managed by the DHCPv6 server. Clients must obtain all configuration information from a DHCPv6 server. The A flag determines whether to use SLAAC.
8. B. For a router to be able to send RA messages, it must be enabled as an IPv6 router using the **ipv6 unicast-routing** global config command.
9. C. When the A flag is set to 1 (default) the client will use SLAAC to configure its GUA address. When M flag is 0 and O flag is 1, a client will look for other configuration parameters (such as DNS server addresses) from a stateless DHCPv6 server.
10. B. Unless a device has been configured statically with a default gateway address, the device can only obtain its default gateway dynamically from the Router Advertisement message. The device will use the link-local address of the router interface, the source IPv6 address of the RA, that is attached to the LAN segment as its IPv6 default gateway address.

Chapter 9

1. B. Hosts send traffic to their default gateways, which is the virtual IP address and the virtual MAC address. The virtual IP address is assigned by the administrator, whereas the virtual MAC address is created automatically by HSRP. The virtual IPv4 and MAC addresses provide consistent default gateway addressing for the end devices. Only the HSRP active router responds to the virtual IP and virtual MAC address.

2. B. Hosts send traffic to their default gateway, which is the virtual IP address and the virtual MAC address. The virtual IP address is assigned by the administrator, whereas the virtual MAC address is created automatically by HSRP. The virtual IPv4 and MAC addresses provide consistent default gateway addressing for the end devices. Only the HSRP active router responds to the virtual IP and virtual MAC address.

3. A. VRRP selects a master router and one or more other routers as backup router. Backup VRRP backup routers monitor the VRRP master router.

4. A. HSRP and GLBP are Cisco proprietary protocols, and VRRP is an IEEE non-proprietary open standard protocol.

5. D. HSRP is a FHRP that provides Layer 3 default gateway redundancy.

6. C. In the Learn state, the router has not determined the virtual IP address and has not yet seen a hello message from the active router. In this state, the router waits to hear from the active router.

7. D. VRRP is a non-proprietary election protocol that dynamically assigns responsibility for one or more virtual routers to the VRRP routers on an IPv4 LAN.

8. D. When frames are sent from HSRP host devices to the default gateway, the destination MAC address of the frame is the virtual router MAC address.

9. D. When the active router fails, the standby router stops seeing hello messages, assumes the role of the forwarding router, and the host devices see no disruption in service.

10. A. GLBP is a Cisco proprietary FHRP protocol that provides redundancy and load balancing (also called load sharing) between a group of redundant routers.

11. C. To force a new HSRP election process when a higher priority router comes online, preemption must be enabled using the **standby preempt** interface command.

Chapter 10

1. B. Ransomware encrypts the data on a host and locks access to it until a ransom is paid.

2. A. An ESA is a network security device that is specifically designed to monitor and secure SMTP traffic.

3. D. Authorization determines which resources the user can access and which operations the user is allowed to perform.

4. A. Local AAA stores usernames and passwords locally in the Cisco router, and users authenticate against the local database. Local AAA is ideal for small networks.

5. A. A switch or wireless access point are 802.1X authenticators in between the client and the authentication server. Authenticators request identifying information from the client, verify that information with the authentication server, and relay a response to the client.

6. D. The supplicant is the client that is requesting network access.

7. C. Port security prevents many types of attacks, including MAC address table overflow.

8. A. Dynamic ARP Inspection (DAI) prevents ARP spoofing and ARP poisoning attacks.

9. D. IP Source Guard (IPSG) prevents MAC and IP address spoofing.

10. C. A MAC address table attack will fill the MAC address table. When the MAC address table is full, the switch treats the frame as an unknown unicast and begins to flood all incoming traffic to all ports only within the local VLAN.

11. A. MAC address table attacks are conducted to overwhelm a switch to disregard the MAC address table entries and instead forward incoming traffic out all ports. Threat actors connected to the LAN can then capture traffic using a protocol analyzer such as Wireshark.

12. D. DHCP starvation attacks occur when a threat actor requests and receives all the available IP addresses for a subnet.

13. E. A threat actor sending BPDU messages with a priority of 0 is trying to become the root bridge in the STP topology.

14. A. Address spoofing attacks occur when the threat actor changes the MAC and/or IP address of the threat actor's device to pose as another legitimate device, such as the default gateway.

15. F. A threat actor can send a gratuitous ARP reply causing all devices to believe that the threat actor's device is a legitimate device, such as the default gateway.

16. C. A threat actor can use packet sniffing software, such as Wireshark, to view the contents of CDP messages, which are sent unencrypted and include a variety of device information, including the IOS version and IP addresses. CDP and LLDP should not be enabled on edge devices and should be disabled globally or on a per-interface basis if not required.

Chapter 11

1. A. Port security can be configured on switches to assist in preventing the MAC address table from being overwhelmed with invalid MAC addresses. ACLs will not assist a switch in filtering broadcast traffic, and increasing the size of the CAM table or the speed of switch ports will not resolve this issue.

2. B. When a violation occurs on a switch port that is configured for port security with the shutdown violation action, it is put into the error-disabled state. It can be brought back up by shutting down the interface and then issuing the **no shutdown** command.

3. B and C. Dynamically learned secure MAC addresses are lost when the switch reboots. Sticky MAC addresses are learned and added to the running config. These addressess can be retained if the configuration is saved and then rebooted. MAC addresses may also be configured statically (that is, manually). If fewer than the maximum number of MAC addresses for a port are configured statically, dynamically learned addresses are added to CAM until the maximum number is reached.

4. B. In port security implementation, an interface can be configured for one of three violation modes: Protect—a port security violation causes the interface to drop packets with unknown source addresses and no notification is sent that a security violation has occurred. Restrict—a port security violation causes the interface to drop packets with unknown source addresses and to send a notification that a security violation has occurred. Shutdown—a port security violation causes the interface to immediately

become error-disabled and turns off the port LED. No notification is sent that a security violation has occurred.

5. A. BPDU guard immediately error-disables a port that receives a BPDU. This prevents rogue switches from being added to the network. BPDU guard should be applied only to all end-user ports.

6. D. With sticky secure MAC addressing, the MAC addresses can be either dynamically learned or manually configured and then stored in the address table and added to the running configuration file. In contrast, dynamic secure MAC addressing provides for dynamically learned MAC addressing that is stored only in the address table.

7. D. BPDU guard can be enabled on all PortFast-enabled ports by using the **spanning-tree portfast bpduguard default** global configuration command. Alternatively, BPDU guard can be enabled on a PortFast-enabled port through the use of the **spanning-tree bpduguard enable** interface configuration command.

8. D. DAI can be configured to check for both destination or source MAC and IPv4 addresses. Destination MAC checks the destination MAC address in the Ethernet header against the target MAC address in the ARP body. Source MAC checks the source MAC address in the Ethernet header against the sender MAC address in the ARP body. IP address checks the ARP body for invalid and unexpected IP addresses including addresses 0.0.0.0, 255.255.255.255, and all IP multicast addresses.

9. A. When DHCP snooping is being configured, the number of DHCP discovery messages that untrusted ports can receive per second should be rate-limited by using the **ip dhcp snooping limit rate** interface configuration command. When a port receives more messages than the rate allows, the extra messages will be dropped.

10. D. If no violation mode is specified when port security is enabled on a switch port, the security violation mode defaults to "shutdown".

11. A, D, and E. Mitigating a VLAN attack can be done by disabling Dynamic Trunking Protocol (DTP), manually setting ports to trunking mode, and by setting the native VLAN of trunk links to VLANs not in use.

12. D. A Port Status of Secure-down means there are no hosts connected. Secure-up means there is at least one host connected to the port. Secure-shutdown means the port is error-disabled.

Chapter 12

1. B. Beacons are the only management frame that may regularly be broadcast by an AP. Probing, authentication, and association frames are used only during the association (or reassociation) process.

2. B. Omnidirectional antennas send the radio signals in a 360 degree pattern around the antenna. This provides coverage to devices situated anywhere around the access point. Dishes, directional, and Yagi antennas focus the radio signals in a single direction, making them less suitable for covering large, open areas.

3. C. SSID cloaking is a weak security feature that is performed by APs and some wireless routers by allowing the SSID beacon frame to be disabled. Although clients have to manually identify the SSID to be connected

to the network, the SSID can be easily discovered.

4. C and E. Two methods can be used by a wireless device to discover and register with an access point: passive mode and active mode. In passive mode, the AP sends a broadcast beacon frame that contains the SSID and other wireless settings. In active mode, the wireless device must be manually configured for the SSID, and then the device broadcasts a probe request.

5. C. Ad hoc mode (also known as independent basic service set or IBSS) is used in a peer-to-peer wireless network, such as when Bluetooth is used. A variation of the ad hoc topology exists when a smart phone or tablet with cellular data access is enabled to create a personal wireless hotspot. Mixed mode allows older wireless NICs to attach to an access point that can use a newer wireless standard.

6. B and C. The 802.11a and 802.11ac standards operate only in the 5 GHZ range. The 802.11b and 802.11g standards operate only in the 2.4 GHz range. The 802.11n standard operates in both the 2.4 and 5 GHz ranges. The 802.11ad standard operates in the 2.4, 5, and 60 GHz ranges.

7. C. 802.11ac provides data rates up to 1.3 Gbps and is still backward compatible with 802.11a/b/g/n devices. 802.11g and 802.11n are older standards that cannot reach speeds over 1 Gbps.

8. A. (MIMO) uses multiple antennas to increase available bandwidth for IEEE 802.11n/ac/ax wireless networks, such as the one in Figure 12-13. Up to eight transmit and receive antennas can be used to increase throughput.

9. B, D, and F. Interference occurs when one signal overlaps a channel reserved for another signal, causing possible distortion. The best practice for 2.4 GHz WLANs that require multiple APs is to use the non-overlapping channels 1, 6, and 11. These are selected because they are 5 channels apart and therefore minimize the interference with adjacent channels.

10. B. WPA and WPA2 Personal are intended for home or small office networks where users authenticate using a pre-shared key (PSK). WPA and WPA2 Enterprise is intended for enterprise networks but requires a RADIUS authentication server which provides additional security. WEP Enterprise is not a valid option.

11. D. When an access point is configured in passive mode, the SSID is broadcast so that the name of wireless network will appear in the listing of available networks for clients. Active is a mode used to configure an access point so that clients must know the SSID to connect to the access point. APs and wireless routers can operate in a mixed mode, meaning that that multiple wireless standards are supported. Open is an authentication mode for an access point that has no impact on the listing of available wireless networks for a client.

Chapter 13

1. A. The first action that should be taken is to secure administrative access to the wireless router. The next action would usually be to configure encryption. Then after the initial group of wireless hosts have connected to the network, MAC address filtering would

be enabled and SSID broadcast disabled. This will prevent new unauthorized hosts from finding and connecting to the wireless network.

2. C. By default, dual-band routers and APs use the same network name on both the 2.4 GHz band and the 5 GHz band. The simplest way to segment traffic is to rename one of the wireless networks.
3. C. The Cisco 3504 WLC dashboard displays when a user logs in to the WLC. It provides some basic settings and menus that users can quickly access to implement a variety of common configurations. The Network Summary page is a dashboard that provides a quick overview of the number of configured wireless networks, associated access points (APs), and active clients. You can also see the number of rogue access points and clients. The Advanced button displays the advanced Summary page providing access to all the features of the WLC.
4. C. Simple Network Management Protocol (SNMP) is used to monitor the network.
5. D. Any private IPv4 address cannot be routed on the Internet. The wireless router will use a service called Network Address Translation (NAT) to convert private IPv4 addresses to Internet-routable IPv4 addresses for wireless devices to gain access to the Internet.
6. D. Many wireless routers have an option for configuring quality of service (QoS). By configuring QoS, certain time-sensitive traffic types, such as voice and video, are prioritized over traffic that is not as time-sensitive, such as email and web browsing.
7. D. Each new WLAN configured on a Cisco 3500 series WLC needs its own VLAN interface. Therefore, it is required that a new VLAN interface be created first before a new WLAN can be created.
8. D. The 2.4 GHz band may be suitable for basic Internet traffic that is not time-sensitive. The 5 GHz band is much less crowded than the 2.4 GHz band; ideal for streaming multimedia. The 5 GHz band has more channels; therefore, the channel chosen is likely interference-free.
9. D. The RADIUS protocol uses security features to protect communications between the RADIUS server and clients. A shared secret is the password used between the WLC and the RADIUS server. It is not for end users.
10. D. Extending a WLAN in a small office or home has become increasingly easier. Manufacturers have made creating a wireless mesh network (WMN) simple through smartphone apps. You buy the system, disperse the access points, plug them in, download the app, and configure your WMN in a few steps.

Chapter 14

1. A and E. Static routing requires a thorough understanding of the entire network for proper implementation. It can be prone to errors and does not scale well for large networks. Static routing uses fewer router resources because no computing is required for updating routes. Static routing can also be more secure because it does not advertise over the network.
2. A. A static default route is a catch-all route for all unmatched networks.

3. C. The route will appear in the routing with a code of S (Static).

4. D. When the interface associated with a static route goes down, the router will remove the route because it is no longer valid.

5. A. A default static route is a route that matches all packets. It identifies the gateway IP address to which the router sends all IP packets for which it does not have a learned or static route. A default static route is simply a static route with 0.0.0.0/0 as the destination IPv4 address or ::/0 for IPv6. Configuring a default static route creates a gateway of last resort.

6. B. Dynamic routing protocols consume more router resources, are suitable for larger networks, and are more useful on networks that are growing and changing.

7. B. A metric is used by a routing protocol to compare routes received from the routing protocol. An exit interface is the interface used to send a packet in the direction of the destination network. A routing protocol is used to exchange routing updates between two or more adjacent routers. The administrative distance represents the trustworthiness of a particular route. The lower an administrative distance, the more trustworthy the learned route is. When a router learns multiple routes toward the same destination, the router uses the administrative distance value to determine which route to place into the routing table.

8. A and C. The code identifies how the route was learned. For instance, **L** identifies the address assigned to a router interface. This allows the router to efficiently determine when it receives a packet for the interface instead of being forwarded. **C** identifies a directly connected network. **S** identifies a static route created to reach a specific network. And **O** identifies a dynamically learned network from another router using the OSPF routing protocol.

9. C. A directly connected network will be added to the routing table when these three conditions are met: (1) the interface is configured with a valid IP address; (2) it is activated with the **no shutdown** command; and (3) it receives a carrier signal from another device that is connected to the interface. An incorrect subnet mask for an IPv4 address will not prevent its appearance in the routing table, although the error may prevent successful communications.

Chapter 15

1. B. A floating static route is a backup route that only appears in the routing table when the interface used with the primary route is down. To test a floating static route, the route must be in the routing table. Therefore, shutting down the interface used as a primary route would allow the floating static route to appear in the routing table.

2. C. The most believable route or the route with the lowest administrative distance is one that is directly connected to a router. In order of trustworthiness is A (AD = 0), D (Static route AD = 1), B (EIGRP AD = 90), and C (OSPF AD = 110). Therefore, the OSPF routes are considered to be the least trustworthy.

3. D. Even though OSPF has a higher administrative distance value (less trustworthy), the best match is the route in the routing

table that has the greatest number of far-left matching bits.

4. B. When only the exit interface is used, the route is a directly connected static route. When the next-hop IP address is used, the route is a recursive static route. When both are used, it is a fully specified static route.

5. A. A fully specified static route can be used to avoid recursive routing table lookups by the router. A fully specified static route contains both the IP address of the next-hop router and the ID of the exit interface.

6. B. By default, dynamic routing protocols have a higher administrative distance than static routes. Configuring a static route with a higher administrative distance than that of the dynamic routing protocol will result in the dynamic route being used instead of the static route. However, should the dynamically learned route fail, the static route will be used as a backup.

7. D. Floating static routes are used as backup routes, often to routes learned from dynamic routing protocols. To be a floating static route, the configured route must have a higher administrative distance than the primary route. For example, if the primary route is learned through OSPF, a floating static route that serves as a backup to the OSPF route must have an administrative distance greater than 110. In this example, the administrative distance of 120 is put at the end of the static route: **ip route 209.165.200.228 255.255.255.248 10.0.0.1 120**.

8. C. A stub router or an edge router connected to an ISP has only one other router as a connection. A default static route works in those situations because all traffic will be sent to one destination. The destination router is the gateway of last resort. The default route is not configured on the gateway, but on the router sending traffic to the gateway.

9. A. A default static route configured for IPv6 is a network prefix of all zeros and a prefix mask of 0, which is expressed as ::/0.

10. B. A floating static route is a backup route that only appears in the routing table when the interface used with the primary route is down. To test a floating static route, the route must be in the routing table. Therefore, shutting down the interface used as a primary route would allow the floating static route to appear in the routing table.

Chapter 16

1. A, C, and D. The **ping**, **show ip route**, and **show ip interface brief** commands provide information to help troubleshoot static routes. The **show version** command does not provide any routing information. The **tracert** command is used at the Windows command prompt and is not an IOS command. The **show arp** command displays learned IP address to MAC address mappings contained in the Address Resolution Protocol (ARP) table.

2. C. When the interface associated with a static route goes down, the router will remove the route because it is no longer valid.

3. C. A router looks up the ARP table entry for the destination IP address to find the Layer 2 Media Access Control (MAC) address of the host. If no entry exists, the router sends an Address Resolution Protocol (ARP) request out of network interface, and the

host responds with an ARP reply, which includes its MAC address.

4. B. The **show cdp neighbors** command provides a list of directly connected Cisco devices. This command validates Layer 2 (and therefore Layer 1) connectivity. For example, if a neighbor device is listed in the command output, but it cannot be pinged, Layer 3 addressing should be investigated.

5. C. The **show ip interface brief** command provides a quick status of all interfaces on the router.

Glossary

Numbers

4G/5G Wireless cellular broadband standards for multiaccess networks carrying both data and voice communications.

802.11 Original IEEE wireless standard supporting speeds of up to 2 Mbps.

802.11a Older IEEE wireless standard that supports speeds of up to 54 Mbps over the 5 GHz frequency range in a small coverage area. However, it is less effective at penetrating building structures and is not interoperable with the 802.11b and 802.11g standards.

802.11ac Modern IEEE wireless standard that supports speeds of up to 1.3 Gbps over the 5 GHz frequency range using MIMO technology. It is backward compatible with 802.11a/n devices.

802.11ax Modern IEEE wireless standard that operates over the 5 GHz frequency range. It is also known as Wi-Fi 6 and high-efficiency wireless (HEW).

802.11b Older IEEE wireless standard that supports speeds of up to 11 Mbps over the 2.4 GHz frequency range. It covers a larger area than 802.11a and is effective at penetrating building structures. It is not interoperable with 802.11a.

802.11g Older IEEE wireless standard that supports speeds of up to 54 Mbps over the 2.4 GHz frequency range. It also is backward compatible with 802.11b devices.

802.11n Current IEEE wireless standard that supports speeds of up to 600 Mbps over the 2.4 and 5 GHz frequency ranges. It is backward compatible with 802.11a/b/g devices. APs and wireless clients require multiple antennas using MIMO technology.

802.1D The original STP standard, which provided a loop-free topology in a network with redundant links. Also called Common Spanning Tree (CST), it assumed one spanning tree instance for the entire bridged network, regardless of the number of VLANs. The updated version of the standard is IEEE 802.1D-2004.

802.1Q Specifies the IEEE networking standard to support VLANs on an Ethernet network.

802.1w The IEEE STP standard for Rapid Spanning Tree Protocol (RSTP), which is an evolution of STP that provides faster convergence than STP.

802.1X This is the IEEE port-based access control and authentication protocol. This protocol restricts unauthorized workstations from connecting to a LAN through publicly accessible switch ports. The authentication server authenticates each workstation that is connected to a switch port before making available any services offered by the switch or the LAN.

802.3ad See *Link Aggregation Control Protocol (LACP)*.

A

active router The name given to the HSRP forwarding router.

ad hoc mode In wireless, when two devices connect wirelessly in a peer-to-peer (P2P) manner without using APs or wireless routers. Examples include wireless clients connecting directly to each other using Bluetooth or Wi-Fi Direct. The IEEE 802.11 standard refers to an ad hoc network as an independent basic service set (IBSS).

address spoofing attacks Consists of attacks where a threat actor uses the IP and or MAC address of another host to impersonate that host.

adjacency table A table in a router that contains a list of the relationships formed between selected neighboring routers and end nodes for the purpose of exchanging routing information. Adjacency is based upon the use of a common media segment.

administrative distance (AD) The feature that routers use to select the best path when there are two or more routes to the same destination from two different routing protocols. The AD represents the "trustworthiness," or reliability, of the route.

Advanced Encryption Standard (AES) This is a very secure commonly used encryption algorithm.

Advanced Malware Protection (AMP) Cisco AMP for Endpoints provides next-generation endpoint protection, scanning files using a variety of antimalware technologies, including the Cisco antivirus engine. Cisco Advanced Malware Protection then goes a step further than most malware detection tools, continuously monitoring every file in your network.

alternate ports Also called a backup port, a switch port in an RSTP topology that offers an alternate path toward the root bridge. An alternate port assumes a discarding state in a stable, active topology. An alternate port is present on nondesignated bridges and makes a transition to a designated port if the current path fails.

application-specific-integrated circuits (ASICs) Electronics added to a switch that allow it to have more ports without degrading performance.

ARP attacks Address resolution protocol (ARP) attacks such as an ARP spoofing and ARP poisoning attacks attempt to divert traffic from intended hosts to an attacker host instead.

ARP poisoning This is a type of man-in-the-middle attack that can be used to intercept, alter, or even stop network traffic. The threat actor creates spoofed ARP messages to make legitimate hosts send frames to them.

ARP spoofing When a threat actor creates spoofed ARP messages to make legitimate hosts send frames to them.

authentication, authorization, and accounting (AAA) A network security service that provides the primary framework to set up access control on a network device (for example, router, switch). AAA is a way to control who is permitted to access a network (authenticate), what they can do while they are there (authorize), and to audit what actions they performed while accessing the network (accounting).

automatic medium-dependent interface crossover (auto-MDIX) A detection on a switch port or hub port to detect the type of cable used between switches or hubs. After the cable type is detected, the port is connected and configured accordingly. With auto-MDIX, a crossover or a straight-through cable can be used for connections to a copper 10/100/1000 port on the switch, regardless of the type of device on the other end of the connection.

autonegotiate Ethernet feature in which two interconnecting devices automatically negotiate duplex and speed settings.

autonomous APs Standalone APs that that are configured locally using the CLI or GUI. Most home and SOHO routers have an autonomous AP. A small network may require a few autonomous APs connected to a Layer 2 switch, and each AP would be individually configured.

autonomous system (AS) Also known as a routing domain, a network of routers under common administration, such as a company or an organization. Typical examples of an AS are a company's internal network and an ISP's network.

B

backup router A VRRP router that monitors the VRRP master router. VRRP can have multiple backup routers. The role of the backup router is similar to that of an HSRP standby router.

Basic Service Area (BSA) In wireless, an IEEE 802.11 name for the wireless coverage area that is provided by a basic service set (BSS).

Basic Service Set (BSS) In wireless, an IEEE 802.11 name for an infrastructure mode wireless network that requires wireless clients to interconnect via a wireless router or AP. APs connect to the network infrastructure using the wired distribution system, such as Ethernet.

Basic Service Set Identifier (BSSID) An IEEE 802.11 name that uniquely identifies each BSS. The BSSID is the formal name of the BSS and is always associated with only one AP.

blacklisting When a firewall or security appliance (for example, IPS, ESA, WSA) are configured with access control rules to deny traffic to and from specific IP addresses.

blocking state A port state for a nondesignated port that does not participate in frame forwarding. The port continues to process received BPDU frames to determine the location and root ID of the root bridge and what port role the switch port should assume in the final active STP topology.

BOOT environment variable A configurable setting on a device that identifies where the IOS image file is located. The boot loader software will use the image file identified by this variable.

boot loader software A small program stored in ROM that runs immediately after POST successfully completes. It is used to initialize a network device such as a router or a switch. The boot loader locates and launches the operating system.

Border Gateway Protocol (BGP) Routing protocol used between Internet service providers (ISPs) and their larger private clients to exchange routing information. An exterior gateway routing protocol that ISPs use to propagate routing information.

BPDU filter A Cisco switch feature used to filter sending or receiving BPDUs on a switch port.

BPDU guard A Cisco switch feature that listens for incoming STP BPDU messages and disables the interface if any are received. The goal is to prevent loops when a switch connects to a port that is expected to have only a host connected to it.

bridge ID (BID) An 8-byte identifier of switches used by STP and RSTP. It consists of a 2-byte bridge priority field and a 6-byte system ID field. The priority field is a configurable bridge priority number, and the system ID is the MAC address of the sending switch.

bridge priority A customizable value between 0 and 65535 (the default is 32768) that can be configured to influence which switch becomes the root bridge.

Bridge Protocol Data Unit (BPDU) A frame used by STP to communicate key information about the avoidance of Layer 2 loops in the network topology.

broadcast domains All nodes that are part of a network segment, VLAN, or subnet, and all devices on the LAN receive broadcast frames from a host within the LAN. A broadcast domain is bounded by a Layer 3 device. A Layer 3 device such as a router sets the boundary of the broadcast domain.

broadcast storm A condition in which broadcasts are flooded endlessly, often due to a looping at Layer 2 (bridge loop). A broadcast storm occurs when there are so many broadcast frames caught in a Layer 2 loop that all available bandwidth is consumed.

buffer Refers to an area of memory used to temporarily store data.

C

Canonical Format Identifier A field in the VLAN tag field consisting of a 1-bit flag. When set to 1, it enables legacy Token Ring frames to be carried across Ethernet links. Other fields in the 802.1Q VLAN tag frame are the Type field, a Priority field, and VLAN ID field.

CAPWAP An IEEE standard protocol that enables a WLC to manage multiple APs and WLANs. CAPWAP is also responsible for the encapsulation and forwarding of WLAN client traffic between an AP and a WLC. CAPWAP is based on LWAPP but adds additional security with Datagram Transport Layer Security (DTLS). CAPWAP establishes tunnels on User Datagram Protocol (UDP) ports. CAPWAP can operate over either IPv4 or IPv6, but uses IPv4 by default.

carrier sense multiple access with collision avoidance (CSMA/CA) How and when data is sent on a wireless network. In a CSMA/CA network, hosts attempt to avoid collisions by beginning transmission only after the channel is sensed to be "idle." When they do transmit, nodes transmit their packet data in its entirety.

child route A route that is a subnet of a classful network address. Also known as a Level 2 route. A child route is an ultimate route.

Cisco Discovery Protocol (CDP) A Cisco proprietary Layer 2 link discovery protocol enabled on all Cisco devices by default. It is used to discover other CDP-enabled devices for autoconfiguring connections and to troubleshoot network devices. Compare with Link Layer Discovery Protocol (LLDP).

Cisco Express Forwarding (CEF) A Cisco proprietary protocol that allows high-speed packet switching in ASICs rather than using CPUs. Cisco Express Forwarding offers "wire speed" routing of packets and load balancing.

Cisco Identity Services Engine (ISE) An example of a NAC device.

Cisco IOS helper address DHCP feature that enables a router to forward DHCPv4 broadcasts to the DHCP IPv4 server. When a router forwards address assignment/parameter requests, it is acting as a DHCPv4 relay agent.

Cisco Talos Intelligence Group One of the largest commercial threat intelligence teams in the world, composed of world-class researchers, analysts, and engineers. Industry-leading visibility, actionable intelligence, and vulnerability research drive rapid detection and protection for Cisco customers against known and emerging threats, and stop threats in the wild to protect the Internet at large.

class of service (CoS) A 3-bit field inserted in a 802.1Q VLAN tagged Ethernet frame to assign a Quality of Service (QoS) marking. The 3-bit field identifies the CoS priority value and is used by Layer 2 switches to specify how the frame should be handled when QoS is enabled.

collision domain A network segment that shares the same bandwidth between the devices, such as between a switch and a PC. Each port on a switch is its own collision domain. Every device connected to a hub is within a single collision domain, meaning that when two devices attempt communication simultaneously, collisions occur.

Common Spanning Tree (CST) The original IEEE 802.1D STP standard, which assumes one spanning tree instance for an entire bridged network, regardless of the number of VLANs.

content addressable memory (CAM) Table in memory that stores source MAC addresses and port numbers learned from frames entering the switch. Also called the MAC address table.

Controller-based APs See *lightweight access points (LAPs).*

converged Convergence means several things in networking: (1) combining voice and video with the traditional data network, (2) providing a loop-free Layer 2 topology for a switched LAN through the use of spanning tree, and (3) providing a stable Layer 3 network where the routers have completed providing each other updates and the routing tables are complete.

CoS priority value This is a CoS value identified using a 3-bit field inserted in a 802.1Q VLAN tagged Ethernet frame. When QoS is enabled, Layer 2 switches use the priority value to identify how to forward frames.

cost An arbitrary value, typically based on hop count, media bandwidth, or other measures, that is assigned by a network administrator and used to compare various paths through an internetwork environment. Routing protocols use cost values to calculate the most favorable path to a particular destination; the lower the cost, the better the path. The OSPF cost is based on the cumulative bandwidth from source to destination.

CPU subsystem Consists of the CPU, DRAM, and the portion of the flash device that makes up the flash file system. POST checks the CPU subsystem upon boot of the device.

CRC This is a process to check for errors within the Layer 2 frame. The device generates a cyclic redundancy check (CRC) and includes this value in the frame check sequence (FCS) field. The receiving device generates a CRC and compares it to the received CRC to look for errors. If the calculations match, no error has occurred. If the calculations do not match, the frame is dropped. CRC errors on Ethernet and serial interfaces usually mean a media or a cable problem.

cut-through switching A method used inside a switch where, after the destination MAC address has been received, the frame is processed without waiting for the complete frame to arrive.

D

data breach An attack in which an organization's data servers or hosts are compromised to steal confidential information (that is, data).

data structures A group of data elements that are stored together under one name. Routing protocols typically use tables or databases for their operations. The adjacency database, link-state database, and forwarding database are all examples of data structures. This information is kept in RAM.

data VLANs VLANs specifically configured to carry user-generated traffic. In particular, a data VLAN does not carry voice-based traffic or traffic used to manage a switch.

Datagram Transport Layer Security (DTLS) Used to add security to CAPWAP. It is a protocol that provides security between the AP and the WLC. It allows them to communicate using encryption and prevents eavesdropping or tampering.

default port cost A measure assigned on a per-link basis in a switched LAN. It is determined by the link bandwidth, with a higher bandwidth having a lower port cost.

default static route A route that matches all packets and identifies the gateway IP address to which the router sends all packets for which it does not have a learned or static route.

default VLAN VLAN that all the ports on a switch are members of when a switch is reset to factory defaults or is new. All switch ports are members of the default VLAN after the initial boot of the switch. On a Cisco switch, VLAN 1 is the default VLAN.

designated ports In spanning tree, a nonroot switch port that is permitted to forward traffic on the network. For a trunk link connecting two switches, one end connects to the designated bridge through the designated port. Only one end of every trunk link in a switched LAN (with spanning tree enabled) connects to a designated port. The selection of designated ports is the last step in the spanning-tree algorithm.

Device Provisioning Protocol (DPP) A replacement to WPS to help authenticate IoT devices.

DHCP Acknowledgment (DHCPACK) A unicast message sent by a DHCP server in response to a device that sends a DHCPREQUEST. The DHCPACK message is used by the DHCP server to complete the DHCP process.

DHCP attacks Dynamic Host Configuration Protocol attacks, such as DHCP spoofing or DHCP starvation, to create a man-in-the-middle or DoS attack.

DHCP Discover (DHCPDISCOVER) A broadcast message sent by a network device to discover an IPv4 DHCP server.

DHCP Offer (DHCPOFFER) A unicast message returned by a DHCP server in response to a client device sending a DHCPDISCOVER broadcast message. The DHCPOFFER message typically contains an IP address, subnet mask, default gateway address, and other information.

DHCP Request (DHCPREQUEST) A broadcast message sent by a network device in response to a DHCPOFFER made by a DHCP server that sent a DHCPOFFER message. The DHCPREQUEST message is used by the device to accept the IP addressing offer made by the DHCP server.

DHCP snooping Cisco switch security feature that is enabled on an interface or VLAN. If a switch receives a DHCP packet on an untrusted port, the switch compares the source packet information with that held in the DHCP Snooping Binding Database. The switch will deny unauthorized DHCP messages incoming on an untrusted port.

DHCP snooping binding table A database that is populated by the DHCP snooping feature. It contains IP to MAC address lease information.

DHCP spoofing Attacks in which a cybercriminal installs a fake DHCP server on the network. Legitimate clients acquire their IP confirmation from the bogus server. These types of attacks force the clients to use both a false Domain Name System (DNS) server and a computer that is under the control of the attacker as their default gateway.

DHCP starvation Type of attack in which the cybercriminal floods the DHCP server with bogus DHCP requests and eventually leases all the available IP addresses in the DHCP server pool. After these IP addresses are issued, the server cannot issue any more addresses, and this situation produces a denial-of-service (DoS) attack because new clients cannot obtain network access.

DHCPv4 The IPv4 version of DHCP. A method of deploying IP address-related information to IPv4 devices. DHCPv4 uses four types of messages: DHCP discover, DHCP offer, DHCP request, and DHCP acknowledgment.

DHCPv4 client A DHCPv4 client requests DHCPv4 services from a DHCPv4 server, which in turn responds with network configuration information for the DHCP client. DHCP clients are typically end devices such as computers, laptops, tablets, smart devices, and more. A Cisco router can also be a DHCP client, as typically implemented in home networks.

DHCPv4 relay agent Allows relaying DHCP messages between a DHCPv4 client and a DHCPv4 server located on a different network.

DHCPv4 server A device can be a dedicated server configured to provide client IPv4 addressing or a Cisco router configured to provide DHCPv4 services without the need for a dedicated server.

DHCPv6 The IPv6 version of DHCP. IPv6 network devices can obtain IPv6 addressing information using one of three options: SLAAC, stateless DHCPv6, and stateful DHCPv6.

DHCPv6 ADVERTISE unicast message A DHCPv6 message sent by a DHCPv6 server to inform the DHCPv6 client that the server is available for DHCPv6 service. It is generated in response to a DHCPv6 SOLICIT message.

DHCPv6 client A DHCPv6 client requests DHCPv6 services from a DHCPv6 server, which in turn responds with network configuration information for the DHCP client. DHCP clients are typically end devices such as computers, laptops, tablets, smart devices, and more. A Cisco router can also be a DHCP client, as typically implemented in home networks.

DHCPv6 INFORMATION-REQUEST message A DHCPv6 message sent by a stateless client to the DHCPv6 server requesting only configuration parameters, such as DNS server address. The client generates its own IPv6 address using the prefix from the RA message and a self-generated Interface ID.

DHCPv6 relay agent Allows relaying DHCP messages between a DHCPv6 client and a DHCPv6 server located on a different network.

DHCPv6 REPLY unicast message The DHCPv6 message unicast message sent by the DHCPv6 server to the client. The DHCPv6 message contains the information requested in the DHCPv6 REQUEST or DHCPv6 INFORMATION-REQUEST message.

DHCPv6 REQUEST message A DHCPv6 message sent by a stateful client to the DHCPv6 server to obtain an IPv6 address and all other configuration parameters from the server.

DHCPv6 server A device that can be a dedicated server configured to provide client IPv6 addressing or a Cisco router configured to provide DHCPv6 services without the need for a dedicated server.

DHCPv6 SOLICIT message A DHCPv6 message sent by the DHCPv6 client to the reserved IPv6 multicast all-DHCPv6-servers address FF02::1:2. This multicast address has link-local scope, which means routers do not forward the messages to other networks.

Direct-Sequence Spread Spectrum (DSSS) This is a wireless modulation technique designed to spread a signal over a larger frequency band. It does this by spreading the signal over a wider frequency, which effectively hides the discernable peak of the signal. A properly configured receiver can reverse the DSSS modulation and reconstruct the original signal. DSSS is used by 802.11b devices to avoid interference from other devices using the same 2.4 GHz frequency.

directional antenna Wireless antenna that focuses the radio signal in a given direction to enhance the signal to and from the AP in the direction the antenna is pointing. This provides a stronger signal strength in one direction and reduced signal strength in all other directions. Examples of directional Wi-Fi antennas include Yagi antenna and parabolic dish antenna.

directly connected interfaces The interfaces on a router.

directly connected networks Networks that are connected to a router's physical Ethernet or serial interfaces.

directly connected static route A static route in which only the router exit interface is specified.

disabled state A spanning tree state for a switch port that is administratively shut down. A disabled port does not function in the spanning-tree process.

discarding state An RSTP state that merges the STP 802.1D disabled, blocking, and listening states together in one state.

distance A measure that identifies how far it is to a destination network, based on hop count, cost, bandwidth, delay, or another metric.

distance vector routing protocol A type of routing protocol in which a router's routing table is based on hop-by-hop metrics and is only aware of the topology from the viewpoint of its directly connected neighbors. EIGRP and RIP are examples of distance vector routing protocols.

Distributed Denial of Service (DDoS) A coordinated attack from many devices, called zombies, with the intention of degrading or halting public access to an organization's website and resources.

distribution system (DS) An IEEE 802.11 term to describe the wired infrastructure to interconnect APs.

dual stack topology An IPv4 to IPv6 migration technique in which a device is enabled for both IPv4 and IPv6 protocols. A transition mechanism used when converting from IPv4 to IPv6. Basically, when using a dual stack, a router runs both IPv4 and IPv6. Other IPv6 migration techniques include translation and tunneling.

duplex mismatch Term used in Ethernet to describe a situation in which one end of the link is set to full duplex and the other end of the link is set to half duplex.

Duplicate Address Detection (DAD) A process used by IPv6 devices using an ICMPv6 Neighbor Solicitation message to verify whether any other device has the same IPv6 address.

Dynamic ARP Inspection (DAI) Cisco Catalyst switch security feature that prevents ARP spoofing and ARP poisoning attacks.

Dynamic Host Configuration Protocol (DHCP) A protocol used to dynamically assign IP configurations to hosts. The services defined by the protocol are used to request and assign an IP address, default gateway, and DNS server address to a network host.

dynamic routing protocols A remote network in a routing table that has been automatically learned using a dynamic routing protocol such as EIGRP or OSPF.

dynamic trunking protocol (DTP) A Cisco proprietary protocol that negotiates both the status and encapsulation of trunk ports.

E

edge router Router connecting the inside network to the outside network (typically the Internet).

egress The exit or the way out.

egress port The port through which a frame exits a switch.

email security appliance (ESA) A mitigation technology device for email-based threats. The Cisco ESA monitors Simple Mail Transfer Protocol (SMTP) traffic using real-time feeds from the Cisco Talos to detect threats, block known threats, remediate against stealth malware that evaded initial detection, discard emails with bad links, block access to newly infected sites, and encrypt content in outgoing email to prevent data loss.

Enhanced Interior Gateway Routing Protocol (EIGRP) An advanced version of IGRP, developed by Cisco. EIGRP provides superior convergence and operating efficiency, and it combines the advantages of link-state protocols with those of distance vector protocols.

equal cost load balancing When a router utilizes multiple paths with the same administrative distance and cost to a destination.

error-disabled state A port that has triggered a port security shutdown violation. The violation must be addressed and the port reenabled using the **shutdown** and **no shutdown** combination commands.

EtherChannel A feature in which up to eight parallel Ethernet segments between the same two devices, each using the same speed, can be combined to act as a single link for forwarding and STP logic.

EUI-64 See *Extended Unique Identifier (EUI-64)*.

exit interface The interface through which frames leave a device.

extended range VLANs Extended range VLANs are numbered 1,006 to 4,094, and they enable service providers to extend their infrastructure to a greater number of customers. Some global enterprises could be large enough to need extended range VLAN IDs. In contrast, normal range VLANs on these switches are numbered 1 to 1,005.

Extended Service Area (ESA) In wireless, an IEEE 802.11 name for the coverage area of an extended service set (BSS).

Extended Service Set (ESS) In wireless, an IEEE 802.11 name when two or more basic service sets (BSSs) are used to create a wireless domain.

extended system ID A VLAN ID or a Multiple Spanning Tree Protocol (MSTP) instance ID. Constitutes 12 bits of the 8-byte BID and contains the ID of the VLAN with which an STP BPDU is associated. The presence of the extended system ID results in bridge priority values incrementing in multiples of 4096.

Extended Unique Identifier method (EUI-64) An IPv6 process that uses a client's 48-bit Ethernet MAC address and inserts another 16 bits in the middle of the 48-bit MAC address to create a 64-bit Interface ID for an IPv6 global unicast address.

Extensible Authentication Protocol (EAP) IEEE 802.1X protocol used between a supplicant and a RADIUS authentication server.

Exterior Gateway Protocol (EGP) A protocol used for routing between ASs. It is also referred to as inter-AS routing. Service providers and large companies may interconnect using an EGP. BGP is the only currently viable EGP, and it is the official routing protocol used on the Internet.

F

fast switching In fast switching, the first packet is copied to packet memory, and the destination network or host is found in the fast-switching cache. The frame is rewritten and sent to the outgoing interface that services the destination. Subsequent packets for the same destination use the same switching path.

fast-switching cache Section of memory used by the fast switching process to temporarily store next-hop frame forwarding information.

firewall A router, dedicated device, or software that denies outside traffic from entering an inside (that is, private) network. However, it permits inside network traffic to exit and return to the inside network. A firewall may use access lists and other methods to ensure the security of the private network.

First Hop Redundancy Protocols (FHRP) A class of protocols that includes HSRP, VRRP, and GLBP, which allows multiple redundant routers on the same subnet to act as a single default router (that is, first-hop router).

FlexConnect A wireless solution for branch office and remote office deployments. It lets you configure and control access points in a branch office from the corporate office through a WAN link, without deploying a controller in each office.

floating static route Static routes used to provide a backup path to a primary static or dynamic route, in the event of a link failure. Used only when the primary route is not available.

Forwarding Information Base (FIB) Used with CEF to provide optimized lookups for more efficient packet forwarding.

forwarding state A state in which an STP port is considered part of the active topology and forwards data frames and sends and receives BPDU frames.

fragment free switching A modified form of cut-through switching in which the switch waits for the collision window (64 bytes) to pass before forwarding the frame. This means each frame will be checked into the data field to make sure no fragmentation has occurred. Fragment free switching provides better error checking than cut-through, with practically no increase in latency.

frame check sequence (FCS) A checksum value found in the last field of a datagram that is used by the switch to validate that the frame is free of errors.

Frequency-Hopping Spread Spectrum (FHSS) A wireless modulation technique designed on a spread spectrum method to communicate. It transmits radio signals by rapidly switching a carrier signal among many frequency channels. FHSS was used by the original 802.11 standard. Walkie-talkies and 900 MHz cordless phones also use FHSS, and Bluetooth uses a variation of FHSS.

full-duplex Operation in which both devices can transmit and receive on the media at the same time.

fully specified static route A static route in which both the output interface and next-hop address are identified.

G

Gateway Load Balancing Protocol (GLBP) A Cisco proprietary protocol that provides both redundancy and load balancing of data, through the use of multiple routers. Routers present a shared GLBP address that end stations use as a default gateway.

giants Problematic Ethernet frames of excess size caused by a malfunctioning NIC or an improperly terminated or unterminated cable.

global unicast addresses (GUA) Routable IPv6 addresses in the IPv6 Internet similar to a public IPv4 address.

gratuitous ARP This is an ARP reply to which no request has been made. Other hosts on the subnet store the MAC address and IP address contained in the gratuitous ARP in their ARP tables. Can be used by threat actors for nefarious reasons.

H

half-duplex Operation in which both devices can transmit and receive on the media but cannot do so simultaneously.

hierarchical network addressing Network addressing scheme where IP network numbers are applied to network segments or VLANs in an orderly fashion that takes the network as a whole into consideration. Blocks of contiguous network addresses are reserved for and configured on devices in a specific area of the network.

High-Level Data Link Control (HDLC) An ISO bit-oriented Layer 2 WAN serial line protocol that supports router-to-router connections. It is the default encapsulation of serial interfaces on Cisco routers. Contrast with Point-to-Point Protocol (PPP).

high-performance computing (HPC) applications Applications that solve complex computational problems using a fast infrastructure connecting super computers and parallel processing techniques.

high port density Switches have high-port densities: 24- and 48-port switches are often just a single rack unit and operate at speeds of 100 Mb/s, 1 Gb/s, and 10 Gb/s. Large enterprise switches may support many hundreds of ports.

High-Speed WAN Interface Card (HWIC) Slot on a router used to install a high-speed WAN interface card.

hop limit Similar to the IPv4 TTL field, IPv6 hop limit is an 8-bit field that indicates the maximum number of links over which the IPv6 packet can travel before being discarded.

host-based intrusion prevention systems (HIPSs) Software used to protect critical computer files and systems from known and unknown malicious attacks.

host route A host route is an IPv4 address with a 32-bit mask or an IPv6 address with a 128-bit mask. Host routes can be added to the routing table.

Hot Standby Router Protocol (HSRP) A Cisco proprietary protocol that allows two (or more) routers to share the duties of being the default router on a subnet, with an active/standby model, with one router acting as the default router and the other sitting by, waiting to take over that role if the first router fails.

hotspot A type of ad hoc network where a cellular device is used to provide a personal Internet connection to other devices. It is also referred to as tethering and provides a temporary solution.

I

ICMP Router Discovery Protocol (IRDP) A legacy FHRP solution that allows IPv4 hosts to locate routers that provide IPv4 connectivity to other (nonlocal) IP networks.

ICMPv6 Neighbor Advertisement (NA) message ICMPv6 message used by an interface to announce its presence on the local link. Similar to an ARP reply for IPv4, ICMPv6 messages are sent by devices in response to an ICMPv6 Neighbor Solicitation message containing the IPv6 address and the corresponding MAC address. NA is used in the IPv6 Neighbor Discovery (ND) feature.

ICMPv6 Neighbor Solicitation (NS) message ICMPv6 message used to discover other IPv6 hosts on the local link. Similar to an ARP request for IPv4, ICMPv6 messages are sent by devices when they know the IPv6 address but need the corresponding MAC address. NS is used in the IPv6 Neighbor Discovery (ND) feature.

ICMPv6 Router Advertisement (RA) message ICMPv6 messages sent by routers to provide addressing information to hosts using SLAAC. A message type used by an IPv6 router to provide IPv6 addressing information to clients. The router sends the message using the IPv6 all-nodes multicast address of FF02::1.

ICMPv6 Router Solicitation (RS) message ICMPv6 messages sent by devices to request an ICMPv6 Router Advertisement message. A message type used by an IPv6 client that sends a multicast to address FF02::2 (all-routers) to obtain an IPv6 address using SLAAC, which does not require the services of a DHCPv6 router.

IEEE 802.1Q header Ethernet header designed to include VLAN related information for the IEEE 802.1Q standard that was developed to add VLAN information to frames as they traverse trunk links.

independent basic service set (IBSS) In wireless, this is an IEEE 802.11 name for an ad hoc wireless network.

infrastructure mode This type of wireless network requires that wireless clients interconnect via a wireless router or AP, such as in WLANs. APs connect to the network infrastructure using the wired distribution system, such as Ethernet.

ingress The entrance or the way in.

ingress port The port through which a frame enters a switch.

input errors These are errors including runts, giants, no buffer, CRC, frame, overrun, and ignored counts reported in the output of the **show interfaces** command.

interface ID Host portion of an IPv6 global unicast address.

Interior Gateway Protocols (IGPs) A protocol used for routing within an AS. It is also referred to as intra-AS routing. Companies, organizations, and even service providers use an IGP on their internal networks. IGPs include RIP, EIGRP, OSPF, and IS-IS.

Interior Gateway Routing Protocol (IGRP) Original routing protocol developed by Cisco Systems. It is a legacy Cisco proprietary distance vector routing protocol. It has been replaced with EIGRP and has not been available since IOS 12.2.

Intermediate System-to-Intermediate System (IS-IS) A routing protocol developed by the ISO.

Internetwork Operating Systems (IOS) The operating system for Cisco devices that provides the majority of a router's or switch's features, with the hardware providing the remaining features.

inter-VLAN routing The process of routing data between VLANs so that communication can occur between the different networks. It can be implemented using legacy inter-VLAN routing, a router-on-a-stick method, or by using a Layer 3 multilayer switch.

IP address spoofing When a threat actor crafts a special packet using false IP address(es) to impersonate another host.

IP Source Guard (IPSG) Cisco Catalyst switch security feature that prevents MAC and IP address spoofing attacks.

IPv6 link-local address Locally unique IPv6 addresses that are used to communicate with other IPv6-enabled devices on the same link and only on that link (subnet). Link-local addresses cannot be routed beyond the local network and are commonly used by IPv6 routing protocols. Every IPv6-enabled interface must have a link-local address. However, a global unicast address is not a requirement.

L

late collisions A collision that occurs after 512 bytes of an Ethernet frame (the preamble) have been transmitted.

Layer 2 loops A loop created by redundant links such as when multiple connections exist between two switches or two ports on the same switch connected to each other. The loop creates broadcast storms as broadcasts and multicasts are continuously forwarded by switches out every port. The switch or switches eventually flood the network. Layer 2 headers do not support a TTL value, so a frame could loop indefinitely.

learning state A state in which a port accepts data frames to populate the MAC address table in an effort to limit flooding of unknown unicast frames. The IEEE 802.1D learning state is seen in both a stable active topology and during topology synchronization changes.

lease A DHCP option identifying the amount of time that an IP address is provided to a host.

legacy inter-VLAN routing An inter-VLAN routing solution performed by connecting different physical router interfaces to different physical switch ports. The switch ports connected to the router are placed in access mode, and each physical interface is assigned to a different VLAN. Each router interface can then accept traffic from the VLAN associated with the switch interface it is connected to, and traffic can be routed to the other VLANs connected to the other interfaces. Compare with router-on-a-stick inter-VLAN routing and Layer 3 inter-VLAN routing.

Lightweight Access Point Protocol (LWAPP) Wireless protocol used to communicate between a lightweight access point (LAP) and a WLAN controller (WLC).

lightweight APs (LAPs) Sometimes abbreviated as LWAP, these are controller-based APs that use the Lightweight Access Point Protocol (LWAPP) to communicate with a WLAN controller (WLC). The WLC serves as the default gateway to all APs. Controller-based APs are useful in situations where many APs are required in the network. As more APs are added, each AP is automatically configured and managed by the WLC. Controller-based WLANs simplify device configuration, troubleshooting, and enhance monitoring and visibility to closely analyze the WLAN.

link aggregation A method of aggregating (that is, combining) multiple links between equipment to increase bandwidth.

Link Aggregation Control Protocol (LACP) An industry-standard protocol that aids in the automatic creation of EtherChannel links.

link aggregation group (LAG) An LACP term used to describe the bundling of several physical ports to form a single logical channel. Cisco uses the term EtherChannel, whereas all other vendors refer to LAG.

Link Layer Discovery Protocol (LLDP) A vendor-neutral neighbor discovery protocol similar to CDP. LLDP works with network devices, such as routers, switches, and wireless LAN access points. Like CDP, LLDP advertises its identity and capabilities to other devices and receives the information from a physically connected Layer 2 device.

link-local address (LLA) These are addresses that are automatically created and used to communicate with devices on the same local link. Link-local addresses are only unique on a given link or network. Refer to IPv6 link-local address.

link-state routing protocol A routing protocol classification in which each router has a topology database based on an SPF tree through the network, with knowledge of all nodes. OSPF and IS-IS are examples of link-state routing protocols.

listening state A state in which a port cannot send or receive data frames, but the port is allowed to receive and send BPDUs. The IEEE 802.1D listening state is seen in both a stable active topology and during topology synchronization changes.

load balancing Load balancing evenly distributes network traffic between multiple links, increasing the utilization of network segments and maximizing available network bandwidth.

local host route When an active interface on a router is configured with an IP address, a local host route is automatically added to the routing table. The local routes are marked with "L" in the output of the routing table.

local route interface An entry in the routing table for a local host route. It is added when an interface is configured and active.

loop guard An STP feature that provides additional protection against Layer 2 forwarding loops (STP loops) caused when a physically redundant port no longer receives STP BPDUs. If BPDUs are not received on a nondesignated port and loop guard is enabled, that port is moved into the STP loop-inconsistent blocking state instead of the listening/learning/forwarding state.

loopback interface A software-only interface that emulates a physical interface. A loopback interface is always up and never goes down.

M

MAC address filtering A wireless security mechanism where an AP is manually configured to permit or deny wireless access to a host based on the MAC hardware address. Devices with different MAC addresses will not be able to join the WLAN.

MAC address flooding A type of LAN attack that occurs when a cybercriminal exploits a default switch behavior to create a MAC address flooding attack. MAC address tables are limited in size. MAC flooding attacks exploit this limitation with fake source MAC addresses until the switch MAC address table is full and the switch is overwhelmed. Compare with CDP reconnaissance attacks, Telnet attacks, VLAN attacks, and DHCP spoofing attacks.

MAC address spoofing This is when a threat actor crafts a special packet using false IP address(es) to impersonate another host.

MAC address table On a switch, a table that lists all known MAC addresses and the bridge/switch port out that the bridge/switch should use to forward frames sent to each MAC address. Also known as a CAM table.

MAC address table overflow See *MAC address table flooding*.

MAC table attacks Network attack such as the MAC address table flooding attack that exploit the default operation of the Layer 2 MAC address table.

malware Software that is designed to exploit or damage end devices and networks. Malware includes computer viruses, Trojan horses, worms, ransomware, spyware, scareware, and adware.

man-in-the-middle attack A type of attack In which a threat actor positions themselves in between a victim and the destination.

Managed Address Configuration flag (M flag) Flag used in DHCPv6 to indicate whether to use stateful DHCPv6. For stateful DHCP, it sets the flag to 1.

management VLAN A VLAN defined by the network administrator as a means of accessing the management capabilities of a switch. The management VLAN SVI is assigned an IP address and subnet mask. It is a security best practice to define the management VLAN to be a VLAN distinct from all other VLANs defined in the switched LAN.

master router A VRRP forwarding router with a role similar to that of the HSRP active router.

Message Integrity Check (MIC) A wireless security protocol used by WPA and WP2 to ensure a message payload and header have not been altered.

metric The quantitative value used by dynamic routing protocols to measure the distance to a given network.

Mode button This is a button on the front of Cisco Catalyst 2960 switches that is used to toggle through port status, port duplex, port speed, and PoE (if supported) status of the port LEDs.

Multiple Input Multiple Output (MIMO) A wireless technology that uses multiple antennas to increase available bandwidth for IEEE 802.1

1n/ac/ax wireless networks. Up to eight transmit and receive antennas can be used to increase throughput.

Multiple Spanning Tree (MST) The Cisco implementation of MSTP.

Multiple Spanning Tree Protocol (MSTP) An evolution of IEEE 802.1D STP and IEEE 802.1w (RSTP) that enables multiple VLANs to be mapped to the same spanning-tree instance, reducing the number of instances needed to support a large number of VLANs. MSTP was introduced as IEEE 802.1s.

N

native VLAN A native VLAN is assigned to an IEEE 802.1Q trunk port. An IEEE 802.1Q trunk port supports tagged and untagged traffic coming from VLANs. The 802.1Q trunk port places untagged traffic on the native VLAN. It is a security best practice to define a native VLAN to be a dummy VLAN distinct from all other VLANs defined in the switched LAN. The native VLAN is not used for any traffic in the switched network.

neighbor A neighbor is another interconnecting router that has an interface on a common network, sometimes referred to as peers. In RIP, neighbors exchange routing information. In EIGRP and OSPF, neighbors exchange routing information and keep in touch by using Hello packets.

network access control (NAC) A NAC device provides authentication, authorization, and accounting (AAA) services for users and administrative users. NAC can manage access policies across a wide variety of users and device types.

next-generation firewall (NGFW) A NGFW provides stateful packet inspection, application visibility and control, next-generation intrusion prevention system (NGIPS), advanced malware protection (AMP), and URL filtering.

next-generation IPS (NGIPS) A NGIPS integrates real-time contextual awareness, security automation, advanced malware protection, and superior threat intelligence with industry-leading network intrusion prevention.

next-hop IP address The next gateway to which a Layer 3 packet is delivered, used to reach its destination.

next-hop router This is the next router to send the packet to reach the destination network.

next-hop static route A static route in which only the next-hop IP address is specified.

normal range VLANs VLANs with VLAN IDs 1 to 1005. VLAN IDs 1 and 1002 to 1005 are automatically created and cannot be removed.

O

omnidirectional antennas Wireless antenna that provide 360-degree doughnut shape wireless coverage. They are commonly used in home and SOHO routers, enterprise networks, and outside areas. Compare with directional antennas.

Open Shortest Path First (OSPF) A popular scalable, link-state routing protocol. It is based on link-state technology and introduced new concepts, such as authentication of routing updates, variable-length subnet masks (VLSMs), and route summarization.

Orthogonal Frequency-Division Multiplexing (OFDM) A wireless modulation technique that is a subset of frequency division multiplexing in which a single channel uses multiple subchannels on adjacent frequencies. OFDM is used by a number of communication systems, including 802.11a/g/n/ac. The new 802.11ax uses a variation of OFDM called Orthogonal Frequency-Division Multiaccess (OFDMA).

Other Configuration flag (O flag) Flag used in DHCPv6 to indicate to use stateless DHCPv6. The O flag value of 1 is used to inform the client that additional configuration information is available from a stateless DHCPv6 server.

out-of-band management A type of management used for the initial configuration of a device or when network access is unavailable that requires direct connection to the console or AUX port and terminal emulation software.

output errors Errors that prevented the final transmission of datagrams out of the interface that is being examined with the **show interfaces** command.

P

parabolic dish antenna A type of directional wireless antenna.

parent route A level 1 route that has been subnetted. A parent route can never be an ultimate route. Parent routes never include an exit interface or a next-hop IP address.

path vector routing protocols A routing protocol that makes routing decisions based on manually configured network policies. BGP is the only path-vector routing protocol and uses configurable attributes to make routing decisions.

Per-VLAN Spanning Tree (PVST+) A Cisco enhancement of STP that provides a separate 802.1D spanning tree instance for each VLAN configured in the network.

phishing A social engineering attack used by threat actors to obtain sensitive victim information. For example, they send fraudulent emails, messages, or use social engineering platforms pretending they are trustworthy entities to trick the victims.

Point-to-Point Protocol (PPP) A Layer 2 WAN protocol that provides router-to-router and host-to-network connections over synchronous and asynchronous circuits. It should be used on Cisco routers when connecting to other vendor routers. It also supports options such as authentication, compression, multilinking, and more. Contrast with High-Level Data Link Control (HDLC).

Port Aggregation Protocol (PAgP) A Cisco proprietary protocol that aids in the automatic creation of EtherChannel links.

port channel interface The virtual interface created when configuring an EtherChannel link on a Cisco Catalyst switch. Configuration tasks are done on the Port Channel interface instead of on each individual port, ensuring configuration consistency throughout the links.

port forwarding Sometimes called tunneling. The act of forwarding a network port from one network node to another. This technique can allow an external user to reach a port on a private IP address (inside a LAN) from the outside through a NAT-enabled router.

port security Switch security feature that limits the number of valid MAC addresses allowed on a port. The MAC addresses of legitimate devices are allowed access, while other MAC addresses are denied.

port triggering A wireless feature that allows the router to temporarily forward data through inbound ports to a specific device. You can use port triggering to forward data to a computer only when a designated port range is used to make an outbound request.

PortFast A switch STP feature in which a port is placed in an STP forwarding state as soon as the interface comes up, bypassing the listening and learning states. This feature is meant for ports connected to end-user devices.

power-on self-test (POST) A series of diagnostic tests performed by a device (router, switch, computer) when booting a computer.

pre-shared key (PSK) When two parties pre-share a secret password that is used to secure communications or authenticate users.

process switching In process switching the first packet is copied to the system buffer. The router looks up the Layer 3 network address in the routing table and initializes the fast-switch cache. The frame is rewritten with the destination address and sent to the outgoing interface that services that destination. Subsequent packets for that destination are sent by the same switching path.

Protected Management Frames (PMF) A wireless security feature that prevents the sending of de-auth frames by threat actors in an attempt to disconnect legitimate clients on the WLAN and then force them to reauthenticate.

public switched telephone network (PSTN) A general term referring to the variety of telephone networks and services in place worldwide. Also called the plain old telephone service (POTS).

PVST+ See *Per-VLAN Spanning Tree (PVST+)*.

Q

quad zero A common phrase used to describe the dotted-decimal address 0.0.0.0 used in default routing.

R

ransomware A type of malware that encrypts the data on a host and locks access to it until a ransom is paid. WannaCry is an example of ransomware.

rapid frame switching A switch forwarding characteristic referring to way the cut-through method makes a forwarding decision as soon as it has looked up the destination MAC address of the frame in its MAC address table. The switch does not have to wait for the rest of the frame to enter the ingress port before making its forwarding decision.

Rapid PVST+ Rapid Per-VLAN Spanning Tree is a Cisco proprietary implementation of RSTP.

Rapid Spanning Tree Protocol (RSTP) The IEEE 802.1w standard that defines an improved version of STP that converges much more quickly and consistently than STP (802.1D).

recursive lookup Occurs when a router has to perform multiple lookups in a routing table before forwarding a packet.

Redundant Power System (RPS) LED Refers to an LED on a Catalyst 2960 switch. An RPS is a device that can provide backup power if the switch power supply fails. The LED displays the status of the RPS.

Remote Authentication Dial-In User Service (RADIUS) An open standard AAA protocol used to provide remote-access authentication, authorization, and accounting. RADIUS encrypts only the password message. RADIUS does not encrypt usernames, accounting information, or any other information carried in the RADIUS message. Compare with Terminal Access Controller Access Control System (TACACS+).

remote networks Networks in the routing table that are not directly connected to the router. Remote networks are connected to other routers. Routes to these networks can either be statically configured or dynamically learned through dynamic routing protocols.

RIPng An IPv6 distance vector routing protocol based on the IPv4 RIPv2 routing protocol. It still has a 15-hop limitation, and the administrative distance is 120.

RIPv1 It is a legacy classful distance vector routing protocol. It has been replaced by RIPv2.

RIPv2 A classless distance vector routing protocol that supports VLSM and is a replacement to RIPv1.

rogue access points Unauthorized access points that have connected to your network.

root bridge The root of a spanning-tree topology. A root bridge exchanges topology information with other bridges in a spanning tree topology to notify all other bridges in the network when topology changes are required. This prevents loops and provides a measure of defense against link failure.

root guard A feature that provides a way to enforce the root bridge placement in the network. The root guard ensures that the port on which root guard is enabled is the designated port.

root path cost A cost of the path from the sending switch to the root bridge that is used to determine STP port roles. It is calculated by summing the individual port costs along the path from the switch to the root bridge.

root port The unique port on a nonroot bridge that has the lowest path cost to the root bridge. Every nonroot bridge in an STP topology must elect a root port. The root port on a switch is used for communication between the switch and the root bridge.

route lookup process The process that a router will use to match a destination IP address route when a packet arrives on an interface.

routed port A Layer 3 switch port configured to be a Layer 3 interface by using the **no switchport** interface configuration command.

router-on-a-stick An inter-VLAN routing solution that requires a router interface to become a trunk link with a switch. The router interface is configured using subinterfaces, and each subinterface is assigned to a specific VLAN. Compare with legacy inter-VLAN routing and Layer 3 inter-VLAN routing.

routing algorithm The process used by a routing protocol to determine the best path routes.

Routing Information Protocol (RIP) Refers to a basic and simple distance vector routing protocol.

routing protocol messages Refers to the message exchange used by different routing protocols. Messages are used to establish neighbor relationships and exchange routing table information.

routing table A data file in RAM that is used to store route information about directly connected and remote networks.

RSTP See *Rapid Spanning Tree Protocol (RSTP).*

runts Any frame less than 64 bytes in length. These frames are automatically discarded by receiving stations. Also called collision fragment. Runts are caused by malfunctioning NICs and improperly terminated Ethernet cables.

S

SANS Institute A private U.S. company that provides information security and cybersecurity training and certification programs.

Secure Copy Protocol (SCP) A secure alternative to TFTP and FTP. It uses SSH to secure traffic.

Secure FTP (SFTP) Also known as SSH File Transfer Protocol. A secure version of FTP. It uses SSH to secure traffic.

Secure Sockets Layer (SSL) A cryptographic protocol designed to provide communication security over the Internet. It has since been replaced by Transport Layer Security (TLS).

service set identifier (SSID) A wireless router advertises its wireless services by sending beacons containing its SSID. The SSID is a name that identifies the wireless domain. To access the local network and Internet, wireless devices associate and authenticate with the AP using the SSID name.

Simultaneous Authentication of Equals (SAE) A new secure password-based authentication and password-authenticated key agreement method to authenticate connecting wireless devices.

solicited-node multicast address This is an IPv6 multicast address associated with an IPv6 unicast address that is mapped to a special Ethernet multicast address.

source IP and destination IP load balancing A type of load balancing used by EtherChannel bundled links. See also *source MAC and destination MAC load balancing*.

source MAC and destination MAC load balancing A type of load balancing used by EtherChannel bundled links. See also *source IP and destination IP load balancing*.

Spanning Tree Algorithm (STA) An algorithm used by STP to calculate the best path to the root switch. It also is used to determine which redundant ports to block.

spanning-tree instance Each STP instance identifies a root bridge that serves as a reference point for all spanning tree calculations to determine which redundant paths to block. PVST+ runs one STP instance for each VLAN.

Spanning Tree Protocol (STP) A protocol defined by IEEE standard 802.1D that allows switches and bridges to create a redundant LAN, with the protocol dynamically causing some ports to block traffic so that the bridge/switch forwarding logic will not cause frames to loop indefinitely around the LAN.

spear phishing A targeted phishing attack focusing on an individual or organization. The goal can be to install malware on a target host, steal information, or encrypt files for ransom.

split MAC Used by CAPWAP to describe how a LAP and WLC divide the MAC functions. For instance, the LAP sends beacons, probe responses, and packet acknowledgments, and the WLC is responsible for authentication, association, and sending wireless traffic on the wired network.

SSID cloaking When an AP disables the SSID beacon frame and therefore wireless clients must manually configure the SSID to connect to the network.

standard static route A static route that routes to a destination network. Other types of static routes include a default static, summary static, and floating static routes.

standby router The HSRP router that monitors the HSRP active router. In the event that the active router fails, the standby router sends a coup message and takes over the active role.

stateful DHCPv6 Similar to DHCP for IPv4, provides IPv6 address, prefix length, and other information, such as DNS server and domain name. Does not provide a default gateway address.

stateful DHCPv6 client An IPv6 client using this option obtains all addressing and configuration information from a stateful DHCPv6 server.

stateful DHCPv6 server Provides all IPv6 configuration information to an IPv6 client.

stateful packet inspection In computing, a stateful network firewall tracks the operating state and characteristics (UDP/TCP) of network connections traversing it. The firewall is configured to distinguish legitimate network packets for different types of connections.

Stateless Address Autoconfiguration (SLAAC) A plug-and-play IPv6 addressing option that allows a device to obtain an IPv6 global unicast address without communicating with a DHCPv6 server. The address is obtained using ICMPv6 RS and RA messages.

stateless DHCPv6 client An IPv6 client using this option automatically obtains some addressing information but contacts a DHCPv6 server for an additional addressing configuration to use, such as DNS addresses.

stateless DHCPv6 server Provides information other than the IPv6 address and prefix length, such as DNS server and domain name. Does not provide a default gateway address.

static route A remote network in a routing table that has been manually entered into the table by a network administrator.

store-and-forward switching A method used inside a switch where the entire frame is received, and the cyclic redundancy check (CRC) is calculated. If valid, the frame is sent to the appropriate port if the destination MAC address was found in the MAC address table, or the frame is broadcasted to all ports except the ingress port.

STP See *Spanning Tree Protocol (STP)*.

STP diameter The maximum number of switches that data must cross to connect any two switches. The IEEE recommends a maximum diameter of seven switches for the default STP timers.

STP manipulation attacks A type of STP that involves manipulating the bridge protocol data unit (BPDU) to disrupt the root bridge election.

stub network A network with only one exit point. In a hub-and-spoke network, the hub network is the stub networks connected to a central hub router.

stub router A router that has only one exit interface from the routing domain and forwards all traffic to a central or distribution router.

subinterfaces These are software-based virtual interfaces associated with a single physical interface. Each subinterface is independently configured with an IP address and VLAN assignment.

summary static route A single static route that can represent multiple contiguous networks to reduce the number of entries in a routing table.

switched virtual interface (SVI) Virtual interfaces for which there is no physical hardware on the device associated. An SVI is created in software. The virtual interfaces are used as a means to remotely manage a switch over a network. They are also used as a method of routing between VLANs.

System LED Shows whether the system is receiving power and is functioning properly on a Catalyst 2960 switch. If the LED is off, it means the system is not powered on. If the LED is green, the system is operating normally. If the LED is amber, the system is receiving power but is not functioning properly.

T

tag protocol ID (TPID) A field in the VLAN tag field called the Type field, which consists of a 16-bit (2-byte) value called the tag protocol ID (TPID) value. For Ethernet, it is set to hexadecimal 0x8100. Other fields in the 802.1Q VLAN tag frame are a Priority field, CFI field, and VLAN ID.

tagged Term to describe an 802.1Q Ethernet frame that has been altered to include a VLAN ID in the packet header. The VLAN ID is used by the receiving switch to identify which port to send a broadcast packet to.

tagged traffic Term to describe traffic that has been tagged to include a VLAN ID in the packet header.

Temporal Key Integrity Protocol (TKIP) A wireless security protocol used by WEP and WPA.

Terminal Access Controller Access Control System (TACACS+) A Cisco proprietary AAA protocol used to provide remote-access authentication, authorization, and accounting. TACACS+ message exchanges are encrypted. Compare with Remote Authentication Dial-In User Service (RADIUS).

tethering A type of ad hoc network where a cellular device is used to provide a personal Internet connection. It is also referred to as a temporary quick solution that enables a smartphone to provide the wireless services of a Wi-Fi router. Other devices can associate and authenticate with the smartphone to use the Internet connection.

threat actor A term used to describe an individual or group of individuals that conduct malicious activities against an individual or organization.

Time to Live (TTL) The field in an IP header that indicates how long a packet is considered valid. Each routing device that an IP packet passes through decrements the TTL by 1.

Transport Layer Security (TLS) A cryptographic protocol that replaces Secure Sockets Layer (SSL) but is still frequently referred to as "SSL." It provides secure communications over a computer network and is commonly used to secure web access with HTTPS.

U

unequal cost load balancing All routing protocols support equal cost load balancing, which enables a router to send packets using multiple routes with the same metric. Only EIGRP supports unequal cost load balancing, which means it can do load balancing across links with different metrics.

untagged frames These are frames that do not originate from a VLAN and are crossing a trunk link. For example, frames generated by a switch such as BPDU, CDP, and more cross the trunk link as untagged frames.

URL filtering Security feature that prevents users from accessing websites based on information contained in a URL list.

user priority A field in the VLAN tag field consisting of a 3-bit value that supports level or service implementation.

V

Virtual LAN (VLAN) A group of hosts with a common set of requirements that communicate as if they were attached to the same network regardless of their physical location. A VLAN has the same attributes as a physical LAN, but it allows for end stations to be grouped together even if they are not located on the same LAN segment.

virtual private network (VPN) Establishes a virtual point-to-point connection through the use of dedicated connections, encryption, or a combination of the two between two endpoints over an unsecured network such as the Internet.

Virtual Router Redundancy Protocol (VRRP) A TCP/IP RFC protocol that allows two (or more) routers to share the duties of being the default router on a subnet, with an active/standby model, with one router acting as the default router and the other sitting by, waiting to take over that role if the first router fails.

VLAN attacks Attacks that specifically target VLANs, such as VLAN hopping and VLAN double-tagging attacks.

VLAN double-tagging When a threat actor embeds a second 802.1Q tag in an existing frame. This enables the frame to be forwarded to a different VLAN.

VLAN hopping An attack that enables traffic from one VLAN to be seen by another VLAN without the aid of a router. The threat actor host attempts to establish a trunk connection with a switch to send and receive traffic from any VLAN.

VLAN tag field The 4-byte field inserted in an Ethernet frame. The VLAN tag field consists of a Type field, a Priority field, a Canonical Format Identifier field, and VLAN ID field.

VLAN Trunking Protocol (VTP) A Cisco proprietary Layer 2 protocol that enables a network manager to configure one or more switches so that they propagate VLAN configuration information to other switches in the network, as well as synchronize the VLAN information with the other switches in the VTP domain.

VLAN trunks The links between switches that support the transmission of traffic associated with more than one VLAN. An 802.1Q trunk port supports traffic coming from many VLANs (tagged traffic), as well as traffic that does not come from a VLAN (untagged traffic).

vlan.dat Cisco switch VLAN configuration information is stored within a VLAN database file called vlan.dat. The file is located in Flash memory of the switch.

voice VLAN Voice VLANs are designed for and dedicated to the transmission of voice traffic involving IP phones or softphones (voice software used instead of a physical phone). QoS configurations are applied to voice VLANs to prioritize voice traffic.

VPN-enabled router A VPN-enabled router provides a secure connection to remote users across a public network and into the enterprise network. VPN services can be integrated into the firewall.

W

WannaCry A type of ransomware that encrypts the data on a host and locks access to it until a ransom is paid.

web security appliance (WSA) A mitigation technology device for web-based threats. The Cisco WSA can perform blacklisting of URLs, URL-filtering, malware scanning, URL categorization, web application filtering, and encryption and decryption of web traffic.

Wi-Fi Protected Access (WPA) Wireless encryption method that is stronger than WEP. It uses the Temporal Key Integrity Protocol (TKIP) encryption algorithm to enhance security.

Wi-Fi Protected Setup (WPS) A wireless network security standard that makes connecting wireless hosts faster and easier on WPA Personal or WPA2 Personal password protected WLANs. To connect a new device, press the WPS router button to enable the discovery of new devices and then on the host connect to the WLAN without entering the network password.

Wi-Fi range extenders A device that boosts the wireless signal to extend Wi-Fi coverage.

WiMAX (Worldwide Interoperability for Microwave Access) Type of access described in the IEEE standard 802.16. WiMAX offers high-speed broadband service with wireless access. It provides broad coverage like a cell phone network rather than using small Wi-Fi hotspots.

Wired Equivalent Privacy (WEP) A legacy method to encrypt wireless traffic between host and destination. WEP is no longer recommended and should never be used.

wireless access point (AP) A device that connects wireless communication devices to form a wireless network, analogous to a hub connecting wired devices to form a LAN. The AP usually connects to a wired network and can relay data between wireless devices and wired devices. Several APs can link together to form a larger network that allows roaming.

Wireless LAN (WLAN) Uses transmitters to cover a medium-sized network, usually up to 300 feet. WLANs are suitable for use in a home, office, and even a campus environment. WLANs are based on the 802.11 standard and a 2.4-GHz or 5-GHz radio frequency.

wireless LAN controller (WLC) A device that controls multiple wireless access points. See *WLAN controller*.

Wireless MANs (WMAN) Uses transmitters to provide wireless service over a larger geographic area. WMANs are suitable for providing wireless access to a metropolitan city or specific district. WMANs use specific licensed frequencies.

wireless mesh network (WMN) Home WLAN solution to extend wireless coverage using a few APs controlled using a phone application.

Wireless Personal-Area Networks (WPAN) Uses low powered transmitters for a short-range network, usually 20 to 30 ft. (6 to 9 meters). Bluetooth and ZigBee based devices are commonly used in WPANs. WPANs are based on the 802.15 standard and a 2.4-GHz radio frequency.

Wireless Wide-Area Networks (WWANs) Uses transmitters to provide coverage over an extensive geographic area. WWANs are suitable for national and global communications. WWANs also use specific licensed frequencies.

WLAN controller (WLC) The central WLAN device that controls multiple lightweight access points (LAPs) using the Lightweight Access Point Protocol (LWAPP). The WLC serves as the default gateway to all LAPs. Controller-based APs are useful in situations where many APs are required in the network. Controller-based WLANs simplify device configuration, troubleshooting, and enhance monitoring and visibility to closely analyze the WLAN.

WPA2 The current recommended wireless encryption method that uses the advanced encryption standard (AES) for encryption.

WPA3 The next generation of Wi-Fi security that will require Protected Management Frames (PMF).

X–Z

Yagi antenna This is a type of directional wireless antenna.

zombies Compromised host devices that are controlled by threat actors for nefarious purposes, including taking part in a coordinated DDoS attack.

Index

SYMBOLS

A

B

C

D

E

F

G

H

I

L

M

N

O

P

Q

R

S

T

U

V

W

Z